BASIC ABSTRACT ALGEBRA

BASIC ABSTRACT ALGEBRA

For Graduate Students and Advanced Undergraduates

Robert B. Ash
Department of Mathematics
University of Illinois

DOVER PUBLICATIONS, INC.
Mineola, New York

Copyright

Bibliographical Note

This Dover edition, first published in 2007, is the first publication in book form of the work originally titled *Abstract Algebra: The Basic Graduate Year,* which is available from the author's website: http://www.math.uiuc.edu/~r-ash/

This edition includes the introductory essay, *Remarks on Expository Writing in Mathematics,* which is available on the same site.

International Standard Book Number: 0-486-45356-1

Manufactured in the United States of America
Dover Publications, Inc., 31 East 2nd Street, Mineola, N.Y. 11501

TABLE OF CONTENTS

CHAPTER 5 SOME BASIC TECHNIQUES OF GROUP THEORY

CHAPTER 6 GALOIS THEORY

CHAPTER 7 INTRODUCING ALGEBRAIC NUMBER THEORY

CHAPTER 8 INTRODUCING ALGEBRAIC GEOMETRY

CHAPTER 9 INTRODUCING NONCOMMUTATIVE ALGEBRA

PREFACE

This is a text for the basic graduate sequence in abstract algebra, offered by most universities. We study fundamental algebraic structures, namely groups, rings, fields and modules, and maps between these structures. The techniques are used in many areas of mathematics, and there are applications to physics, engineering and computer science as well. In addition, I have attempted to communicate the intrinsic beauty of the subject. Ideally, the reasoning underlying each step of a proof should be completely clear, but the overall argument should be as brief as possible, allowing a sharp overview of the result. These two requirements are in opposition, and it is my job as expositor to try to resolve the conflict.

My primary goal is to help the reader learn the subject, and there are times when informal or intuitive reasoning leads to greater understanding than a formal proof. In the text, there are three types of informal arguments:

1. The concrete or numerical example with all features of the general case. Here, the example indicates how the proof should go, and the formalization amounts to substituting Greek letters for numbers. There is no essential loss of rigor in the informal version.
2. Brief informal surveys of large areas. There are two of these, p-adic numbers and group representation theory. References are given to books accessible to the beginning graduate student.
3. Intuitive arguments that replace lengthy formal proofs which do not reveal why a result is true. In this case, explicit references to a precise formalization are given. I am not saying that the formal proof should be avoided, just that the basic graduate year, where there are many pressing matters to cope with, may not be the appropriate place, especially when the result rather than the proof technique is used in applications.

I would estimate that about 90 percent of the text is written in conventional style, and I hope that the book will be used as a classroom text as well as a supplementary reference.

Solutions to all problems are included in the text; in my experience, most students find this to be a valuable feature. The writing style for the solutions is similar to that of the main text, and this allows for wider coverage as well as reinforcement of the basic ideas.

Chapters 1–4 cover basic properties of groups, rings, fields and modules. The typical student will have seen some but not all of this material in an undergraduate algebra course. [It should be possible to base an undergraduate course on Chapters 1–4, traversed at a suitable pace with detailed coverage of the exercises.] In Chapter 4, the fundamental structure theorems for finitely generated modules over a principal ideal domain are developed concretely with the aid of the Smith normal form. Students will undoubtedly be comfortable with elementary row and column operations, and this will significantly aid the learning process.

In Chapter 5, the theme of groups acting on sets leads to a nice application to combinatorics as well as the fundamental Sylow theorems and some results on simple groups. Analysis of normal and subnormal series leads to the Jordan-Hölder theorem and to solvable and nilpotent groups. The final section, on defining a group by generators and relations, concentrates on practical cases where the structure of a group can be deduced from its presentation. Simplicity of the alternating groups and semidirect products are covered in the exercises.

Chapter 6 goes quickly to the fundamental theorem of Galois theory; this is possible because the necessary background has been covered in Chapter 3. After some examples of direct calculation of a Galois group, we proceed to finite fields, which are of great importance in applications, and cyclotomic fields, which are fundamental in algebraic number theory. The Galois group of a cubic is treated in detail, and the quartic is covered in an appendix. Sections on cyclic and Kummer extensions are followed by Galois' fundamental theorem on solvability by radicals. The last section of the chapter deals with transcendental extensions and transcendence bases.

In the remaining chapters, we begin to apply the results and methods of abstract algebra to related areas. The title of each chapter begins with "Introducing ... ", and the areas to be introduced are algebraic number theory, algebraic geometry, noncommutative algebra and homological algebra (including categories and functors).

Algebraic number theory and algebraic geometry are the two major areas that use the tools of commutative algebra (the theory of commutative rings). In Chapter 7, after an example showing how algebra can be applied in number theory, we assemble some algebraic equipment: integral extensions, norms, traces, discriminants, Noetherian and Artinian modules and rings. We then prove the fundamental theorem on unique factorization of ideals in a Dedekind domain. The chapter concludes with an informal introduction to p-adic numbers and some ideas from valuation theory.

Chapter 8 begins geometrically with varieties in affine space. This provides motivation for Hilbert's fundamental theorems, the basis theorem and the Nullstellensatz. Several equivalent versions of the Nullstellensatz are given, as well as some corollaries with geometric significance. Further geometric considerations lead to the useful algebraic techniques of localization and primary decomposition. The remainder of the chapter is concerned with the tensor product and its basic properties.

Chapter 9 begins the study of noncommutative rings and their modules. The basic theory of simple and semisimple rings and modules, along with Schur's lemma and Jacobson's theorem, combine to yield Wedderburn's theorem on the structure of semisimple rings. We indicate the precise connection between the two popular definitions of simple ring in the literature. After an informal introduction to group representations, Maschke's theorem on semisimplicity of modules over the group algebra is proved. The introduction of the Jacobson radical gives more insight into the structure of rings and modules. The chapter ends with the Hopkins-Levitzki theorem that an Artinian ring is Noetherian, and the useful lemma of Nakayama.

In Chapter 10, we introduce some of the tools of homological algebra. Waiting until the last chapter for this is a deliberate decision. Students need as much exposure as possible to specific algebraic systems before they can appreciate the broad viewpoint of category theory. Even experienced students may have difficulty absorbing the abstract definitions of kernel, cokernel, product, coproduct, direct and inverse limit. To aid the reader, functors are introduced via the familiar examples of hom and tensor. No attempt is made to work with general abelian categories. Instead, we stay within the category of modules and study projective, injective and flat modules.

In a supplement, we go much farther into homological algebra than is usual in the basic algebra sequence. We do this to help students cope with the massive formal machinery that makes it so difficult to gain a working knowledge of this area. We concentrate on the results that are most useful in applications: the long exact homology sequence and the properties of the derived functors Tor and Ext. There is a complete proof of the snake lemma, a rarity in the literature. In this case, going through a long formal proof is entirely appropriate, because doing so will help improve algebraic skills. The point is not to avoid difficulties, but to make most efficient use of the finite amount of time available.

Robert B. Ash
October 2000

Further Remarks

Many mathematicians believe that formalism aids understanding, but I believe that when one is learning a subject, formalism often prevents understanding. The most important skill is the ability to think intuitively. This is true even in a highly abstract field such as homological algebra. My writing style reflects this view.

Classroom lectures are inherently inefficient. If the pace is slow enough to allow comprehension as the lecture is delivered, then very little can be covered. If the pace is fast enough to allow decent coverage, there will unavoidably be large gaps. Thus the student must depend on the textbook, and the current trend in algebra is to produce massive encyclopedias, which are likely to be quite discouraging to the beginning graduate student. Instead, I have attempted to write a text of manageable size, which can be read by students, including those working independently.

Another goal is to help the student reach an advanced level as quickly and efficiently as possible. When I omit a lengthy formal argument, it is because I judge that the increase in algebraic skills is insufficient to justify the time and effort involved in going through the formal proof. In all cases, I give explicit references where the details can be found. One can argue that learning to write formal proofs is an essential part of the student's mathematical training. I agree, but the ability to think intuitively is fundamental and must come first. I would add that the way things are today, there is absolutely no danger that the student will be insufficiently exposed to formalism and abstraction. In fact there is quite a bit of it in this book, although not 100 percent.

I offer this text in the hope that it will make the student's trip through algebra more enjoyable. I have done my best to avoid gaps in the reasoning. I never use the phrase "it is easy to see" under any circumstances. I welcome comments and suggestions for improvement.

Remarks on Expository Writing in Mathematics

Robert B. Ash

Successful graduate students in mathematics are able to reach an advanced level in one or more areas. Textbooks are an important part of this process. A skilled lecturer is able to illuminate and clarify many ideas, but if the pace of a course is fast enough to allow decent coverage, gaps will inevitably result. Students will depend on the text to fill these gaps, but the experience of most students is that the usual text is difficult for the novice to read. At one extreme, the text is a thousand page, twenty pound encyclopedia which cannot be read linearly in a finite amount of time. At the other extreme, the presentation in the book is essentially a seminar lecture with huge gaps.

So it seems that improvements in readability of textbooks would be highly desirable, and the natural question is "What makes a text readable?" Is it possible to answer such a question concretely? I am going to try.

First, we need to be clear on exactly who is trying to read these books. Textbooks that are opaque for students may turn out to be quite useful to the research specialist. I will assume that the reader of the text is not already an expert in the area.

The path to readability is certainly not unique, but here is some advice that may be useful.

1. Adopt a Linear Style

The idea is to help the student move from point A to point B as quickly and efficiently as possible. When learning a subject, I don't like lengthy detours, digressions or interruptions. I don't like having my path blocked by exercises that are used in the text but are not accompanied by solutions. It is fine to say that the reader should be challenged to participate actively in the learning process, but if I have to do 100 exercises to get through Chapter 1, and I get stuck on exercise 3, the chances are that I will abandon the project. This leads naturally to the next category.

2. Include Solutions to Exercises

There is an enormous increase in content when solutions are included. I trust my readers to decide which barriers they will attempt to leap over and which obstacles they will walk around. This often invites the objection that I am spoon-feeding my readers. My reply is that I would love to be spoon-fed class field theory, if only it were possible. Abstract mathematics is difficult enough without introducing gratuitous roadblocks.

3. Discuss the Intuitive Content of Results

A formal proof of the fundamental theorem of calculus should be accompanied by some intuition: If x changes by a small amount dx, then the area under the curve $y = f(x)$ changes by $dA = f(x)\,dx$, hence $dA/dx = f(x)$. Engineers and physicists will appreciate this viewpoint. They are trying to explain and predict the behavior of physical systems. This is a legitimate use of mathematics, and for this purpose, a formal proof may very well be unnecessary or even counterproductive.

Here is an example at a more advanced level. If $X = \text{Spec } A$ is the set of all prime ideals of the commutative ring A, topologized with the Zariski topology, it is useful to think informally of the elements $f \in A$ as functions on X. The value of f at the point P is the coset $f + P \in A/P$. Thus $f(P) = 0$ iff $f \in P$. This suggests an abstract analog of the zero set of a collection of polynomials. For any subset S of A, we define $V(S) = \{P \in \text{Spec } A : P \supseteq S\}$. In other words, $V(S)$ consists of all $P \in X$ such that every $f \in S$ vanishes at P.

4. Replace Abstract Arguments by Algorithmic Procedures if Possible

One formal proof that a countable union of countable sets A_n is countable goes like this. Let f_n be a surjective mapping of the positive integers N onto A_n. If we define $h : N \times N \to \cup_{n-1}^{\infty} A_n$ by $h(n,m) = f_n(m)$, then h is surjective and therefore the union of the A_n is countable (because $N \times N$ is countable). But a proof that actually gives an algorithm has more impact. List the sets A_n as an array:

$$
\begin{aligned}
A_1 &: a_{11}\ a_{12}\ a_{13} \cdots \\
A_2 &: a_{21}\ a_{22}\ a_{23} \cdots \\
A_3 &: a_{31}\ a_{32}\ a_{33} \cdots \\
&\ \vdots
\end{aligned}
$$

Then count the union by Cantor's diagonal method, for example,
$a_{11}, a_{12}, a_{21}, a_{13}, a_{22}, a_{31}, a_{14}, a_{23}, a_{32}, a_{41}$, and so on.

5. Use the Concrete Example with All the Features of the General Case.

Consider the Euclidean algorithm and its corollary that if d is the greatest common divisor of a and b, there exist integers s and t such that $sa + tb = d$. I like to take a numerical example such as $a = 123$ and $b = 54$ and carry out all the details. In this case, a carefully chosen example will have all the features of the general case. What this means is that a formal proof essentially involves substituting Greek letters for numbers. The concrete example instructs you on how to write a formal proof, and is easier to follow.

6. Avoid Serious Gaps in the Reasoning

This may be the most important component of readability. Nothing wears out the reader more than a proof in which step n+1 does not follow from step n. If the conclusions are not justified by the arguments given, the student trying to learn the subject is in for a rough ride.

Gaps can take several forms. An algebraic computation may be difficult to follow because not enough details are provided. A result from a later chapter can be tacitly assumed. Knowledge of a subject not covered in the text might be required. But perhaps the most common gap occurs when steps that are essential for understanding the proof are simply omitted. One can argue that it may be difficult to determine how much detail to give. After all, we are not going to write proofs in the formal language of set theory. We are not going to expand the text to ten thousand pages. We want the steps in a proof to be as clear as possible, and at the same we want the argument to be as brief as possible. My job as expositor is to resolve the conflict between these two objectives. If I go too far in one direction or the other, my readers will let me know about it.

Closing Comments

I hope to see a change in the reward structure and system of values at research-oriented universities so that teaching and expository writing become legitimate as a specialty. This will help to improve the current situation in which many advanced areas of mathematics are inaccessible to most students because no satisfactory exposition exists. I hope to see more mathematicians write lecture notes for their courses and post the results on the web for all to use.

BASIC ABSTRACT ALGEBRA

Chapter 0

Prerequisites

All topics listed in this chapter are covered in *A Primer of Abstract Mathematics* by Robert B. Ash, MAA 1998.

0.1 Elementary Number Theory

The greatest common divisor of two integers can be found by the Euclidean algorithm, which is reviewed in the exercises in Section 2.5. Among the important consequences of the algorithm are the following three results.

0.1.1

If d is the greatest common divisor of a and b, then there are integers s and t such that $sa + tb = d$. In particular, if a and b are relatively prime, there are integers s and t such that $sa + tb = 1$.

0.1.2

If a prime p divides a product $a_1 \cdots a_n$ of integers, then p divides at least one a_i

0.1.3 Unique Factorization Theorem

If a is an integer, not 0 or ± 1, then

(1) a can be written as a product $p_1 \cdots p_n$ of primes.

(2) If $a = p_1 \cdots p_n = q_1 \cdots q_m$, where the p_i and q_j are prime, then $n = m$ and, after renumbering, $p_i = \pm q_i$ for all i.

[We allow negative primes, so that, for example, -17 is prime. This is consistent with the general definition of prime element in an integral domain; see Section 2.6.]

0.1.4 The Integers Modulo m

If a and b are integers and m is a positive integer ≥ 2, we write $a \equiv b \bmod m$, and say that a is *congruent* to b modulo m, if $a - b$ is divisible by m. Congruence modulo m is an equivalence relation, and the resulting equivalence classes are called *residue classes* mod m. Residue classes can be added, subtracted and multiplied consistently by choosing a representative from each class, performing the appropriate operation, and calculating the residue class of the result. The collection $\mathbb{Z}_m$ of residue classes mod m forms a commutative ring under addition and multiplication. $\mathbb{Z}_m$ is a field if and only if m is prime. (The general definitions of ring, integral domain and field are given in Section 2.1.)

0.1.5

(1) The integer a is relatively prime to m if and only if a is a unit mod m, that is, a has a multiplicative inverse mod m.

(2) If c divides ab and a and c are relatively prime, then c divides b.

(3) If a and b are relatively prime to m, then ab is relatively prime to m.

(4) If $ax \equiv ay \bmod m$ and a is relatively prime to m, then $x \equiv y \bmod m$.

(5) If $d = \gcd(a, b)$, the greatest common divisor of a and b, then a/d and b/d are relatively prime.

(6) If $ax \equiv ay \bmod m$ and $d = \gcd(a, m)$, then $x \equiv y \bmod m/d$.

(7) If a_i divides b for $i = 1, \dots, r$, and a_i and a_j are relatively prime whenever $i \neq j$, then the product $a_1 \cdots a_r$ divides b.

(8) The product of two integers is their greatest common divisor times their least common multiple.

0.1.6 Chinese Remainder Theorem

If $m_1, \dots, m_r$ are relatively prime in pairs, then the system of simultaneous equations $x \equiv b_j \bmod m_j, j = 1, \dots, r$, has a solution for arbitrary integers b_j. The set of solutions forms a single residue class mod m=$m_1 \cdots m_r$, so that there is a unique solution mod m.

This result can be derived from the abstract form of the Chinese remainder theorem; see Section 2.3.

0.1.7 Euler's Theorem

The *Euler phi function* is defined by $\varphi(n)$ = the number of integers in $\{1, \dots, n\}$ that are relatively prime to n. For an explicit formula for $\varphi(n)$, see Section 1.1, Problem 13. Euler's theorem states that if $n \geq 2$ and a is relatively prime to n, then $a^{\varphi(n)} \equiv 1 \bmod n$.

0.1.8 Fermat's Little Theorem

If a is any integer and p is a prime not dividing a, then $a^{p-1} \equiv 1 \bmod p$. Thus for any integer a and prime p, whether or not p divides a, we have $a^p \equiv a \bmod p$.

For proofs of (0.1.7) and (0.1.8), see (1.3.4).

0.2 Set Theory

0.2.1

A *partial ordering* on a set S is a relation on S that is reflexive ($x \leq x$ for all $x \in S$), antisymmetric ($x \leq y$ and $y \leq x$ implies $x = y$), and transitive ($x \leq y$ and $y \leq z$ implies $x \leq z$). If for all $x, y \in S$, either $x \leq y$ or $y \leq x$, the ordering is *total.*

0.2.2

A *well-ordering* on S is a partial ordering such that every nonempty subset A of S has a smallest element a. (Thus $a \leq b$ for every $b \in A$).

0.2.3 Well-Ordering Principle

Every set can be well-ordered.

0.2.4 Maximum Principle

If T is any chain (totally ordered subset) of a partially ordered set S, then T is contained in a maximal chain M. (Maximal means that M is not properly contained in a larger chain.)

0.2.5 Zorn's Lemma

If S is a nonempty partially ordered set such that every chain of S has an upper bound in S, then S has a maximal element.

(The element x is an upper bound of the set A if $a \leq x$ for every $a \in A$. Note that x need not belong to A, but in the statement of Zorn's lemma, we require that if A is a chain of S, then A has an upper bound that actually belongs to S.)

0.2.6 Axiom of Choice

Given any family of nonempty sets S_i, $i \in I$, we can choose an element of each S_i. Formally, there is a function f whose domain is I such that $f(i) \in S_i$ for all $i \in I$.

The well-ordering principle, the maximum principle, Zorn's lemma, and the axiom of choice are equivalent in the sense that if any one of these statements is added to the basic axioms of set theory, all the others can be proved. The statements themselves cannot be proved from the basic axioms. Constructivist mathematics rejects the axiom of choice and its equivalents. In this philosophy, an assertion that we can choose an element from each S_i must be accompanied by an explicit algorithm. The idea is appealing, but its acceptance results in large areas of interesting and useful mathematics being tossed onto the scrap heap. So at present, the mathematical mainstream embraces the axiom of choice, Zorn's lemma et al.

0.2.7 Proof by Transfinite Induction

To prove that statement P_i holds for all i in the well-ordered set I, we do the following:

1. Prove the basis step P_0, where 0 is the smallest element of I.

2. If $i > 0$ and we assume that P_j holds for all $j < i$ (the transfinite induction hypothesis), prove P_i.

It follows that P_i is true for all i.

0.2.8

We say that the size of the set A is less than or equal to the size of B (notation $A \leq_s B$) if there is an injective map from A to B. We say that A and B have the same size ($A =_s B$) if there is a bijection between A and B.

0.2.9 Schröder-Bernstein Theorem

If $A \leq_s B$ and $B \leq_s A$, then $A =_s B$. (This can be proved without the axiom of choice.)

0.2.10

Using (0.2.9), one can show that if sets of the same size are called equivalent, then $\leq_s$ on equivalence classes is a partial ordering. It follows with the aid of Zorn's lemma that the ordering is total. The equivalence class of a set A, written $|A|$, is called the *cardinal number* or *cardinality* of A. In practice, we usually identify $|A|$ with any convenient member of the equivalence class, such as A itself.

0.2.11

For any set A, we can always produce a set of greater cardinality, namely the *power set* 2^A, that is, the collection of all subsets of A.

0.2.12

Define addition and multiplication of cardinal numbers by $|A|+|B| = |A \cup B|$ and $|A||B| = |A \times B|$. In defining addition, we assume that A and B are disjoint. (They can always be disjointized by replacing $a \in A$ by $(a, 0)$ and $b \in B$ by $(b, 1)$.)

0.2.13

If $\aleph_0$ is the cardinal number of a countably infinite set, then $\aleph_0 + \aleph_0 = \aleph_0\aleph_0 = \aleph_0$. More generally,

(a) If α and β are cardinals, with $\alpha \leq \beta$ and β infinite, then $\alpha + \beta = \beta$.

(b) If $\alpha \neq 0$ (i.e., α is nonempty), $\alpha \leq \beta$ and β is infinite, then $\alpha\beta = \beta$.

0.2.14

If A is an infinite set, then A and the set of all finite subsets of A have the same cardinality.

0.3 Linear Algebra

It is not feasible to list all results presented in an undergraduate course in linear algebra. Instead, here is a list of topics that are covered in a typical course.

1. Sums, products, transposes, inverses of matrices; symmetric matrices.
2. Elementary row and column operations; reduction to echelon form.
3. Determinants: evaluation by Laplace expansion and Cramer's rule.
4. Vector spaces over a field; subspaces, linear independence and bases.
5. Rank of a matrix; homogeneous and nonhomogeneous linear equations.
6. Null space and range of a matrix; the dimension theorem.
7. Linear transformations and their representation by matrices.
8. Coordinates and matrices under change of basis.
9. Inner product spaces and the projection theorem.
10. Eigenvalues and eigenvectors; diagonalization of matrices with distinct eigenvalues, symmetric and Hermitian matrices.
11. Quadratic forms.

A more advanced course might cover the following topics:

12. Generalized eigenvectors and the Jordan canonical form.
13. The minimal and characteristic polynomials of a matrix; Cayley-Hamilton theorem.
14. The adjoint of a linear operator.
15. Projection operators.
16. Normal operators and the spectral theorem.

Chapter 1

Group Fundamentals

1.1 Groups and Subgroups

1.1.1 Definition

A *group* is a nonempty set G on which there is defined a binary operation $(a,b) \to ab$ satisfying the following properties.

Closure: If a and b belong to G, then ab is also in G;

Associativity: $a(bc) = (ab)c$ for all $a, b, c \in G$;

Identity: There is an element $1 \in G$ such that $a1 = 1a = a$ for all a in G;

Inverse: If a is in G, then there is an element a^{-1} in G such that $aa^{-1} = a^{-1}a = 1$.

A group G is *abelian* if the binary operation is commutative, i.e., $ab = ba$ for all a, b in G. In this case the binary operation is often written additively ($(a,b) \to a+b$), with the identity written as 0 rather than 1.

There are some very familiar examples of abelian groups under addition, namely the integers $\mathbb{Z}$, the rationals $\mathbb{Q}$, the real numbers $\mathbb{R}$, the complex numers $\mathbb{C}$, and the integers $\mathbb{Z}_m$ modulo m. Nonabelian groups will begin to appear in the next section.

The associative law generalizes to products of any finite number of elements, for example, $(ab)(cde) = a(bcd)e$. A formal proof can be given by induction. If two people A and B form $a_1 \cdots a_n$ in different ways, the last multiplication performed by A might look like $(a_1 \cdots a_i)(a_{i+1} \cdots a_n)$, and the last multiplication by B might be $(a_1 \cdots a_j)(a_{j+1} \cdots a_n)$. But if (without loss of generality) $i < j$, then (induction hypothesis)

$$(a_1 \cdots a_j) = (a_1 \cdots a_i)(a_{i+1} \cdots a_j)$$

and

$$(a_{i+1} \cdots a_n) = (a_{i+1} \cdots a_j)(a_{j+1} \cdots a_n).$$

By the $n = 3$ case, i.e., the associative law as stated in the definition of a group, the products computed by A and B are the same.

The identity is unique ($1' = 1'1 = 1$), as is the inverse of any given element (if b and b' are inverses of a, then $b = 1b = (b'a)b = b'(ab) = b'1 = b'$). Exactly the same argument shows that if b is a right inverse, and a a left inverse, of a, then $b = b'$.

1.1.2 Definitions and Comments

A *subgroup* H of a group G is a nonempty subset of G that forms a group under the binary operation of G. Equivalently, H is a nonempty subset of G such that if a and b belong to H, so does ab^{-1}. (Note that $1 = aa^{-1} \in H$; also, $ab = a((b^{-1})^{-1}) \in H$.)

If A is any subset of a group G, the *subgroup generated by* A is the smallest subgroup containing A, often denoted by $\langle A \rangle$. Formally, $\langle A \rangle$ is the intersection of all subgroups containing A. More explicitly, $\langle A \rangle$ consists of all finite products $a_1 \cdots a_n$, $n = 1, 2, \ldots$, where for each i, either a_i or a_i^{-1} belongs to A. To see this, note that all such products belong to any subgroup containing A, and the collection of all such products forms a subgroup. In checking that the inverse of an element of $\langle A \rangle$ also belongs to $\langle A \rangle$, we use the fact that

$$(a_1 \cdots a_n)^{-1} = a_n^{-1} \cdots a_1^{-1}$$

which is verified directly: $(a_1 \cdots a_n)(a_n^{-1} \cdots a_1^{-1}) = 1$.

1.1.3 Definitions and Comments

The groups G_1 and G_2 are said to be *isomorphic* if there is a bijection $f \colon G_1 \to G_2$ that preserves the group operation, in other words, $f(ab) = f(a)f(b)$. Isomorphic groups are essentially the same; they differ only notationally. Here is a simple example. A group G is *cyclic* if G is generated by a single element: $G = \langle a \rangle$. A finite cyclic group generated by a is necessarily abelian, and can be written as $\{1, a, a^2, \ldots, a^{n-1}\}$ where $a^n = 1$, or in additive notation, $\{0, a, 2a, \ldots, (n-1)a\}$, with $na = 0$. Thus a finite cyclic group with n elements is isomorphic to the additive group $\mathbb{Z}_n$ of integers modulo n. Similarly, if G is an infinite cyclic group generated by a, then G must be abelian and can be written as $\{1, a^{\pm 1}, a^{\pm 2}, \ldots\}$, or in additive notation as $\{0, \pm a, \pm 2a, \ldots\}$. In this case, G is isomorphic to the additive group $\mathbb{Z}$ of all integers.

The *order* of an element a in a group G (denoted $|a|$) is the least positive integer n such that $a^n = 1$; if no such integer exists, the order of a is infinite. Thus if $|a| = n$, then the cyclic subgroup $\langle a \rangle$ generated by a has exactly n elements, and $a^k = 1$ iff k is a multiple of n. (Concrete examples are more illuminating than formal proofs here. Start with 0 in the integers modulo 4, and continually add 1; the result is $0, 1, 2, 3, 0, 1, 2, 3, 0, 1, 2, 3, \ldots$.)

The *order of the group* G, denoted by $|G|$, is simply the number of elements in G.

1.1.4 Proposition

If G is a finite cyclic group of order n, then G has exactly one (necessarily cyclic) subgroup of order n/d for each positive divisor d of n, and G has no other subgroups. If G is an infinite cyclic group, the (necessarily cyclic) subgroups of G are of the form $\{1, b^{\pm 1}, b^{\pm 2}, \ldots\}$, where b is an arbitrary element of G, or, in additive notation, $\{0, \pm b, \pm 2b, \ldots\}$.

Proof. Again, an informal argument is helpful. Suppose that H is a subgroup of $\mathbb{Z}_{20}$ (the integers with addition modulo 20). If the smallest positive integer in H is 6 (a non-divisor of 20) then H contains $6, 12, 18, 4$ (oops, a contradiction, 6 is supposed to be the smallest positive integer). On the other hand, if the smallest positive integer in H is 4, then $H = \{4,8,12,16,0\}$. Similarly, if the smallest positive integer in a subgroup H of the additive group of integers $\mathbb{Z}$ is 5, then $H = \{0, \pm 5, \pm 10, \pm 15, \pm 20, \dots\}$. ♣

If $G = \{1, a, \dots, a^{n-1}\}$ is a cyclic group of order n, when will an element a^r also have order n? To discover the answer, let's work in $\mathbb{Z}_{12}$. Does 8 have order 12? We compute $8, 16, 24$ $(= 0)$, so the order of 8 is 3. But if we try 7, we get $7, 14, 21, \dots, 77, 84 = 7 \times 12$, so 7 does have order 12. The point is that the least common multiple of 7 and 12 is simply the product, while the lcm of 8 and 12 is smaller than the product. Equivalently, the greatest common divisor of 7 and 12 is 1, while the gcd of 8 and 12 is $4 > 1$. We have the following result.

1.1.5 Proposition

If G is a cyclic group of order n generated by a, the following conditions are equivalent:

(a) $|a^r| = n$.

(b) r and n are relatively prime.

(c) r is a *unit* mod n, in other words, r has an *inverse* mod n (an integer s such that $rs \equiv 1 \bmod n$).

Furthermore, the set U_n of units mod n forms a group under multiplication. The order of this group is $\varphi(n)$ = the number of positive integers less than or equal to n that are relatively prime to n; φ is the familiar *Euler* φ *function.*

Proof. The equivalence of (a) and (b) follows from the discussion before the statement of the proposition, and the equivalence of (b) and (c) is handled by a similar argument. For example, since there are 12 distinct multiples of 7 mod 12, one of them must be 1; specifically, $7 \times 7 \equiv 1 \bmod 12$. But since 8×3 is 0 mod 12, no multiple of 8 can be 1 mod 12. (If $8x \equiv 1$, multiply by 3 to reach a contradiction.) Finally, U_n is a group under multiplication because the product of two integers relatively prime to n is also relatively prime to n. ♣

Problems For Section 1.1

1. A *semigroup* is a nonempty set with a binary operation satisfying closure and associativity (we drop the identity and inverse properties from the definition of a group). A *monoid* is a semigroup with identity (so that only the inverse property is dropped). Give an example of a monoid that is not a group, and an example of a semigroup that is not a monoid.
2. In $\mathbb{Z}_6$, the group of integers modulo 6, find the order of each element.
3. List all subgroups of $\mathbb{Z}_6$.

4. Let S be the set of all n by n matrices with real entries. Does S form a group under matrix addition?

5. Let S^* be the set of all nonzero n by n matrices with real entries. Does S^* form a group under matrix multiplication?

6. If H is a subgroup of the integers $\mathbb{Z}$ and $H \neq \{0\}$, what does H look like?

7. Give an example of an infinite group that has a nontrivial finite subgroup (trivial means consisting of the identity alone).

8. Let a and b belong to the group G. If $ab = ba$ and $|a| = m$, $|b| = n$, where m and n are relatively prime, show that $|ab| = mn$ and that $\langle a \rangle \cap \langle b \rangle = \{1\}$.

9. If G is a finite abelian group, show that G has an element g such that $|g|$ is the least common multiple of $\{|a|: a \in G\}$.

10. Show that a group G cannot be the union of two proper subgroups, in other words, if $G = H \cup K$ where H and K are subgroups of G, then $H = G$ or $K = G$. Equivalently, if H and K are subgroups of a group G, then $H \cup K$ cannot be a subgroup unless $H \subseteq K$ or $K \subseteq H$.

11. In an arbitrary group, let a have finite order n, and let k be a positive integer. If (n, k) is the greatest common divisor of n and k, and $[n, k]$ the least common multiple, show that the order of a^k is $n/(n, k) = [n, k]/k$.

12. Suppose that the prime factorization of the positive integer n is

$$n = p_1^{e_1} p_2^{e_2} \cdots p_r^{e_r}$$

and let A_i be the set of all positive integers $m \in \{1, 2, \ldots, n\}$ such that p_i divides m. Show that if $|S|$ is the number of elements in the set S, then

$$\begin{aligned} |A_i| &= \frac{n}{p_i}, \\ |A_i \cap A_j| &= \frac{n}{p_i p_j} \quad \text{for } i \neq j, \\ |A_i \cap A_j \cap A_k| &= \frac{n}{p_i p_j p_k} \quad \text{for } i, j, k \text{ distinct}, \end{aligned}$$

and so on.

13. Continuing Problem 12, show that the number of positive integers less than or equal to n that are relatively prime to n is

$$\varphi(n) = n(1 - \frac{1}{p_1})(1 - \frac{1}{p_2}) \cdots (1 - \frac{1}{p_r}).$$

14. Give an example of a finite group G (of order at least 3) with the property that the only subgroups of G are $\{1\}$ and G itself.

15. Does an infinite group with the property of Problem 14 exist?

1.2 Permutation Groups

1.2.1 Definition

A *permutation* of a set S is a bijection on S, that is, a function $\pi\colon S \to S$ that is one-to-one and onto. (If S is finite, then π is one-to-one if and only if it is onto.) If S is not too large, it is feasible to describe a permutation by listing the elements $x \in S$ and the corresponding values $\pi(x)$. For example, if $S = \{1,2,3,4,5\}$, then

$$\begin{bmatrix} 1 & 2 & 3 & 4 & 5 \\ 3 & 5 & 4 & 1 & 2 \end{bmatrix}$$

is the permutation such that $\pi(1) = 3$, $\pi(2) = 5$, $\pi(3) = 4$, $\pi(4) = 1$, $\pi(5) = 2$. If we start with any element $x \in S$ and apply π repeatedly to obtain $\pi(x), \pi(\pi(x))$, $\pi(\pi(\pi(x)))$, and so on, eventually we must return to x, and there are no repetitions along the way because π is one-to-one. For the above example, we obtain $1 \to 3 \to 4 \to 1$, $2 \to 5 \to 2$. We express this result by writing

$$\pi = (1,3,4)(2,5)$$

where the *cycle* $(1,3,4)$ is the permutation of S that maps 1 to 3, 3 to 4 and 4 to 1, leaving the remaining elements 2 and 5 fixed. Similarly, $(2,5)$ maps 2 to 5, 5 to 2, 1 to 1, 3 to 3 and 4 to 4. The product of $(1,3,4)$ and $(2,5)$ is interpreted as a composition, with the right factor $(2,5)$ applied first, as with composition of functions. In this case, the cycles are disjoint, so it makes no difference which mapping is applied first.

The above analysis illustrates the fact that *any permutation can be expressed as a product of disjoint cycles, and the cycle decomposition is unique.*

1.2.2 Definitions and Comments

A permutation π is said to be *even* if its cycle decomposition contains an even number of even cycles (that is, cycles of even length); otherwise π is *odd.* A cycle can be decomposed further into a product of (not necessarily disjoint) two-element cycles, called *transpositions.* For example,

$$(1,2,3,4,5) = (1,5)(1,4)(1,3)(1,2)$$

where the order of application of the mappings is from right to left.

Multiplication by a transposition changes the parity of a permutation (from even to odd, or vice versa). For example,

$$(2,4)(1,2,3,4,5) = (2,3)(1,4,5)$$
$$(2,6)(1,2,3,4,5) = (1,6,2,3,4,5);$$

$(1,2,3,4,5)$ has no cycles of even length, so is even; $(2,3)(1,4,5)$ and $(1,6,2,3,4,5)$ each have one cycle of even length, so are odd.

Since a cycle of even length can be expressed as the product of an odd number of transpositions, we can build an even permutation using an even number of transpositions, and an odd permutation requires an odd number of transpositions. A decomposition into transpositions is not unique; for example, $(1,2,3,4,5) = (1,4)(1,5)(1,4)(1,3)(1,2)(3,5)$, but as mentioned above, the cycle decomposition is unique. Since multiplication by a transposition changes the parity, it follows that if a permutation is expressed in two different ways as a product of transpositions, the number of transpositions will agree in parity (both even or both odd).

Consequently, *the product of two even permutations is even; the product of two odd permutations is even*; and *the product of an even and an odd permutation is odd.* To summarize very compactly, define the *sign* of the permutation π as

$$\operatorname{sgn}(\pi) = \begin{cases} +1 & \text{if } \pi \text{ is even} \\ -1 & \text{if } \pi \text{ is odd} \end{cases}$$

Then for arbitrary permutations π_1 and π_2 we have

$$\operatorname{sgn}(\pi_1\pi_2) = \operatorname{sgn}(\pi_1)\operatorname{sgn}(\pi_2).$$

1.2.3 Definitions and Comments

There are several permutation groups that are of major interest. The set S_n of *all* permutations of $\{1, 2, \ldots, n\}$ is called the *symmetric group on n letters*, and its subgroup A_n of all *even* permutations of $\{1, 2, \ldots, n\}$ is called the *alternating group on n letters.* (The group operation is composition of functions.) Since there are as many even permutations as odd ones (any transposition, when applied to the members of S_n, produces a one-to-one correspondence between even and odd permutations), it follows that A_n is half the size of S_n. Denoting the size of the set S by $|S|$, we have

$$|S_n| = n!, \quad |A_n| = \tfrac{1}{2}n!$$

We now define and discuss informally D_{2n}, the *dihedral group of order 2n.* Consider a regular polygon with center O and vertices $V_1, V_2, \ldots, V_n$, arranged so that as we move counterclockwise around the figure, we encounter $V_1, V_2, \ldots$ in turn. To eliminate some of the abstraction, let's work with a regular pentagon with vertices A, B, C, D, E, as shown in Figure 1.2.1.

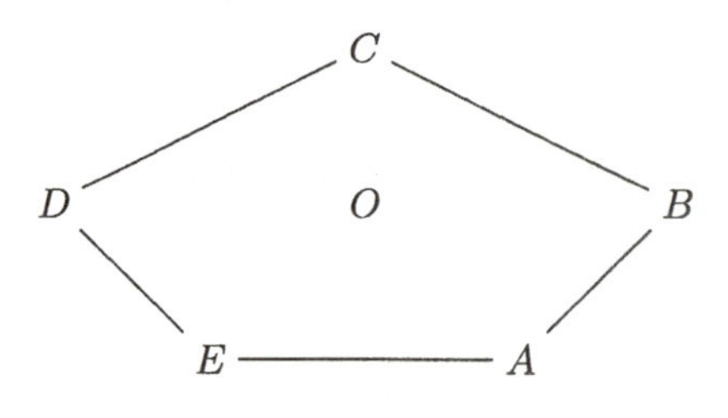

Figure 1.2.1

The group D_{10} consists of the *symmetries* of the pentagon, i.e., those permutations that can be realized via a rigid motion (a combination of rotations and reflections). All symmetries can be generated by two basic operations R and F:

R is counterclockwise rotation by $\frac{360}{n} = \frac{360}{5} = 72$ degrees,

F ("flip") is reflection about the line joining the center O to the first vertex (A in this case).

The group D_{2n} contains $2n$ elements, namely, I (the identity), $R, R^2, \dots, R^{n-1}$, F, RF, R^2F, ... , $R^{n-1}F$ (RF means F followed by R). For example, in the case of the pentagon, $F = (B, E)(C, D)$ and $R = (A, B, C, D, E)$, so $RF = (A, B)(C, E)$, which is the reflection about the line joining O to D; note that RF can also be expressed as FR^{-1}. In visualizing the effect of a permutation such as F, interpret F's taking B to E as vertex B moving to where vertex E was previously.

D_{2n} will contain exactly n rotations $I, R, \dots, R^{n-1}$ and n reflections $F, RF, \dots, R^{n-1}F$. If n is odd, each reflection is determined by a line joining the center to a vertex (and passing through the midpoint of the opposite side). If n is even, half the reflections are determined by a line passing through two vertices (as well as the center), and the other half by a line passing through the midpoints of two opposite sides (as well as the center).

1.2.4 An Abstract Characterization of the Dihedral Group

Consider the *free group with generators R and F*, in other words all finite sequences whose components are R, R^{-1}, F and F^{-1}. The group operation is concatenation, subject to the constraint that if a symbol and its inverse occur consecutively, they may be cancelled. For example, $RFFFF^{-1}RFR^{-1}RFF$ is identified with $RFFRFFF$, also written as RF^2RF^3. If we add further restrictions (so the group is no longer "free"), we can obtain D_{2n}. Specifically, D_{2n} is the group defined by *generators* R and F, subject to the *relations*

$$R^n = I, \quad F^2 = I, \quad RF = FR^{-1}.$$

The relations guarantee that there are only $2n$ distinct group elements $I, R, \dots, R^{n-1}$ and $F, RF, \dots, R^{n-1}F$. For example, with $n = 5$ we have

$$F^2R^2F = FFRRF = FFRFR^{-1} = FFFR^{-1}R^{-1} = FR^{-2} = FR^3;$$

also, R cannot be the same as R^2F, since this would imply that $I = RF$, or $F = R^{-1} = R^4$, and there is no way to get this using the relations. Since the product of two group elements is completely determined by the defining relations, it follows that there cannot be more than one group with the given generators and relations. (This statement is true "up to isomorphism"; it is always possible to create lots of isomorphic copies of any given group.) The symmetries of the regular n-gon provide a concrete realization.

Later we will look at more systematic methods of analyzing groups defined by generators and relations.

Problems For Section 1.2

1. Find the cycle decomposition of the permutation

$$\begin{bmatrix} 1 & 2 & 3 & 4 & 5 & 6 \\ 4 & 6 & 3 & 1 & 2 & 5 \end{bmatrix}$$

and determine whether the permutation is even or odd.

2. Consider the dihedral group D_8 as a group of permutations of the square. Assume that as we move counterclockwise around the square, we encounter the vertices A, B, C, D in turn. List all the elements of D_8.
3. In S_5, how many 5-cycles are there; that is, how many permutations are there with the same cycle structure as $(1, 2, 3, 4, 5)$?
4. In S_5, how many permutations are products of two disjoint transpositions, such as $(1, 2)(3, 4)$?
5. Show that if $n \geq 3$, then S_n is not abelian.
6. Show that the products of two disjoint transpositions in S_4, together with the identity, form an abelian subgroup V of S_4. Describe the multiplication table of V (known as the *four group*).
7. Show that the cycle structure of the inverse of a permutation π coincides with that of π. In particular, the inverse of an even permutation is even (and the inverse of an odd permutation is odd), so that A_n is actually a group.
8. Find the number of 3-cycles, i.e., permutations consisting of exactly one cycle of length 3, in S_4.
9. Suppose H is a subgroup of A_4 with the property that for every permutation π in A_4, π^2 belongs to H. Show that H contains all 3-cycles in A_4. (Since 3-cycles are even, H in fact contains all 3-cycles in S_4.)
10. Consider the permutation

$$\pi = \begin{bmatrix} 1 & 2 & 3 & 4 & 5 \\ 2 & 4 & 5 & 1 & 3 \end{bmatrix}.$$

Count the number of *inversions* of π, that is, the number of pairs of integers that are out of their natural order in the second row of π. For example, 2 and 5 are in natural order, but 4 and 3 are not. Compare your result with the parity of π.

11. Show that the parity of any permutation π is the same as the parity of the number of inversions of π.

1.3 Cosets, Normal Subgroups, and Homomorphisms

1.3.1 Definitions and Comments

Let H be a subgroup of the group G. If $g \in G$, the *right coset* of H generated by g is

$$Hg = \{hg \colon h \in H\};$$

similarly, the *left coset* of H generated by g is

$$gH = \{gh \colon h \in H\}.$$

It follows (Problem 1) that if $a, b \in G$, then

$$Ha = Hb \quad \text{if and only if} \quad ab^{-1} \in H$$

and

$$aH = bH \quad \text{if and only if} \quad a^{-1}b \in H.$$

Thus if we define a and b to be equivalent iff $ab^{-1} \in H$, we have an equivalence relation (Problem 2), and (Problem 3) the equivalence class of a is

$$\{b\colon ab^{-1} \in H\} = Ha.$$

Therefore *the right cosets partition G (similarly for the left cosets)*. Since $h \to ha$, $h \in H$, is a one-to-one correspondence, each coset has $|H|$ elements. There are as many right cosets as left cosets, since the map $aH \to Ha^{-1}$ is a one-to-one correspondence (Problem 4). If $[G : H]$, the *index* of H in G, denotes the number of right (or left) cosets, we have the following basic result.

1.3.2 Lagrange's Theorem

If H is a subgroup of G, then $|G| = |H|[G : H]$. In particular, if G is finite then $|H|$ divides $|G|$, and

$$\frac{|G|}{|H|} = [G : H].$$

Proof. There are $[G : H]$ cosets, each with $|H|$ members. ♣

1.3.3 Corollary

Let G be a finite group.

(i) If $a \in G$ then $|a|$ divides $|G|$; in particular, $a^{|G|} = 1$. Thus $|G|$ is a multiple of the order of each of its elements, so if we define the *exponent* of G to be the least common multiple of $\{|a|\colon a \in G\}$, then $|G|$ is a multiple of the exponent.

(ii) If G has prime order, then G is cyclic.

Proof. If the element $a \in G$ has order n, then $H = \{1, a, a^2, \dots, a^{n-1}\}$ is a cyclic subgroup of G with $|H| = n$. By Lagrange's theorem, n divides $|G|$, proving (i). If $|G|$ is prime then we may take $a \neq 1$, and consequently $n = |G|$. Thus H is a subgroup with as many elements as G, so in fact H and G coincide, proving (ii). ♣

Here is another corollary.

1.3.4 Euler's Theorem

If a and n are relatively prime positive integers, with $n \geq 2$, then $a^{\varphi(n)} \equiv 1 \mod n$. A special case is *Fermat's Little Theorem*: If p is a prime and a is a positive integer not divisible by p, then $a^{p-1} \equiv 1 \mod p$.

Proof. The group of units mod n has order $\varphi(n)$, and the result follows from (1.3.3). ♣

We will often use the notation $H \leq G$ to indicate that H is a subgroup of G. If H is a proper subgroup, i.e., $H \leq G$ but $H \neq G$, we write $H < G$.

1.3.5 The Index is Multiplicative

If $K \leq H \leq G$, then $[G : K] = [G : H][H : K]$.

Proof. Choose representatives a_i from each left coset of H in G, and representatives b_j from each left coset of K in H. If cK is any left coset of K in G, then $c \in a_iH$ for some unique i, and if $c = a_ih, h \in H$, then $h \in b_jK$ for some unique j, so that c belongs to a_ib_jK. The map $(a_i, b_j) \rightarrow a_ib_jK$ is therefore onto, and it is one-to-one by the uniqueness of i and j. We therefore have a bijection between a set of size $[G : H][H : K]$ and a set of size $[G : K]$, as asserted. ♣

Now suppose that H and K are subgroups of G, and define HK to be the set of all products $hk, h \in H, k \in K$. Note that HK need not be a group, since $h_1k_1h_2k_2$ is not necessarily equal to $h_1h_2k_1k_2$. If G is abelian, then HK will be a group, and we have the following useful generalization of this observation.

1.3.6 Proposition

If $H \leq G$ and $K \leq G$, then $HK \leq G$ if and only if $HK = KH$. In this case, HK is the subgroup generated by $H \cup K$.

Proof. If HK is a subgroup, then $(HK)^{-1}$, the collection of all inverses of elements of HK, must coincide with HK. But $(HK)^{-1} = K^{-1}H^{-1} = KH$. Conversely, if $HK = KH$, then the inverse of an element in HK also belongs to HK, because $(HK)^{-1} = K^{-1}H^{-1} = KH = HK$. The product of two elements in HK belongs to HK, because $(HK)(HK) = HKHK = HHKK = HK$. The last statement follows from the observation that any subgroup containing H and K must contain HK. ♣

The set product HK defined above suggests a multiplication operation on cosets. If H is a subgroup of G, we can multiply aH and bH, and it is natural to hope that we get abH. This does not always happen, but here is one possible criterion.

1.3.7 Lemma

If $H \leq G$, then $(aH)(bH) = abH$ for all $a, b \in G$ iff $cHc^{-1} = H$ for all $c \in G$. (Equivalently, $cH = Hc$ for all $c \in G$.

Proof. If the second condition is satisfied, then $(aH)(bH) = a(Hb)H = abHH = abH$. Conversely, if the first condition holds, then $cHc^{-1} \subseteq cHc^{-1}H$ since $1 \in H$, and $(cH)(c^{-1}H) = cc^{-1}H(= H)$ by hypothesis. Thus $cHc^{-1} \subseteq H$, which implies that $H \subseteq c^{-1}Hc$. Since this holds for all $c \in G$, we have $H \subseteq cHc^{-1}$, and the result follows. ♣

Notice that we have proved that if $cHc^{-1} \subseteq H$ for all $c \in G$, then in fact $cHc^{-1} = H$ for all $c \in G$.

1.3.8 Definition

Let H be a subgroup of G. If any of the following equivalent conditions holds, we say that H is a *normal subgroup* of G, or that H is *normal* in G:

(1) $cHc^{-1} \subseteq H$ for all $c \in G$ (equivalently, $c^{-1}Hc \subseteq H$ for all $c \in G$).

(2) $cHc^{-1} = H$ for all $c \in G$ (equivalently, $c^{-1}Hc = H$ for all $c \in G$).

(3) $cH = Hc$ for all $c \in G$.

(4) Every left coset of H in G is also a right coset.

(5) Every right coset of H in G is also a left coset.

We have established the equivalence of (1), (2) and (3) above, and (3) immediately implies (4). To show that (4) implies (3), suppose that $cH = Hd$. Then since c belongs to both cH and Hc, i.e., to both Hd and Hc, we must have $Hd = Hc$ because right cosets partition G, so that any two right cosets must be either disjoint or identical. The equivalence of (5) is proved by a symmetrical argument.

Notation: $H \trianglelefteq G$ indicates that H is a normal subgroup of G; if H is a proper normal subgroup, we write $H \triangleleft G$.

1.3.9 Definition of the Quotient Group

If H is normal in G, we may define a group multiplication on cosets, as follows. If aH and bH are (left) cosets, let

$$(aH)(bH) = abH;$$

by (1.3.7), $(aH)(bH)$ is simply the set product. If a_1 is another member of aH and b_1 another member of bH, then $a_1H = aH$ and $b_1H = bH$ (Problem 5). Therefore the set product of a_1H and b_1H is also abH. The point is that the product of two cosets does not depend on which representatives we select.

To verify that cosets form a group under the above multiplication, we consider the four defining requirements.

Closure: The product of two cosets is a coset.

Associativity: This follows because multiplication in G is associative.

Identity: The coset $1H = H$ serves as the identity.

Inverse: The inverse of aH is $a^{-1}H$.

The group of cosets of a normal subgroup N of G is called the *quotient group* of G by N; it is denoted by G/N.

Since the identity in G/N is $1N = N$, we have, intuitively, "set everything in N equal to 1".

1.3.10 Example

Let $GL(n, \mathbb{R})$ be the set of all nonsingular n by n matrices with real coefficients, and let $SL(n, \mathbb{R})$ be the subgroup formed by matrices whose determinant is 1 (GL stands for "general linear" and SL for "special linear"). Then $SL(n, \mathbb{R}) \triangleleft GL(n, R)$, because if A is a nonsingular n by n matrix and B is n by n with determinant 1, then $\det(ABA^{-1}) = \det A \det B \det A^{-1} = \det B = 1$.

1.3.11 Definition

If $f\colon G \to H$, where G and H are groups, then f is said to be a *homomorphism* if for all a, b in G, we have

$$f(ab) = f(a)f(b).$$

This idea will look familiar if G and H are abelian, in which case, using additive notation, we write

$$f(a + b) = f(a) + f(b);$$

thus a linear transformation on a vector space is, in particular, a homomorphism on the underlying abelian group. If f is a homomorphism from G to H, it must map the identity of G to the identity of H, since $f(a) = f(a1_G) = f(a)f(1_G)$; multiply by $f(a)^{-1}$ to get $1_H = f(1_G)$. Furthermore, the inverse of $f(a)$ is $f(a^{-1})$, because

$$1 = f(aa^{-1}) = f(a)f(a^{-1}),$$

so that $[f(a)]^{-1} = f(a^{-1})$.

1.3.12 The Connection Between Homomorphisms and Normal Subgroups

If $f\colon G \to H$ is a homomorphism, define the *kernel* of f as

$$\ker f = \{a \in G\colon f(a) = 1\};$$

then $\ker f$ is a normal subgroup of G. For if $a \in G$ and $b \in \ker f$, we must show that aba^{-1} belongs to $\ker f$. But $f(aba^{-1}) = f(a)f(b)f(a^{-1}) = f(a)(1)f(a)^{-1} = 1$.

Conversely, every normal subgroup is the kernel of a homomorphism. To see this, suppose that $N \trianglelefteq G$, and let H be the quotient group G/N. Define the map $\pi\colon G \to G/N$ by $\pi(a) = aN$; π is called the *natural* or *canonical* map. Since

$$\pi(ab) = abN = (aN)(bN) = \pi(a)\pi(b),$$

π is a homomorphism. The kernel of π is the set of all $a \in G$ such that $aN = N(= 1N)$, or equivalently, $a \in N$. Thus $\ker \pi = N$.

1.3.13 Proposition

A homomorphism f is injective if and only if its kernel K is trivial, that is, consists only of the identity.

Proof. If f is injective and $a \in K$, then $f(a) = 1 = f(1)$, hence $a = 1$. Conversely, if K is trivial and $f(a) = f(b)$, then $f(ab^{-1}) = f(a)f(b^{-1}) = f(a)[f(b)]^{-1} = f(a)[f(a)]^{-1} = 1$, so $ab^{-1} \in K$. Thus $ab^{-1} = 1$, i.e., $a = b$, proving f injective. ♣

1.3.14 Some Standard Terminology

A *monomorphism* is an injective homomorphism

An *epimorphism* is a surjective homomorphism

An *isomorphism* is a bijective homomorphism

An *endomorphism* is a homomorphism of a group to itself

An *automorphism* is an isomorphism of a group with itself

We close the section with a result that is applied frequently.

1.3.15 Proposition

Let $f: G \to H$ be a homomorphism.

(i) If K is a subgroup of G, then $f(K)$ is a subgroup of H. If f is an epimorphism and K is normal, then $f(K)$ is also normal.

(ii) If K is a subgroup of H, then $f^{-1}(K)$ is a subgroup of G. If K is normal, so is $f^{-1}(K)$.

Proof. (i) If $f(a)$ and $f(b)$ belong to $f(K)$, so does $f(a)f(b)^{-1}$, since this element coincides with $f(ab^{-1})$. If K is normal and $c \in G$, we have $f(c)f(K)f(c)^{-1} = f(cKc^{-1}) = f(K)$, so if f is surjective, then $f(K)$ is normal.

(ii) If a and b belong to $f^{-1}(K)$, so does ab^{-1}, because $f(ab^{-1}) = f(a)f(b)^{-1}$, which belongs to K. If $c \in G$ and $a \in f^{-1}(K)$ then $f(cac^{-1}) = f(c)f(a)f(c)^{-1}$, so if K is normal, we have $cac^{-1} \in f^{-1}(K)$, proving $f^{-1}(K)$ normal. ♣

Problems For Section 1.3

In Problems 1–6, H is a subgroup of the group G, and a and b are elements of G.

1. Show that $Ha = Hb$ iff $ab^{-1} \in H$.
2. Show that "$a \sim b$ iff $ab^{-1} \in H$" defines an equivalence relation.
3. If we define a and b to be equivalent iff $ab^{-1} \in H$, show that the equivalence class of a is Ha.
4. Show that $aH \to Ha^{-1}$ is a one-to-one correspondence between left and right cosets of H.

5. If aH is a left coset of H in G and $a_1 \in aH$, show that the left coset of H generated by a_1 (i.e., a_1H), is also aH.
6. If $[G : H] = 2$, show that H is a normal subgroup of G.
7. Let S_3 be the group of all permutations of $\{1, 2, 3\}$, and take a to be permutation $(1, 2, 3)$, b the permutation $(1, 2)$, and e the identity permutation. Show that the elements of S_3 are, explicitly, e, a, a^2, b, ab and a^2b.
8. Let H be the subgroup of S_3 consisting of the identity e and the permutation $b = (1, 2)$. Compute the left cosets and the right cosets of H in S_3.
9. Continuing Problem 8, show that H is not a normal subgroup of S_3.
10. Let f be an endomorphism of the integers $\mathbb{Z}$. Show that f is completely determined by its action on 1. If $f(1) = r$, then f is multiplication by r; in other words, $f(n) = rn$ for every integer n.
11. If f is an automorphism of $\mathbb{Z}$, and I is the identity function on $\mathbb{Z}$, show that f is either I or $-I$.
12. Since the composition of two automorphisms is an automorphism, and the inverse of an automorphism is an automorphism, it follows that the set of automorphisms of a group is a group under composition. In view of Problem 11, give a simple description of the group of automorphisms of $\mathbb{Z}$.
13. Let H and K be subgroups of the group G. If $x, y \in G$, define $x \sim y$ iff x can be written as hyk for some $h \in H$ and $k \in K$. Show that $\sim$ is an equivalence relation.
14. The equivalence class of $x \in G$ is $HxK = \{hxk : h \in H, k \in K\}$, called a *double coset* associated with the subgroups H and K. Thus the double cosets partition G. Show that any double coset can be written as a union of right cosets of H, or equally well as a union of left cosets of K.

1.4 The Isomorphism Theorems

Suppose that N is a normal subgroup of G, f is a homomorphism from G to H, and π is the natural map from G to G/N, as pictured in Figure 1.4.1.

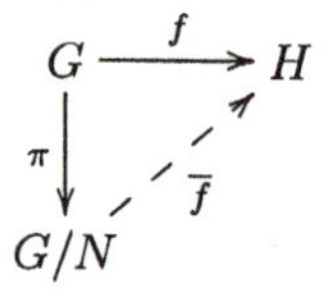

Figure 1.4.1

We would like to find a homomorphism $\overline{f} : G/N \to H$ that makes the diagram *commutative.* Commutativity means that we get the same result by traveling directly from G to H via f as we do by taking the roundabout route via π followed by $\overline{f}$. This requirement translates to $\overline{f}(aN) = f(a)$. Here is the key result for finding such an $\overline{f}$.

1.4.1 Factor Theorem

Any homomorphism f whose kernel K contains N can be factored through G/N. In other words, in Figure 1.4.1 there is a unique homomorphism $\overline{f}\colon G/N \to H$ such that $\overline{f} \circ \pi = f$. Furthermore:

(i) $\overline{f}$ is an epimorphism if and only if f is an epimorphism;

(ii) $\overline{f}$ is a monomorphism if and only if $K = N$;

(iii) $\overline{f}$ is an isomorphism if and only if f is an epimorphism and $K = N$.

Proof. If the diagram is to commute, then $\overline{f}(aN)$ must be $f(a)$, and it follows that $\overline{f}$, if it exists, is unique. The definition of $\overline{f}$ that we have just given makes sense, because if $aN = bN$, then $a^{-1}b \in N \subseteq K$, so $f(a^{-1}b) = 1$, and therefore $f(a) = f(b)$. Since

$$\overline{f}(aNbN) = \overline{f}(abN) = f(ab) = f(a)f(b) = \overline{f}(aN)\overline{f}(bN),$$

$\overline{f}$ is a homomorphism. By construction, $\overline{f}$ has the same image as f, proving (i). Now the kernel of $\overline{f}$ is

$$\{aN\colon f(a) = 1\} = \{aN\colon a \in K\} = K/N.$$

By (1.3.13), a homomorphism is injective, i.e., a monomorphism, if and only if its kernel is trivial. Thus $\overline{f}$ is a monomorphism if and only if K/N consists only of the identity element N. This means that if a is any element of K, then the coset aN coincides with N, which forces a to belong to N. Thus $\overline{f}$ is a monomorphism if and only if $K = N$, proving (ii). Finally, (iii) follows immediately from (i) and (ii). ♣

The factor theorem yields a fundamental result.

1.4.2 First Isomorphism Theorem

If $f\colon G \to H$ is a homomorphism with kernel K, then the image of f is isomorphic to G/K.

Proof. Apply the factor theorem with $N = K$, and note that f must be an epimorphism of G onto its image. ♣

If we are studying a subgroup K of a group G, or perhaps the quotient group G/K, we might try to construct a homomorphism f whose kernel is K and whose image H has desirable properties. The first isomorphism theorem then gives $G/K \cong H$ (where $\cong$ is our symbol for isomorphism). If we know something about H, we may get some insight into K and G/K.

We will prove several other isomorphism theorems after the following preliminary result.

1.4.3 Lemma

Let H and N be subgroups of G, with N normal in G. Then:

(i) $HN = NH$, and therefore by (1.3.6), HN is a subgroup of G.

(ii) N is a normal subgroup of HN.

(iii) $H \cap N$ is a normal subgroup of H.

Proof. (i) We have $hN = Nh$ for every $h \in G$, in particular for every $h \in H$.
 (ii) Since N is normal in G, it must be normal in the subgroup HN.
 (iii) $H \cap N$ is the kernel of the canonical map $\pi\colon G \to G/N$, restricted to H. ♣

The subgroups we are discussing are related by a "parallelogram" or "diamond", as Figure 1.4.2 suggests.

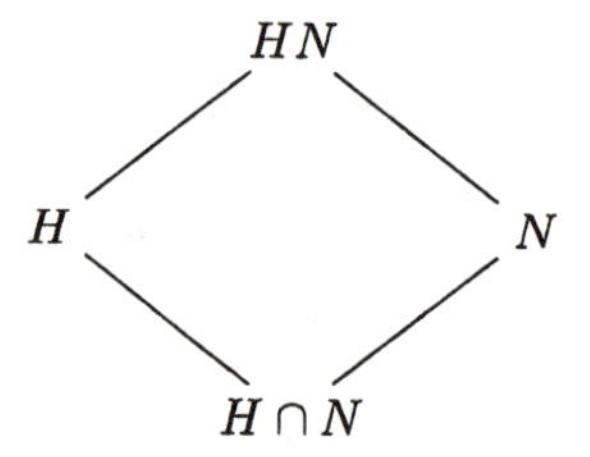

Figure 1.4.2

1.4.4 Second Isomorphism Theorem

If H and N are subgroups of G, with N normal in G, then

$$H/(H \cap N) \cong HN/N.$$

Note that we write HN/N rather than H/N, since N need not be a subgroup of H.

Proof. Let π be the canonical epimorphism from G to G/N, and let π_0 be the restriction of π to H. Then the kernel of π_0 is $H \cap N$, so by the first isomorphism theorem, $H/(H \cap N)$ is isomorphic to the image of π_0, which is $\{hN\colon h \in H\} = HN/N$. (To justify the last equality, note that for any $n \in N$ we have $hnN = hN$). ♣

1.4.5 Third Isomorphism Theorem

If N and H are normal subgroups of G, with N contained in H, then

$$G/H \cong (G/N)/(H/N),$$

a "cancellation law".

Proof. This will follow directly from the first isomorphism theorem if we can find an epimorphism of G/N onto G/H with kernel H/N, and there is a natural candidate: $f(aN) = aH$. To check that f is well-defined, note that if $aN = bN$ then $a^{-1}b \in N \subseteq H$, so $aH = bH$. Since a is an arbitrary element of G, f is surjective, and by definition of coset multiplication, f is a homomorphism. But the kernel of f is

$$\{aN\colon aH = H\} = \{aN\colon a \in H\} = H/N. \quad ♣$$

Now suppose that N is a normal subgroup of G. If H is a subgroup of G containing N, there is a natural analog of H in the quotient group G/N, namely, the subgroup H/N. In fact we can make this correspondence very precise. Let

$$\psi(H) = H/N$$

be a map from the set of subgroups of G containing N to the set of subgroups of G/N. We claim that ψ is a bijection. For if $H_1/N = H_2/N$ then for any $h_1 \in H_1$, we have $h_1N = h_2N$ for some $h_2 \in H_2$, so that $h_2^{-1}h_1 \in N$, which is contained in H_2. Thus $H_1 \subseteq H_2$, and by symmetry the reverse inclusion holds, so that $H_1 = H_2$ and ψ is injective. Now if Q is a subgroup of G/N and $\pi\colon G \to G/N$ is canonical, then

$$\pi^{-1}(Q) = \{a \in G\colon aN \in Q\},$$

a subgroup of G containing N, and

$$\psi(\pi^{-1}(Q)) = \{aN\colon aN \in Q\} = Q,$$

proving ψ surjective.

The map ψ has a number of other interesting properties, summarized in the following result, sometimes referred to as the fourth isomorphism theorem.

1.4.6 Correspondence Theorem

If N is a normal subgroup of G, then the map $\psi\colon H \to H/N$ sets up a one-to-one correspondence between subgroups of G containing N and subgroups of G/N. The inverse of ψ is the map $\tau\colon Q \to \pi^{-1}(Q)$, where π is the canonical epimorphism of G onto G/N. Furthermore:

(i) $H_1 \leq H_2$ if and only if $H_1/N \leq H_2/N$, and, in this case,

$$[H_2\colon H_1] = [H_2/N\colon H_1/N].$$

(ii) H is a normal subgroup of G if and only if H/N is a normal subgroup of G/N.

More generally,

(iii) H_1 is a normal subgroup of H_2 if and only if H_1/N is a normal subgroup of H_2/N, and in this case, $H_2/H_1 \cong (H_2/N)/H_1/N)$.

Proof. We have established that ψ is a bijection with inverse τ. If $H_1 \leq H_2$, we have $H_1/N \leq H_2/N$ immediately, and the converse follows from the above proof that ψ is injective. To prove the last statement of (i), let η map the left coset $aH_1, a \in H_2$, to the left coset $(aN)(H_1/N)$. Then η is a well-defined injective map because

$$\begin{aligned} aH_1 = bH_1 \quad &\text{iff} \quad a^{-1}b \in H_1 \\ &\text{iff} \quad (aN)^{-1}(bN) = a^{-1}bN \in H_1/N \\ &\text{iff} \quad (aN)(H_1/N) = (bN)(H_1/N); \end{aligned}$$

η is surjective because a ranges over all of H_2.

To prove (ii), assume that $H \trianglelefteq G$; then for any $a \in G$ we have

$$(aN)(H/N)(aN)^{-1} = (aHa^{-1})/N = H/N$$

so that $H/N \trianglelefteq G/N$. Conversely, suppose that H/N is normal in G/N. Consider the homomorphism $a \rightarrow (aN)(H/N)$, the composition of the canonical map of G onto G/N and the canonical map of G/N onto $(G/N)/(H/N)$. The element a will belong to the kernel of this map if and only if $(aN)(H/N) = H/N$, which happens if and only if $aN \in H/N$, that is, $aN = hN$ for some $h \in H$. But since N is contained in H, this statement is equivalent to $a \in H$. Thus H is the kernel of a homomorphism, and is therefore a normal subgroup of G.

Finally, the proof of (ii) also establishes the first part of (iii); just replace H by H_1 and G by H_2. The second part of (iii) follows from the third isomorphism theorem (with the same replacement). ♣

We conclude the section with a useful technical result.

1.4.7 Proposition

If H is a subgroup of G and N is a normal subgroup of G, we know by (1.4.3) that HN, the subgroup generated by $H \cup N$, is a subgroup of G. If H is also a normal subgroup of G, then HN is normal in G as well. More generally, if for each i in the index set I, we have $H_i \trianglelefteq G$, then $\langle H_i, i \in I \rangle$, the subgroup generated by the H_i (technically, by the set $\cup_{i \in I} H_i$) is a normal subgroup of G.

Proof. A typical element in the subgroup generated by the H_i is $a = a_1 a_2 \cdots a_n$ where a_k belongs to H_{i_k}. If $g \in G$ then

$$g(a_1 a_2 \cdots a_n)g^{-1} = (ga_1 g^{-1})(ga_2 g^{-1}) \cdots (ga_n g^{-1})$$

and $ga_k g^{-1} \in H_{i_k}$ because $H_{i_k} \trianglelefteq G$. Thus gag^{-1} belongs to $\langle H_i, i \in I \rangle$. ♣

Problems For Section 1.4

1. Let $\mathbb{Z}$ be the integers, and $n\mathbb{Z}$ the set of integer multiples of n. Show that $\mathbb{Z}/n\mathbb{Z}$ is isomorphic to $\mathbb{Z}_n$, the additive group of integers modulo n. (This is not quite a tautology if we view $\mathbb{Z}_n$ concretely as the set $\{0, 1, \ldots, n-1\}$, with sums and differences reduced modulo n.)

2. If m divides n then $\mathbb{Z}_m \leq \mathbb{Z}_n$; for example, we can identify $\mathbb{Z}_4$ with the subgroup $\{0, 3, 6, 9\}$ of $\mathbb{Z}_{12}$. Show that $\mathbb{Z}_n/\mathbb{Z}_m \cong \mathbb{Z}_{n/m}$.
3. Let a be an element of the group G, and let $f_a\colon G \to G$ be "conjugation by a", that is, $f_a(x) = axa^{-1}, x \in G$. Show that f_a is an automorphism of G.
4. An *inner automorphism* of G is an automorphism of the form f_a (defined in Problem 3) for some $a \in G$. Show that the inner automorphisms of G form a group under composition of functions (a subgroup of the group of all automorphisms of G).
5. Let $Z(G)$ be the *center* of G, that is, the set of all x in G such that $xy = yx$ for all y in G. Thus $Z(G)$ is the set of elements that commute with everything in G. Show that $Z(G)$ is a normal subgroup of G, and that the group of inner automorphisms of G is isomorphic to $G/Z(G)$.
6. If f is an automorphism of Z_n, show that f is multiplication by m for some m relatively prime to n. Conclude that the group of automorphisms of $\mathbb{Z}_n$ can be identified with the group of units mod n.
7. The diamond diagram associated with the second isomorphism theorem (1.4.4) illustrates least upper bounds and greatest lower bounds in a lattice. Verify that HN is the smallest subgroup of G containing both H and N, and $H \cap N$ is the largest subgroup of G contained in both H and N.
8. Let g be an automorphism of the group G, and f_a an inner automorphism, as defined in Problems 3 and 4. Show that $g \circ f_a \circ g^{-1}$ is an inner automorphism. Thus the group of inner automorphisms of G is a normal subgroup of the group of all automorphisms.
9. Identify a large class of groups for which the only inner automorphism is the identity mapping.

1.5 Direct Products

1.5.1 External and Internal Direct Products

In this section we examine a popular construction. Starting with a given collection of groups, we build a new group with the aid of the cartesian product. Let's start with two given groups H and K, and let $G = H \times K$, the set of all ordered pairs (h, k), $h \in H, k \in K$. We define multiplication on G componentwise:

$$(h_1, k_1)(h_2, k_2) = (h_1 h_2, k_1 k_2).$$

Since $(h_1 h_2, k_1 k_2)$ belongs to G, it follows that G is closed under multiplication. The multiplication operation is associative because the individual products on H and K are associative. The identity element in G is $(1_H, 1_K)$, and the inverse of (h, k) is (h^{-1}, k^{-1}). Thus G is a group, called the *external direct product* of H and K.

We may regard H and K as subgroups of G. More precisely, G contains isomorphic copies of H and K, namely

$$\overline{H} = \{(h, 1_K)\colon h \in H\} \text{ and } \overline{K} = \{(1_H, k)\colon k \in K\}.$$

Furthermore, $\overline{H}$ and $\overline{K}$ are normal subgroups of G. (Note that $(h,k)(h_1,1_K)(h^{-1},k^{-1}) = (hh_1h^{-1},1_K)$, with $hh_1h^{-1} \in H$.) Also, from the definitions of $\overline{H}$ and $\overline{K}$, we have

$$G = \overline{H}\,\overline{K} \text{ and } \overline{H} \cap \overline{K} = \{1\}, \text{ where } 1 = (1_H, 1_K).$$

If a group G contains normal subgroups H and K such that $G = HK$ and $H \cap K = \{1\}$, we say that G is the *internal direct product* of H and K.

Notice the key difference between external and internal direct products. We *construct* the external direct product from the component groups H and K. On the other hand, starting with a given group we *discover* subgroups H and K such that G is the internal direct product of H and K. Having said this, we must admit that in practice the distinction tends to be blurred, because of the following result.

1.5.2 Proposition

If G is the internal direct product of H and K, then G is isomorphic to the external direct product $H \times K$.

Proof. Define $f\colon H \times K \to G$ by $f(h,k) = hk$; we will show that f is an isomorphism. First note that if $h \in H$ and $k \in K$ then $hk = kh$. (Consider $hkh^{-1}k^{-1}$, which belongs to K since $hkh^{-1} \in K$, and also belongs to H since $kh^{-1}k^{-1} \in H$; thus $hkh^{-1}k^{-1} = 1$, so $hk = kh$.)

(a) f is a homomorphism, since

$$f((h_1,k_1)(h_2,k_2)) = f(h_1h_2,k_1k_2) = h_1h_2k_1k_2 = (h_1k_1)(h_2k_2) = f(h_1,k_1)f(h_2,k_2).$$

(b) f is surjective, since by definition of internal direct product, $G = HK$.

(c) f is injective, for if $f(h,k) = 1$ then $hk = 1$, so that $h = k^{-1}$.Thus h belongs to both H and K, so by definition of internal direct product, h is the identity, and consequently so is k. The kernel of f is therefore trivial. ♣

External and internal direct products may be defined for any number of factors. We will restrict ourselves to a finite number of component groups, but the generalization to arbitrary cartesian products with componentwise multiplication is straightforward.

1.5.3 Definitions and Comments

If $H_1, H_2, \dots H_n$ are arbitrary groups, the *external direct product* of the H_i is the cartesian product $G = H_1 \times H_2 \times \cdots \times H_n$, with componentwise multiplication:

$$(h_1, h_2, \dots, h_n)(h_1', h_2', \dots h_n') = (h_1h_1', h_2h_2', \dots h_nh_n');$$

G contains an isomorphic copy of each H_i, namely

$$\overline{H}_i = \{(1_{H_1}, \dots, 1_{H_{i-1}}, h_i, 1_{H_{i+1}}, \dots, 1_{H_n})\colon h_i \in H_i\}.$$

As in the case of two factors, $G = \overline{H}_1\overline{H}_2 \cdots \overline{H}_n$, and $\overline{H}_i \trianglelefteq G$ for all i; furthermore, if $g \in G$ then g has a unique representation of the form

$$g = \overline{h}_1\,\overline{h}_2 \cdots \overline{h}_n \text{ where } \overline{h}_i \in \overline{H}_i.$$

Specifically, $g = (h_1, \dots, h_n) = (h_1, 1, \dots, 1) \dots (1, \dots, 1, h_n)$. The representation is unique because the only way to produce the i-th component h_i of g is for h_i to be the ith component of the factor from $\overline{H}_i$. If a group G contains normal subgroups $H_1, \dots, H_n$ such that $G = H_1 \cdots H_n$, and each $g \in G$ can be uniquely represented as $h_1 \cdots h_n$ with $h_i \in H_i, i = 1, 2, \dots, n$, we say that G is the *internal direct product* of the H_i. As in the case of two factors, if G is the internal direct product of the H_i, then G is isomorphic to the external direct product $H_1 \times \cdots \times H_n$; the isomorphism $f \colon H_1 \times \cdots \times H_n \to G$ is given by $f(h_1, \dots, h_n) = h_1 \cdots h_n$. The next result frequently allows us to recognize when a group is an internal direct product.

1.5.4 Proposition

Suppose that $G = H_1 \cdots H_n$, where each H_i is a normal subgroup of G. The following conditions are equivalent:

(1) G is the internal direct product of the H_i.

(2) $H_i \bigcap \prod_{j \neq i} H_j = \{1\}$ for $i = 1, \dots, n$; thus it does not matter in which order the H_i are listed.

(3) $H_i \bigcap \prod_{j=1}^{i-1} H_j = \{1\}$ for $i = 1, \dots, n$.

Proof. (1) implies (2): If g belongs to the product of the $H_j, j \neq i$, then g can be written as $h_1 \cdots h_n$ where $h_i = 1$ and $h_j \in H_j$ for $j \neq i$. But if g also belongs to H_i then g can be written as $k_1 \cdots k_n$ where $k_i = g$ and $k_j = 1$ for $j \neq i$. By uniqueness of representation in the internal direct product, $h_i = k_i = 1$ for all i, so $g = 1$.

(2) implies (3): If g belongs to H_i and, in addition, $g = h_1 \cdots h_{i-1}$ with $h_j \in H_j$, then $g = h_1 \cdots h_{i-1} 1_{H_{i+1}} \cdots 1_{H_n}$, hence $g = 1$ by (2).

(3) implies (1): If $g \in G$ then since $G = H_1 \cdots H_n$ we have $g = h_1 \cdots h_n$ with $h_i \in H_i$. Suppose that we have another representation $g = k_1 \cdots k_n$ with $k_i \in H_i$. Let i be the largest integer such that $h_i \neq k_i$. If $i < n$ we can cancel the $h_t (= k_t), t > i$, to get $h_1 \cdots h_i = k_1 \cdots k_i$. If $i = n$ then $h_1 \cdots h_i = k_1 \cdots k_i$ by assumption. Now *any product of the H_i is a subgroup of G* (as in (1.5.2), $h_i h_j = h_j h_i$ for $i \neq j$, and the result follows from (1.3.6)). Therefore

$$h_i k_i^{-1} \in \prod_{j=1}^{i-1} H_j,$$

and since $h_i k_i^{-1} \in H_i$, we have $h_i k_i^{-1} = 1$ by (3). Therefore $h_i = k_i$, which is a contradiction. ♣

Problems For Section 1.5

In Problems 1–5, C_n is a cyclic group of order n, for example, $C_n = \{1, a, \dots, a^{n-1}\}$ with $a^n = 1$.

1. Let C_2 be a cyclic group of order 2. Describe the multiplication table of the direct product $C_2 \times C_2$. Is $C_2 \times C_2$ cyclic?

2. Show that $C_2 \times C_2$ is isomorphic to the four group (Section 1.2, Problem 6).
3. Show that the direct product $C_2 \times C_3$ is cyclic, in particular, it is isomorphic to C_6.
4. If n and m are relatively prime, show that $C_n \times C_m$ is isomorphic to C_{nm}, and is therefore cyclic.
5. If n and m are not relatively prime, show that $C_n \times C_m$ is not cyclic.
6. If p and q are distinct primes and $|G| = p, |H| = q$, show that the direct product $G \times H$ is cyclic.
7. If H and K are arbitrary groups, show that $H \times K \cong K \times H$.
8. If G, H and K are arbitrary groups, show that $G \times (H \times K) \cong (G \times H) \times K$. In fact, both sides are isomorphic to $G \times H \times K$.

Chapter 2

Ring Fundamentals

2.1 Basic Definitions and Properties

2.1.1 Definitions and Comments

A *ring* R is an abelian group with a multiplication operation $(a, b) \to ab$ that is associative and satisfies the distributive laws: $a(b+c) = ab+ac$ and $(a+b)c = ab+ac$ for all $a, b, c \in R$. We will always assume that R has at least two elements, including a multiplicative identity 1_R satisfying $a1_R = 1_Ra = a$ for all a in R. The multiplicative identity is often written simply as 1, and the additive identity as 0. If $a, b,$ and c are arbitrary elements of R, the following properties are derived quickly from the definition of a ring; we sketch the technique in each case.

(1) $a0 = 0a = 0$ $[a0 + a0 = a(0+0) = a0;\ 0a + 0a = (0+0)a = 0a]$

(2) $(-a)b = a(-b) = -(ab)$ $[0 = 0b = (a + (-a))b = ab + (-a)b,\ \text{so}\ (-a)b = -(ab)]$
$[0 = a0 = a(b + (-b)) = ab + a(-b),\ \text{so}\ a(-b) = -(ab)]$

(3) $(-1)(-1) = 1$ [take $a = 1, b = -1$ in (2)]

(4) $(-a)(-b) = ab$ [replace b by $-b$ in (2)]

(5) $a(b - c) = ab - ac$ $[a(b + (-c)) = ab + a(-c) = ab + (-(ac)) = ab - ac]$

(6) $(a - b)c = ac - bc$ $[(a + (-b))c = ac + (-b)c) = ac - (bc) = ac - bc]$

(7) $1 \neq 0$ [If $1 = 0$ then for all a we have $a = a1 = a0 = 0$, so $R = \{0\}$, contradicting the assumption that R has at least two elements]

(8) The multiplicative identity is unique [If $1'$ is another multiplicative identity then $1 = 11' = 1'$]

2.1.2 Definitions and Comments

If a and b are nonzero but $ab = 0$, we say that a and b are *zero divisors*; if $a \in R$ and for some $b \in R$ we have $ab = ba = 1$, we say that a is a *unit* or that a is *invertible*.

Note that ab need not equal ba; if this holds for all $a, b \in R$, we say that R is a *commutative ring*.

An *integral domain* is a commutative ring with no zero divisors.

A *division ring* or *skew field* is a ring in which every nonzero element a has a multiplicative inverse a^{-1} (i.e., $aa^{-1} = a^{-1}a = 1$). Thus the nonzero elements form a group under multiplication.

A *field* is a commutative division ring. Intuitively, in a ring we can do addition, subtraction and multiplication without leaving the set, while in a field (or skew field) we can do division as well.

Any finite integral domain is a field. To see this, observe that if $a \neq 0$, the map $x \to ax$, $x \in R$, is injective because R is an integral domain. If R is finite, the map is surjective as well, so that $ax = 1$ for some x.

The *characteristic* of a ring R (written Char R) is the smallest positive integer such that $n1 = 0$, where $n1$ is an abbreviation for $1 + 1 + \dots 1$ (n times). If $n1$ is never 0, we say that R has *characteristic 0*. Note that the characteristic can never be 1, since $1_R \neq 0$. If R is an integral domain and Char $R \neq 0$, then Char R must be a prime number. For if Char$R = n = rs$ where r and s are positive integers greater than 1, then $(r1)(s1) = n1 = 0$, so either $r1$ or $s1$ is 0, contradicting the minimality of n.

A *subring* of a ring R is a subset S of R that forms a ring under the operations of addition and multiplication defined on R. In other words, S is an additive subgroup of R that contains 1_R and is closed under multiplication. Note that 1_R is automatically the multiplicative identity of S, since the multiplicative identity is unique (see (8) of 2.1.1).

2.1.3 Examples

1. The integers $\mathbb{Z}$ form an integral domain that is not a field.

2. Let $\mathbb{Z}_n$ be the integers modulo n, that is, $\mathbb{Z}_n = \{0, 1, \dots, n-1\}$ with addition and multiplication mod n. (If $a \in \mathbb{Z}_n$ then a is identified with all integers $a + kn$, $k = 0, \pm 1, \pm 2, \dots \}$. Thus, for example, in $\mathbb{Z}_9$ the multiplication of 3 by 4 results in 3 since $12 \equiv 3 \bmod 9$, and therefore 12 is identified with 3.

$\mathbb{Z}_n$ is a ring, which is an integral domain (and therefore a field, since $\mathbb{Z}_n$ is finite) if and only if n is prime. For if $n = rs$ then $rs = 0$ in $\mathbb{Z}_n$; if n is prime then every nonzero element in $\mathbb{Z}_n$ has a multiplicative inverse, by Fermat's little theorem 1.3.4.

Note that by definition of characteristic, any field of prime characteristic p contains an isomorphic copy of $\mathbb{Z}_p$. Any field of characteristic 0 contains a copy of $\mathbb{Z}$, hence a copy of the rationals $\mathbb{Q}$.

3. If $n \geq 2$, then the set $M_n(R)$ of all n by n matrices with coefficients in a ring R forms a noncommutative ring, with the identity matrix I_n as multiplicative identity. If we identify the element $c \in R$ with the diagonal matrix cI_n, we may regard R as a subring of $M_n(R)$. It is possible for the product of two nonzero matrices to be zero, so that $M_n(R)$ is not an integral domain. (To generate a large class of examples, let E_{ij} be the matrix with 1 in row i, column j, and 0's elsewhere. Then $E_{ij}E_{kl} = \delta_{jk}E_{il}$, where δ_{jk} is 1 when $j = k$, and 0 otherwise.)

4. Let $1, i, j$ and k be basis vectors in 4-dimensional Euclidean space, and define multiplication of these vectors by

$$i^2 = j^2 = k^2 = -1, \quad ij = k, \quad jk = i, \quad ki = j, \quad ji = -ij, \quad kj = -jk, \quad ik = -ki \quad (1)$$

Let H be the set of all linear combinations $a + bi + cj + dk$ where a, b, c and d are real numbers. Elements of H are added componentwise and multiplied according to the above rules, i.e.,

$$(a + bi + cj + dk)(x + yi + zj + wk) = (ax - by - cz - dw) + (ay + bx + cw - dz)i \\ + (az + cx + dy - bw)j + (aw + dx + bz - cy)k.$$

H (after Hamilton) is called the ring of *quaternions*. In fact H is a division ring; the inverse of $a + bi + cj + dk$ is $(a^2 + b^2 + c^2 + d^2)^{-1}(a - bi - cj - dk)$.

H can also be represented by 2 by 2 matrices with complex entries, with multiplication of quaternions corresponding to ordinary matrix multiplication. To see this, let

$$\mathbf{1} = \begin{bmatrix} 1 & 0 \\ 0 & 1 \end{bmatrix}, \quad \mathbf{i} = \begin{bmatrix} i & 0 \\ 0 & -i \end{bmatrix}, \quad \mathbf{j} = \begin{bmatrix} 0 & 1 \\ -1 & 0 \end{bmatrix}, \quad \mathbf{k} = \begin{bmatrix} 0 & i \\ i & 0 \end{bmatrix};$$

a direct computation shows that $\mathbf{1}, \mathbf{i}, \mathbf{j}$ and $\mathbf{k}$ obey the multiplication rules (1) given above. Thus we may identify the quaternion $a + bi + cj + dk$ with the matrix

$$a\mathbf{1} + b\mathbf{i} + c\mathbf{j} + d\mathbf{k} = \begin{bmatrix} a + bi & c + di \\ -c + di & a - bi \end{bmatrix}$$

(where in the matrix, i is $\sqrt{-1}$, not the quaternion i).

The set of 8 elements ± 1, $\pm i$, $\pm j$, $\pm k$ forms a group under multiplication; it is called the *quaternion group*.

5. If R is a ring, then $R[X]$, the set of all polynomials in X with coefficients in R, is also a ring under ordinary polynomial addition and multiplication, as is $R[X_1, \dots, X_n]$, the set of polynomials in n variables $X_i, 1 \leq i \leq n$, with coefficients in R. Formally, the polynomial $A(X) = a_0 + a_1X + \cdots + a_nX^n$ is simply the sequence $(a_0, \dots, a_n)$; the symbol X is a placeholder. The product of two polynomials $A(X)$ and $B(X)$ is a polynomial whose X^k-coefficient is $a_0b_k + a_1b_{k-1} + \cdots + a_kb_0$. If we wish to evaluate a polynomial on R, we use the *evaluation map*

$$a_0 + a_1X + \cdots + a_nX^n \to a_0 + a_1x + \cdots + a_nx^n$$

where x is a particular element of R. A nonzero polynomial can evaluate to 0 at all points of R. For example, $X^2 + X$ evaluates to 0 on $\mathbb{Z}_2$, the field of integers modulo 2, since $1 + 1 = 0 \bmod 2$. We will say more about evaluation maps in Section 2.5, when we study polynomial rings.

6. If R is a ring, then $R[[X]]$, the set of *formal power series*

$$a_0 + a_1X + a_2X^2 + \dots$$

with coefficients in R, is also a ring under ordinary addition and multiplication of power series. The definition of multiplication is purely formal and convergence is never mentioned; we simply define the coefficient of X^n in the product of $a_0 + a_1X + a_2X^2 + \dots$ and $b_0 + b_1X + b_2X^2 + \dots$ to be $a_0b_n + a_1b_{n-1} + \cdots + a_{n-1}b_1 + a_nb_0$.

In Examples 5 and 6, if R is an integral domain, so are $R[X]$ and $R[[X]]$. In Example 5, look at leading coefficients to show that if $f(X) \neq 0$ and $g(X) \neq 0$, then $f(X)g(X) \neq 0$. In Example 6, if $f(X)g(X) = 0$ with $f(X) \neq 0$, let a_i be the first nonzero coefficient of $f(X)$. Then $a_ib_j = 0$ for all j, and therefore $g(X) = 0$.

2.1.4 Lemma

The *generalized associative law* holds for multiplication in a ring. There is also a *generalized distributive law*:

$$(a_1 + \cdots + a_m)(b_1 + \cdots + b_n) = \sum_{i=1}^{m}\sum_{j=1}^{n} a_i b_j.$$

Proof. The argument for the generalized associative law is exactly the same as for groups; see the beginning of Section 1.1. The generalized distributive law is proved in two stages. First set $m = 1$ and work by induction on n, using the left distributive law $a(b + c) = ab + ac$. Then use induction on m and the right distributive law $(a + b)c = ac + bc$ on $(a_1 + \cdots + a_m + a_{m+1})(b_1 + \cdots + b_n)$. ♣

2.1.5 Proposition

$(a + b)^n = \sum_{k=0}^{n} \binom{n}{k} a^k b^{n-k}$, the *binomial theorem*, is valid in any ring, if $ab = ba$.

Proof. The standard proof via elementary combinatorial analysis works. Specifically, $(a + b)^n = (a + b) \dots (a + b)$, and we can expand this product by multiplying an element (a or b) from object 1 (the first $(a + b)$) times an element from object 2 times ... times an element from object n, in all possible ways. Since $ab = ba$, these terms are of the form $a^k b^{n-k}, 0 \le k \le n$. The number of terms corresponding to a given k is the number of ways of selecting k objects from a collection of n, namely $\binom{n}{k}$. ♣

Problems For Section 2.1

1. If R is a field, is $R[X]$ a field always? sometimes? never?
2. If R is a field, what are the units of $R[X]$?
3. Consider the ring of formal power series with rational coefficients.
   (a) Give an example of a nonzero element that does not have a multiplicative inverse, and thus is not a unit.
   (b) Give an example of a nonconstant element (one that is not simply a rational number) that does have a multiplicative inverse, and therefore is a unit.
4. Let $\mathbb{Z}[i]$ be the ring of *Gaussian integers* $a + bi$, where $i = \sqrt{-1}$ and a and b are integers. Show that $\mathbb{Z}[i]$ is an integral domain that is not a field.
5. Continuing Problem 4, what are the units of $\mathbb{Z}[i]$?
6. Establish the following quaternion identities:
   (a) $(x_1 + y_1 i + z_1 j + w_1 k)(x_2 - y_2 i - z_2 j - w_2 k)$
   $= (x_1x_2 + y_1y_2 + z_1z_2 + w_1w_2) + (-x_1y_2 + y_1x_2 - z_1w_2 + w_1z_2)i$
   $+ (-x_1z_2 + z_1x_2 - w_1y_2 + y_1w_2)j + (-x_1w_2 + w_1x_2 - y_1z_2 + z_1y_2)k$
   (b) $(x_2 + y_2 i + z_2 j + w_2 k)(x_1 - y_1 i - z_1 j - w_1 k)$
   $= (x_1x_2 + y_1y_2 + z_1z_2 + w_1w_2) + (x_1y_2 - y_1x_2 + z_1w_2 - w_1z_2)i$
   $+ (x_1z_2 - z_1x_2 + w_1y_2 - y_1w_2)j + (x_1w_2 - w_1x_2 + y_1z_2 - z_1y_2)k$

(c) The product of a quaternion $h = a + bi + cj + dk$ and its *conjugate* $h^* = a - bi - cj - dk$ is $a^2 + b^2 + c^2 + d^2$. If q and t are quaternions, then $(qt)^* = t^* q^*$.

7. Use Problem 6 to establish *Euler's Identity* for real numbers x_r, y_r, z_r, w_r, $r = 1, 2$:

$$\begin{aligned}(x_1^2 + y_1^2 + z_1^2 + w_1^2)(x_2^2 + y_2^2 + z_2^2 + w_2^2) &= (x_1x_2 + y_1y_2 + z_1z_2 + w_1w_2)^2 \\ &+ (x_1y_2 - y_1x_2 + z_1w_2 - w_1z_2)^2 \\ &+ (x_1z_2 - z_1x_2 + w_1y_2 - y_1w_2)^2 \\ &+ (x_1w_2 - w_1x_2 + y_1z_2 - z_1y_2)^2\end{aligned}$$

8. Recall that an endomorphism of a group G is a homomorphism of G to itself. Thus if G is abelian, an endomorphism is a function $f\colon G \to G$ such that $f(a + b) = f(a) + f(b)$ for all $a, b \in G$. Define addition of endomorphisms in the natural way, $(f+g)(a) = f(a)+g(a)$, and define multiplication as functional composition, $(fg)(a) = f(g(a))$. Show that the set End G of endomorphisms of G becomes a ring under these operations.

9. Continuing Problem 8, what are the units of End G?

10. It can be shown that every positive integer is the sum of 4 squares. A key step is to prove that if n and m can be expressed as sums of 4 squares, so can nm. Do this using Euler's identity (see Problem 7), and illustrate for the case $n = 34, m = 54$.

11. Which of the following collections of n by n matrices form a ring under matrix addition and multiplication?

    (a) symmetric matrices

    (b) matrices whose entries are 0 except possibly in column 1

    (c) lower triangular matrices ($a_{ij} = 0$ for $i < j$)

    (d) upper triangular matrices ($a_{ij} = 0$ for $i > j$)

2.2 Ideals, Homomorphisms, and Quotient Rings

Let $f\colon R \to S$, where R and S are rings. Rings are, in particular, abelian groups under addition, so we know what it means for f to be a group homomorphism: $f(a + b) = f(a) + f(b)$ for all a, b in R. It is then automatic that $f(0_R) = 0_S$ (see (1.3.11)). It is natural to consider mappings f that preserve multiplication as well as addition, i.e.,

$$f(a + b) = f(a) + f(b) \text{ and } f(ab) = f(a)f(b) \text{ for all } a, b \in R.$$

But here it does not follow that f maps the multiplicative identity 1_R to the multiplicative identity 1_S. We have $f(a) = f(a1_R) = f(a)f(1_R)$, but we cannot multiply on the left by $f(a)^{-1}$, which might not exist. We avoid this difficulty by only considering functions f that have the desired behavior.

2.2.1 Definition

If $f\colon R \to S$, where R and S are rings, we say that f is a *ring homomorphism* if $f(a+b) = f(a) + f(b)$ and $f(ab) = f(a)f(b)$ for all $a, b \in R$, and $f(1_R) = 1_S$.

2.2.2 Example

Let $f\colon \mathbb{Z} \to M_n(R)$, $n \geq 2$, be defined by $f(n) = nE_{11}$ (see 2.1.3, Example 3). Then we have $f(a+b) = f(a) + f(b), f(ab) = f(a)f(b)$, but $f(1) \neq I_n$. Thus f is not a ring homomorphism.

In Chapter 1, we proved the basic isomorphism theorems for groups, and a key observation was the connection between group homomorphisms and normal subgroups. We can prove similar theorems for rings, but first we must replace the normal subgroup by an object that depends on multiplication as well as addition.

2.2.3 Definitions and Comments

Let I be a subset of the ring R, and consider the following three properties:

(1) I is an additive subgroup of R.

(2) If $a \in I$ and $r \in R$ then $ra \in I$; in other words, $rI \subseteq I$ for every $r \in R$.

(3) If $a \in I$ and $r \in R$ then $ar \in I$; in other words, $Ir \subseteq I$ for every $r \in R$.

If (1) and (2) hold, I is said to be a *left ideal* of R. If (1) and (3) hold, I is said to be a *right ideal* of R. If all three properties are satisfied, I is said to be an *ideal* (or *two-sided ideal*) of R, a *proper ideal* if $I \neq R$, a *nontrivial ideal* if I is neither R nor $\{0\}$.

If $f\colon R \to S$ is a ring homomorphism, its *kernel* is

$$\ker f = \{r \in R\colon f(r) = 0\};$$

exactly as in (1.3.13), f is injective if and only if $\ker f = \{0\}$.

Now it follows from the definition of ring homomorphism that $\ker f$ is an ideal of R. The kernel must be a proper ideal because if $\ker f = R$ then f is identically 0, in particular, $f(1_R) = 1_S = 0_S$, a contradiction (see (7) of 2.1.1). Conversely, every proper ideal is the kernel of a ring homomorphism, as we will see in the discussion to follow.

2.2.4 Construction of Quotient Rings

Let I be a proper ideal of the ring R. Since I is a subgroup of the additive group of R, we can form the quotient group R/I, consisting of cosets $r + I$, $r \in R$. We define multiplication of cosets in the natural way:

$$(r + I)(s + I) = rs + I.$$

To show that multiplication is well-defined, suppose that $r + I = r' + I$ and $s + I = s' + I$, so that $r' - r$ is an element of I, call it a, and $s' - s$ is an element of I, call it b. Thus

$$r's' = (r + a)(s + b) = rs + as + rb + ab,$$

and since I is an ideal, we have $as \in I$, $rb \in I$, and $ab \in I$. Consequently, $r's' + I = rs + I$, so the multiplication of two cosets is independent of the particular representatives r and s that we choose. From our previous discussion of quotient groups, we know that the cosets of the ideal I form a group under addition, and the group is abelian because R

itself is an abelian group under addition. Since multiplication of cosets $r + I$ and $s + I$ is accomplished simply by multiplying the coset representatives r and s in R and then forming the coset $rs + I$, we can use the ring properties of R to show that the cosets of I form a ring, called the *quotient ring* of R by I. The identity element of the quotient ring is $1_R + I$, and the zero element is $0_R + I$. Furthermore, if R is a commutative ring, so is R/I. The fact that I is proper is used in verifying that R/I has at least two elements. For if $1_R + I = 0_R + I$, then $1_R = 1_R - 0_R \in I$; thus for any $r \in R$ we have $r = r1_R \in I$, so that $R = I$, a contradiction.

2.2.5 Proposition

Every proper ideal I is the kernel of a ring homomorphism.

Proof. Define the *natural* or *canonical* map $\pi \colon R \to R/I$ by $\pi(r) = r + I$. We already know that π is a homomorphism of abelian groups and its kernel is I (see (1.3.12)). To verify that π preserves multiplication, note that

$$\pi(rs) = rs + I = (r + I)(s + I) = \pi(r)\pi(s);$$

since

$$\pi(1_R) = 1_R + I = 1_{R/I},$$

π is a ring homomorphism. ♣

2.2.6 Proposition

Suppose $f \colon R \to S$ is a ring homomorphism and the only ideals of R are $\{0\}$ and R. (In particular, if R is a division ring, then R satisfies this hypothesis.) Then f is injective.

Proof. Let $I = \ker f$, an ideal of R (see 2.2.3). If $I = R$ then f is identically zero, and is therefore not a legal ring homomorphism since $f(1_R) = 1_S \neq 0_S$. Thus $I = \{0\}$, so that f is injective.

If R is a division ring, then in fact R has no nontrivial left or right ideals. For suppose that I is a left ideal of R and $a \in I$, $a \neq 0$. Since R is a division ring, there is an element $b \in R$ such that $ba = 1$, and since I is a left ideal, we have $1 \in I$, which implies that $I = R$. If I is a right ideal, we choose the element b such that $ab = 1$. ♣

2.2.7 Definitions and Comments

If X is a nonempty subset of the ring R, then $\langle X \rangle$ will denote the *ideal generated by* X, that is, the smallest ideal of R that contains X. Explicitly,

$$\langle X \rangle = RXR = \text{the collection of all finite sums of the form } \sum_i r_i x_i s_i$$

with $r_i, s_i \in R$ and $x_i \in X$. To show that this is correct, verify that the finite sums of the given type form an ideal containing X. On the other hand, if J is any ideal containing X, then all finite sums $\sum_i r_i x_i s_i$ must belong to J.

If R is commutative, then $rxs = rsx$, and we may as well drop the s. In other words:

In a commutative ring, $\langle X \rangle = RX =$ all finite sums $\sum_i r_i x_i,\ r_i \in R, x_i \in X$.

An ideal generated by a single element a is called a *principal ideal* and is denoted by $\langle a \rangle$ or (a). In this case, $X = \{a\}$, and therefore:

In a commutative ring, the principal ideal generated by a is $\langle a \rangle = \{ra \colon r \in R\}$,

the set of all multiples of a, sometimes denoted by Ra.

2.2.8 Definitions and Comments

In an arbitrary ring, we will sometimes need to consider the *sum* of two ideals I and J, defined as $\{x + y \colon x \in I, y \in J\}$. It follows from the distributive laws that $I + J$ is also an ideal. Similarly, the sum of two left [resp. right] ideals is a left [resp. right] ideal.

Problems For Section 2.2

1. What are the ideals in the ring of integers?
2. Let $M_n(R)$ be the ring of n by n matrices with coefficients in the ring R. If C_k is the subset of $M_n(R)$ consisting of matrices that are 0 except perhaps in column k, show that C_k is a left ideal of $M_n(R)$. Similarly, if R_k consists of matrices that are 0 except perhaps in row k, then R_k is a right ideal of $M_n(R)$.
3. In Problem 2, assume that R is a division ring, and let E_{ij} be the matrix with 1 in row i, column j, and 0's elsewhere.
   (a) If $A \in M_n(R)$, show that $E_{ij}A$ has row j of A as its i^{th}row, with 0's elsewhere.
   (b) Now suppose that $A \in C_k$. Show that $E_{ij}A$ has a_{jk} in the ik position, with 0's elsewhere, so that if a_{jk} is not zero, then $a_{jk}^{-1} E_{ij} A = E_{ik}$.
   (c) If A is a nonzero matrix in C_k with $a_{jk} \neq 0$, and C is any matrix in C_k, show that
   $$\sum_{i=1}^{n} c_{ik} a_{jk}^{-1} E_{ij} A = C.$$
4. Continuing Problem 3, if a nonzero matrix A in C_k belongs to the left ideal I of $M_n(R)$, show that every matrix in C_k belongs to I. Similarly, if a nonzero matrix A in R_k belongs to the right ideal I of $M_n(R)$, every matrix in R_k belongs to I.
5. Show that if R is a division ring, then $M_n(R)$ has no nontrivial two-sided ideals.
6. In $R[X]$, express the set I of polynomials with no constant term as $\langle f \rangle$ for an appropriate f, and thus show that I is a principal ideal.
7. Let R be a commutative ring whose only proper ideals are $\{0\}$ and R. Show that R is a field.
8. Let R be the ring $\mathbb{Z}_n$ of integers modulo n, where n may be prime or composite. Show that every ideal of R is principal.

2.3 The Isomorphism Theorems For Rings

The basic ring isomorphism theorems may be proved by adapting the arguments used in Section 1.4 to prove the analogous theorems for groups. Suppose that I is an ideal of the ring R, f is a ring homomorphism from R to S with kernel K, and π is the natural map, as indicated in Figure 2.3.1. To avoid awkward analysis of special cases, let us make a blanket assumption that any time a quotient ring R_0/I_0 appears in the statement of a theorem, the ideal I_0 is proper.

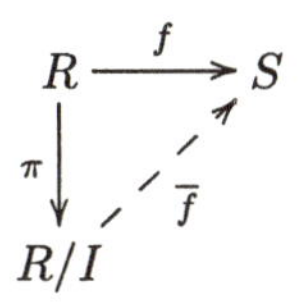

Figure 2.3.1

2.3.1 Factor Theorem For Rings

Any ring homomorphism whose kernel contains I can be factored through R/I. In other words, in Figure 2.3.1 there is a unique ring homomorphism $\overline{f}\colon R \to S$ that makes the diagram commutative. Furthermore,

(i) $\overline{f}$ is an epimorphism if and only if f is an epimorphism;

(ii) $\overline{f}$ is a monomorphism if and only if $\ker f = I$;

(iii) $\overline{f}$ is an isomorphism if and only if f is an epimorphism and $\ker f = I$.

Proof. The only possible way to define $\overline{f}$ is $\overline{f}(a + I) = f(a)$. To verify that $\overline{f}$ is well-defined, note that if $a + I = b + I$, then $a - b \in I \subseteq K$, so $f(a - b) = 0$, i.e.,$f(a) = f(b)$. Since f is a ring homomorphism, so is $\overline{f}$. To prove (i), (ii) and (iii), the discussion in (1.4.1) may be translated into additive notation and copied. ♣

2.3.2 First Isomorphism Theorem For Rings

If $f\colon R \to S$ is a ring homomorphism with kernel K, then the image of f is isomorphic to R/K.

Proof. Apply the factor theorem with $I = K$, and note that f is an epimorphism onto its image. ♣

2.3.3 Second Isomorphism Theorem For Rings

Let I be an ideal of the ring R, and let S be a subring of R. Then:

(a) $S + I (= \{x + y\colon x \in S, y \in I\})$ is a subring of R;

(b) I is an ideal of $S + I$;

(c) $S \cap I$ is an ideal of S;

(d) $(S+I)/I$ is isomorphic to $S/(S \cap I)$, as suggested by the "parallelogram" or "diamond" diagram in Figure 2.3.2.

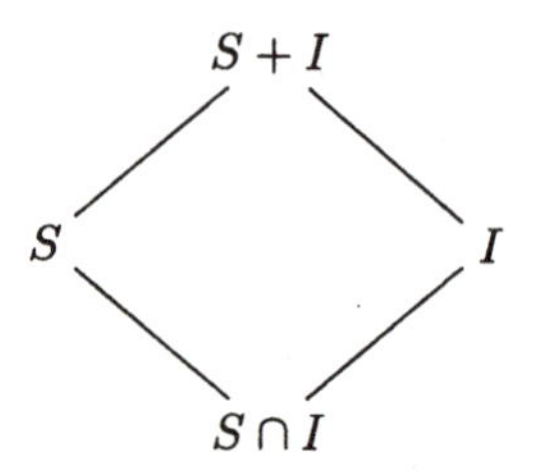

Figure 2.3.2

Proof. (a) Verify directly that $S+I$ is an additive subgroup of R that contains 1_R (since $1_R \in S$ and $0_R \in I$) and is closed under multiplication. For example, if $a \in S$, $x \in I$, $b \in S$, $y \in I$, then $(a+x)(b+y) = ab + (ay + xb + xy) \in S+I$.

(b) Since I is an ideal of R, it must be an ideal of the subring $S+I$.

(c) This follows from the definitions of subring and ideal.

(d) Let $\pi \colon R \to R/I$ be the natural map, and let π_0 be the restriction of π to S. Then π_0 is a ring homomorphism whose kernel is $S \cap I$ and whose image is $\{a+I\colon a \in S\} = (S+I)/I$. (To justify the last equality, note that if $s \in S$ and $x \in I$ we have $(s+x)+I = s+I$.) By the first isomorphism theorem for rings, $S/\ker \pi_0$ is isomorphic to the image of π_0, and (d) follows. ♣

2.3.4 Third Isomorphism Theorem For Rings

Let I and J be ideals of the ring R, with $I \subseteq J$. Then J/I is an ideal of R/I, and $R/J \cong (R/I)/(J/I)$.

Proof. Define $f\colon R/I \to R/J$ by $f(a+I) = a+J$. To check that f is well-defined, suppose that $a+I = b+I$. Then $a-b \in I \subseteq J$, so $a+J = b+J$. By definition of addition and multiplication of cosets in a quotient ring, f is a ring homomorphism. Now

$$\ker f = \{a+I\colon a+J = J\} = \{a+I\colon a \in J\} = J/I$$

and

$$\operatorname{im} f = \{a+J\colon a \in R\} = R/J$$

(where im denotes image). The result now follows from the first isomorphism theorem for rings. ♣

2.3.5 Correspondence Theorem For Rings

If I is an ideal of the ring R, then the map $S \to S/I$ sets up a one-to-one correspondence between the set of all subrings of R containing I and the set of all subrings of R/I, as well as a one-to-one correspondence between the set of all ideals of R containing I and the set of all ideals of R/I. The inverse of the map is $Q \to \pi^{-1}(Q)$, where π is the canonical map: $R \to R/I$.

Proof. The correspondence theorem for groups yields a one-to-one correspondence between additive subgroups of R containing I and additive subgroups of R/I. We must check that subrings correspond to subrings and ideals to ideals. If S is a subring of R then S/I is closed under addition, subtraction and multiplication. For example, if s and s' belong to S, we have $(s+I)(s'+I) = ss'+I \in S/I$. Since $1_R \in S$ we have $1_R+I \in S/I$, proving that S/I is a subring of R/I. Conversely, if S/I is a subring of R/I, then S is closed under addition, subtraction and multiplication, and contains the identity, hence is a subring or R. For example, if $s, s' \in S$ then $(s+I)(s'+I) \in S/I$, so that $ss'+I = t+I$ for some $t \in S$, and therefore $ss' - t \in I$. But $I \subseteq S$, so $ss' \in S$.

Now if J is an ideal of R containing I, then J/I is an ideal of R/I by the third isomorphism theorem for rings. Conversely, let J/I be an ideal of R/I. If $r \in R$ and $x \in J$ then $(r+I)(x+I) \in J/I$, that is, $rx + I \in J/I$. Thus for some $j \in J$ we have $rx - j \in I \subseteq J$, so $rx \in J$. A similar argument shows that $xr \in J$, and that J is an additive subgroup of R. It follows that J is an ideal of R. ♣

We now consider the Chinese remainder theorem, which is an abstract version of a result in elementary number theory. Along the way, we will see a typical application of the first isomorphism theorem for rings; in fact the development of any major theorem of algebra is likely to include an appeal to one or more of the isomorphism theorems. The following observations may make the ideas easier to visualize.

2.3.6 Definitions and Comments

(i) If a and b are integers that are congruent modulo n, then $a - b$ is a multiple of n. Thus $a - b$ belongs to the ideal I_n consisting of all multiples of n in the ring $\mathbb{Z}$ of integers. Thus we may say that a is congruent to b modulo I_n. In general, if $a, b \in R$ and I is an ideal of R, we say that $a \equiv b \bmod I$ if $a - b \in I$.

(ii) The integers a and b are relatively prime if and only if the integer 1 can be expressed as a linear combination of a and b. Equivalently, the sum of the ideals I_a and I_b is the entire ring $\mathbb{Z}$. In general, we say that the ideals I and J in the ring R are *relatively prime* if $I + J = R$.

(iii) If I_{n_i} consists of all multiples of n_i in the ring of integers $(i = 1, \dots k)$, then the intersection $\cap_{i=1}^k I_{n_i}$ is I_r where r is the least common multiple of the n_i. If the n_i are relatively prime in pairs, then r is the product of the n_i.

(iv) If $R_1, \dots, R_n$ are rings, the *direct product* of the R_i is defined as the ring of n-tuples $(a_1, \dots, a_n)$, $a_i \in R_i$, with componentwise addition and multiplication, that is, with

$$(a_1, \dots, a_n) + (b_1, \dots, b_n) = (a_1 + b_1, \dots, a_n + b_n),$$
$$(a_1, \dots, a_n)(b_1, \dots, b_n) = (a_1 b_1, \dots, a_n b_n).$$

The zero element is $(0, \dots, 0)$ and the multiplicative identity is $(1, \dots, 1)$.

2.3.7 Chinese Remainder Theorem

Let R be an arbitrary ring, and let $I_1, \dots, I_n$ be ideals in R that are relatively prime in pairs, that is, $I_i + I_j = R$ for all $i \neq j$.

(1) If $a_1 = 1$ (the multiplicative identity of R) and $a_j = 0$ (the zero element of R) for $j = 2, \dots, n$, then there is an element $a \in R$ such that $a \equiv a_i \bmod I_i$ for all $i = 1, \dots, n$.

(2) More generally, if $a_1, \dots, a_n$ are arbitrary elements of R, there is an element $a \in R$ such that $a \equiv a_i \bmod I_i$ for all $i = 1, \dots, n$.

(3) If b is another element of R such that $b \equiv a_i \bmod I_i$ for all $i = 1, \dots, n$, then $b \equiv a \bmod I_1 \cap I_2 \cap \dots \cap I_n$. Conversely, if $b \equiv a \bmod \cap_{i=1}^n I_i$, then $b \equiv a_i \bmod I_i$ for all i.

(4) $R/\bigcap_{i=1}^n I_i$ is isomorphic to the direct product $\prod_{i=1}^n R/I_i$.

Proof. (1) If $j > 1$ we have $I_1 + I_j = R$, so there exist elements $b_j \in I_1$ and $c_j \in I_j$ such that $b_j + c_j = 1$; thus

$$\prod_{j=2}^{n} (b_j + c_j) = 1.$$

Expand the left side and observe that any product containing at least one b_j belongs to I_1, while $c_2 \dots c_n$ belongs to $\prod_{j=2}^n I_j$, the collection of all finite sums of products $x_2 \dots x_n$ with $x_j \in I_j$. Thus we have elements $b \in I_1$ and $a \in \prod_{j=2}^n I_j$ (a subset of each I_j) with $b + a = 1$. Consequently, $a \equiv 1 \bmod I_1$ and $a \equiv 0 \bmod I_j$ for $j > 1$, as desired.

(2) By the argument of part (1), for each i we can find c_i with $c_i \equiv 1 \bmod I_i$ and $c_i \equiv 0 \bmod I_j, j \neq i$. If $a = a_1 c_1 + \dots + a_n c_n$, then a has the desired properties. To see this, write $a - a_i = a - a_i c_i + a_i(c_i - 1)$, and note that $a - a_i c_i$ is the sum of the $a_j c_j, j \neq i$, and is therefore congruent to 0 mod I_i.

(3) We have $b \equiv a_i \bmod I_i$ for all i iff $b - a \equiv 0 \bmod I_i$ for all i, that is, iff $b - a \in \cap_{i=1}^n I_i$, and the result follows.

(4) Define $f \colon R \to \prod_{i=1}^n R/I_i$ by $f(a) = (a + I_1, \dots, a + I_n)$. If $a_1, \dots, a_n \in R$, then by part (2) there is an element $a \in R$ such that $a \equiv a_i \bmod I_i$ for all i. But then $f(a) = (a_1 + I_1, \dots, a_n + I_n)$, proving that f is surjective. Since the kernel of f is the intersection of the ideals I_j, the result follows from the first isomorphism theorem for rings. ♣

The concrete version of the Chinese remainder theorem can be recovered from the abstract result; see Problems 3 and 4.

Problems For Section 2.3

1. Show that the group isomorphisms of Section 1.4, Problems 1 and 2, are ring isomorphisms as well.
2. Give an example of an ideal that is not a subring, and a subring that is not an ideal.
3. If the integers $m_i, i = 1, \dots, n$, are relatively prime in pairs, and $a_1, \dots, a_n$ are arbitrary integers, show that there is an integer a such that $a \equiv a_i \bmod m_i$ for all i, and that any two such integers are congruent modulo $m_1 \dots m_n$.
4. If the integers $m_i, i = 1, \dots, n$, are relatively prime in pairs and $m = m_1 \dots m_n$, show that there is a ring isomorphism between $\mathbb{Z}_m$ and the direct product $\prod_{i=1}^n \mathbb{Z}_{m_i}$. Specifically, $a \bmod m$ corresponds to $(a \bmod m_1, \dots, a \bmod m_n)$.
5. Suppose that $R = R_1 \times R_2$ is a direct product of rings. Let R_1' be the ideal $R_1 \times \{0\} = \{(r_1, 0) : r_1 \in R_1\}$, and let R_2' be the ideal $\{(0, r_2) : r_2 \in R_2\}$. Show that $R/R_1' \cong R_2$ and $R/R_2' \cong R_1$.
6. If $I_1, \dots, I_n$ are ideals, the *product* $I_1 \dots I_n$ is defined as the set of all finite sums $\sum_i a_{1i} a_{2i} \dots a_{ni}$, where $a_{ki} \in I_k$, $k = 1, \dots, n$. [See the proof of part (1) of 2.3.7; a brief check shows that the product of ideals is an ideal.]

   Assume that R is a commutative ring. Under the hypothesis of the Chinese remainder theorem, show that the intersection of the ideals I_i coincides with their product.
7. Let $I_1, \dots, I_n$ be ideals in the ring R. Suppose that $R/\cap_i I_i$ is isomorphic to $\prod_i R/I_i$ via $a + \cap_i I_i \to (a + I_1, \dots, a + I_n)$. Show that the ideals I_i are relatively prime in pairs.

2.4 Maximal and Prime Ideals

If I is an ideal of the ring R, we might ask"What is the smallest ideal containing I" and "What is the largest ideal containing I". Neither of these questions is challenging; the smallest ideal is I itself, and the largest ideal is R. But if I is a proper ideal and we ask for a maximal proper ideal containing I, the question is much more interesting.

2.4.1 Definition

A *maximal ideal* in the ring R is a proper ideal that is not contained in any strictly larger proper ideal.

2.4.2 Theorem

Every proper ideal I of the ring R is contained in a maximal ideal. Consequently, every ring has at least one maximal ideal.

Proof. The argument is a prototypical application of Zorn's lemma. Consider the collection of all proper ideals containing I, partially ordered by inclusion. Every chain $\{J_t, t \in T\}$ of proper ideals containing I has an upper bound, namely the union of the chain. (Note that the union is still a proper ideal, because the identity 1_R belongs to none of the ideals J_t.) By Zorn, there is a maximal element in the collection, that is, a maximal ideal containing I. Now take $I = \{0\}$ to conclude that every ring has at least one maximal ideal. ♣

We have the following characterization of maximal ideals.

2.4.3 Theorem

Let M be an ideal in the commutative ring R. Then M is a maximal ideal if and only if R/M is a field.

Proof. Suppose M is maximal. We know that R/M is a ring (see 2.2.4); we need to find the multiplicative inverse of the element $a + M$ of R/M, where $a + M$ is not the zero element, i.e., $a \notin M$. Since M is maximal, the ideal $Ra + M$, which contains a and is therefore strictly larger than M, must be the ring R itself. Consequently, the identity element 1 belongs to $Ra + M$. If $1 = ra + m$ where $r \in R$ and $m \in M$, then

$$(r + M)(a + M) = ra + M = (1 - m) + M = 1 + M \text{ (since } m \in M),$$

proving that $r + M$ is the multiplicative inverse of $a + M$.

Conversely, if R/M is a field, then M must be a proper ideal. If not, then $M = R$, so that R/M contains only one element, contradicting the requirement that $1 \neq 0$ in R/M (see (7) of 2.1.1). By (2.2.6), the only ideals of R/M are $\{0\}$ and R/M, so by the correspondence theorem 2.3.5, there are no ideals properly between M and R. Therefore M is a maximal ideal. ♣

If in (2.4.3) we relax the requirement that R/M be a field, we can identify another class of ideals.

2.4.4 Definition

A *prime ideal* in a commutative ring R is a proper ideal P such that for any two elements a, b in R,

$$ab \in P \text{ implies that } a \in P \text{ or } b \in P.$$

We can motivate the definition by looking at the ideal (p) in the ring of integers. In this case, $a \in (p)$ means that p divides a, so that (p) will be a prime ideal if and only if

$$p \text{ divides } ab \text{ implies that } p \text{ divides } a \text{ or } p \text{ divides } b,$$

which is equivalent to the requirement that p be a prime number.

2.4.5 Theorem

If P is an ideal in the commutative ring R, then P is a prime ideal if and only if R/P is an integral domain.

Proof. Suppose P is prime. Since P is a proper ideal, R/P is a ring. We must show that if $(a+P)(b+P)$ is the zero element P in R/P, then $a+P=P$ or $b+P=P$, i.e., $a \in P$ or $b \in P$. This is precisely the definition of a prime ideal.

Conversely, if R/P is an integral domain, then, as in (2.4.3), P is a proper ideal. If $ab \in P$, then $(a+P)(b+P)$ is zero in R/P, so that $a+P=P$ or $b+P=P$, i.e., $a \in P$ or $b \in P$. ♣

2.4.6 Corollary

In a commutative ring, a maximal ideal is prime.

Proof. This is immediate from (2.4.3) and (2.4.5). ♣

2.4.7 Corollary

Let $f\colon R \to S$ be an epimorphism of commutative rings. Then:

(i) If S is a field then $\ker f$ is a maximal ideal of R;

(ii) If S is an integral domain then $\ker f$ is a prime ideal of R.

Proof. By the first isomorphism theorem (2.3.2), S is isomorphic to $R/\ker f$, and the result now follows from (2.4.3) and (2.4.5). ♣

2.4.8 Example

Let $\mathbb{Z}[X]$ be the set of all polynomials $f(X) = a_0 + a_1X + \cdots + a_nX^n, n = 0, 1, \ldots$ in the indeterminate X, with integer coefficients. The ideal generated by X, that is, the collection of all multiples of X, is

$$\langle X \rangle = \{f(X) \in \mathbb{Z}[X]\colon a_0 = 0\}.$$

The ideal generated by 2 is

$$\langle 2 \rangle = \{f(X) \in \mathbb{Z}[X]\colon \text{all } a_i \text{ are even integers.}\}$$

Both $\langle X \rangle$ and $\langle 2 \rangle$ are proper ideals, since $2 \notin \langle X \rangle$ and $X \notin \langle 2 \rangle$. In fact we can say much more. Consider the ring homomorphisms $\varphi\colon \mathbb{Z}[X] \to \mathbb{Z}$ and $\psi\colon \mathbb{Z}[X] \to \mathbb{Z}_2$ given by $\varphi(f(X)) = a_0$ and $\psi(f(X)) = \overline{a}_0$, where $\overline{a}_0$ is a_0 reduced modulo 2. We will show that both $\langle X \rangle$ and $\langle 2 \rangle$ are prime ideals that are not maximal.

First note that by (2.4.7), $\langle X \rangle$ is prime because it is the kernel of φ. Then observe that $\langle X \rangle$ is not maximal because it is properly contained in $\langle 2, X \rangle$, the ideal generated by 2 and X.

To verify that $\langle 2 \rangle$ is prime, note that it is the kernel of the homomorphism from $\mathbb{Z}[X]$ onto $\mathbb{Z}_2[X]$ that takes $f(X)$ to $\overline{f}(X)$, where the overbar indicates that the coefficients of $f(X)$ are to be reduced modulo 2. Since $\mathbb{Z}_2[X]$ is an integral domain (see the comment at the end of 2.1.3), $\langle 2 \rangle$ is a prime ideal. Since $\langle 2 \rangle$ is properly contained in $\langle 2, X \rangle$, $\langle 2 \rangle$ is not maximal.

Finally, $\langle 2, X\rangle$ is a maximal ideal, since

$$\ker\psi = \{a_0 + Xg(X)\colon a_0 \text{ is even and } g(X) \in \mathbb{Z}[X]\} = \langle 2, X\rangle.$$

Thus $\langle 2, X\rangle$ is the kernel of a homomorphism onto a field, and the result follows from (2.4.7).

2.4.9 Problems For Section 2.4

1. We know from Problem 1 of Section 2.2 that in the ring of integers, all ideals I are of the form $\langle n\rangle$ for some $n \in \mathbb{Z}$, and since $n \in I$ implies $-n \in I$, we may take n to be nonnegative. Let $\langle n\rangle$ be a nontrivial ideal, so that n is a positive integer greater than 1. Show that $\langle n\rangle$ is a prime ideal if and only if n is a prime number.
2. Let I be a nontrivial prime ideal in the ring of integers. Show that in fact I must be maximal.
3. Let $F[[X]]$ be the ring of formal power series with coefficients in the field F (see (2.1.3), Example 6). Show that $\langle X\rangle$ is a maximal ideal.
4. Perhaps the result of Problem 3 is a bit puzzling. Why can't we argue that just as in (2.4.8), $\langle X\rangle$ is properly contained in $\langle 2, X\rangle$, and therefore $\langle X\rangle$ is not maximal?
5. Let I be a proper ideal of $F[[X]]$. Show that $I \subseteq \langle X\rangle$, so that $\langle X\rangle$ is the unique maximal ideal of $F[[X]]$. (A commutative ring with a unique maximal ideal is called a *local ring*.)
6. Show that every ideal of $F[[X]]$ is principal, and specifically every nonzero ideal is of the form (X^n) for some $n = 0, 1, \dots$.
7. Let $f\colon R \to S$ be a ring homomorphism, with R and S commutative. If P is a prime ideal of S, show that the preimage $f^{-1}(P)$ is a prime ideal of R.
8. Show that the result of Problem 7 does not hold in general when P is a maximal ideal.
9. Show that a prime ideal P cannot be the intersection of two strictly larger ideals I and J.

2.5 Polynomial Rings

In this section, *all rings are assumed commutative.* To see a good reason for this restriction, consider the *evaluation map* (also called the *substitution map*) E_x, where x is a fixed element of the ring R. This map assigns to the polynomial $a_0 + a_1X + \cdots + a_nX^n$ in $R[X]$ the value $a_0 + a_1x + \cdots + a_nx^n$ in R. It is tempting to say that "obviously", E_x is a ring homomorphism, but we must be careful. For example,

$$E_x[(a + bX)(c + dX)] = E_x(ac + (ad + bc)X + bdX^2) = ac + (ad + bc)x + bdx^2,$$
$$E_x(a + bX)E_x(c + dX) = (a + bx)(c + dx) = ac + adx + bxc + bxdx,$$

and these need not be equal if R is not commutative.

The *degree*, abbreviated deg, of a polynomial $a_0 + a_1X + \cdots + a_nX^n$ (with *leading coefficient* $a_n \neq 0$) is n; it is convenient to define the degree of the zero polynomial as $-\infty$. If f and g are polynomials in $R[X]$, where R is a *field*, ordinary long division allows us to express f as $qg + r$, where the degree of r is less than the degree of g. We have a similar result over an arbitrary commutative ring, if g is *monic*, i.e., the leading coefficient of g is 1. For example (with $R = \mathbb{Z}$), we can divide $2X^3 + 10X^2 + 16X + 10$ by $X^2 + 3X + 5$:

$$2X^3 + 10X^2 + 16X + 10 = 2X(X^2 + 3X + 5) + 4X^2 + 6X + 10.$$

The remainder $4X^2 + 6X + 10$ does not have degree less than 2, so we divide it by $X^2 + 3X + 5$:

$$4X^2 + 6X + 10 = 4(X^2 + 3X + 5) - 6X - 10.$$

Combining the two calculations, we have

$$2X^3 + 10X^2 + 16X + 10 = (2X + 4)(X^2 + 3X + 5) + (-6X - 10)$$

which is the desired decomposition.

2.5.1 Division Algorithm

If f and g are polynomials in $R[X]$, with g monic, there are unique polynomials q and r in $R[X]$ such that $f = qg + r$ and deg $r <$ deg g. If R is a field, g can be any nonzero polynomial.

Proof. The above procedure, which works in any ring R, shows that q and r exist.If $f = qg + r = q_1g + r_1$ where r and r_1 are of degree less than deg g, then $g(q - q_1) = r_1 - r$. But if $q - q_1 \neq 0$, then, since g is monic, the degree of the left side is at least deg g, while the degree of the right side is less than deg g, a contradiction. Therefore $q = q_1$, and consequently $r = r_1$. ♣

2.5.2 Remainder Theorem

If $f \in R[X]$ and $a \in R$, then for some unique polynomial $q(X)$ in $R[X]$ we have

$$f(X) = q(X)(X - a) + f(a);$$

hence $f(a) = 0$ if and only if $X - a$ divides $f(X)$.

Proof. By the division algorithm, we may write $f(X) = q(X)(X - a) + r(X)$ where the degree of r is less than 1, i.e., r is a constant. Apply the evaluation homomorphism $X \to a$ to show that $r = f(a)$. ♣

2.5.3 Theorem

If R is an integral domain, then a nonzero polynomial f in $R[X]$ of degree n has at most n roots in R, counting multiplicity.

Proof. If $f(a_1) = 0$, then by (2.5.2), possibly applied several times, we have $f(X) = q_1(X)(X - a_1)^{n_1}$, where $q_1(a_1) \neq 0$ and the degree of q_1 is $n - n_1$. If a_2 is another root of f, then $0 = f(a_2) = q_1(a_2)(a_2 - a_1)^{n_1}$. But $a_1 \neq a_2$ and R is an integral domain, so $q_1(a_2)$ must be 0, i.e. a_2 is a root of $q_1(X)$. Repeating the argument, we have $q_1(X) = q_2(X)(X - a_2)^{n_2}$, where $q_2(a_2) \neq 0$ and $\deg q_2 = n - n_1 - n_2$. After n applications of (2.5.2), the quotient becomes constant, and we have $f(X) = c(X - a_1)^{n_1} \dots (X - a_k)^{n_k}$ where $c \in R$ and $n_1 + \dots + n_k = n$. Since R is an integral domain, the only possible roots of f are $a_1, \dots, a_k$. ♣

2.5.4 Example

Let $R = \mathbb{Z}_8$, which is not an integral domain. The polynomial $f(X) = X^3$ has four roots in R, namely 0, 2, 4 and 6.

Problems For Section 2.5

In Problems 1-4, we review the Euclidean algorithm. Let a and b be positive integers, with $a > b$. Divide a by b to obtain

$$a = bq_1 + r_1 \text{ with } 0 \le r_1 < b,$$

then divide b by r_1 to get

$$b = r_1 q_2 + r_2 \text{ with } 0 \le r_2 < r_1,$$

and continue in this fashion until the process terminates:

$$\begin{aligned} r_1 &= r_2 q_3 + r_3, \quad 0 \le r_3 < r_2, \\ &\vdots \\ r_{j-2} &= r_{j-1} q_j + r_j, \quad 0 \le r_j < r_{j-1}, \\ r_{j-1} &= r_j q_{j+1} \end{aligned}$$

1. Show that the greatest common divisor of a and b is the last remainder r_j.
2. If d is the greatest common divisor of a and b, show that there are integers x and y such that $ax + by = d$.
3. Define three sequences by

$$\begin{aligned} r_i &= r_{i-2} - q_i r_{i-1} \\ x_i &= x_{i-2} - q_i x_{i-1} \\ y_i &= y_{i-2} - q_i y_{i-1} \end{aligned}$$

for $i = -1, 0, 1, \ldots$ with initial conditions $r_{-1} = a$, $r_0 = b$, $x_{-1} = 1$, $x_0 = 0$, $y_{-1} = 0$, $y_0 = 1$. (The q_i are determined by dividing r_{i-2} by r_{i-1}.) Show that we can generate all steps of the algorithm, and at each stage, $r_i = ax_i + by_i$.

4. Use the procedure of Problem 3 (or any other method) to find the greatest common divisor d of $a = 123$ and $b = 54$, and find integers x and y such that $ax + by = d$.
5. Use Problem 2 to show that $\mathbb{Z}_p$ is a field if and only if p is prime.
6. If $a(X)$ and $b(X)$ are polynomials with coefficients in a field F, the Euclidean algorithm can be used to find their greatest common divisor. The previous discussion can be taken over verbatim, except that instead of writing

$$a = q_1 b + r_1 \text{ with } 0 \le r_1 < b,$$

we write

$$a(X) = q_1(X)b(X) + r_1(X) \text{ with } \deg r_1(X) < \deg b(X).$$

The greatest common divisor can be defined as the monic polynomial of highest degree that divides both $a(X)$ and $b(X)$.

Let $f(X)$ and $g(X)$ be polynomials in $F[X]$, where F is a field. Show that the ideal I generated by $f(X)$ and $g(X)$, i.e., the set of all linear combinations $a(X)f(X) + b(X)g(X)$, with $a(X), b(X) \in F[X]$, is the principal ideal $J = \langle d(X) \rangle$ generated by the greatest common divisor $d(X)$ of $f(X)$ and $g(X)$.

7. (*Lagrange Interpolation Formula*) Let $a_0, a_1, \ldots, a_n$ be distinct points in the field F, and define

$$P_i(X) = \prod_{j \ne i} \frac{X - a_j}{a_i - a_j}, \quad i = 0, 1, \ldots, n;$$

then $P_i(a_i) = 1$ and $P_i(a_j) = 0$ for $j \ne i$. If $b_0, b_1, \ldots, b_n$ are arbitrary elements of F (not necessarily distinct), use the P_i to find a polynomial $f(X)$ of degree n or less such that $f(a_i) = b_i$ for all i.

8. In Problem 7, show that $f(X)$ is the unique polynomial of degree n or less such that $f(a_i) = b_i$ for all i.
9. Suppose that f is a polynomial in $F[X]$, where F is a field. If $f(a) = 0$ for every $a \in F$, it does not in general follow that f is the zero polynomial. Give an example.
10. Give an example of a field F for which it does follow that $f = 0$.

2.6 Unique Factorization

If we are asked to find the greatest common divisor of two integers, say 72 and 60, one method is to express each integer as a product of primes; thus $72 = 2^3 \times 3^2$, $60 = 2^2 \times 3 \times 5$. The greatest common divisor is the product of terms of the form p^e, where for each prime appearing in the factorization, we use the minimum exponent. Thus $\gcd(72, 60) = 2^2 \times 3^1 \times 5^0 = 12$. To find the least common multiple, we use the maximum exponent: $\text{lcm}(72, 60) = 2^3 \times 3^2 \times 5^1 = 360$. The key idea is that every integer (except 0, 1 and -1) can be uniquely represented as a product of primes. It is natural to ask whether there are integral domains other than the integers in which unique factorization is possible. We now begin to study this question; throughout this section, *all rings are assumed to be integral domains.*

2.6.1 Definitions

Recall from (2.1.2) that a *unit* in a ring R is an element with a multiplicative inverse. The elements a and b are *associates* if $a = ub$ for some unit u.

Let a be a nonzero nonunit; a is said to be *irreducible* if it cannot be represented as a product of nonunits. In other words, if $a = bc$, then either b or c must be a unit.

Again let a be a nonzero nonunit; a is said to be *prime* if whenever a divides a product of terms, it must divide one of the factors. In other words, if a divides bc, then a divides b or a divides c (a divides b means that $b = ar$ for some $r \in R$). It follows from the definition that if p is any nonzero element of R, then p is prime if and only if $\langle p \rangle$ is a prime ideal.

The units of $\mathbb{Z}$ are 1 and -1, and the irreducible and the prime elements coincide. But these properties are not the same in an arbitrary integral domain.

2.6.2 Proposition

If a is prime, then a is irreducible, but not conversely.

Proof. We use the standard notation $r|s$ to indicate that r divides s. Suppose that a is prime, and that $a = bc$. Then certainly $a|bc$, so by definition of prime, $a|b$ or $a|c$, say $a \mid b$. If $b = ad$ then $b = bcd$, so $cd = 1$ and therefore c is a unit. (Note that b cannot be 0, for if so, $a = bc = 0$, which is not possible since a is prime.) Similarly, if $a|c$ with $c = ad$ then $c = bcd$, so $bd = 1$ and b is a unit. Therefore a is irreducible.

To give an example of an irreducible element that is not prime, consider $R = \mathbb{Z}[\sqrt{-3}\,] = \{a + ib\sqrt{3}\colon a, b \in \mathbb{Z}\}$; in R, 2 is irreducible but not prime. To see this, first suppose that we have a factorization of the form

$$2 = (a + ib\sqrt{3}\,)(c + id\sqrt{3}\,);$$

take complex conjugates to get

$$2 = (a - ib\sqrt{3}\,)(c - id\sqrt{3}\,).$$

Now multiply these two equations to obtain

$$4 = (a^2 + 3b^2)(c^2 + 3d^2).$$

Each factor on the right must be a divisor of 4, and there is no way that $a^2 + 3b^2$ can be 2. Thus one of the factors must be 4 and the other must be 1. If, say, $a^2 + 3b^2 = 1$, then $a = \pm 1$ and $b = 0$. Thus in the original factorization of 2, one of the factors must be a unit, so 2 is irreducible. Finally, 2 divides the product $(1 + i\sqrt{3}\,)(1 - i\sqrt{3}\,)$ $(= 4)$, so if 2 were prime, it would divide one of the factors, which means that 2 divides 1, a contradiction since $1/2$ is not an integer. ♣

The distinction between irreducible and prime elements disappears in the presence of unique factorization.

2.6.3 Definition

A *unique factorization domain* (UFD) is an integral domain R satisfying the following properties:

(UF1) Every nonzero element a in R can be expressed as $a = up_1 \dots p_n$, where u is a unit and the p_i are irreducible.

(UF2): If a has another factorization, say $a = vq_1 \dots q_m$, where v is a unit and the q_i are irreducible, then $n = m$ and, after reordering if necessary, p_i and q_i are associates for each i.

Property UF1 asserts the existence of a factorization into irreducibles, and UF2 asserts uniqueness.

2.6.4 Proposition

In a unique factorization domain, a is irreducible if and only if a is prime.

Proof. By (2.6.2), prime implies irreducible, so assume a irreducible, and let a divide bc. Then we have $ad = bc$ for some $d \in R$. We factor d, b and c into irreducibles to obtain

$$aud_1 \dots d_r = vb_1 \dots b_s wc_1 \dots c_t$$

where u, v and w are units and the d_i, b_i and c_i are irreducible. By uniqueness of factorization, a, which is irreducible, must be an associate of some b_i or c_i. Thus a divides b or a divides c. ♣

2.6.5 Definitions and Comments

Let A be a nonempty subset of R, with $0 \notin A$. The element d is a *greatest common divisor* (gcd) of A if d divides each a in A, and whenever e divides each a in A, we have $e|d$.

If d' is another gcd of A, we have $d|d'$ and $d'|d$, so that d and d' are associates. We will allow ourselves to speak of "the" greatest common divisor, suppressing but not forgetting that the gcd is determined up to multiplication by a unit.

The elements of A are said to be *relatively prime* (or the set A is said to be relatively prime) if 1 is a greatest common divisor of A.

The nonzero element m is a *least common multiple* (lcm) of A if each a in A divides m, and whenever $a|e$ for each a in A, we have $m|e$.

Greatest common divisors and least common multiples always exist for finite subsets of a UFD; they may be found by the technique discussed at the beginning of this section.

We will often use the fact that for any $a, b \in R$, we have $a|b$ if and only if $\langle b \rangle \subseteq \langle a \rangle$. This follows because $a|b$ means that $b = ac$ for some $c \in R$. For short, *divides means contains.*

It would be useful to be able to recognize when an integral domain is a UFD. The following criterion is quite abstract, but it will help us to generate some explicit examples.

2.6.6 Theorem

Let R be an integral domain.

(1) If R is a UFD then R satisfies the *ascending chain condition* (acc) *on principal ideals*: If $a_1, a_2, \dots$ belong to R and $\langle a_1 \rangle \subseteq \langle a_2 \rangle \subseteq \dots$, then the sequence eventually stabilizes, that is, for some n we have $\langle a_n \rangle = \langle a_{n+1} \rangle = \langle a_{n+2} \rangle = \dots$.

(2) If R satisfies the ascending chain condition on principal ideals, then R satisfies UF1, that is, every nonzero element of R can be factored into irreducibles.

(3) If R satisfies UF1 and in addition, every irreducible element of R is prime, then R is a UFD.

Thus R is a UFD if and only if R satisfies the ascending chain condition on principal ideals and every irreducible element of R is prime.

Proof. (1) If $\langle a_1 \rangle \subseteq \langle a_2 \rangle \subseteq \dots$ then $a_{i+1} | a_i$ for all i. Therefore the prime factors of a_{i+1} consist of some (or all) of the prime factors of a_i. Multiplicity is taken into account here; for example, if p^3 is a factor of a_i, then p^k will be a factor of a_{i+1} for some $k \in \{0, 1, 2, 3\}$. Since a_1 has only finitely many prime factors, there will come a time when the prime factors are the same from that point on, that is, $\langle a_n \rangle = \langle a_{n+1} \rangle = \dots$ for some n.

(2) Let a_1 be any nonzero element. If a_1 is irreducible, we are finished, so let $a_1 = a_2 b_2$ where neither a_2 nor b_2 is a unit. If both a_2 and b_2 are irreducible, we are finished, so we can assume that one of them, say a_2, is not irreducible. Since a_2 divides a_1 we have $\langle a_1 \rangle \subseteq \langle a_2 \rangle$, and in fact the inclusion is proper because $a_2 \notin \langle a_1 \rangle$. (If $a_2 = c a_1$ then $a_1 = a_2 b_2 = c a_1 b_2$, so b_2 is a unit, a contradiction.) Continuing, we have $a_2 = a_3 b_3$ where neither a_3 nor b_3 is a unit, and if, say, a_3 is not irreducible, we find that $\langle a_2 \rangle \subset \langle a_3 \rangle$. If a_1 cannot be factored into irreducibles, we obtain, by an inductive argument, a strictly increasing chain $\langle a_1 \rangle \subset \langle a_2 \rangle \subset \dots$ of principal ideals.

(3) Suppose that $a = u p_1 p_2 \dots p_n = v q_1 q_2 \dots q_m$ where the p_i and q_i are irreducible and u and v are units. Then p_1 is a prime divisor of $v q_1 \dots q_m$, so p_1 divides one of the q_i, say q_1. But q_1 is irreducible, and therefore p_1 and q_1 are associates. Thus we have, up to multiplication by units, $p_2 \dots p_n = q_2 \dots q_m$. By an inductive argument, we must have $m = n$, and after reordering, p_i and q_i are associates for each i. ♣

We now give a basic sufficient condition for an integral domain to be a UFD.

2.6.7 Definition

A *principal ideal domain* (PID) is an integral domain in which every ideal is principal, that is, generated by a single element.

2.6.8 Theorem

Every principal ideal domain is a unique factorization domain. For short, PID implies UFD.

Proof. If$\langle a_1\rangle \subseteq \langle a_2\rangle \subseteq \dots$, let $I = \cup_i \langle a_i\rangle$. Then I is an ideal, necessarily principal by hypothesis. If $I = \langle b\rangle$ then b belongs to some $\langle a_n\rangle$, so $I \subseteq \langle a_n\rangle$. Thus if $i \geq n$ we have $\langle a_i\rangle \subseteq I \subseteq \langle a_n\rangle \subseteq \langle a_i\rangle$. Therefore $\langle a_i\rangle = \langle a_n\rangle$ for all $i \geq n$, so that R satisfies the acc on principal ideals.

Now suppose that a is irreducible. Then $\langle a\rangle$ is a proper ideal, for if $\langle a\rangle = R$ then $1 \in \langle a\rangle$, so that a is a unit. By the acc on principal ideals, $\langle a\rangle$ is contained in a maximal ideal I. (Note that we need not appeal to the general result (2.4.2), which uses Zorn's lemma.) If $I = \langle b\rangle$ then b divides the irreducible element a, and b is not a unit since I is proper. Thus a and b are associates, so $\langle a\rangle = \langle b\rangle = I$. But I, a maximal ideal, is prime by (2.4.6), hence a is prime. The result follows from (2.6.6). ♣

The following result gives a criterion for a UFD to be a PID. (Terminology: the *zero ideal* is $\{0\}$; a *nonzero ideal* is one that is not $\{0\}$.)

2.6.9 Theorem

R is a PID if and only if R is a UFD and every nonzero prime ideal of R is maximal.

Proof. Assume R is a PID; then R is a UFD by (2.6.8). If $\langle p\rangle$ is a nonzero prime ideal of R, then $\langle p\rangle$ is contained in the maximal ideal $\langle q\rangle$, so that q divides the prime p. Since a maximal ideal must be proper, q cannot be a unit, so that p and q are associates. But then $\langle p\rangle = \langle q\rangle$ and $\langle p\rangle$ is maximal.

The proof of the converse is given in the exercises. ♣

Problems For Section 2.6

Problems 1-6 form a project designed to prove that if R is a UFD and every nonzero prime ideal of R is maximal, then R is a PID.

Let I be an ideal of R; since $\{0\}$ is principal, we can assume that $I \neq \{0\}$. Since R is a UFD, every nonzero element of I can be written as $up_1 \dots p_t$ where u is a unit and the p_i are irreducible, hence prime. Let $r = r(I)$ be the minimum such t. We are going to prove by induction on r that I is principal.

1. If $r = 0$, show that $I = \langle 1\rangle = R$.
2. If the result holds for all $r < n$, let $r = n$, with $up_1 \dots p_n \in I$, hence $p_1 \dots p_n \in I$. Since p_1 is prime, $\langle p_1\rangle$ is a prime ideal, necessarily maximal by hypothesis. By (2.4.3), $R/\langle p_1\rangle$ is a field. If b belongs to I but not to $\langle p_1\rangle$, show that for some $c \in R$ we have $bc - 1 \in \langle p_1\rangle$.
3. By Problem 2, $bc - dp_1 = 1$ for some $d \in R$. Show that this implies that $p_2 \dots p_n \in I$, which contradicts the minimality of n. Thus if b belongs to I, it must also belong to $\langle p_1\rangle$, that is, $I \subseteq \langle p_1\rangle$.
4. Let $J = \{x \in R\colon xp_1 \in I\}$. Show that J is an ideal.
5. Show that $Jp_1 = I$.
6. Since $p_1 \dots p_n = (p_2 \dots p_n)p_1 \in I$, we have $p_2 \dots p_n \in J$. Use the induction hypothesis to conclude that I is principal.

7. Let p and q be prime elements in the integral domain R, and let $P = \langle p \rangle$ and $Q = \langle q \rangle$ be the corresponding prime ideals. Show that it is not possible for P to be a proper subset of Q.

8. If R is a UFD and P is a nonzero prime ideal of R, show that P contains a nonzero principal prime ideal.

2.7 Principal Ideal Domains and Euclidean Domains

In Section 2.6, we found that a principal ideal domain is a unique factorization domain, and this exhibits a class of rings in which unique factorization occurs. We now study some properties of PID's, and show that any integral domain in which the Euclidean algorithm works is a PID. If I is an ideal in $\mathbb{Z}$, in fact if I is simply an additive subgroup of $\mathbb{Z}$, then I consists of all multiples of some positive integer n; see Section 1.1, Problem 6. Thus $\mathbb{Z}$ is a PID.

Now suppose that A is a nonempty subset of the PID R. The ideal $\langle A \rangle$ generated by A consists of all finite sums $\sum r_i a_i$ with $r_i \in R$ and $a_i \in A$; see (2.2.7). We show that if d is a greatest common divisor of A, then d generates A, and conversely.

2.7.1 Proposition

Let R be a PID, with A a nonempty subset of R. Then d is a greatest common divisor of A if and only if d is a generator of $\langle A \rangle$.

Proof. Let d be a gcd of A, and assume that $\langle A \rangle = \langle b \rangle$. Then d divides every $a \in A$, so d divides all finite sums $\sum r_i a_i$. In particular d divides b, hence $\langle b \rangle \subseteq \langle d \rangle$; that is, $\langle A \rangle \subseteq \langle d \rangle$. But if $a \in A$ then $a \in \langle b \rangle$, so b divides a. Since d is a gcd of A, it follows that b divides d, so $\langle d \rangle$ is contained in $\langle b \rangle = \langle A \rangle$. We conclude that $\langle A \rangle = \langle d \rangle$, proving that d is a generator of $\langle A \rangle$.

Conversely, assume that d generates $\langle A \rangle$. If $a \in A$ then a is a multiple of d, so $d \mid a$. By (2.2.7), d can be expressed as $\sum r_i a_i$, so any element that divides everything in A divides d. Therefore d is a gcd of A. ♣

2.7.2 Corollary

If d is a gcd of A, where A is a nonempty subset of the PID R, then d can be expressed as a finite linear combination $\sum r_i a_i$ of elements of A with coefficients in R.

Proof. By (2.7.1), $d \in \langle A \rangle$, and the result follows from (2.2.7). ♣

As a special case, we have the familiar result that the greatest common divisor of two integers a and b can be expressed as $ax + by$ for some integers x and y.

The Euclidean algorithm in $\mathbb{Z}$ is based on the division algorithm: if a and b are integers and $b \neq 0$, then a can be divided by b to produce a quotient and remainder. Specifically, we have $a = bq + r$ for some $q, r \in \mathbb{Z}$ with $|r| < |b|$. The Euclidean algorithm performs equally well for polynomials with coefficients in a field; the absolute value of an integer is replaced by the degree of a polynomial. It is possible to isolate the key property that makes the Euclidean algorithm work.

2.7.3 Definition

Let R be an integral domain. R is said to be a *Euclidean domain* (ED) if there is a function Ψ from $R \setminus \{0\}$ to the nonnegative integers satisfying the following property:

If a and b are elements of R, with $b \neq 0$, then a can be expressed as $bq + r$ for some $q, r \in R$, where either $r = 0$ or $\Psi(r) < \Psi(b)$.

We can replace"$r = 0$ or $\Psi(r) < \Psi(b)$" by simply "$\Psi(r) < \Psi(b)$" if we define $\Psi(0)$ to be $-\infty$.

In any Euclidean domain, we may use the Euclidean algorithm to find the greatest common divisor of two elements; see the problems in Section 2.5 for a discussion of the procedure in $\mathbb{Z}$ and in $F[X]$, where F is a field.

A Euclidean domain is automatically a principal ideal domain, as we now prove.

2.7.4 Theorem

If R is a Euclidean domain, then R is a principal ideal domain. For short, ED implies PID.

Proof. Let I be an ideal of R. If $I = \{0\}$ then I is principal, so assume $I \neq \{0\}$. Then $\{\Psi(b) \colon b \in I, b \neq 0\}$ is a nonempty set of nonnegative integers, and therefore has a smallest element n. Let b be any nonzero element of I such that $\Psi(b) = n$; we claim that $I = \langle b \rangle$. For if a belongs to I then we have $a = bq + r$ where $r = 0$ or $\Psi(r) < \Psi(b)$. Now $r = a - bq \in I$ (because a and b belong to I), so if $r \neq 0$ then $\Psi(r) < \Psi(b)$ is impossible by minimality of $\Psi(b)$. Thus b is a generator of I. ♣

The most familiar Euclidean domains are $\mathbb{Z}$ and $F[X]$, with F a field. We now examine some less familiar cases.

2.7.5 Example

Let $\mathbb{Z}[\sqrt{d}]$ be the ring of all elements $a + b\sqrt{d}$, where $a, b \in \mathbb{Z}$. If $d = -2, -1, 2$ or 3, we claim that $\mathbb{Z}[\sqrt{d}]$ is a Euclidean domain with

$$\Psi(a + b\sqrt{d}) = |a^2 - db^2|.$$

Since the $a + b\sqrt{d}$ are real or complex numbers, there are no zero divisors, and $\mathbb{Z}[\sqrt{d}]$ is an integral domain. Let $\alpha, \beta \in \mathbb{Z}[\sqrt{d}], \beta \neq 0$, and divide α by β to get $x + y\sqrt{d}$. Unfortunately, x and y need not be integers, but at least they are rational numbers. We can find integers reasonably close to x and y; let x_0 and y_0 be integers such that $|x - x_0|$ and $|y - y_0|$ are at most $1/2$. Let

$$q = x_0 + y_0\sqrt{d}, \quad r = \beta((x - x_0) + (y - y_0)\sqrt{d});$$

then

$$\beta q + r = \beta(x + y\sqrt{d}) = \alpha.$$

We must show that $\Psi(r) < \Psi(\beta)$. Now

$$\Psi(a + b\sqrt{d}) = |(a + b\sqrt{d})(a - b\sqrt{d})|,$$

and it follows from (Problem 4 that for all $\gamma, \delta \in \mathbb{Z}[\sqrt{d}]$ we have

$$\Psi(\gamma\delta) = \Psi(\gamma)\Psi(\delta).$$

(When $d = -1$, this says that the magnitude of the product of two complex numbers is the product of the magnitudes.) Thus $\Psi(r) = \Psi(\beta)[(x - x_0)^2 - d(y - y_0)^2]$, and the factor in brackets is at most $\frac{1}{4} + |d|(\frac{1}{4}) \leq \frac{1}{4} + \frac{3}{4} = 1$. The only possibility for equality occurs when $d = 3$ ($d = -3$ is excluded by hypothesis) and $|x - x_0| = |y - y_0| = \frac{1}{2}$. But in this case, the factor in brackets is $|\frac{1}{4} - 3(\frac{1}{4})| = \frac{1}{2} < 1$. We have shown that $\Psi(r) < \Psi(\beta)$, so $\mathbb{Z}[\sqrt{d}]$ is a Euclidean domain.

When $d = -1$, we obtain the *Gaussian integers* $a + bi$, $a, b \in \mathbb{Z}$, $i = \sqrt{-1}$.

Problems For Section 2.7

1. Let $A = \{a_1, \dots, a_n\}$ be a finite subset of the PID R. Show that m is a least common multiple of A iff m is a generator of the ideal $\cap_{i=1}^n \langle a_i \rangle$.
2. Find the gcd of $11 + 3i$ and $8 - i$ in the ring of Gaussian integers.
3. Suppose that R is a Euclidean domain in which $\Psi(a) \leq \Psi(ab)$ for all nonzero elements $a, b \in R$. Show that $\Psi(a) \geq \Psi(1)$, with equality if and only if a is a unit in R.
4. Let $R = \mathbb{Z}[\sqrt{d}]$, where d is any integer, and define $\Psi(a + b\sqrt{d}) = |a^2 - db^2|$. Show that for all nonzero α and β, $\Psi(\alpha\beta) = \Psi(\alpha)\Psi(\beta)$, and if d is not a perfect square, then $\Psi(\alpha) \leq \Psi(\alpha\beta)$.
5. Let $R = \mathbb{Z}[\sqrt{d}]$ where d is not a perfect square. Show that 2 is not prime in R. (Show that 2 divides $d^2 - d$.)
6. If d is a negative integer with $d \leq -3$, show that 2 is irreducible in $\mathbb{Z}[\sqrt{d}]$.
7. Let $R = \mathbb{Z}[\sqrt{d}]$ where d is a negative integer. We know (see (2.7.5)) that R is an ED, hence a PID and a UFD, for $d = -1$ and $d = -2$. Show that for $d \leq -3$, R is not a UFD.
8. Find the least common multiple of $11 + 3i$ and $8 - i$ in the ring of Gaussian integers.
9. If $\alpha = a + bi$ is a Gaussian integer, let $\Psi(\alpha) = a^2 + b^2$ as in (2.7.5). If $\Psi(\alpha)$ is prime in $\mathbb{Z}$, show that α is prime in $\mathbb{Z}[i]$.

2.8 Rings of Fractions

It was recognized quite early in mathematical history that the integers have a significant defect: the quotient of two integers need not be an integer. In such a situation a mathematician is likely to say "I want to be able to divide one integer by another, and I will". This will be legal if the computation takes place in a field F containing the integers $\mathbb{Z}$. Any such field will do, since if a and b belong to F and $b \neq 0$, then $a/b \in F$. How do we know that a suitable F exists? With hindsight we can take F to be the rationals $\mathbb{Q}$, the reals $\mathbb{R}$, or the complex numbers $\mathbb{C}$. In fact, $\mathbb{Q}$ is the smallest field containing $\mathbb{Z}$, since any field $F \supseteq \mathbb{Z}$ contains a/b for all $a, b \in \mathbb{Z}$, $b \neq 0$, and consequently $F \supseteq \mathbb{Q}$.

For an arbitrary integral domain D, the same process that leads from the integers to the rationals allows us to construct a field F whose elements are (essentially) fractions a/b, $a, b \in D$, $b \neq 0$. F is called the *field of fractions* or *quotient field* of D. The mathematician's instinct to generalize then leads to the following question: If R is an arbitrary commutative ring (not necessarily an integral domain), can we still form fractions with numerator and denominator in R? Difficulties quickly arise; for example, how do we make sense of $\frac{a}{b}\frac{c}{d}$ when $bd = 0$? Some restriction must be placed on the allowable denominators, and we will describe a successful approach shortly. Our present interest is in the field of fractions of an integral domain, but later we will need the more general development. Since the ideas are very similar, we will give the general construction now.

2.8.1 Definitions and Comments

Let S be a subset of the ring R; we say that S is *multiplicative* if (a) $0 \notin S$, (b) $1 \in S$, and (c) whenever a and b belong to S, we have $ab \in S$. We can merge (b) and (c) by stating that S is closed under multiplication, if we regard 1 as the empty product. Here are some standard examples of multiplicative sets.

(1) The set of all nonzero elements of an integral domain

(2) The set of all nonzero elements of a commutative ring R that are not zero divisors

(3) $R \setminus P$, where P is a prime ideal of the commutative ring R

If S is a multiplicative subset of the commutative ring R, we define the following equivalence relation on $R \times S$:

$$(a, b) \sim (c, d) \text{ if and only if for some } s \in S \text{ we have } s(ad - bc) = 0.$$

If we are constructing the field of fractions of an integral domain, then (a, b) is our first approximation to a/b. Also, since the elements $s \in S$ are never 0 and R has no zero divisors, we have $(a, b) \sim (c, d)$ iff $ad = bc$, and this should certainly be equivalent to $a/b = c/d$.

Let us check that we have a legal equivalence relation. (Commutativity of multiplication will be used many times to slide an element to a more desirable location in a formula. There is also a theory of rings of fractions in the *non*commutative case, but we will not need the results, and in view of the serious technical difficulties that arise, we will not discuss this area.)

Reflexivity and symmetry follow directly from the definition. For transitivity, suppose that $(a, b) \sim (c, d)$ and $(c, d) \sim (e, f)$. Then for some elements s and t in S we have

$$s(ad - bc) = 0 \text{ and } t(cf - de) = 0.$$

Multiply the first equation by tf and the second by sb, and add the results to get

$$std(af - be) = 0,$$

which implies that $(a, b) \sim (e, f)$, proving transitivity.

If $a \in R$ and $b \in S$, we define the fraction $\frac{a}{b}$ to be the equivalence class of the pair (a, b). The set of all equivalence classes is denoted by $S^{-1}R$, and in view of what we are about to prove, is called the *ring of fractions of R by S.* The term *localization of R by S* is also used, because it will turn out that in Examples (1) and (3) above, $S^{-1}R$ is a local ring (see Section 2.4, Problem 5).

We now make the set of fractions into a ring in a natural way.

addition $\dfrac{a}{b} + \dfrac{c}{d} = \dfrac{ad + bc}{bd}$

multiplication $\dfrac{a}{b}\dfrac{c}{d} = \dfrac{ac}{bd}$

additive identity $\dfrac{0}{1}$ $(= \dfrac{0}{s}$ for any $s \in S)$

additive inverse $-(\dfrac{a}{b}) = \dfrac{-a}{b}$

multiplicative identity $\dfrac{1}{1}$ $(= \dfrac{s}{s}$ for any $s \in S)$

2.8.2 Theorem

Let S be a multiplicative subset of the commutative ring R. With the above definitions, $S^{-1}R$ is a commutative ring. If R is an integral domain, so is $S^{-1}R$. If R is an integral domain and $S = R\setminus\{0\}$, then $S^{-1}R$ is a field (the *field of fractions* or *quotient field* of R).

Proof. First we show that addition is well-defined. If $a_1/b_1 = c_1/d_1$ and $a_2/b_2 = c_2/d_2$, then for some $s, t \in S$ we have

$$s(a_1d_1 - b_1c_1) = 0 \quad \text{and} \quad t(a_2d_2 - b_2c_2) = 0 \tag{1}$$

Multiply the first equation of (1) by tb_2d_2 and the second equation by sb_1d_1, and add the results to get

$$st[(a_1b_2 + a_2b_1)d_1d_2 - (c_1d_2 + c_2d_1)b_1b_2] = 0.$$

Thus

$$\frac{a_1b_2 + a_2b_1}{b_1b_2} = \frac{c_1d_2 + c_2d_1}{d_1d_2},$$

in other words,

$$\frac{a_1}{b_1} + \frac{a_2}{b_2} = \frac{c_1}{d_1} + \frac{c_2}{d_2}$$

so that addition of fractions does not depend on the particular representative of an equivalence class.

Now we show that multiplication is well-defined. We follow the above computation as far as (1), but now we multiply the first equation by ta_2d_2, the second by sc_1b_1, and add. The result is

$$st[a_1a_2d_1d_2 - b_1b_2c_1c_2] = 0$$

which implies that

$$\frac{a_1}{b_1}\frac{a_2}{b_2} = \frac{c_1}{d_1}\frac{c_2}{d_2},$$

as desired. We now know that the fractions in $S^{-1}R$ can be added and multiplied in exactly the same way as ratios of integers, so checking the defining properties of a commutative ring essentially amounts to checking that the rational numbers form a commutative ring; see Problems 3 and 4 for some examples.

Now assume that R is an integral domain. It follows that if a/b is zero in $S^{-1}R$, i.e., $a/b = 0/1$, then $a = 0$ in R. (For some $s \in S$ we have $sa = 0$, and since R is an integral domain and $s \neq 0$, we must have $a = 0$.) Thus if $\frac{a}{b}\frac{c}{d} = 0$, then $ac = 0$, so either a or c is 0, and consequently either a/b or c/d is zero. Therefore $S^{-1}R$ is an integral domain.

If R is an integral domain and $S = R \setminus \{0\}$, let a/b be a nonzero element of $S^{-1}R$. Then both a and b are nonzero, so $a, b \in S$. By definition of multiplication we have $\frac{a}{b}\frac{b}{a} = \frac{1}{1}$. Thus a/b has a multiplicative inverse, so $S^{-1}R$ is a field. ♣

When we go from the integers to the rational numbers, we don't lose the integers in the process, in other words, the rationals contain a copy of the integers, namely, the rationals of the form $a/1$, $a \in \mathbb{Z}$. So a natural question is whether $S^{-1}R$ contains a copy of R.

2.8.3 Proposition

Define $f\colon R \to S^{-1}R$ by $f(a) = a/1$. Then f is a ring homomorphism. If S has no zero divisors then f is a monomorphism, and we say that R can be *embedded* in $S^{-1}R$. In particular:

(i) A commutative ring R can be embedded in its *complete* (or *full*) ring of fractions ($S^{-1}R$, where S consists of all non-divisors of zero in R).

(ii) An integral domain can be embedded in its quotient field.

Proof. We have $f(a+b) = \frac{a+b}{1} = \frac{a}{1} + \frac{b}{1} = f(a) + f(b)$, $f(ab) = \frac{ab}{1} = \frac{a}{1}\frac{b}{1} = f(a)f(b)$, and $f(1) = \frac{1}{1}$, proving that f is a ring homomorphism. If S has no zero divisors and $f(a) = a/1 = 0/1$, then for some $s \in S$ we have $sa = 0$, and since s cannot be a zero divisor, we have $a = 0$. Thus f is a monomorphism. ♣

2.8.4 Corollary

The quotient field F of an integral domain R is the smallest field containing R.

Proof. By (2.8.3), we may regard R as a subset of F, so that F is a field containing R. But if L is any field containing R, then all fractions a/b, $a, b \in R$, must belong to L. Thus $F \subseteq L$. ♣

Problems For Section 2.8

1. If the integral domain D is in fact a field, what is the quotient field of D?
2. If D is the set $F[X]$ of all polynomials over a field F, what is the quotient field of D?
3. Give a detailed proof that addition in a ring of fractions is associative.
4. Give a detailed proof that the distributive laws hold in a ring of fractions.
5. Let R be an integral domain with quotient field F, and let h be a ring monomorphism from R to a field L. Show that h has a unique extension to a monomorphism from F to L.
6. Let h be the ring homomorphism from $\mathbb{Z}$ to $\mathbb{Z}_p$, p prime, given by $h(x) = x \bmod p$. Why can't the analysis of Problem 5 be used to show that h extends to a monomorphism of the rationals to $\mathbb{Z}_p$? (This can't possibly work since $\mathbb{Z}_p$ is finite, but what goes wrong?)
7. Let S be a multiplicative subset of the commutative ring R, with $f\colon R \to S^{-1}R$ defined by $f(a) = a/1$. If g is a ring homomorphism from R to a commutative ring R' and $g(s)$ is a unit in R' for each $s \in S$, we wish to find a ring homomorphism $\overline{g}\colon S^{-1}R \to R'$ such that $\overline{g}(f(a)) = g(a)$ for every $a \in R$, i.e., such that the diagram below is commutative. Show that there is only one conceivable way to define $\overline{g}$.

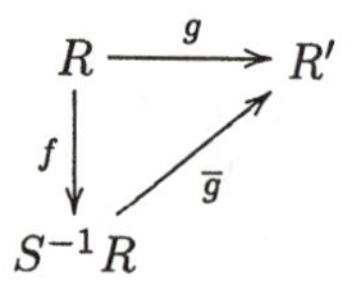

8. Show that the mapping you have defined in Problem 7 is a well-defined ring homomorphism.

2.9 Irreducible Polynomials

2.9.1 Definitions and Comments

In (2.6.1) we defined an irreducible element of a ring; it is a nonzero nonunit which cannot be represented as a product of nonunits. If R is an integral domain, we will refer to an irreducible element of $R[X]$ as an *irreducible polynomial.* Now in $F[X]$, where F is a field, the units are simply the nonzero elements of F (Section 2.1, Problem 2). Thus in this case, an irreducible element is a polynomial of degree at least 1 that cannot be factored into two polynomials of lower degree. A polynomial that is not irreducible is said to be *reducible* or *factorable.* For example, $X^2 + 1$, regarded as an element of $\mathbb{R}[X]$, where $\mathbb{R}$ is the field of real numbers, is irreducible, but if we replace $\mathbb{R}$ by the larger field $\mathbb{C}$ of complex numbers, $X^2 + 1$ is factorable as $(X - i)(X + i)$, $i = \sqrt{-1}$. We say that $X^2 + 1$ is *irreducible over* $\mathbb{R}$ but not *irreducible over* $\mathbb{C}$.

Now consider $D[X]$, where D is a unique factorization domain but not necessarily a field, for example, $D = \mathbb{Z}$. The polynomial $12X + 18$ is not an irreducible element of $\mathbb{Z}[X]$ because it can be factored as the product of the two nonunits 6 and $2X + 3$.

It is convenient to factor out the greatest common divisor of the coefficients (6 in this case). The result is a *primitive polynomial*, one whose *content* (gcd of coefficients) is 1. A primitive polynomial will be irreducible if and only if it cannot be factored into two polynomials of lower degree.

In this section, we will compare irreducibility over a unique factorization domain D and irreducibility over the quotient field F of D. Here is the key result.

2.9.2 Proposition

Let D be a unique factorization domain with quotient field F. Suppose that f is a nonzero polynomial in $D[X]$ and that f can be factored as gh, where g and h belong to $F[X]$. Then there is a nonzero element $\lambda \in F$ such that $\lambda g \in D[X]$ and $\lambda^{-1}h \in D[X]$. Thus if f is factorable over F, then it is factorable over D. Equivalently, if f is irreducible over D, then f is irreducible over F.

Proof. The coefficients of g and h are quotients of elements of D. If a is the least common denominator for g (technically, the least common multiple of the denominators of the coefficients of g), let $g^* = ag \in D[X]$. Similarly, let $h^* = bh \in D[X]$. Thus $abf = g^*h^*$ with $g^*, h^* \in D[X]$ and $c = ab \in D$.

Now if p is a prime factor of c, we will show that either p divides all coefficients of g^* or p divides all coefficients of h^*. We do this for all prime factors of c to get $f = g_0h_0$ with $g_0, h_0 \in D[X]$. Since going from g to g_0 involves only multiplication or division by nonzero constants in D, we have $g_0 = \lambda g$ for some nonzero $\lambda \in F$. But then $h_0 = \lambda^{-1}h$, as desired.

Now let

$$g^*(X) = g_0 + g_1X + \cdots + g_sX^s, \quad h^*(X) = h_0 + h_1X + \cdots + h_tX^t.$$

Since p is a prime factor of $c = ab$ and $abf = g^*h^*, p$ must divide all coefficients of g^*h^*. If p does not divide every g_i and p does not divide every h_i, let g_u and h_v be the coefficients of minimum index not divisible by p. Then the coefficient of X^{u+v} in g^*h^* is

$$g_0h_{u+v} + g_1h_{u+v-1} + \cdots + g_uh_v + \cdots + g_{u+v-1}h_1 + g_{u+v}h_0.$$

But by choice of u and v, p divides every term of this expression except g_uh_v, so that p cannot divide the entire expression. So there is a coefficient of g^*h^* not divisible by p, a contradiction. ♣

The technique of the above proof yields the following result.

2.9.3 Gauss' Lemma

Let f and g be nonconstant polynomials in $D[X]$, where D is a unique factorization domain. If c denotes content, then $c(fg) = c(f)c(g)$, up to associates. In particular, the product of two primitive polynomials is primitive.

Proof. By definition of content we may write $f = c(f)f^*$ and $g = c(g)g^*$ where f^* and g^* are primitive. Thus $fg = c(f)c(g)f^*g^*$. It follows that $c(f)c(g)$ divides every coefficient of fg, so $c(f)c(g)$ divides $c(fg)$. Now let p be any prime factor of $c(fg)$; then p divides $c(f)c(g)f^*g^*$, and the proof of (2.9.2) shows that either p divides every coefficient of $c(f)f^*$ or p divides every coefficient of $c(g)g^*$. If, say, p divides every coefficient of $c(f)f^*$, then (since p is prime) either p divides $c(f)$ or p divides every coefficient of f^*. But f^* is primitive, so that p divides $c(f)$, hence p divides $c(f)c(g)$. We conclude that $c(fg)$ divides $c(f)c(g)$, and the result follows. ♣

2.9.4 Corollary of the Proof of (2.9.3)

If h is a nonconstant polynomial in $D[X]$ and $h = ah^*$ where h^* is primitive and $a \in D$, then a must be the content of h.

Proof. Since a divides every coefficient of h, a must divide $c(h)$. If p is any prime factor of $c(h)$, then p divides every coefficient of ah^*, and as in (2.9.3), either p divides a or p divides every coefficient of h^*, which is impossible by primitivity of h^*. Thus $c(h)$ divides a, and the result follows. ♣

Proposition 2.9.2 yields a precise statement comparing irreducibility over D with irreducibility over F.

2.9.5 Proposition

Let D be a unique factorization domain with quotient field F. If f is a nonconstant polynomial in $D[X]$, then f is irreducible over D if and only if f is primitive and irreducible over F.

Proof. If f is irreducible over D, then f is irreducible over F by (2.9.2). If f is not primitive, then $f = c(f)f^*$ where f^* is primitive and $c(f)$ is not a unit. This contradicts the irreducibility of f over D. Conversely, if $f = gh$ is a factorization of the primitive polynomial f over D, then g and h must be of degree at least 1. Thus neither g nor h is a unit in $F[X]$, so $f = gh$ is a factorization of f over F. ♣

Here is another basic application of (2.9.2).

2.9.6 Theorem

If R is a unique factorization domain, so is $R[X]$.

Proof. If $f \in R[X]$, $f \neq 0$, then f can be factored over the quotient field F as $f = f_1 f_2 \cdots f_k$, where the f_i are irreducible polynomials in $F[X]$. (Recall that $F[X]$ is a Euclidean domain, hence a unique factorization domain.) By (2.9.2), for some nonzero $\lambda_1 \in F$ we may write $f = (\lambda_1 f_1)(\lambda_1^{-1} f_2 \cdots f_k)$ with $\lambda_1 f_1$ and $\lambda_1^{-1} f_2 \cdots f_k$ in $R[X]$. Again by (2.9.2), we have

$$\lambda_1^{-1} f_2 \cdots f_k = f_2 \lambda_1^{-1} f_3 \cdots f_k = (\lambda_2 f_2)(\lambda_2^{-1} \lambda_1^{-1} f_3 \cdots f_k)$$

with $\lambda_2 f_2$ and $\lambda_2^{-1}\lambda_1^{-1} f_3 \ldots f_k \in R[X]$. Continuing inductively, we express f as $\prod_{i=1}^k \lambda_i f_i$ where the $\lambda_i f_i$ are in $R[X]$ and are irreducible over F. But $\lambda_i f_i$ is the product of its content and a primitive polynomial (which is irreducible over F, hence over R by (2.9.5)). Furthermore, the content is either a unit or a product of irreducible elements of the UFD R, and these elements are irreducible in $R[X]$ as well. This establishes the existence of a factorization into irreducibles.

Now suppose that $f = g_1 \cdots g_r = h_1 \cdots h_s$, where the g_i and h_i are nonconstant irreducible polynomials in $R[X]$. (Constant polynomials cause no difficulty because R is a UFD.) By (2.9.5), the g_i and h_i are irreducible over F, and since $F[X]$ is a UFD, we have $r = s$ and, after reordering if necessary, g_i and h_i are associates (in $F[X]$) for each i. Now $g_i = c_i h_i$ for some constant $c_i \in F$, and we have $c_i = a_i/b_i$ with $a_i, b_i \in R$. Thus $b_i g_i = a_i h_i$, with g_i and h_i primitive by (2.9.5). By (2.9.4), $b_i g_i$ has content b_i and $a_i h_i$ has content a_i. Therefore a_i and b_i are associates, which makes c_i a unit in R, which in turn makes g_i and h_i associates in $R[X]$, proving uniqueness of factorization. ♣

The following result is often used to establish irreducibility of a polynomial.

2.9.7 Eisenstein's Irreducibility Criterion

Let R be a UFD with quotient field F, and let $f(X) = a_n X^n + \cdots + a_1 X + a_0$ be a polynomial in $R[X]$, with $n \geq 1$ and $a_n \neq 0$. If p is prime in R and p divides a_i for $0 \leq i < n$, but p does not divide a_n and p^2 does not divide a_0, then f is irreducible over F. Thus by (2.9.5), if f is primitive then f is irreducible over R.

Proof. If we divide f by its content to produce a primitive polynomial f^*, the hypothesis still holds for f^*. (Since p does not divide a_n, it is not a prime factor of $c(f)$, so it must divide the ith coefficient of f^* for $0 \leq i < n$.) If we can prove that f^* is irreducible over R, then by (2.9.5), f^* is irreducible over F, and therefore so is f. Thus we may assume without loss of generality that f is primitive, and prove that f is irreducible over R.

Assume that $f = gh$, with $g(X) = g_0 + \cdots + g_r X^r$ and $h(X) = h_0 + \cdots + h_s X^s$. If $r = 0$ then g_0 divides all coefficients a_i of f, so g_0 divides $c(f)$, hence $g(= g_0)$ is a unit. Thus we may assume that $r \geq 1$, and similarly $s \geq 1$. By hypothesis, p divides $a_0 = g_0 h_0$ but p^2 does not divide a_0, so p cannot divide both g_0 and h_0. Assume that p fails to divide h_0, so that p divides g_0; the argument is symmetrical in the other case. Now $g_r h_s = a_n$, and by hypothesis, p does not divide a_n, so that p does not divide g_r. Let i be the smallest integer such that p does not divide g_i; then $1 \leq i \leq r < n$ (since $r + s = n$ and $s \geq 1$). Now

$$a_i = g_0 h_i + g_1 h_{i-1} + \cdots + g_i h_0$$

and by choice of i, p divides $g_0, \ldots, g_{i-1}$. But p divides the entire sum a_i, so p must divide the last term $g_i h_0$. Consequently, either p divides g_i, which contradicts the choice of i, or p divides h_0, which contradicts our earlier assumption. Thus there can be no factorization of f as a product of polynomials of lower degree; in other words, f is irreducible over R. ♣

Problems For Section 2.9

1. (The *rational root test*, which can be useful in factoring a polynomial over $\mathbb{Q}$.) Let $f(X) = a_nX^n + \cdots + a_1X + a_0 \in \mathbb{Z}[X]$. If f has a rational root u/v where u and v are relatively prime integers and $v \neq 0$, show that v divides a_n and u divides a_0.
2. Show that for every positive integer n, there is at least one irreducible polynomial of degree n over the integers.
3. If $f(X) \in \mathbb{Z}[X]$ and p is prime, we can reduce all coefficients of f modulo p to obtain a new polynomial $f_p(X) \in \mathbb{Z}_p[X]$. If f is factorable over $\mathbb{Z}$, then f_p is factorable over $\mathbb{Z}_p$. Therefore if f_p is irreducible over $\mathbb{Z}_p$, then f is irreducible over $\mathbb{Z}$. Use this idea to show that the polynomial $X^3 + 27X^2 + 5X + 97$ is irreducible over $\mathbb{Z}$. (Note that Eisenstein does not apply.)
4. If we make a change of variable $X = Y + c$ in the polynomial $f(X)$, the result is a new polynomial $g(Y) = f(Y + c)$. If g is factorable over $\mathbb{Z}$, so is f since $f(X) = g(X - c)$. Thus if f is irreducible over $\mathbb{Z}$, so is g. Use this idea to show that $X^4+4X^3+6X^2+4X+4$ is irreducible over $\mathbb{Z}$.
5. Show that in $\mathbb{Z}[X]$, the ideal $\langle n, X\rangle, n \geq 2$, is not principal, and therefore $\mathbb{Z}[X]$ is a UFD that is not a PID.
6. Show that if F is a field, then $F[X, Y]$, the set of all polynomials $\sum a_{ij}X^iY^j$, $a_{ij} \in F$, is not a PID since the ideal $\langle X, Y\rangle$ is not principal.
7. Let $f(X, Y) = X^2 + Y^2 + 1 \in \mathbb{C}[X, Y]$, where $\mathbb{C}$ is the field of complex numbers. Write f as $Y^2 + (X^2 + 1)$ and use Eisenstein's criterion to show that f is irreducible over $\mathbb{C}$.
8. Show that $f(X, Y) = X^3 + Y^3 + 1$ is irreducible over $\mathbb{C}$.

Chapter 3

Field Fundamentals

3.1 Field Extensions

If F is a field and $F[X]$ is the set of all *polynomials over* F, that is, polynomials with coefficients in F, we know that $F[X]$ is a Euclidean domain, and therefore a principal ideal domain and a unique factorization domain (see Sections 2.6 and 2.7). Thus any nonzero polynomial f in $F[X]$ can be factored uniquely as a product of irreducible polynomials. Any root of f must be a root of one of the irreducible factors, but at this point we have no concrete information about the existence of roots and how they might be found. For example, X^2+1 has no real roots, but if we consider the larger field of complex numbers, we get two roots, $+i$ and $-i$. It appears that the process of passing to a larger field may help produce roots, and this turns out to be correct.

3.1.1 Definitions

If F and E are fields and $F \subseteq E$, we say that E is an *extension* of F, and we write $F \leq E$, or sometimes E/F.

If E is an extension of F, then in particular E is an abelian group under addition, and we may multiply the "vector" $x \in E$ by the "scalar" $\lambda \in F$, and the axioms of a vector space are satisfied. Thus if $F \leq E$, then E is a vector space over F. The dimension of this vector space is called the *degree* of the extension, written $[E : F]$. If $[E : F] = n < \infty$, we say that E is a *finite extension* of F, or that *the extension* E/F *is finite*, or that E is *of degree* n *over* F.

If f is a nonconstant polynomial over the field F, and f has no roots in F, we can always produce a root of f in an extension field of F. We do this after a preliminary result.

3.1.2 Lemma

Let $f\colon F \to E$ be a homomorphism of fields, i.e., $f(a+b) = f(a)+f(b)$, $f(ab) = f(a)f(b)$ (all $a, b \in F$), and $f(1_F) = 1_E$. Then f is a monomorphism.

Proof. First note that *a field F has no ideals except $\{0\}$ and F*. For if a is a nonzero member of the ideal I, then $ab = 1$ for some $b \in F$, hence $1 \in I$, and therefore $I = F$. Taking I to be the kernel of f, we see that I cannot be all of F because $f(1) \neq 0$. Thus I must be $\{0\}$, so that f is injective. ♣

3.1.3 Theorem

Let f be a nonconstant polynomial over the field F. Then there is an extension E/F and an element $\alpha \in E$ such that $f(\alpha) = 0$.

Proof. Since f can be factored into irreducibles, we may assume without loss of generality that f itself is irreducible. The ideal $I = \langle f(X) \rangle$ in $F[X]$ is prime (see (2.6.1)), in fact maximal (see (2.6.9)). Thus $E = F[X]/I$ is a field by (2.4.3). We have a problem at this point because F need not be a subset of E, but we can place an isomorphic copy of F inside E via the homomorphism $h\colon a \to a + I$; by (3.1.2), h is a monomorphism, so we may identify F with a subfield of E. Now let $\alpha = X + I$; if $f(X) = a_0 + a_1X + \cdots + a_nX^n$, then

$$\begin{aligned} f(\alpha) &= (a_0 + I) + a_1(X + I) + \cdots + a_n(X + I)^n \\ &= (a_0 + a_1X + \cdots + a_nX^n) + I \\ &= f(X) + I \end{aligned}$$

which is zero in E. ♣

The extension E is sometimes said to be obtained from F by *adjoining a root α* of f. Here is a further connection between roots and extensions.

3.1.4 Proposition

Let f and g be polynomials over the field F. Then f and g are relatively prime if and only if f and g have no common root in any extension of F.

Proof. If f and g are relatively prime, their greatest common divisor is 1, so there are polynomials $a(X)$ and $b(X)$ over F such that $a(X)f(X) + b(X)g(X) = 1$. If α is a common root of f and g, then the substitution of α for X yields $0 = 1$, a contradiction. Conversely, if the greatest common divisor $d(X)$ of $f(X)$ and $g(X)$ is nonconstant, let E be an extension of F in which $d(X)$ has a root α (E exists by (3.1.3)). Since $d(X)$ divides both $f(X)$ and $g(X)$, α is a common root of f and g in E. ♣

3.1.5 Corollary

If f and g are distinct monic irreducible polynomials over F, then f and g have no common roots in any extension of F.

Proof. If h is a nonconstant divisor of the irreducible polynomials f and g, then up to multiplication by constants, h coincides with both f and g, so that f is a constant multiple of g. This is impossible because f and g are monic and distinct. Thus f and g are relatively prime, and the result follows from (3.1.4). ♣

If E is an extension of F and $\alpha \in E$ is a root of a polynomial $f \in F[X]$, it is often of interest to examine the field $F(\alpha)$ generated by F and α, in other words the smallest subfield of E containing F and α (more precisely, containing all elements of F along with α). The field $F(\alpha)$ can be described abstractly as the intersection of all subfields of E containing F and α, and more concretely as the collection of all rational functions

$$\frac{a_0 + a_1\alpha + \cdots + a_m\alpha^m}{b_0 + b_1\alpha + \cdots + b_n\alpha^n}$$

with $a_i, b_j \in F, m, n = 0, 1, \ldots,$ and $b_0 + b_1\alpha + \cdots + b_n\alpha^n \neq 0$. In fact there is a much less complicated description of $F(\alpha)$, as we will see shortly.

3.1.6 Definitions and Comments

If E is an extension of F, the element $\alpha \in E$ is said to be *algebraic* over F is there is a nonconstant polynomial $f \in F[X]$ such that $f(\alpha) = 0$; if α is not algebraic over F, it is said to be *transcendental* over F. If every element of E is algebraic over F, then E is said to be an *algebraic extension* of F.

Suppose that $\alpha \in E$ is algebraic over F, and let I be the set of all polynomials g over F such that $g(\alpha) = 0$. If g_1 and g_2 belong to I, so does $g_1 \pm g_2$, and if $g \in I$ and $c \in F[X]$, then $cg \in I$. Thus I is an ideal of $F[X]$, and since $F[X]$ is a PID, I consists of all multiples of some $m(X) \in F[X]$. Any two such generators must be multiples of each other, so if we require that $m(X)$ be monic, then $m(X)$ is unique. The polynomial $m(X)$ has the following properties:

(1) If $g \in F[X]$, then $g(\alpha) = 0$ if and only if $m(X)$ divides $g(X)$.

(2) $m(X)$ is the monic polynomial of least degree such that $m(\alpha) = 0$.

(3) $m(X)$ is the unique monic irreducible polynomial such that $m(\alpha) = 0$.

Property (1) follows because $g(\alpha) = 0$ iff $g(X) \in I$, and $I = \langle m(X) \rangle$, the ideal generated by $m(X)$. Property (2) follows from (1). To prove (3), note that if $m(X) = h(X)k(X)$ with $\deg h$ and $\deg k$ less than $\deg m$, then either $h(\alpha) = 0$ or $k(\alpha) = 0$, so that by (1), either $h(X)$ or $k(X)$ is a multiple of $m(X)$, which is impossible. Thus $m(X)$ is irreducible, and uniqueness of $m(X)$ follows from (3.1.5).

The polynomial $m(X)$ is called the *minimal polynomial* of α over F, sometimes written as $\min(\alpha, F)$.

3.1.7 Theorem

If $\alpha \in E$ is algebraic over F and the minimal polynomial $m(X)$ of α over F has degree n, then $F(\alpha) = F[\alpha]$, the set of polynomials in α with coefficients in F. In fact, $F[\alpha]$ is the set $F_{n-1}[\alpha]$ of all polynomials *of degree at most* $n-1$ with coefficients in F, and $1, \alpha, \ldots, \alpha^{n-1}$ form a basis for the vector space $F[\alpha]$ over the field F. Consequently, $[F(\alpha) : F] = n$.

Proof. Let $f(X)$ be any nonzero polynomial over F of degree $n-1$ or less. Then since $m(X)$ is irreducible and $\deg f < \deg m$, $f(X)$ and $m(X)$ are relatively prime, and there are polynomials $a(X)$ and $b(X)$ over F such that $a(X)f(X) + b(X)m(X) = 1$. But then $a(\alpha)f(\alpha) = 1$, so that any nonzero element of $F_{n-1}[\alpha]$ has a multiplicative inverse. It follows that $F_{n-1}[\alpha]$ is a field. (This may not be obvious, since the product of two polynomials of degree $n-1$ or less can have degree greater than $n-1$, but if $\deg g > n-1$, then divide g by m to get $g(X) = q(X)m(X) + r(X)$ where $\deg r(X) < \deg m(X) = n$. Replace X by α to get $g(\alpha) = r(\alpha) \in F_{n-1}[\alpha]$. Less abstractly, if $m(\alpha) = \alpha^3 + \alpha + 1 = 0$, then $\alpha^3 = -\alpha - 1$, $\alpha^4 = -\alpha^2 - \alpha$, and so on.)

Now any field containing F and α must contain all polynomials in α, in particular all polynomials of degree at most $n-1$. Therefore $F_{n-1}[\alpha] \subseteq F[\alpha] \subseteq F(\alpha)$. But $F(\alpha)$ is the smallest field containing F and α, so $F(\alpha) \subseteq F_{n-1}[\alpha]$, and we conclude that $F(\alpha) = F[\alpha] = F_{n-1}[\alpha]$. Finally, the elements $1, \alpha, \dots, \alpha^{n-1}$ certainly span $F_{n-1}[\alpha]$, and they are linearly independent because if a nontrivial linear combination of these elements were zero, we would have a nonzero polynomial of degree less than that of $m(X)$ with α as a root, contradicting (2) of (3.1.6). ♣

We now prove a basic multiplicativity result for extensions, after a preliminary discussion.

3.1.8 Lemma

Suppose that $F \leq K \leq E$, the elements $\alpha_i, i \in I$, form a basis for E over K, and the elements $\beta_j, j \in J$, form a basis for K over F. (I and J need not be finite.) Then the products $\alpha_i\beta_j, i \in I, j \in J$, form a basis for E over F.

Proof. If $\gamma \in E$, then γ is a linear combination of the α_i with coefficients $a_i \in K$, and each a_i is a linear combination of the β_j with coefficients $b_{ij} \in F$. It follows that the $\alpha_i\beta_j$ span E over F. Now if $\sum_{i,j} \lambda_{ij}\alpha_i\beta_j = 0$, then $\sum_i \lambda_{ij}\alpha_i = 0$ for all j, and consequently $\lambda_{ij} = 0$ for all i, j, and the $\alpha_i\beta_j$ are linearly independent. ♣

3.1.9 The Degree is Multiplicative

If $F \leq K \leq E$, then $[E : F] = [E : K][K : F]$. In particular, $[E : F]$ is finite if and only if $[E : K]$ and $[K : F]$ are both finite.

Proof. In (3.1.8), we have $[E : K] = |I|$, $[K : F] = |J|$, and $[E : F] = |I||J|$. ♣

We close this section by proving that every finite extension is algebraic.

3.1.10 Theorem

If E is a finite extension of F, then E is an algebraic extension of F.

Proof. Let $\alpha \in E$, and let $n = [E : F]$. Then $1, \alpha, \alpha^2, \dots, \alpha^n$ are $n + 1$ vectors in an n-dimensional vector space, so they must be linearly dependent. Thus α is a root of a nonzero polynomial with coefficients in F, which means that α is algebraic over F. ♣

Problems For Section 3.1

1. Let E be an extension of F, and let S be a subset of E. If $F(S)$ is the subfield of E generated by S over F, in other words, the smallest subfield of E containing F and S, describe $F(S)$ explicitly, and justify your characterization.
2. If for each $i \in I$, K_i is a subfield of the field E, the *composite* of the K_i (notation $\bigvee_i K_i$) is the smallest subfield of E containing every K_i. As in Problem 1, describe the composite explicitly.
3. Assume that α is algebraic over F, with $[F[\alpha] : F] = n$. If $\beta \in F[\alpha]$, show that $[F[\beta] : F] \leq n$, in fact $[F[\beta] : F]$ divides n.
4. The minimal polynomial of $\sqrt{2}$ over the rationals $\mathbb{Q}$ is $X^2 - 2$, by (3) of (3.1.6). Thus $\mathbb{Q}[\sqrt{2}]$ consists of all numbers of the form $a_0 + a_1\sqrt{2}$, where a_0 and a_1 are rational. By Problem 3, we know that $-1 + \sqrt{2}$ has a minimal polynomial over $\mathbb{Q}$ of degree at most 2. Find this minimal polynomial.
5. If α is algebraic over F and β belongs to $F[\alpha]$, describe a systematic procedure for finding the minimal polynomial of β over F.
6. If E/F and the element $\alpha \in E$ is transcendental over F, show that $F(\alpha)$ is isomorphic to $F(X)$, the field of rational functions with coefficients in F.
7. Theorem 3.1.3 gives one method of adjoining a root of a polynomial, and in fact there is essentially only one way to do this. If E is an extension of F and $\alpha \in E$ is algebraic over F with minimal polynomial $m(X)$, let I be the ideal $\langle m(X) \rangle \subseteq F[X]$. Show that $F(\alpha)$ is isomorphic to $F[X]/I$. [Define $\varphi\colon F[X] \to E$ by $\varphi(f(X)) = f(\alpha)$, and use the first isomorphism theorem for rings.]
8. In the proof of (3.1.3), we showed that if f is irreducible in $F[X]$, then $I = \langle f \rangle$ is a maximal ideal. Show that conversely, if I is a maximal ideal, then f is irreducible.
9. Suppose that $F \leq E \leq L$, with $\alpha \in L$. What is the relation between the minimal polynomial of α over F and the minimal polynomial of α over E?
10. If $\alpha_1, \dots, \alpha_n$ are algebraic over F, we can successively adjoin the α_i to F to obtain the field $F[\alpha_1, \dots, \alpha_n]$ consisting of all polynomials over F in the α_i. Show that

$$[F[\alpha_1, \dots, \alpha_n] : F] \leq \prod_{i=1}^{n} [F(\alpha_i) : F] < \infty$$

3.2 Splitting Fields

If f is a polynomial over the field F, then by (3.1.3) we can find an extension E_1 of F containing a root α_1 of f. If not all roots of f lie in E_1, we can find an extension E_2 of E_1 containing another root α_2 of f. If we continue the process, eventually we reach a complete factorization of f. In this section we examine this idea in detail.

If E is an extension of F and $\alpha_1, \dots, \alpha_k \in E$, we will use the notation $F(\alpha_1, \dots, \alpha_k)$ for the subfield of E generated by F and the α_i. Thus $F(\alpha_1, \dots, \alpha_k)$ is the smallest subfield of E containing all elements of F along with the α_i. ("Smallest" means that $F(\alpha_1, \dots, \alpha_k)$ is the intersection of all such subfields.) Explicitly, $F(\alpha_1, \dots, \alpha_k)$ is the collection of all rational functions in the α_i with nonzero denominators.

3.2.1 Definitions and Comments

If E is an extension of F and $f \in F[X]$, we say that f *splits* over E if f can be written as $\lambda(X-\alpha_1)\cdots(X-\alpha_k)$ for some $\alpha_1,\dots,\alpha_k \in E$ and $\lambda \in F$.

(There is a subtle point that should be mentioned. We would like to refer to the α_i as "the" roots of f, but in doing so we are implicitly assuming that if β is an element of some extension E' of E and $f(\beta)=0$, then β must be one of the α_i. This follows upon substituting β into the equation $f(X)=\lambda(X-\alpha_1)\cdots(X-\alpha_k)=0$.)

If K is an extension of F and $f \in F[X]$, we say that K is a *splitting field* for f over F if f splits over K but not over any proper subfield of K containing F.

Equivalently, K is a splitting field for f over F if f splits over K and K is generated over F by the roots $\alpha_1,\dots,\alpha_k$ of f, in other words, $F(\alpha_1,\dots,\alpha_k)=K$. For if K is a splitting field for f, then since f splits over K we have all $\alpha_j \in K$, so $F(\alpha_1,\dots,\alpha_k)\subseteq K$. But f splits over $F(\alpha_1,\dots,\alpha_k)$, and it follows that $F(\alpha_1,\dots,\alpha_k)$ cannot be a proper subfield; it must coincide with K. Conversely, if f splits over K and $F(\alpha_1,\dots,\alpha_k)=K$, let L be a subfield of K containing F. If f splits over L then all α_i belong to L, so $K=F(\alpha_1,\dots,\alpha_k)\subseteq L\subseteq K$, so $L=K$.

If $f \in F[X]$ and f splits over the extension E of F, then E contains a unique splitting field for f, namely $F(\alpha_1,\dots,\alpha_k)$.

3.2.2 Proposition

If $f \in F[X]$ and $\deg f = n$, then f has a splitting field K over F with $[K:F]\le n!$.

Proof. We may assume that $n \ge 1$. (If f is constant, take $K=F$.) By (3.1.3), F has an extension E_1 containing a root α_1 of f, and the extension $F(\alpha_1)/F$ has degree at most n. (Since $f(\alpha_1)=0$, the minimal polynomial of α_1 divides f; see (3.1.6) and (3.1.7).) We may then write $f(X)=(X-\alpha_1)^{r_1}g(X)$, where α_1 is not a root of g and $\deg g \le n-1$. If g is nonconstant, we can find an extension of $F(\alpha_1)$ containing a root α_2 of g, and the extension $F(\alpha_1,\alpha_2)$ will have degree at most $n-1$ over $F(\alpha_1)$. Continue inductively and use (3.1.9) to reach an extension of degree at most $n!$ containing all the roots of f. ♣

If $f \in F[X]$ and f splits over E, then we may pick any root α of f and adjoin it to F to obtain the extension $F(\alpha)$. Roots of the same irreducible factor of f yield essentially the same extension, as the next result shows.

3.2.3 Theorem

If α and β are roots of the irreducible polynomial $f \in F[X]$ in an extension E of F, then $F(\alpha)$ is isomorphic to $F(\beta)$ via an isomorphism that carries α into β and is the identity on F.

Proof. Without loss of generality we may assume f monic (if not, divide f by its leading coefficient). By (3.1.6), part (3), f is the minimal polynomial of both α and β. By (3.1.7), the elements of $F(\alpha)$ can be expressed uniquely as $a_0+a_1\alpha+\cdots+a_{n-1}\alpha^{n-1}$, where the a_i belong to F and n is the degree of f. The desired isomorphism is given by

$$a_0+a_1\alpha+\cdots+a_{n-1}\alpha^{n-1}\to a_0+a_1\beta+\cdots+a_{n-1}\beta^{n-1}. \quad ♣$$

If f is a polynomial in $F[X]$ and F is isomorphic to the field F' via the isomorphism i, we may regard f as a polynomial over F'. We simply use i to transfer f. Thus if $f = a_0 + a_1X + \cdots a_nX^n$, then $f' = i(f) = i(a_0) + i(a_1)X + \cdots + i(a_n)X^n$. There is only a notational difference between f and f', and we expect that splitting fields for f and f' should also be essentially the same. We prove this after the following definition.

3.2.4 Definition

If E and E' are extensions of F and i is an isomorphism of E and E', we say that i is an F-*isomorphism* if i fixes F, that is, if $i(a) = a$ for every $a \in F$. F-homomorphisms, F-monomorphisms, etc., are defined similarly.

3.2.5 Isomorphism Extension Theorem

Suppose that F and F' are isomorphic, and the isomorphism i carries the polynomial $f \in F[X]$ to $f' \in F'[X]$. If K is a splitting field for f over F and K' is a splitting field for f' over F', then i can be extended to an isomorphism of K and K'. In particular, if $F = F'$ and i is the identity function, we conclude that any two splitting fields of f are F-isomorphic.

Proof. Carry out the construction of a splitting field for f over F as in (3.2.2), and perform exactly the same steps to construct a splitting field for f' over F'. At every stage, there is only a notational difference between the fields obtained. Furthermore, we can do the first construction inside K and the second inside K'. But the comment at the end of (3.2.1) shows that the splitting fields that we have constructed coincide with K and K'. ♣

3.2.6 Example

We will find a splitting field for $f(X) = X^3 - 2$ over the rationals $\mathbb{Q}$.

If α is the positive cube root of 2, then the roots of f are $\alpha, \alpha(-\frac{1}{2} + i\frac{1}{2}\sqrt{3})$ and $\alpha(-\frac{1}{2} - i\frac{1}{2}\sqrt{3})$. The polynomial f is irreducible, either by Eisenstein's criterion or by the observation that if f were factorable, it would have a linear factor, and there is no rational number whose cube is 2. Thus f is the minimal polynomial of α, so $[\mathbb{Q}(\alpha) : \mathbb{Q}] = 3$. Now since α and $i\sqrt{3}$ generate all the roots of f, the splitting field is $K = \mathbb{Q}(\alpha, i\sqrt{3})$. (We regard all fields in this example as subfields of the complex numbers $\mathbb{C}$.) Since $i\sqrt{3} \notin \mathbb{Q}(\alpha)$ (because $\mathbb{Q}(\alpha)$ is a subfield of the reals), $[\mathbb{Q}(\alpha, i\sqrt{3}) : \mathbb{Q}(\alpha)]$ is at least 2. But $i\sqrt{3}$ is a root of $X^2 + 3 \in \mathbb{Q}(\alpha)[X]$, so the degree of $\mathbb{Q}(\alpha, i\sqrt{3})$ over $\mathbb{Q}(\alpha)$ is a most 2, and therefore is exactly 2. Thus

$$[K : \mathbb{Q}] = [Q(\alpha, i\sqrt{3}) : \mathbb{Q}] = [Q(\alpha, i\sqrt{3}) : \mathbb{Q}(\alpha)][\mathbb{Q}(\alpha) : \mathbb{Q}] = 2 \times 3 = 6.$$

Problems For Section 3.2

1. Find a splitting field for $f(X) = X^2 - 4X + 4$ over $\mathbb{Q}$.
2. Find a splitting field K for $f(X) = X^2 - 2X + 4$ over $\mathbb{Q}$, and determine the degree of K over $\mathbb{Q}$.

3. Find a splitting field K for $f(X) = X^4 - 2$ over $\mathbb{Q}$, and determine $[K : \mathbb{Q}]$.
4. Let $\mathcal{C}$ be a *family* of polynomials over F, and let K be an extension of F. Show that the following two conditions are equivalent:

   (a) Each $f \in \mathcal{C}$ splits over K, but if $F \leq K' < K$, then it is not true that each $f \in \mathcal{C}$ splits over K'.

   (b) Each $f \in \mathcal{C}$ splits over K, and K is generated over F by the roots of all the polynomials in $\mathcal{C}$.

   If one, and hence both, of these conditions are satisfied, we say that K is a *splitting field for* $\mathcal{C}$ over F.
5. Suppose that K is a splitting field for the finite set of polynomials $\{f_1, \ldots, f_r\}$ over F. Express K as a splitting field for a single polynomial f over F.
6. If m and n are distinct square-free positive integers greater than 1, show that the splitting field $\mathbb{Q}(\sqrt{m}, \sqrt{n})$ of $(X^2 - m)(X^2 - n)$ has degree 4 over $\mathbb{Q}$.

3.3 Algebraic Closures

If f is a polynomial of degree n over the rationals or the reals, or more generally over the complex numbers, then f need not have any rational roots, or even real roots, but we know that f always has n complex roots, counting multiplicity. This favorable situation can be duplicated for any field F, that is, we can construct an algebraic extension C of F with the property that any polynomial in $C[X]$ splits over C. There are many ways to express this idea.

3.3.1 Proposition

If C is a field, the following conditions are equivalent:

(1) Every nonconstant polynomial $f \in C[X]$ has at least one root in C.

(2) Every nonconstant polynomial $f \in C[X]$ splits over C.

(3) Every irreducible polynomial $f \in C[X]$ is linear.

(4) C has no proper algebraic extensions.

If any (and hence all) of these conditions are satisfied, we say that C is *algebraically closed.*

Proof. (1) implies (2): By (1) we may write $f = (X - \alpha_1)g$. Proceed inductively to show that any nonconstant polynomial is a product of linear factors.

(2) implies (3): If f is an irreducible polynomial in $C[X]$, then by (2.9.1), f is nonconstant. By (2), f is a product of linear factors. But f is irreducible, so there can be only one such factor.

(3) implies (4): Let E be an algebraic extension of C. If $\alpha \in E$, let f be the minimal polynomial of α over C. Then f is irreducible and by (3), f is of the form $X - \alpha$. But then $\alpha \in C$, so $E = C$.

(4) implies (1): Let f be a nonconstant polynomial in $C[X]$, and adjoin a root α of f to obtain $C(\alpha)$, as in (3.1.3). But then $C(\alpha)$ is an algebraic extension of C, so by (4), $\alpha \in C$. ♣

It will be useful to embed an arbitrary field F in an algebraically closed field.

3.3.2 Definitions and Comments

An extension C of F is an *algebraic closure* of F if C is algebraic over F and C is algebraically closed.

Note that C is minimal among algebraically closed extensions of F. For if $F \leq K \leq C$ and $\alpha \in C, \alpha \notin K$, then since α is algebraic over F it is algebraic over K. But since $\alpha \notin K$, the minimal polynomial of α over K is a nonlinear irreducible polynomial in $K[X]$. By (3) of (3.3.1), K cannot be algebraically closed.

If C is an algebraic extension of F, then in order for C to be an algebraic closure of F it is sufficient that every polynomial in $F[X]$ (rather than $C[X]$) splits over C. To prove this, we will need the following result.

3.3.3 Proposition

If E is generated over F by finitely many elements $\alpha_1, \dots, \alpha_n$ algebraic over F (so that $E = F(\alpha_1, \dots, \alpha_n)$), then E is a finite extension of F.

Proof. Set $E_0 = F$ and $E_k = F(\alpha_1, \dots, \alpha_k)$, $1 \leq k \leq n$ (so $E_n = E$). Then $E_k = E_{k-1}(\alpha_k)$, where α_k is algebraic over F and hence over E_{k-1}. But by (3.1.7), $[E_k : E_{k-1}]$ is the degree of the minimal polynomial of α_k over E_{k-1}, which is finite. By (3.1.9), $[E : F] = \prod_{k=1}^{n} [E_k : E_{k-1}] < \infty$. ♣

3.3.4 Corollary

If E is an extension of F and A is the set of all elements in E that are algebraic over F (the *algebraic closure of F in E*), then A is a subfield of E.

Proof. If $\alpha, \beta \in A$, then the sum, difference, product and quotient (if $\beta \neq 0$) of α and β belong to $F(\alpha, \beta)$, which is a finite extension of F by (3.3.3), and therefore an algebraic extension of F by (3.1.10). But then $\alpha + \beta, \alpha - \beta, \alpha\beta$ and α/β belong to A, proving that A is a field. ♣

3.3.5 Corollary (Transitivity of Algebraic Extensions)

If E is algebraic over K (in other words, every element of E is algebraic over K), and K is algebraic over F, then E is algebraic over F.

Proof. Let $\alpha \in E$, and let $m(X) = b_0 + b_1 X + \cdots + b_{n-1} X^{n-1} + X^n$ be the minimal polynomial of α over K. The b_i belong to K and are therefore algebraic over F. If $L = F(b_0, b_1, \dots, b_{n-1})$, then by (3.3.3), L is a finite extension of F. Since the coefficients of $m(X)$ belong to L, α is algebraic over L, so by (3.1.7), $L(\alpha)$ is a finite extension of L. By (3.1.9), $L(\alpha)$ is a finite extension of F. By (3.1.10), α is algebraic over F. ♣

Now we can add another condition to (3.3.1).

3.3.6 Proposition

Let C be an algebraic extension of F. Then C is an algebraic closure of F if and only if every nonconstant polynomial in $F[X]$ splits over C.

Proof. The "only if" part follows from (2) of (3.3.1), since $F \subseteq C$. Thus assume that every nonconstant polynomial in $F[X]$ splits over C. If f is a nonconstant polynomial in $C[X]$, we will show that f has at least one root in C, and it will follow from (1) of (3.3.1) that C is algebraically closed. Adjoin a root α of f to obtain the extension $C(\alpha)$. Then $C(\alpha)$ is algebraic over C by (3.1.7), and C is algebraic over F by hypothesis. By (3.3.5), $C(\alpha)$ is algebraic over F, so α is algebraic over F. But then α is a root of some polynomial $g \in F[X]$, and by hypothesis, g splits over C. By definition of "splits" (see (3.2.1)), all roots of g lie in C, in particular $\alpha \in C$. Thus f has at least one root in C. ♣

To avoid a lengthy excursion into formal set theory, we argue intuitively to establish the following three results. (For complete proofs, see the appendix to Chapter 3.)

3.3.7 Theorem

Every field F has an algebraic closure.

Informal argument. Well-order $F[X]$ and use transfinite induction, beginning with the field $F_0 = F$. At stage f we adjoin all roots of the polynomial f by constructing a splitting field for f over the field $F_{<f}$ that has been generated so far by the recursive procedure. When we reach the end of the process, we will have a field C such that every polynomial f in $F[X]$ splits over C. By (3.3.6), C is an algebraic closure of F. ♣

3.3.8 Theorem

Any two algebraic closures C and C' of F are F-isomorphic.

Informal argument. Carry out the recursive procedure described in (3.3.7) in both C and C'. At each stage we may use the fact that any two splitting fields of the same polynomial are F-isomorphic; see (3.2.5). When we finish, we have F-isomorphic algebraic closures of F, say $D \subseteq C$ and $D' \subseteq C'$. But an algebraic closure is a minimal algebraically closed extension by (3.3.2), and therefore $D = C$ and $D' = C'$. ♣

3.3.9 Theorem

If E is an algebraic extension of F, C is an algebraic closure of F, and i is an embedding (that is, a monomorphism) of F into C, then i can be extended to an embedding of E into C.

Informal argument. Each $\alpha \in E$ is a root of some polynomial in $F[X]$, so if we allow α to range over all of E, we get a collection S of polynomials in $F[X]$. Within C, carry out the recursive procedure of (3.3.7) on the polynomials in S. The resulting field lies inside C and contains an F-isomorphic copy of E. ♣

Problems For Section 3.3

1. Show that the converse of (3.3.3) holds, that is, if E is a finite extension of F, then E is generated over F by finitely many elements that are algebraic over F.
2. An *algebraic number* is a complex number that is algebraic over the rational field $\mathbb{Q}$. A *transcendental number* is a complex number that is not algebraic over $\mathbb{Q}$. Show that there only countably many algebraic numbers, and consequently there are uncountably many transcendental numbers.
3. Give an example of an extension C/F such that C is algebraically closed but C is not an algebraic extension of F.
4. Give an example of an extension E/F such that E is an algebraic but not a finite extension of F.
5. In the proof of (3.3.7), why is C algebraic over F?
6. Show that the set A of algebraic numbers is an algebraic closure of $\mathbb{Q}$.
7. If E is an algebraic extension of the infinite field F, show that $|E| = |F|$.
8. Show that any set S of nonconstant polynomials in $F[X]$ has a splitting field over F.
9. Show that an algebraically closed field must be infinite.

3.4 Separability

If f is a polynomial in $F[X]$, we can construct a splitting field K for f over F, and all roots of f must lie in K. In this section we investigate the multiplicity of the roots.

3.4.1 Definitions and Comments

An irreducible polynomial $f \in F[X]$ is *separable* if f has no repeated roots in a splitting field; otherwise f is *inseparable*. If f is an arbitrary polynomial, not necessarily irreducible, then we call f separable if each of its irreducible factors is separable.

Thus if $f(X) = (X-1)^2(X-3)$ over $\mathbb{Q}$, then f is separable, because the irreducible factors $(X-1)$ and $(X-3)$ do not have repeated roots. We will see shortly that over a field of characteristic 0 (for example, the rationals), every polynomial is separable. Here is a method for testing for multiple roots.

3.4.2 Proposition

If

$$f(X) = a_0 + a_1X + \cdots + a_nX^n \in F[X],$$

let f' be the *derivative* of f, defined by

$$f'(X) = a_1 + 2a_2X + \cdots + na_nX^{n-1}.$$

[Note that the derivative is a purely formal expression; we completely ignore questions about existence of limits. One can check by brute force that the usual rules for differentiating a sum and product apply].

If g is the greatest common divisor of f and f', then f has a repeated root in a splitting field if and only if the degree of g is at least 1.

Proof. If f has a repeated root, we can write $f(X) = (X-\alpha)^r h(X)$ where $r \geq 2$. Applying the product rule for derivatives, we see that $(X - \alpha)$ is a factor of both f and f', and consequently $\deg g \geq 1$. Conversely, if $\deg g \geq 1$, let α be a root of g in some splitting field. Then $(X - \alpha)$ is a factor of both f and f'. We will show that α is a repeated root of f. If not, we may write $f(X) = (X - \alpha)h(X)$ where $h(\alpha) \neq 0$. Differentiate to obtain $f'(X) = (X - \alpha)h'(X) + h(X)$, hence $f'(\alpha) = h(\alpha) \neq 0$. This contradicts the fact that $(X - \alpha)$ is a factor of f'. ♣

3.4.3 Corollary

(1) Over a field of characteristic zero, every polynomial is separable.

(2) Over a field F of prime characteristic p, the irreducible polynomial f is inseparable if and only if f' is the zero polynomial. Equivalently, f is a polynomial in X^p; we abbreviate this as $f \in F[X^p]$.

Proof. (1) Without loss of generality, we can assume that we are testing an irreducible polynomial f. The derivative of X^n is nX^{n-1}, and in a field of characteristic 0, n cannot be 0. Thus f' is a nonzero polynomial whose degree is less than that of f. Since f is irreducible, the gcd of f and f' is either 1 or f, and the latter is excluded because f cannot possibly divide f'. By (3.4.2), f is separable.

(2) If $f' \neq 0$, the argument of (1) shows that f is separable. If $f' = 0$, then $\gcd(f, f') = f$, so by (3.4.2), f is inseparable. In characteristic p, an integer n is zero if and only if n is a multiple of p, and it follows that $f' = 0$ iff $f \in F[X^p]$. ♣

By (3.4.3), part (1), every polynomial over the rationals (or the reals or the complex numbers) is separable. This pleasant property is shared by finite fields as well. First note that a finite field F cannot have characteristic 0, since a field of characteristic 0 must contain a copy of the integers (and the rationals as well), and we cannot squeeze infinitely many integers into a finite set. Now recall the binomial expansion modulo p, which is simply $(a + b)^p = a^p + b^p$, since p divides $\binom{p}{k}$ for $1 \leq k \leq p - 1$. [By induction, $(a + b)^{p^n} = a^{p^n} + b^{p^n}$ for every positive integer n.] Here is the key step in the analysis.

3.4.4 The Frobenius Automorphism

Let F be a finite field of characteristic p, and define $f: F \to F$ by $f(\alpha) = \alpha^p$. Then f is an automorphism. In particular, if $\alpha \in F$ then $\alpha = \beta^p$ for some $\beta \in F$.

Proof. We have $f(1) = 1$ and

$$f(\alpha + \beta) = (\alpha + \beta)^p = \alpha^p + \beta^p = f(\alpha) + f(\beta),$$
$$f(\alpha\beta) = (\alpha\beta)^p = \alpha^p\beta^p = f(\alpha)f(\beta)$$

so f is a monomorphism. But an injective function from a finite set to itself is automatically surjective, and the result follows. ♣

3.4.5 Proposition

Over a finite field, every polynomial is separable.

Proof. Suppose that f is an irreducible polynomial over the finite field F with repeated roots in a splitting field. By (3.4.3), part (2), $f(X)$ has the form $a_0 + a_1X^p + \cdots + a_nX^{np}$ with the $a_i \in F$. By (3.4.4), for each i there is an element $b_i \in F$ such that $b_i^p = a_i$. But then

$$(b_0 + b_1X + \cdots + b_nX^n)^p = b_0^p + b_1^pX^p + \cdots + b_n^pX^{np} = f(X)$$

which contradicts the irreducibility of f. ♣

Separability of an element can be defined in terms of its minimal polynomial.

3.4.6 Definitions and Comments

If E is an extension of F and $\alpha \in E$, then α is *separable over* F if α is algebraic over F and $\min(\alpha, F)$ is a separable polynomial. If every element of E is separable over F, we say that *E is a separable extension of F* or *the extension E/F is separable* or *E is separable over F*. By (3.4.3) and (3.4.5), every algebraic extension of a field of characteristic zero or a finite field is separable.

3.4.7 Lemma

If $F \leq K \leq E$ and E is separable over F, then K is separable over F and E is separable over K.

Proof. Since K is a subfield of E, K/F is separable. If $\alpha \in E$, then since α is a root of $\min(\alpha, F)$, it follows from (1) of (3.1.6) that $\min(\alpha, K)$ divides $\min(\alpha, F)$. By hypothesis, $\min(\alpha, F)$ has no repeated roots in a splitting field, so neither does $\min(\alpha, K)$. Thus E/K is separable. ♣

The converse of (3.4.7) is also true: If K/F and E/K are separable, then E/F is separable. Thus we have *transitivity of separable extensions.* We will prove this (for finite extensions) in the exercises.

In view of (3.4.6), we can produce many examples of separable extensions. Inseparable extensions are less common, but here is one way to construct them.

3.4.8 Example

Let $F = \mathbb{F}_p(t)$ be the set of rational functions (in the indeterminate t) with coefficients in the field with p elements (the integers mod p). Thus an element of F looks like

$$\frac{a_0 + a_1 t + \cdots + a_m t^m}{b_0 + b_1 t + \cdots + b_n t^n}.$$

with the a_i and b_j in $\mathbb{F}_p$. Adjoin $\sqrt[p]{t}$, that is, a root of $X^p - t$, to create the extension E. Note that $X^p - t$ is irreducible by Eisenstein, because t is irreducible in $\mathbb{F}_p[t]$. (The product of two nonconstant polynomials in t cannot possibly be t.) The extension E/F is inseparable, since

$$X^p - t = X^p - (\sqrt[p]{t})^p = (X - \sqrt[p]{t})^p,$$

which has multiple roots.

Problems For Section 3.4

1. Give an example of a separable polynomial f whose derivative is zero. (In view of (3.4.3), f cannot be irreducible.)
2. Let $\alpha \in E$, where E is an algebraic extension of a field F of prime characteristic p. Let $m(X)$ be the minimal polynomial of α over the field $F(\alpha^p)$. Show that $m(X)$ splits over E, and in fact α is the only root, so that $m(X)$ is a power of $(X - \alpha)$.
3. Continuing Problem 2, if α is separable over the field $F(\alpha^p)$, show that $\alpha \in F(\alpha^p)$.
4. A field F is said to be *perfect* if every polynomial over F is separable. Equivalently, every algebraic extension of F is separable. Thus fields of characteristic zero and finite fields are perfect. Show that if F has prime characteristic p, then F is perfect if and only if every element of F is the pth power of some element of F. For short we write $F = F^p$.

   In Problems 5-8, we turn to transitivity of separable extensions.
5. Let E be a finite extension of a field F of prime characteristic p, and let $K = F(E^p)$ be the subfield of E obtained from F by adjoining the pth powers of all elements of E. Show that $F(E^p)$ consists of all finite linear combinations of elements in E^p with coefficients in F.
6. Let E be a finite extension of the field F of prime characteristic p, and assume that $E = F(E^p)$. If the elements $y_1, \dots, y_r \in E$ are linearly independent over F, show that $y_1^p, \dots, y_r^p$ are linearly independent over F.
7. Let E be a finite extension of the field F of prime characteristic p. Show that the extension is separable if and only if $E = F(E^p)$.
8. If $F \le K \le E$ with $[E : F] < \infty$, with E separable over K and K separable over F, show that E is separable over F.
9. Let f be an irreducible polynomial in $F[X]$, where F has characteristic $p > 0$. Express $f(X)$ as $g(X^{p^m})$, where the nonnegative integer m is a large as possible. (This makes sense because $X^{p^0} = X$, so $m = 0$ always works, and f has finite degree, so m is bounded above.) Show that g is irreducible and separable.

10. Continuing Problem 9, if f has only one distinct root α, show that $\alpha^{p^m} \in F$.
11. If E/F, where char $F = p > 0$, and the element $\alpha \in E$ is algebraic over F, show that the minimal polynomial of α over F has only one distinct root if and only if $\alpha^{p^n} \in F$ for some nonnegative integer n. (In this case we say that α is *purely inseparable* over F.)

3.5 Normal Extensions

Let E/F be a field extension. In preparation for Galois theory, we are going to look at monomorphisms defined on E, especially those which fix F. First we examine what an F-monomorphism does to the roots of a polynomial in $F[X]$.

3.5.1 Lemma

Let $\sigma: E \to E$ be an F-monomorphism, and assume that the polynomial $f \in F[X]$ splits over E. If α is a root of f in E, then so is $\sigma(\alpha)$. Thus σ permutes the roots of f.

Proof. If $b_0 + b_1\alpha + \cdots + b_n\alpha^n = 0$, with the $b_i \in F$, apply σ and note that since σ is an F-monomorphism, $\sigma(b_i) = b_i$ and $\sigma(\alpha^i) = (\sigma(\alpha))^i$. Thus

$$b_0 + b_1\sigma(\alpha) + \cdots + b_n(\sigma(\alpha))^n = 0. \quad \clubsuit$$

Now let C be an algebraic closure of E. It is convenient to have C available because it will contain all the roots of a polynomial $f \in E[X]$, even if f does not split over E. We are going to count the number of embeddings of E in C that fix F, that is, the number of F-monomorphisms of E into C. Here is the key result.

3.5.2 Theorem

Let E/F be a finite separable extension of degree n, and let σ be an embedding of F in C. Then σ extends to exactly n embeddings of E in C; in other words, there are exactly n embeddings τ of E in C such that the restriction $\tau|_F$ of τ to F coincides with σ. In particular, taking σ to be the identity function on F, there are exactly n F-monomorphisms of E into C.

Proof. An induction argument works well. If $n = 1$ then $E = F$ and there is nothing to prove, so assume $n > 1$ and choose an element α that belongs to E but not to F. If f is the minimal polynomial of α over F, let $g = \sigma(f)$. (This is a useful shorthand notation, indicating that if a_i is one of the coefficients of f, the corresponding coefficient of g is $\sigma(a_i)$.) Any factorization of g can be translated via the inverse of σ to a factorization of f, so g is separable and irreducible over the field $\sigma(F)$. If β is any root of g, then there is a unique isomorphism of $F(\alpha)$ and $(\sigma(F))(\beta)$ that carries α into β and coincides with σ on F. Explicitly,

$$b_0 + b_1\alpha + \cdots + b_r\alpha^r \to \sigma(b_0) + \sigma(b_1)\beta + \cdots + \sigma(b_r)\beta^r.$$

Now if $\deg g = r$, then $[F(\alpha) : F] = \deg f = \deg g = r$ as well, so by (3.1.9), $[E : F(\alpha)] = n/r < n$. By separability, g has exactly r distinct roots in C, so there are exactly r possible choices of β. In each case, by the induction hypothesis,the resulting embedding of $F(\alpha)$ in C has exactly n/r extensions to embeddings of E in C. This produces n distinct embeddings of E in C extending σ. But if τ is any embedding of F in C that extends σ, then just as in (3.5.1), τ must take α to a root of g, i.e., to one of the β's. If there were more than n possible τ's, there would have to be more than n/r possible extensions of at least one of the embeddings of $F(\alpha)$ in C. This would contradict the induction hypothesis. ♣

3.5.3 Example

Adjoin the positive cube root of 2 to the rationals to get $E = \mathbb{Q}(\sqrt[3]{2})$. The roots of the irreducible polynomial $f(X) = X^3 - 2$ are $\sqrt[3]{2}$, $\omega\sqrt[3]{2}$ and $\omega^2\sqrt[3]{2}$, where $\omega = e^{i2\pi/3} = -\frac{1}{2} + \frac{1}{2}i\sqrt{3}$ and $\omega^2 = e^{i4\pi/3} = -\frac{1}{2} - \frac{1}{2}i\sqrt{3}$.

Notice that the polynomial f has a root in E but does not split in E (because the other two roots are complex and E consists entirely of real numbers). We give a special name to extensions that do not have this annoying drawback.

3.5.4 Definition

The algebraic extension E/F is *normal* (we also say that E is *normal over* F) if every irreducible polynomial over F that has at least one root in E splits over E. In other words, if $\alpha \in E$, then all *conjugates* of α over F (i.e., all roots of the minimal polynomial of α over F) belong to E.

Here is an equivalent condition.

3.5.5 Theorem

The finite extension E/F is normal if and only if every F-monomorphism of E into an algebraic closure C is actually an F-automorphism of E. (The hypothesis that E/F is finite rather than simply algebraic can be removed, but we will not need the more general result.)

Proof. If E/F is normal, then as in (3.5.1), an F-monomorphism τ of E into C must map each element of E to one of its conjugates. Thus by hypothesis, $\tau(E) \subseteq E$. But $\tau(E)$ is an isomorphic copy of E , so it must have the same degree as E over F. Since the degree is assumed finite, we have $\tau(E) = E$. (All we are saying here is that an m-dimensional subspace of an m-dimensional vector space is the entire space.) Conversely, let $\alpha \in E$, and let β be any conjugate of α over F. As in the proof of (3.5.2), there is an F-monomorphism of E into C that carries α to β. If all such embeddings are F-automorphisms of E, we must have $\beta \in E$, and we conclude that E is normal over F. ♣

3.5.6 Remarks

In (3.5.2) and (3.5.5), the algebraic closure can be replaced by any fixed normal extension of F containing E; the proof is the same. Also, the implication $\tau(E) \subseteq E \Rightarrow \tau(E) = E$

holds for any F-monomorphism τ and any finite extension E/F; normality is not involved.

The next result yields many explicit examples of normal extensions.

3.5.7 Theorem

The finite extension E/F is normal if and only if E is a splitting field for some polynomial $f \in F[X]$.

Proof. Assume that E is normal over F. Let $\alpha_1, \dots, \alpha_n$ be a basis for E over F, and let f_i be the minimal polynomial of α_i over $F, i = 1, \dots, n$. Since f_i has a root α_i in E, f_i splits over E, hence so does $f = f_1 \cdots f_n$. If f splits over a field K with $F \subseteq K \subseteq E$, then each α_i belongs to K, and therefore K must coincide with E. Thus E is a splitting field for f over F. Conversely, let E be a splitting field for f over F, where the roots of f are α_i, $i = 1, \dots, n$. Let τ be an F-monomorphism of E into an algebraic closure. As in (3.5.1), τ takes each α_i into another root of f, and therefore τ takes a polynomial in the α_i to another polynomial in the α_i. But $F(\alpha_1, \dots, \alpha_n) = E$, so $\tau(E) \subseteq E$. By (3.5.6), τ is an automorphism of E, so by (3.5.5), E/F is normal. ♣

3.5.8 Corollary

Let $F \leq K \leq E$, where E is a finite extension of F. If E/F is normal, so is E/K.

Proof. By (3.5.7), E is a splitting field for some polynomial $f \in F[X]$, so that E is generated over F by the roots of f. But then $f \in K[X]$ and E is generated over K by the roots of f. Again by (3.5.7), E/K is normal. ♣

3.5.9 Definitions and Comments

If E/F is normal and separable, it is said to be a *Galois extension*; we also say that E is *Galois over* F. It follows from (3.5.2) and (3.5.5) that if E/F is a finite Galois extension, then there are exactly $[E : F]$ F-automorphisms of E. If E/F is finite and separable but not normal, then at least one F-embedding of E into an algebraic closure must fail to be an automorphism of E. Thus in this case, the number of F-automorphisms of E is less than the degree of the extension.

If E/F is an arbitrary extension, the *Galois group* of the extension, denoted by $\text{Gal}(E/F)$, is the set of F-automorphisms of E. (The set is a group under composition of functions.)

3.5.10 Example

Let $E = \mathbb{Q}(\sqrt[3]{2})$, as in (3.5.3). The Galois group of the extension consists of the identity automorphism alone. For any $\mathbb{Q}$-monomorphism σ of E must take $\sqrt[3]{2}$ into a root of $X^3 - 2$. Since the other two roots are complex and do not belong to E, $\sqrt[3]{2}$ must map to itself. But σ is completely determined by its action on $\sqrt[3]{2}$, and the result follows.

If E/F is not normal, we can always enlarge E to produce a normal extension of F. If C is an algebraic closure of E, then C contains all the roots of every polynomial in $F[X]$, so C/F is normal. Let us try to look for a smaller normal extension.

3.5.11 The Normal Closure

Let E be a finite extension of F, say $E = F(\alpha_1, \dots, \alpha_n)$. If $N \supseteq E$ is any normal extension of F, then N must contain the α_i along with all conjugates of the α_i, that is, all roots of $\min(\alpha_i, F), i = 1, \dots, n$. Thus if f is the product of these minimal polynomials, then N must contain the splitting field K for f over F. But K/F is normal by (3.5.7), so K must be the smallest normal extension of F that contains E. It is called the *normal closure* of E over F.

We close the section with an important result on the structure of finite separable extensions.

3.5.12 Theorem of the Primitive Element

If E/F is a finite separable extension, then $E = F(\alpha)$ for some $\alpha \in E$. We say that α is a *primitive element* of E over F.

Proof. We will argue by induction on $n = [E : F]$. If $n = 1$ then $E = F$ and we can take α to be any member of F. If $n > 1$, choose $\alpha \in E \setminus F$. By the induction hypothesis, there is a primitive element β for E over $F(\alpha)$, so that $E = F(\alpha, \beta)$. We are going to show that if $c \in F$ is properly chosen, then $E = F(\alpha + c\beta)$. Now by (3.5.2), there are exactly n F-monomorphisms of E into an algebraic closure C, and each of these maps restricts to an F-monomorphism of $F(\alpha + c\beta)$ into C. If $F(\alpha + c\beta) \neq E$, then $[F(\alpha + c\beta) : F] < n$, and it follows from (3.5.2) that at least two embeddings of E, say σ and τ, must coincide when restricted. Therefore

$$\sigma(\alpha) + c\sigma(\beta) = \tau(\alpha) + c\tau(\beta),$$

hence

$$c = \frac{\sigma(\alpha) - \tau(\alpha)}{\tau(\beta) - \sigma(\beta)}. \tag{1}$$

(If $\tau(\beta) = \sigma(\beta)$ then by the previous equation, $\tau(\alpha) = \sigma(\alpha)$. But an F-embedding of E is determined by what it does to α and β, hence $\sigma = \tau$, a contradiction.) Now an F-monomorphism must map α to one of its conjugates over F, and similarly for β. Thus there are only finitely many possible values for the ratio in (1). If we select c to be different from each of these values, we reach a contradiction of our assumption that $F(\alpha + c\beta) \neq E$. The proof is complete if F is an infinite field. We must leave a gap here, to be filled later (see (6.4.4)). If F is finite, then so is E (since E is a finite-dimensional vector space over F). We will show that the multiplicative group of nonzero elements of a finite field E is cyclic, so if α is a generator of this group, then $E = F(\alpha)$. ♣

Problems For Section 3.5

1. Give an example of fields $F \leq K \leq E$ such that E/F is normal but K/F is not.
2. Let $E = \mathbb{Q}(\sqrt{a})$, where a is an integer that is not a perfect square. Show that $E/\mathbb{Q}$ is normal.

3. Give an example of fields $F \le K \le E$ such that E/K and K/F are normal, but E/F is not. Thus transitivity fails for normal extensions.
4. Suppose that in (3.5.2), the hypothesis of separability is dropped. State and prove an appropriate conclusion.
5. Show that $E = \mathbb{Q}(\sqrt{2}, \sqrt{3})$ is a Galois extension of $\mathbb{Q}$.
6. In Problem 5, find the Galois group of $E/\mathbb{Q}$.
7. Let E be a finite extension of F, and let K be a normal closure (= minimal normal extension) of E over F, as in (3.5.11). Is K unique?
8. If E_1 and E_2 are normal extensions of F, show that $E_1 \cap E_2$ is normal over F.

Appendix To Chapter 3

In this appendix, we give a precise development of the results on algebraic closure treated informally in the text.

A3.1 Lemma

Let E be an algebraic extension of F, and let $\sigma\colon E \to E$ be an F-monomorphism. Then σ is surjective, hence σ is an automorphism of E.

Proof. Let $\alpha \in E$, and let $f(X)$ be the minimal polynomial of α over F. We consider the subfield L of E generated over F by the roots of f that lie in E. Then L is an extension of F that is finitely generated by algebraic elements, so by (3.3.3), L/F is finite. As in (3.5.1), σ takes a root of f to a root of f, so $\sigma(L) \subseteq L$. But $[L : F] = [\sigma(L) : F] < \infty$ (σ maps a basis to a basis), and consequently $\sigma(L) = L$. But $\alpha \in L$, so $\alpha \in \sigma(L)$. ♣

The following result, due to Artin, is crucial.

A3.2 Theorem

If F is any field, there is an algebraically closed field E containing F.

Proof. For each nonconstant polynomial f in $F[X]$, we create a variable $X(f)$. If T is the collection of all such variables, we can form the ring $F[T]$ of all polynomials in all possible finite sets of variables in T, with coefficients in F. Let I be the ideal of $F[T]$ generated by the polynomials $f(X(f))$, $f \in F[X]$. We claim that I is a proper ideal. If not, then $1 \in I$, so there are finitely many polynomials $f_1, \dots, f_n$ in $F[X]$ and polynomials $h_1, \dots, h_n$ in $F[T]$ such that $\sum_{i=1}^{n} h_i f_i(X(f_i)) = 1$. Now only finitely many variables $X_i = X(f_i), i = 1, \dots, m$, can possibly appear in the h_i, so we have an equation of the form

$$\sum_{i=1}^{n} h_i(X_1, \dots, X_m) f_i(X_i) = 1 \tag{1}$$

where $m \geq n$. Let L be the extension of F formed by successively adjoining the roots of $f_1, \ldots, f_n$. Then each f_i has a root $\alpha_i \in L$. If we set $\alpha_i = 0$ for $n \leq i < m$ and then set $X_i = \alpha_i$ for each i in (1), we get $0 = 1$, a contradiction.

Thus the ideal I is proper, and is therefore contained in a maximal ideal $\mathcal{M}$. Let E_1 be the field $F[T]/\mathcal{M}$. Then E_1 contains an isomorphic copy of F, via the map taking $a \in F$ to $a + \mathcal{M} \in E_1$. (Note that if $a \in \mathcal{M}, a \neq 0$, then $1 = a^{-1}a \in \mathcal{M}$, a contradiction.) Consequently, we can assume that $F \leq E_1$. If f is any nonconstant polynomial in $F[X]$, then $X(f) + \mathcal{M} \in E_1$ and $f(X(f) + \mathcal{M}) = f(X(f)) + \mathcal{M} = 0$ because $f(X(f)) \in I \subseteq \mathcal{M}$.

Iterating the above procedure, we construct a chain of fields $F \leq E_1 \leq E_2 \leq \cdots$ such that every polynomial of degree at least 1 in $E_n[X]$ has a root in E_{n+1}. The union E of all the E_n is a field, and every nonconstant polynomial f in $E[X]$ has all its coefficients in some E_n. Therefore f has a root in $E_{n+1} \subseteq E$. ♣

A3.3 Theorem

Every field F has an algebraic closure.

Proof. By (3.5.12), F has an algebraically closed extension L. If E is the algebraic closure of F in L (see 3.3.4), then E/F is algebraic. Let f be a nonconstant polynomial in $E[X]$. Then f has a root α in L (because L is algebraically closed). We now have α algebraic over E (because $f \in E[X]$), and E algebraic over F. As in (3.3.5), α is algebraic over F, hence $\alpha \in E$. By (3.3.1), E is algebraically closed. ♣

A3.4 Problem

Suppose that σ is a monomorphism of F into the algebraically closed field L. Let E be an algebraic extension of F, and α an element of E with minimal polynomial f over F. We wish to extend σ to a monomorphism from $F(\alpha)$ to L. In how many ways can this be done?

Let σf be the polynomial in $(\sigma F)[X]$ obtained from f by applying σ to the coefficients of f. Any extension of f is determined by what it does to α, and as in (3.5.1), the image of α is a root of σf. Now the number of distinct roots of f in an algebraic closure of F, call it t, is the same as the number of distinct roots of σf in L; this follows from the isomorphism extension theorem (3.2.5). Thus the number of extensions is at most t. But if β is any root of σf, we can construct an extension of σ by mapping the element $h(\alpha) \in F(\alpha)$ to $(\sigma h)(\beta)$; in particular, α is mapped to β. To show that the definition makes sense, suppose that $h_1(\alpha) = h_2(\alpha)$. Then $(h_1 - h_2)(\alpha) = 0$, so f divides $h_1 - h_2$ in $F[X]$. Consequently, σf divides $\sigma h_1 - \sigma h_2$ in $(\sigma F)[X]$, so $(\sigma h_1)(\beta) = (\sigma h_2)(\beta)$.

We conclude that the number of extensions of σ is the number of distinct roots of f in an algebraic closure of F.

Rather than extend σ one element at a time, we now attempt an extension to all of E.

A3.5 Theorem

Let $\sigma \colon F \to L$ be a monomorphism, with L algebraically closed. If E is an algebraic extension of F, then σ has an extension to a monomorphism $\tau \colon E \to L$.

Proof. Let $\mathcal{G}$ be the collection of all pairs (K, μ) where K is an intermediate field between F and E and μ is an extension of σ to a monomorphism from K to L. We partially order $\mathcal{G}$ by $(K_1, \mu) \leq (K_2, \rho)$ iff $K_1 \subseteq K_2$ and ρ restricted to K_1 coincides with μ. Since $(F, \sigma) \in \mathcal{G}$, we have $\mathcal{G} \neq \emptyset$. If the pairs $(K_i, \mu_i), i \in I$, form a chain, there is an upper bound (K, μ) for the chain, where K is the union of the K_i and μ coincides with μ_i on each K_i. By Zorn's lemma, $\mathcal{G}$ has a maximal element (K_0, τ). If $K_0 \subset E$, let $\alpha \in E \setminus K_0$. By (3.5.12), τ has an extension to $K_0(\alpha)$, contradicting maximality of (K_0, τ). ♣

A3.6 Corollary

In (3.5.12), if E is algebraically closed and L is algebraic over $\sigma(F)$, then τ is an isomorphism.

Proof. Since E is algebraically closed, so is $\tau(E)$. Since L is algebraic over $\sigma(F)$, it is algebraic over the larger field $\tau(E)$. By (1) $\iff$ (4) in (3.3.1), $L = \tau(E)$. ♣

A3.7 Theorem

Any two algebraic closures L and E of a field F are F-isomorphic.

Proof. We can assume that F is a subfield of L and $\sigma \colon F \to L$ is the inclusion map. By (3.5.12), σ extends to an isomorphism τ of E and L, and since τ is an extension of σ, it is an F-monomorphism. ♣

A3.8 Theorem (=Theorem 3.3.9)

If E is an algebraic extension of F and C is an algebraic closure of F, then any embedding of F into C can be extended to an embedding of E into C.

Proof. Repeat the proof of (3.5.12), with the mapping μ required to be an embedding. ♣

A3.9 Remark

The argument just given assumes that E is a subfield of C. This can be assumed without loss of generality, by (3.5.12), (3.5.12) and (3.5.12). In other words, we can assume that an algebraic closure of F contains a specified algebraic extension of F.

A3.10 Theorem

Let E be an algebraic extension of F, and let L be the algebraic closure of F containing E (see 3.5.12). If σ is an F-monomorphism from E to L, then σ can be extended to an automorphism of L.

Proof. We have L algebraically closed and L/E algebraic, so by (3.5.12) with E replaced by L and F by E, σ extends to a monomorphism from L to L, an F-monomorphism by hypothesis. The result follows from (3.5.12). ♣

Chapter 4

Module Fundamentals

4.1 Modules and Algebras

4.1.1 Definitions and Comments

A vector space M over a field R is a set of objects called vectors, which can be added, subtracted and multiplied by scalars (members of the underlying field). Thus M is an abelian group under addition, and for each $r \in R$ and $x \in M$ we have an element $rx \in M$. Scalar multiplication is distributive and associative, and the multiplicative identity of the field acts as an identity on vectors. Formally,

$$r(x+y) = rx + ry; \quad (r+s)x = rx + sx; \quad r(sx) = (rs)x; \quad 1x = x$$

for all $x, y \in M$ and $r, s \in R$. A module is just a vector space over a ring. The formal definition is exactly as above, but we relax the requirement that R be a field, and instead allow an arbitrary ring. We have written the product rx with the scalar r on the left, and technically we get a *left R-module* over the ring R. The axioms of a *right R-module* are

$$(x+y)r = xr + yr; \quad x(r+s) = xr + xs; \quad (xs)r = x(sr), \quad x1 = x.$$

"Module" will always mean left module unless stated otherwise. Most of the time, there is no reason to switch the scalars from one side to the other (especially if the underlying ring is commutative). But there are cases where we must be very careful to distinguish between left and right modules (see Example 6 of (4.1.3)).

4.1.2 Some Basic Properties of Modules

Let M be an R-module. The technique given for rings in (2.1.1) can be applied to establish the following results, which hold for any $x \in M$ and $r \in R$. We distinguish the zero vector 0_M from the zero scalar 0_R.

(1) $r0_M = 0_M$ $[r0_M = r(0_M + 0_M) = r0_M + r0_M]$

(2) $0_R x = 0_M$ $[0_R x = (0_R + 0_R)x = 0_R x + 0_R x]$

(3) $(-r)x = r(-x) = -(rx)$ [as in (2) of (2.1.1) with a replaced by r and b by x]

(4) If R is a field, or more generally a division ring, then $rx = 0_M$ implies that either $r = 0_R$ or $x = 0_M$. [If $r \neq 0$, multiply the equation $rx = 0_M$ by r^{-1}.]

4.1.3 Examples

1. If M is a vector space over the field R, then M is an R-module.

2. Any ring R is a module over itself. Rather than check all the formal requirements, think intuitively: Elements of a ring can be added and subtracted, and we can certainly multiply $r \in R$ by $x \in R$, and the usual rules of arithmetic apply.

3. If R is any ring, then R^n, the set of all n-tuples with components in R, is an R-module, with the usual definitions of addition and scalar multiplication (as in Euclidean space, e.g., $r(x_1, \dots, x_n) = (rx_1, \dots, rx_n)$, etc).

4. Let $M = M_{mn}(R)$ be the set of all $m \times n$ matrices with entries in R. Then M is an R-module, where addition is ordinary matrix addition, and multiplication of the scalar c by the matrix A means multiplication of each entry of A by c.

5. Every abelian group A is a $\mathbb{Z}$-module. Addition and subtraction is carried out according to the group structure of A; the key point is that we can multiply $x \in A$ by the integer n. If $n > 0$, then $nx = x + x + \cdots + x$ (n times); if $n < 0$, then $nx = -x - x - \cdots - x$ ($|n|$ times).

In all of these examples, we can switch from left to right modules by a simple notational change. This is definitely not the case in the next example.

6. Let I be a left ideal of the ring R; then I is a left R-module. (If $x \in I$ and $r \in R$ then rx (but not necessarily xr) belongs to I.) Similarly, a right ideal is a right R-module, and a two-sided ideal is both a left and a right R-module.

An R-module M permits addition of vectors and scalar multiplication. If multiplication of vectors is allowed, we have an R-algebra.

4.1.4 Definitions and Comments

Let R be a commutative ring. We say that M is an *algebra over* R, or that M is an R-*algebra*, if M is an R-module that is also a ring (not necessarily commutative), and the ring and module operations are compatible, i.e.,

$$r(xy) = (rx)y = x(ry) \text{ for all } x, y \in M \text{ and } r \in R.$$

4.1.5 Examples

1. Every commutative ring R is an algebra over itself (see Example 2 of (4.1.3)).

2. An arbitrary ring R is always a $\mathbb{Z}$-algebra (see Example 5 of (4.1.3)).

3. If R is a commutative ring, then $M_n(R)$, the set of all $n \times n$ matrices with entries in R, is an R-algebra (see Example 4 of (4.1.3)).

4. If R is a commutative ring, then the polynomial ring $R[X]$ is an R-algebra, as is the ring $R[[X]]$ of formal power series; see Examples 5 and 6 of (2.1.3). The compatibility condition is satisfied because an element of R can be regarded as a polynomial of degree 0.

5. If E/F is a field extension, then E is an algebra over F. This continues to hold if E is a division ring, and in this case we say that E is a *division algebra* over F.

To check that a subset S of a vector space is a subspace, we verify that S is closed under addition of vectors and multiplication of a vector by a scalar. Exactly the same idea applies to modules and algebras.

4.1.6 Definitions and Comments

If N is a nonempty subset of the R-module M, we say that N is a *submodule* of M (notation $N \leq M$) if for every $x, y \in N$ and $r, s \in R$, we have $rx + sy \in N$. If M is an R-algebra, we say that N is a *subalgebra* if N is a submodule that is also a subring.

For example, if A is an abelian group ($= \mathbb{Z}$-module), the submodules of A are the subsets closed under addition and multiplication by an integer (which amounts to addition also). Thus the submodules of A are simply the subgroups. If R is a ring, hence a module over itself, the submodules are those subsets closed under addition and also under multiplication by any $r \in R$, in other words, the left ideals. (If we take R to be a right R-module, then the submodules are the right ideals.)

We can produce many examples of subspaces of vector spaces by considering kernels and images of linear transformations. A similar idea applies to modules.

4.1.7 Definitions and Comments

Let M and N be R-modules. A *module homomorphism* (also called an *R-homomorphism)* from M to N is a map $f\colon M \to N$ such that

$$f(rx + sy) = rf(x) + sf(y) \text{ for all } x, y \in M \text{ and } r, s \in R.$$

Equivalently, $f(x+y) = f(x) + f(y)$ and $f(rx) = rf(x)$ for all $x, y \in M$ and $r \in R$.

The *kernel* of a homomorphism f is $\ker f = \{x \in M\colon f(x) = 0\}$, and the *image* of f is $\{f(x)\colon x \in M\}$.

If follows from the definitions that the kernel of f is a submodule of M, and the image of f is a submodule of N.

If M and N are R-algebras, an *algebra homomorphism* or *homomorphism of algebras* from M to N is an R-module homomorphism that is also a ring homomorphism.

4.1.8 Another Way to Describe an Algebra

Assume that A is an algebra over the commutative ring R, and consider the map $r \to r1$ of R into A. The commutativity of R and the compatibility of the ring and module operations imply that the map is a ring homomorphism. To see this, note that if $r, s \in R$ then

$$(rs)1 = (sr)1 = s(r1) = s[(r1)1] = (r1)(s1).$$

Furthermore, if $y \in A$ then

$$(r1)y = r(1y) = r(y1) = y(r1)$$

so that $r1$ belongs to the *center* of A, i.e., the set of elements that commute with everything in A.

Conversely, if f is a ring homomorphism from the commutative ring R to the center of the ring A, we can make A into an R-module via $rx = f(r)x$. The compatibility conditions are satisfied because

$$r(xy) = f(r)(xy) = (f(r)x)y = (rx)y$$

and

$$(f(r)x)y = (xf(r))y = x(f(r)y) = x(ry).$$

Because of this result, the definition of an R-algebra is sometimes given as follows. The ring A is an algebra over the commutative ring R if there exists a ring homomorphism of R into the center of A. For us at this stage, such a definition would be a severe overdose of abstraction.

Notational Convention: We will often write the module $\{0\}$ (and the ideal $\{0\}$ in a ring) simply as 0.

Problems For Section 4.1

1. If I is an ideal of the ring R, show how to make the quotient ring R/I into a left R-module, and also show how to make R/I into a right R-module.
2. Let A be a commutative ring and F a field. Show that A is an algebra over F if and only if A contains (an isomorphic copy of) F as a subring.

Problems 3, 4 and 5 illustrate that familiar properties of vector spaces need not hold for modules.

3. Give an example of an R-module M with nonzero elements $r \in R$ and $x \in M$ such that $rx = 0$.
4. Let M be the additive group of rational numbers. Show that any two elements of M are linearly dependent (over the integers $\mathbb{Z}$).
5. Continuing Problem 4, show that M cannot have a basis, that is, a linearly independent spanning set over $\mathbb{Z}$.
6. Prove the *modular law* for subgroups of a given group G: With the group operation written multiplicatively,

$$A(B \cap C) = (AB) \cap C$$

if $A \subseteq C$. Switching to additive notation, we have, for submodules of a given R-module,

$$A + (B \cap C) = (A + B) \cap C,$$

again if $A \subseteq C$.

7. Let T be a linear transformation on the vector space V over the field F. Show how to make V into an R-module in a natural way, where R is the polynomial ring $F[X]$.

4.2 The Isomorphism Theorems For Modules

If N is a submodule of the R-module M (notation $N \leq M$), then in particular N is an additive subgroup of M, and we may form the quotient group M/N in the usual way. In fact M/N becomes an R-module if we define $r(x + N) = rx + N$. (This makes sense because if x belongs to the submodule N, so does rx.) Since scalar multiplication in the quotient module M/N is carried out via scalar multiplication in the original module M, we can check the module axioms without difficulty. The canonical map $\pi\colon M \to M/N$ is a module homomorphism with kernel N. Just as with groups and rings, we can establish the basic isomorphism theorems for modules.

4.2.1 Factor Theorem For Modules

Any module homomorphism $f\colon M \to M'$ whose kernel contains N can be factored through M/N. In other words, there is a unique module homomorphism $\overline{f}\colon M/N \to M'$ such that $\overline{f}(x+N) = f(x)$. Furthermore, (i) $\overline{f}$ is an epimorphism if and only if f is an epimorphism; (ii) $\overline{f}$ is a monomorphism if and only if $\ker f = N$; (iii) $\overline{f}$ is an isomorphism if and only if f is a epimorphism and $\ker f = N$.

Proof. Exactly as in (2.3.1), with appropriate notational changes. (In Figure 2.3.1, replace R by M, S by M' and I by N.) ♣

4.2.2 First Isomorphism Theorem For Modules

If $f\colon M \to M'$ is a module homomorphism with kernel N, then the image of f is isomorphic to M/N.

Proof. Apply the factor theorem, and note that f is an epimorphism onto its image. ♣

4.2.3 Second Isomorphism Theorem For Modules

Let S and T be submodules of M, and let $S + T = \{x + y\colon x \in S, y \in T\}$. Then $S + T$ and $S \cap T$ are submodules of M and

$$(S + T)/T \cong S/(S \cap T).$$

Proof. The module axioms for $S + T$ and $S \cap T$ can be checked in routine fashion. Define a map $f\colon S \to M/T$ by $f(x) = x + T$. Then f is a module homomorphism whose kernel is $S \cap T$ and whose image is $\{x + T\colon x \in S\} = (S + T)/T$. The first isomorphism theorem for modules gives the desired result. ♣

4.2.4 Third Isomorphism Theorem For Modules

If $N \leq L \leq M$, then

$$M/L \cong (M/N)/(L/N).$$

Proof. Define $f\colon M/N \to M/L$ by $f(x+N) = x+L$. As in (2.3.4), the kernel of f is $\{x+N\colon x \in L\} = L/N$, and the image of f is $\{x+L\colon x \in M\} = M/L$. The result follows from the first isomorphism theorem for modules. ♣

4.2.5 Correspondence Theorem For Modules

Let N be a submodule of the R-module M. The map $S \to S/N$ sets up a one-to-one correspondence between the set of all submodules of M containing N and the set of all submodules of M/N. The inverse of the map is $T \to \pi^{-1}(T)$, where π is the canonical map: $M \to M/N$.

Proof. The correspondence theorem for groups yields a one-to-one correspondence between additive subgroups of M containing N and additive subgroups of M/N. We must check that submodules correspond to submodules, and it is sufficient to show that if $S_1/N \leq S_2/N$, then $S_1 \leq S_2$ (the converse is immediate). If $x \in S_1$, then $x+N \in S_1/N \subseteq S_2/N$, so $x+N = y+N$ for some $y \in S_2$. Thus $x-y \in N \subseteq S_2$, and since $y \in S_2$ we must have $x \in S_2$ as well. The ore $S_1 \leq S_2$. ♣

We now look at modules that have a particularly simple structure, and can be used as building blocks for more complicated modules.

4.2.6 Definitions and Comments

An R-module M is *cyclic* if it is generated by a single element x. In other words,

$$M = Rx = \{rx\colon r \in R\}.$$

Thus every element of M is a scalar multiple of x. (If $x = 0$, then $M = \{0\}$, which is called the *zero module* and is often written simply as 0.) A cyclic vector space over a field is a one-dimensional space, assuming that $x \neq 0$.

The *annihilator* of an *element* y in the R-module M is $I_y = \{r \in R\colon ry = 0\}$, a left ideal of R. If R is commutative, and M is cyclic with generator x, then $M \cong R/I_x$. To see this, apply the first isomorphism theorem for modules to the map $r \to rx$ of R onto M. The *annihilator* of the *module* M is $I_o = \{r \in R\colon ry = 0 \text{ for every } y \in M\}$. Note that I_o is a two-sided ideal, because if $r \in I_0$ and $s \in R$, then for every $y \in M$ we have $(rs)y = r(sy) = 0$. When R is commutative, annihilating the generator of a cyclic module is equivalent to annihilating the entire module.

4.2.7 Lemma

(a) If x generates a cyclic module M over the commutative ring R, then $I_x = I_o$, so that $M \cong R/I_o$. (In this situation, I_o is frequently referred to as the *order ideal* of M.)
(b) Two cyclic R-modules over a commutative ring are isomorphic if and only if they have the same annihilator.

Proof. (a) If $rx = 0$ and $y \in M$, then $y = sx$ for some $s \in R$, so $ry = r(sx) = s(rx) = s0 = 0$. Conversely, if r annihilates M, then in particular, $rx = 0$.

(b) The "if" part follows from (a), so assume that $g\colon Rx \to Ry$ is an isomorphism of cyclic R-modules. Since g is an isomorphism, $g(x)$ must be a generator of Ry, so we may as well take $g(x) = y$. Then $g(rx) = rg(x) = ry$, so rx corresponds to ry under the isomorphism. Therefore r belongs to the annihilator of Rx if and only if r belongs to the annihilator of Ry. ♣

Problems For Section 4.2

1. Show that every submodule of the quotient module M/N can be expressed as $(L+N)/N$ for some submodule L of M.
2. In Problem 1, must L contain N?
3. In the matrix ring $M_n(R)$, let M be the submodule generated by E_{11}, the matrix with 1 in row 1, column 1, and 0's elsewhere. Thus $M = \{AE_{11}\colon A \in M_n(R)\}$. Show that M consists of all matrices whose entries are zero except perhaps in column 1.
4. Continuing Problem 3, show that the annihilator of E_{11} consists of all matrices whose first column is zero, but the annihilator of M is $\{0\}$.
5. If I is an ideal of the ring R, show that R/I is a cyclic R-module.
6. Let M be an R-module, and let I be an ideal of R. We wish to make M into an R/I-module via $(r + I)m = rm, r \in R, m \in M$. When will this be legal?
7. Assuming legality in Problem 6, let M_1 be the resulting R/I-module, and note that as sets, $M_1 = M$. Let N be a subset of M and consider the following two statements:

   (a) N is an R-submodule of M;

   (b) N is an R/I-submodule of M_1.

   Can one of these statements be true and the other false?

4.3 Direct Sums and Free Modules

4.3.1 Direct Products

In Section 1.5, we studied direct products of groups, and the basic idea seems to carry over to modules. Suppose that we have an R-module M_i for each i in some index set I (possibly infinite). The members of the *direct product* of the M_i, denoted by $\prod_{i\in I} M_i$, are all families $(a_i, i \in I)$, where $a_i \in M_i$. (A family is just a function on I whose value at the element i is a_i.) Addition is described by $(a_i) + (b_i) = (a_i + b_i)$ and scalar multiplication by $r(a_i) = (ra_i)$.

There is nothing wrong with this definition, but the resulting mathematical object has some properties that are undesirable. If (e_i) is the family with 1 in position i and zeros elsewhere, then (thinking about vector spaces) it would be useful to express an arbitrary element of the direct product in terms of the e_i. But if the index set I is infinite, we will need concepts of limit and convergence, and this will take us out of algebra and into analysis. Another approach is to modify the definition of direct product.

4.3.2 Definitions

The *external direct sum* of the modules $M_i, i \in I$, denoted by $\oplus_{i \in I} M_i$, consists of all families $(a_i, i \in I)$ with $a_i \in M_i$, such that $a_i = 0$ for all but finitely many i. Addition and scalar multiplication are defined exactly as for the direct product, so that the external direct sum coincides with the direct product when the index set I is finite.

The R-module M is the *internal direct sum* of the submodules M_i if each $x \in M$ can be expressed uniquely as $x_{i_1} + \cdots + x_{i_n}$ where $0 \neq x_{i_k} \in M_{i_k}, k = 1, \ldots, n$. (The positive integer n and the elements x_{i_k} depend on x. In any expression of this type, the indices i_k are assumed distinct.)

Just as with groups, the internal and external direct sums are isomorphic. To see this without a lot of formalism, let the element $x_{i_k} \in M_{i_k}$ correspond to the family that has x_{i_k} in position i_k and zeros elsewhere. We will follow standard practice and refer to the "direct sum" without the qualifying adjective. Again as with groups, the next result may help us to recognize when a module can be expressed as a direct sum.

4.3.3 Proposition

The module M is the direct sum of submodules M_i if and only if both of the following conditions are satisfied:

(1) $M = \sum_i M_i$, that is, each $x \in M$ is a finite sum of the form $x_{i_1} + \cdots + x_{i_n}$, where $x_{i_k} \in M_{i_k}$;

(2) For each i, $M_i \cap \sum_{j \neq i} M_j = 0$.

(Note that in condition (1), we do *not* assume that the representation is unique. Observe also that another way of expressing (2) is that if $x_{i_1} + \cdots + x_{i_n} = 0$, with $x_{i_k} \in M_{i_k}$, then $x_{i_k} = 0$ for all k.)

Proof. The necessity of the conditions follows from the definition of external direct sum, so assume that (1) and (2) hold. If $x \in M$ then by (1), x is a finite sum of elements from various M_i's. For convenience in notation, say $x = x_1 + x_2 + x_3 + x_4$ with $x_i \in M_i$, $i = 1, 2, 3, 4$. If the representation is not unique, say $x = y_1 + y_2 + y_4 + y_5 + y_6$ with $y_i \in M_i$, $i = 1, 2, 4, 5, 6$. Then x_3 is a sum of terms from modules other than M_3, so by (2), $x_3 = 0$. Similarly, $y_5 = y_6 = 0$ and we have $x_1 + x_2 + x_4 = y_1 + y_2 + y_4$. But then $x_1 - y_1$ is a sum of terms from modules other than M_1, so by (2), $x_1 = y_1$. Similarly $x_2 = y_2$, $x_4 = y_4$, and the result follows. ♣

A basic property of the direct sum $M = \oplus_{i \in I} M_i$ is that homomorphisms $f_i \colon M_i \to N$ can be "lifted" to M. In other words, there is a unique homomorphism $f \colon M \to N$ such that for each i, $f = f_i$ on M_i. Explicitly,

$$f(x_{i_1} + \cdots + x_{i_r}) = f_{i_1}(x_{i_1}) + \cdots + f_{i_r}(x_{i_r}).$$

[No other choice is possible for f, and since each f_i is a homomorphism, so is f.]

We know that every vector space has a basis, but not every module is so fortunate; see Section 4.1, Problem 5. We now examine modules that have this feature.

4.3.4 Definitions and Comments

Let S be a subset of the R-module M. We say that S is *linearly independent over* R if $\lambda_1 x_1 + \cdots + \lambda_k x_k = 0$ implies that all $\lambda_i = 0$ $(\lambda_i \in R, x_i \in S, k = 1, 2, \dots)$. We say that S is a *spanning* (or *generating*) *set for* M *over* R, or that S *spans (generates)* M *over* R if each $x \in M$ can be written as a finite linear combination of elements of S with coefficients in R. We will usually omit "over R" if the underlying ring R is clearly identified. A *basis* is a linearly independent spanning set, and a module that has a basis is said to be *free*.

Suppose that M is a free module with basis $(b_i, i \in I)$, and we look at the submodule M_i spanned by the basis element b_i. (In general, the submodule spanned (or generated) by a subset T of M consists of all finite linear combinations of elements of T with coefficients in R. Thus the submodule spanned by b_i is the set of all $rb_i, r \in R$.) If R is regarded as a module over itself, then the map $r \to rb_i$ is an R-module isomorphism of R and M_i, because $\{b_i\}$ is a linearly independent set. Since the b_i span M, it follows that M is the sum of the submodules M_i, and by linear independence of the b_i, the sum is direct. Thus we have an illuminating interpretation of a free module:

A free module is a direct sum of isomorphic copies of the underlying ring R.

Conversely, a direct sum of copies of R is a free R-module. If e_i has 1 as its ith component and zeros elsewhere, the e_i form a basis.

This characterization allows us to recognize several examples of free modules.

1. For any positive integer n, R^n is a free R-module.

2. The matrix ring $M_{mn}(R)$ is a free R-module with basis E_{ij}, $i = 1, \dots, m$, $j = 1, \dots, n$.

3. The polynomial ring $R[X]$ is a free R-module with basis $1, X, X^2, \dots$.

We will adopt the standard convention that the zero module is free with the empty set as basis.

Any two bases for a vector space over a field have the same cardinality. This property does not hold for arbitrary free modules, but the following result covers quite a few cases.

4.3.5 Theorem

Any two bases for a free module M over a commutative ring R have the same cardinality.

Proof. If I is a maximal ideal of R, then $k = R/I$ is a field, and $V = M/IM$ is a vector space over k. [By IM we mean all finite sums $\sum a_i x_i$ with $a_i \in I$ and $x_i \in M$; thus IM is a submodule of M. If $r + I \in k$ and $x + IM \in M/IM$, we take $(r + I)(x + IM)$ to be $rx + IM$. The scalar multiplication is well-defined because $r \in I$ or $x \in IM$ implies that $rx \in IM$. We can express this in a slightly different manner by saying that I annihilates M/IM. The requirements for a vector space can be checked routinely.]

Now if (x_i) is a basis for M, let $\overline{x}_i = x_i + IM$. Since the x_i span M, the $\overline{x}_i$ span M/IM. If $\sum \overline{a}_i \overline{x}_i = 0$, where $\overline{a}_i = a_i + I, a_i \in R$, then $\sum a_i x_i \in IM$. Thus $\sum a_i x_i = \sum b_j x_j$ with $b_j \in I$. Since the x_i form a basis, we must have $a_i = b_j$ for some j. Consequently $a_i \in I$, so that $\overline{a}_i = 0$ in k. We conclude that the $\overline{x}_i$ form a basis for V over k, and since the dimension of V over k depends only on M, R and I, and not on a particular basis for M, the result follows. ♣

4.3.6 Some Key Properties of Free Modules

Suppose that M is a free module with basis (x_i), and we wish to construct a module homomorphism f from M to an arbitrary module N. Just as with vector spaces, we can specify $f(x_i) = y_i \in N$ arbitrarily on basis elements, and extend by linearity. Thus if $x = \sum a_i x_i \in M$, we have $f(x) = \sum a_i y_i$. (The idea should be familiar; for example, a linear transformation on Euclidean 3-space is determined by what it does to the three standard basis vectors.) Now let's turn this process around:

> If N is an arbitrary module, we can express N as a homomorphic image of a free module.

All we need is a set $(y_i, i \in I)$ of generators for N. (If all else fails, we can take the y_i to be all the elements of N.) We then construct a free module with basis $(x_i, i \in I)$. (To do this, take the direct sum of copies of R, as many copies as there are elements of I.) Then map x_i to y_i for each i.

Note that by the first isomorphism theorem, every module is a quotient of a free module.

Problems For Section 4.3

1. Show that in Proposition 4.3.3, (2) can be replaced by the weaker condition that for each i, $M_i \cap \sum_{j<i} M_j = 0$. (Assume a fixed total ordering on the index set.)
2. Let A be a finite abelian group. Is it possible for A to be a free $\mathbb{Z}$-module?
3. Let r and s be elements in the ideal I of the commutative ring R. Show that r and s are linearly dependent over R.
4. In Problem 3, regard I as an R-module. Can I be free?
5. Give an example of an infinite abelian group that is a free $\mathbb{Z}$-module, and an example of an infinite abelian group that is not free.
6. Show that a module M is free if and only if M has a subset S such that any function f from S to a module N can be extended uniquely to a module homomorphism from M to N.

7. Let M be a free module, expressed as the direct sum of α copies of the underlying ring R, where α and $|R|$ are infinite cardinals. Find the cardinality of M.
8. In Problem 7, assume that all bases B have the same cardinality, e.g., R is commutative. Find the cardinality of B.

4.4 Homomorphisms and Matrices

Suppose that M is a free R-module with a finite basis of n elements $v_1, \dots, v_n$, sometimes called a free module of *rank n*. We know from Section 4.3 that M is isomorphic to the direct sum of n copies of R. Thus we can regard M as R^n, the set of all n-tuples with components in R. Addition and scalar multiplication are performed componentwise, as in (4.1.3), Example 3. Note also that the direct sum coincides with the direct product, since we are summing only finitely many modules.

Let N be a free R-module of rank m, with basis $w_1, \dots, w_m$, and suppose that f is a module homomorphism from M to N. Just as in the familiar case of a linear transformation on a finite-dimensional vector space, we are going to represent f by a matrix. For each j, $f(v_j)$ is a linear combination of the basis elements w_j, so that

$$f(v_j) = \sum_{i=1}^{m} a_{ij} w_i, \; j = 1, \dots, n \tag{1}$$

where the a_{ij} belong to R.

It is natural to associate the $m \times n$ matrix A with the homomorphism f, and it appears that we have an isomorphism of some sort, but an isomorphism of what? If f and g are homomorphisms of M into N, then f and g can be added (and subtracted): $(f+g)(x) = f(x) + g(x)$. If f is represented by the matrix A and g by B, then $f+g$ corresponds to $A+B$. This gives us an abelian group isomorphism of $\text{Hom}_R(M, N)$, the set of all R-module homomorphisms from M to N, and $M_{mn}(R)$, the set of all $m \times n$ matrices with entries in R. In addition, $M_{mn}(R)$ is an R-module, so it is tempting to say "obviously, we have an R-module isomorphism". But we must be very careful here. If $f \in \text{Hom}_R(M, N)$ and $s \in R$, we can define sf in the natural way: $(sf)(x) = sf(x)$. However, if we carry out the "routine" check that $sf \in \text{Hom}_R(M, N)$, there is one step that causes alarm bells to go off:

$$(sf)(rx) = sf(rx) = srf(x), \text{ but } r(sf)(x) = rsf(x)$$

and the two expressions can disagree if R is not commutative. Thus $\text{Hom}_R(M, N)$ need not be an R-module. Let us summarize what we have so far.

4.4.1 The Correspondence Between Homomorphisms and Matrices

Associate with each $f \in \text{Hom}_R(M, N)$ a matrix A as in (1) above. This yields an abelian group isomorphism, and also an R-module isomorphism if R is commutative.

Now let $m = n$, so that the dimensions are equal and the matrices are square, and take $v_i = w_i$ for all i. A homomorphism from M to itself is called an *endomorphism* of

M, and we use the notation $\text{End}_R(M)$ for $\text{Hom}_R(M,M)$. Since $\text{End}_R(M)$ is a ring under composition of functions, and $M_n(R)$ is a ring under matrix multiplication, it is plausible to conjecture that we have a ring isomorphism. If f corresponds to A and g to B, then we can apply g to both sides of (1) to obtain

$$g(f(v_j)) = \sum_{i=1}^{n} a_{ij} \sum_{k=1}^{n} b_{ki} v_k = \sum_{k=1}^{n} (\sum_{i=1}^{n} a_{ij} b_{ki}) v_k. \tag{2}$$

If R is commutative, then $a_{ij}b_{ki} = b_{ki}a_{ij}$, and the matrix corresponding to $gf = g \circ f$ is BA, as we had hoped. In the noncommutative case, we will not be left empty-handed if we define the *opposite ring* R^o, which has exactly the same elements as R and the same addition structure. However, multiplication is done backwards, i.e., ab in R^o is ba in R. It is convenient to attach a superscript ^{o} to the elements of R^o, so that

$$a^o b^o = ba \text{ (more precisely, } a^o b^o = (ba)^o).$$

Thus in (2) we have $a_{ij}b_{ki} = b^o_{ki}a^o_{ij}$. To summarize,

> The endomorphism ring $\text{End}_R(M)$ is isomorphic to the ring of $n \times n$ matrices with coefficients in the opposite ring R^o. If R is commutative, then $\text{End}_R(M)$ is ring-isomorphic to $M_n(R)$.

4.4.2 Preparation For The Smith Normal Form

We now set up some basic machinery to be used in connection with the Smith normal form and its applications. Assume that M is a free $\mathbb{Z}$-module of rank n, with basis $x_1, \dots, x_n$, and that K is a submodule of M with finitely many generators $u_1, \dots, u_m$. (We say that K is *finitely generated.*) We change to a new basis $y_1, \dots, y_n$ via $Y = PX$, where X [resp. Y] is a column vector with components x_i [resp. y_i]. Since X and Y are bases, the $n \times n$ matrix P must be invertible, and we need to be very clear on what this means. If the determinant of P is nonzero, we can construct P^{-1}, for example by the "adjoint divided by determinant" formula given in Cramer's rule. But the underlying ring is $\mathbb{Z}$, not $\mathbb{Q}$, so we require that the coefficients of P^{-1} be integers. (For a more transparent equivalent condition, see Problem 1.) Similarly, we are going to change generators of K via $V = QU$, where Q is an invertible $m \times m$ matrix and U is a column vector with components u_i.

The generators of K are linear combinations of basis elements, so we have an equation of the form $U = AX$, where A is an $m \times n$ matrix called the *relations matrix.* Thus

$$V = QU = QAX = QAP^{-1}Y.$$

so the new relations matrix is

$$B = QAP^{-1}.$$

Thus B is obtained from A by pre-and postmultiplying by invertible matrices, and we say that A and B are *equivalent.* We will see that two matrices are equivalent iff they have the same Smith normal form. The point we wish to emphasize now is that if we know the matrix P, we can compute the new basis Y, and if we know the matrix Q, we can compute the new system of generators V. In our applications, P and Q will be constructed by elementary row and column operations.

Problems For Section 4.4

1. Show that a square matrix P over the integers has an inverse with integer entries if and only if P is *unimodular*, that is, the determinant of P is ± 1.
2. Let V be the direct sum of the R-modules $V_1, \dots, V_n$, and let W be the direct sum of R-modules $W_1, \dots, W_m$. Indicate how a module homomorphism from V to W can be represented by a matrix. (The entries of the matrix need not be elements of R.)
3. Continuing Problem 2, show that if V^n is the direct sum of n copies of the R-module V, then we have a ring isomorphism

$$\text{End}_R(V^n) \cong M_n(\text{End}_R(V)).$$

4. Show that if R is regarded as an R-module, then $\text{End}_R(R)$ is isomorphic to the opposite ring R^o.
5. Let R be a ring, and let $f \in \text{End}_R(R)$. Show that for some $r \in R$ we have $f(x) = xr$ for all $x \in R$.
6. Let M be a free R-module of rank n. Show that $\text{End}_R(M) \cong M_n(R^o)$, a ring isomorphism.
7. Continuing Problem 6, if R is commutative, show that the ring isomorphism is in fact an R-algebra isomorphism.

4.5 Smith Normal Form

We are going to describe a procedure that is very similar to reduction of a matrix to echelon form. The result is that every matrix over a principal ideal domain is equivalent to a matrix in Smith normal form. Explicitly, the Smith matrix has nonzero entries only on the main diagonal. The main diagonal entries are, from the top, $a_1, \dots, a_r$ (possibly followed by zeros), where the a_i are nonzero and a_i divides a_{i+1} for all i.

We will try to convey the basic ideas via a numerical example. This will allow us to give informal but convincing proofs of some major theorems. A formal development is given in Jacobson, Basic Algebra I, Chapter 3. All our computations will be in the ring of integers, but we will indicate how the results can be extended to an arbitrary principal ideal domain. Let's start with the following matrix:

$$\begin{bmatrix} 0 & 0 & 22 & 0 \\ -2 & 2 & -6 & -4 \\ 2 & 2 & 6 & 8 \end{bmatrix}$$

As in (4.4.2), we assume a free $\mathbb{Z}$-module M with basis x_1, x_2, x_3, x_4, and a submodule K generated by u_1, u_2, u_3, where $u_1 = 22x_3, u_2 = -2x_1 + 2x_2 - 6x_3 - 4x_4, u_3 = 2x_1 + 2x_2 + 6x_3 + 8x_4$. The first step is to bring the smallest positive integer to the 1-1 position. Thus interchange rows 1 and 3 to obtain

$$\begin{bmatrix} 2 & 2 & 6 & 8 \\ -2 & 2 & -6 & -4 \\ 0 & 0 & 22 & 0 \end{bmatrix}$$

Since all entries in column 1, and similarly in row 1, are divisible by 2, we can pivot about the 1-1 position, in other words, use the 1-1 entry to produce zeros. Thus add row 1 to row 2 to get

$$\begin{bmatrix} 2 & 2 & 6 & 8 \\ 0 & 4 & 0 & 4 \\ 0 & 0 & 22 & 0 \end{bmatrix}$$

Add -1 times column 1 to column 2, then add -3 times column 1 to column 3, and add -4 times column 1 to column 4. The result is

$$\begin{bmatrix} 2 & 0 & 0 & 0 \\ 0 & 4 & 0 & 4 \\ 0 & 0 & 22 & 0 \end{bmatrix}$$

Now we have "peeled off" the first row and column, and we bring the smallest positive integer to the 2-2 position. It's already there, so no action is required. Furthermore, the 2-2 element is a multiple of the 1-1 element, so again no action is required. Pivoting about the 2-2 position, we add -1 times column 2 to column 4, and we have

$$\begin{bmatrix} 2 & 0 & 0 & 0 \\ 0 & 4 & 0 & 0 \\ 0 & 0 & 22 & 0 \end{bmatrix}$$

Now we have peeled off the first two rows and columns, and we bring the smallest positive integer to the 3-3 position; again it's already there. But 22 is not a multiple of 4, so we have more work to do. Add row 3 to row 2 to get

$$\begin{bmatrix} 2 & 0 & 0 & 0 \\ 0 & 4 & 22 & 0 \\ 0 & 0 & 22 & 0 \end{bmatrix}$$

Again we pivot about the 2-2 position; 4 does not divide 22, but if we add -5 times column 2 to column 3, we have

$$\begin{bmatrix} 2 & 0 & 0 & 0 \\ 0 & 4 & 2 & 0 \\ 0 & 0 & 22 & 0 \end{bmatrix}$$

Interchange columns 2 and 3 to get

$$\begin{bmatrix} 2 & 0 & 0 & 0 \\ 0 & 2 & 4 & 0 \\ 0 & 22 & 0 & 0 \end{bmatrix}$$

Add -11 times row 2 to row 3 to obtain

$$\begin{bmatrix} 2 & 0 & 0 & 0 \\ 0 & 2 & 4 & 0 \\ 0 & 0 & -44 & 0 \end{bmatrix}$$

Finally, add -2 times column 2 to column 3, and then (as a convenience to get rid of the minus sign) multiply row (or column) 3 by -1; the result is

$$\begin{bmatrix} 2 & 0 & 0 & 0 \\ 0 & 2 & 0 & 0 \\ 0 & 0 & 44 & 0 \end{bmatrix}$$

which is the Smith normal form of the original matrix. Although we had to backtrack to produce a new pivot element in the 2-2 position, the new element is smaller than the old one (since it is a remainder after division by the original number). Thus we cannot go into an infinite loop, and the algorithm will indeed terminate in a finite number of steps. In view of (4.4.2), we have the following interpretation.

We have a new basis y_1, y_2, y_3, y_4 for M, and new generators v_1, v_2, v_3 for K, where $v_1 = 2y_1, v_2 = 2y_2$, and $v_3 = 44y_3$. In fact since the v_j's are nonzero multiples of the corresponding y_j's, they are linearly independent, and consequently form a basis of K. The new basis and set of generators can be expressed in terms of the original sets; see Problems 1–3 for the technique.

The above discussion indicates that the Euclidean algorithm guarantees that the Smith normal form can be computed in finitely many steps. Therefore the Smith procedure can be carried out in any Euclidean domain. In fact we can generalize to a principal ideal domain. Suppose that at a particular stage of the computation, the element a occupies the 1-1 position of the Smith matrix S, and the element b is in row 1, column 2. To use a as a pivot to eliminate b, let d be the greatest common divisor of a and b, and let r and s be elements of R such that $ar + bs = d$ (see (2.7.2)). We postmultiply the Smith matrix by a matrix T of the following form (to aid in the visualization, we give a concrete 5×5 example):

$$\begin{bmatrix} r & b/d & 0 & 0 & 0 \\ s & -a/d & 0 & 0 & 0 \\ 0 & 0 & 1 & 0 & 0 \\ 0 & 0 & 0 & 1 & 0 \\ 0 & 0 & 0 & 0 & 1 \end{bmatrix}$$

The 2×2 matrix in the upper left hand corner has determinant -1, and is therefore invertible over R. The element in the 1-1 position of ST is $ar + bs = d$, and the element in the 1-2 position is $ab/d - ba/d = 0$, as desired. We have replaced the pivot element a by a divisor d, and this will decrease the number of prime factors, guaranteeing the finite termination of the algorithm. Similarly, if b were in the 2-1 position, we would premultiply S by the transpose of T; thus in the upper left hand corner we would have

$$\begin{bmatrix} r & s \\ b/d & -a/d \end{bmatrix}$$

Problems For Section 4.5

1. Let A be the matrix

$$\begin{bmatrix} -2 & 3 & 0 \\ -3 & 3 & 0 \\ -12 & 12 & 6 \end{bmatrix}$$

over the integers. Find the Smith normal form of A. (It is convenient to begin by adding column 2 to column 1.)

2. Continuing Problem 1, find the matrices P and Q, and verify that QAP^{-1} is the Smith normal form.
3. Continuing Problem 2, if the original basis for M is $\{x_1, x_2, x_3\}$ and the original set of generators of K is $\{u_1, u_2, u_3\}$, find the new basis and set of generators.

It is intuitively reasonable, but a bit messy to prove, that if a matrix A over a PID is multiplied by an invertible matrix, then the greatest common divisor of all the $i \times i$ minors of A is unchanged. Accept this fact in doing Problems 4 and 5.

4. The nonzero components a_i of the Smith normal form S of A are called the *invariant factors* of A. Show that the invariant factors of A are unique (up to associates).
5. Show that two $m \times n$ matrices are equivalent if and only if they have the same invariant factors, i.e. (by Problem 4), if and only if they have the same Smith normal form.
6. Recall that when a matrix over a field is reduced to row-echelon form (only row operations are involved), a pivot column is followed by non-pivot columns whose entries are zero in all rows below the pivot element. When a similar computation is carried out over the integers, or more generally over a Euclidean domain, the resulting matrix is said to be in *Hermite normal form*. We indicate the procedure in a typical example. Let

$$A = \begin{bmatrix} 6 & 4 & 13 & 5 \\ 9 & 6 & 0 & 7 \\ 12 & 8 & -1 & 12 \end{bmatrix}.$$

Carry out the following sequence of steps:

1. Add -1 times row 1 to row 2
2. Interchange rows 1 and 2
3. Add -2 times row 1 to row 2, and then add -4 times row 1 to row 3
4. Add -1 times row 2 to row 3
5. Interchange rows 2 and 3
6. Add -3 times row 2 to row 3
7. Interchange rows 2 and 3
8. Add -4 times row 2 to row 3
9. Add 5 times row 2 to row 1 (this corresponds to choosing $0, 1, \ldots, m-1$ as a complete system of residues mod m)
10. Add 2 times row 3 to row 1, and then add row 3 to row 2

We now have reduced A to Hermite normal form.

7. Continuing Problem 6, consider the simultaneous equations

$$6x + 4y + 13z \equiv 5,\ 9x + 6y \equiv 7,\ 12x + 8y - z \equiv 12 \pmod{m}$$

For which values of $m \geq 2$ will the equations be consistent?

4.6 Fundamental Structure Theorems

The Smith normal form yields a wealth of information about modules over a principal ideal domain. In particular, we will be able to see exactly what finitely generated abelian groups must look like.

Before we proceed, we must mention a result that we will use now but not prove until later (see (7.5.5), Example 1, and (7.5.9)). If M is a finitely generated module over a PID R, then every submodule of M is finitely generated. [R is a Noetherian ring, hence M is a Noetherian R-module.] To avoid gaps in the current presentation, we can restrict our attention to finitely generated submodules.

4.6.1 Simultaneous Basis Theorem

Let M be a free module of finite rank $n \geq 1$ over the PID R, and let K be a submodule of M. Then there is a basis $\{y_1, \dots, y_n\}$ for M and nonzero elements $a_1, \dots, a_r \in R$ such that $r \leq n$, a_i divides a_{i+1} for all i, and $\{a_1 y_1, \dots, a_r y_r\}$ is a basis for K.

Proof. This is a corollary of the construction of the Smith normal form, as explained in Section 4.5. ♣

4.6.2 Corollary

Let M be a free module of finite rank n over the PID R. Then every submodule of M is free of rank at most n.

Proof. By (4.6.1), the submodule K has a basis with $r \leq n$ elements. ♣

In (4.6.2), the hypothesis that M has finite rank can be dropped, as the following sketch suggests. We can well-order the generators u_α of K, and assume as a transfinite induction hypothesis that for all $\beta < \alpha$, the submodule K_β spanned by all the generators up to u_β is free of rank at most that of M, and that if $\gamma < \beta$, then the basis of K_γ is contained in the basis of K_β. The union of the bases S_β of the K_β is a basis S_α for K_α. Furthermore, the inductive step preserves the bound on rank. This is because $|S_\beta| \leq$ rank M for all $\beta < \alpha$, and $|S_\alpha|$ is the smallest cardinal bounded below by all $|S_\beta|, \beta < \alpha$. Thus $|S_\alpha| \leq$ rank M.

4.6.3 Fundamental Decomposition Theorem

Let M be a finitely generated module over the PID R. Then there are ideals $I_1 = \langle a_1 \rangle$, $I_2 = \langle a_2 \rangle, \dots, I_n = \langle a_n \rangle$ of R such that $I_1 \supseteq I_2 \supseteq \cdots \supseteq I_n$ (equivalently, $a_1 \mid a_2 \mid \cdots \mid a_n$) and

$$M \cong R/I_1 \oplus R/I_2 \oplus \cdots \oplus R/I_n.$$

Thus M is a direct sum of cyclic modules.

Proof. By (4.3.6), M is the image of a free module R^n under a homomorphism f. If K is the kernel of f, then by (4.6.1) we have a basis $y_1, \dots, y_n$ for R^n and a corresponding basis $a_1y_1, \dots, a_ry_r$ for K. We set $a_i = 0$ for $r < i \le n$. Then

$$M \cong R^n/K \cong \frac{Ry_1 \oplus \cdots \oplus Ry_n}{Ra_1y_1 \oplus \cdots \oplus Ra_ny_n} \cong \bigoplus_{i=1}^{n} Ry_i/Ra_iy_i.$$

(To justify the last step, apply the first isomorphism theorem to the map

$$r_1y_1 + \cdots + r_ny_n \to (r_1y_1 + Ra_1y_1, \dots, r_ny_n + Ra_ny_n.)$$

But

$$Ry_i/Ra_iy_i \cong R/Ra_i,$$

as can be seen via an application of the first isomorphism theorem to the map $r \to ry_i + Ra_iy_i$. Thus if $I_i = Ra_i$, $i = 1, \dots, n$, we have

$$M \cong \bigoplus_{i=1}^{n} R/I_i$$

and the result follows. ♣

Remark It is plausible, and can be proved formally, that the uniqueness of invariant factors in the Smith normal form implies the uniqueness of the decomposition (4.6.3). Intuitively, the decomposition is completely specified by the sequence $a_1, \dots, a_n$, as the proof of (4.6.3) indicates.

4.6.4 Finite Abelian Groups

Suppose that G is a finite abelian group of order 1350; what can we say about G? In the decomposition theorem (4.6.3), the components of G are of the form $\mathbb{Z}/\mathbb{Z}a_i$, that is, cyclic groups of order a_i. We must have $a_i \mid a_{i+1}$ for all i, and since the order of a direct sum is the product of the orders of the components, we have $a_1 \cdots a_r = 1350$.

The first step in the analysis is to find the prime factorization of 1350, which is $(2)(3^3)(5^2)$. One possible choice of the a_i is $a_1 = 3$, $a_2 = 3$, $a_3 = 150$. It is convenient to display the prime factors of the a_i, which are called *elementary divisors*, as follows:

$$\begin{aligned} a_1 &= 3 = 2^0 3^1 5^0 \\ a_2 &= 3 = 2^0 3^1 5^0 \\ a_3 &= 150 = 2^1 3^1 5^2 \end{aligned}$$

Since $a_1a_2a_3 = 2^1 3^3 5^2$, the sum of the exponents of 2 must be 1, the sum of the exponents of 3 must be 3, and the sum of the exponents of 5 must be 2. A particular distribution of exponents of a prime p corresponds to a partition of the sum of the exponents. For example, if the exponents of p were 0, 1, 1 and 2, this would correspond to a partition of 4 as $1+1+2$. In the above example, the partitions are $1 = 1$, $3 = 1+1+1$, $2 = 2$. We can count the number of abelian groups of order 1350 (up to isomorphism) by counting partitions. There is only one partition of 1, there are two partitions of 2 (2 and $1+1$) and three partitions of 3 (3, $1+2$ and $1+1+1$). [This pattern does not continue; there are five partitions of 4, namely 4, $1+3$, $1+1+2$, $1+1+1+1$, $2+2$, and seven partitions of 5, namely 5, $1+4$, $1+1+3$, $1+1+1+2$, $1+1+1+1+1$, $1+2+2$, $2+3$.] We specify a group by choosing a partition of 1, a partition of 3 and a partition of 2, and the number of possible choices is $(1)(3)(2) = 6$. Each choice of a sequence of partitions produces a different sequence of invariant factors. Here is the entire list; the above example appears as entry (5).

(1) $a_1 = 2^1 3^3 5^2 = 1350$, $G \cong \mathbb{Z}_{1350}$

(2) $a_1 = 2^0 3^0 5^1 = 5$, $a_2 = 2^1 3^3 5^1 = 270$, $G \cong \mathbb{Z}_5 \oplus \mathbb{Z}_{270}$

(3) $a_1 = 2^0 3^1 5^0 = 3$, $a_2 = 2^1 3^2 5^2 = 450$, $G \cong \mathbb{Z}_3 \oplus \mathbb{Z}_{450}$

(4) $a_1 = 2^0 3^1 5^1 = 15$, $a_2 = 2^1 3^2 5^1 = 90$, $G \cong \mathbb{Z}_{15} \oplus \mathbb{Z}_{90}$

(5) $a_1 = 2^0 3^1 5^0 = 3$, $a_2 = 2^0 3^1 5^0 = 3$, $a_3 = 2^1 3^1 5^2 = 150$, $G \cong \mathbb{Z}_3 \oplus \mathbb{Z}_3 \oplus \mathbb{Z}_{150}$

(6) $a_1 = 2^0 3^1 5^0 = 3$, $a_2 = 2^0 3^1 5^1 = 15$, $a_3 = 2^1 3^1 5^1 = 30$, $G \cong \mathbb{Z}_3 \oplus \mathbb{Z}_{15} \oplus \mathbb{Z}_{30}$.

In entry (6) for example, the maximum number of summands in a partition is 3 ($= 1 + 1 + 1$), and this reveals that there will be three invariant factors. The partition $2 = 1 + 1$ has only two summands, and it is "pushed to the right" so that 5^1 appears in a_2 and a_3 but not a_1. (Remember that we must have $a_1 \mid a_2 \mid a_3$.). Also, we can continue to decompose some of the components in the direct sum representation of G. (If m and n are relatively prime, then $\mathbb{Z}_{mn} \cong \mathbb{Z}_m \oplus \mathbb{Z}_n$ by the Chinese remainder theorem.) However, this does not change the conclusion that there are only 6 mutually nonisomorphic abelian groups of order 1350.

Before examining infinite abelian groups, let's come back to the fundamental decomposition theorem.

4.6.5 Definitions and Comments

If x belongs to the R-module M, where R is any integral domain, then x is a *torsion element* if $rx = 0$ for some nonzero $r \in R$. The *torsion submodule* T of M is the set of torsion elements. (T is indeed a submodule; if $rx = 0$ and $sy = 0$, then $rs(x+y) = 0$.) M is a *torsion module* if T is all of M, and M is *torsion-free* if T consists of 0 alone, in other words, $rx = 0$ implies that either $r = 0$ or $x = 0$. A free module must be torsion-free, by definition of linear independence. Now assume that R is a PID, and decompose M as in (4.6.3), where $a_1, \dots, a_r$ are nonzero and $a_{r+1} = \cdots = a_n = 0$. Each module $R/\langle a_i \rangle$, $1 \le i \le r$, is torsion (it is annihilated by a_i), and the $R/\langle a_i \rangle$, $r+1 \le i \le n$, are copies of R. Thus $\oplus_{i=r+1}^n R/\langle a_i \rangle$ is free. We conclude that

(*) every finitely generated module over a PID is the direct sum of its torsion submodule and a free module

and

(**) every finitely generated torsion-free module over a PID is free.

In particular, a finitely generated abelian group is the direct sum of a number (possibly zero) of finite cyclic groups and a free abelian group (possibly $\{0\}$).

4.6.6 Abelian Groups Specified by Generators and Relations

Suppose that we have a free abelian group F with basis x_1, x_2, x_3, and we impose the following constraints on the x_i:

$$2x_1 + 2x_2 + 8x_3 = 0, \quad -2x_1 + 2x_2 + 4x_3 = 0. \tag{1}$$

What we are doing is forming a "submodule of relations" K with generators

$$u_1 = 2x_1 + 2x_2 + 8x_3 \quad \text{and} \quad u_2 = -2x_1 + 2x_2 + 4x_3 \tag{2}$$

and we are identifying every element in K with zero. This process yields the abelian group $G = F/K$, which is generated by $x_1 + K$, $x_2 + K$ and $x_3 + K$. The matrix associated with (2) is

$$\begin{bmatrix} 2 & 2 & 8 \\ -2 & 2 & 4 \end{bmatrix}$$

and a brief computation gives the Smith normal form

$$\begin{bmatrix} 2 & 0 & 0 \\ 0 & 4 & 0 \end{bmatrix}.$$

Thus we have a new basis y_1, y_2, y_3 for F and new generators $2y_1, 4y_2$ for K. The quotient group F/K is generated by y_1+K, y_2+K and y_3+K, with $2(y_1+K) = 4(y_2+K) = 0+K$. In view of (4.6.3) and (4.6.5), we must have

$$F/K \cong \mathbb{Z}_2 \oplus \mathbb{Z}_4 \oplus \mathbb{Z}.$$

Canonical forms of a square matrix A can be developed by reducing the matrix $xI - A$ to Smith normal form. In this case, R is the polynomial ring $F[X]$ where F is a field. But the analysis is quite lengthy, and I prefer an approach in which the Jordan canonical form is introduced at the very beginning, and then used to prove some basic results in the theory of linear operators; see Ash, A Primer of Abstract Mathematics, MAA 1998.

Problems For Section 4.6

1. Classify all abelian groups of order 441.
2. Classify all abelian groups of order 40.
3. Identify the abelian group given by generators x_1, x_2, x_3 and relations

$$x_1 + 5x_2 + 3x_3 = 0, \; 2x_1 - x_2 + 7x_3 = 0, \; 3x_1 + 4x_2 + 2x_3 = 0.$$

4. In (4.6.6), suppose we cancel a factor of 2 in Equation (1). This changes the matrix associated with (2) to

$$\begin{bmatrix} 1 & 1 & 4 \\ -1 & 1 & 2 \end{bmatrix},$$

whose Smith normal form differs from that given in the text. What's wrong?

5. Let M, N and P be abelian groups. If $M \oplus N \cong M \oplus P$, show by example that N need not be isomorphic to P.

6. In Problem 5, show that $M \oplus N \cong M \oplus P$ does imply $N \cong P$ if M, N and P are finitely generated.

4.7 Exact Sequences and Diagram Chasing

4.7.1 Definitions and Comments

Suppose that the R-module M is the direct sum of the submodules A and B. Let f be the *inclusion* or *injection* map of A into M (simply the identity function on A), and let g be the *natural projection* of M on B, given by $g(a+b) = b$, $a \in A$, $b \in B$. The image of f, namely A, coincides with the kernel of g, and we say that the sequence

$$A \xrightarrow{f} M \xrightarrow{g} B \tag{1}$$

is *exact* at M. A longer (possibly infinite) sequence of homomorphisms is said to be exact if it is exact at each junction, that is, everywhere except at the left and right endpoints, if they exist.

There is a natural exact sequence associated with any module homomorphism $g\colon M \to N$, namely

$$0 \longrightarrow A \xrightarrow{f} M \xrightarrow{g} B \longrightarrow 0 \tag{2}$$

In the diagram, A is the kernel of g, f is the injection map, and B is the image of g. A five term exact sequence with zero modules at the ends, as in (2), is called a *short exact sequence.* Notice that exactness at A is equivalent to $\ker f = 0$, i.e., injectivity of f. Exactness at B is equivalent to $\operatorname{im} g = B$, i.e., surjectivity of g. Notice also that by the first isomorphism theorem, we may replace B by M/A and g by the canonical map of M onto M/A, while preserving exactness.

Now let's come back to (1), where M is the direct sum of A and B, and attach zero modules to produce the short exact sequence (2). If we define h as the injection of B into M and e as the projection of M on A, we have (see (3) below) $g \circ h = 1$ and $e \circ f = 1$, where 1 stands for the identity map.

$$0 \longrightarrow A \underset{f}{\overset{e}{\leftrightarrows}} M \underset{g}{\overset{h}{\leftrightarrows}} B \longrightarrow 0 \tag{3}$$

The short exact sequence (2) is said to *split on the right* if there is a homomorphism $h\colon B \to M$ such that $g \circ h = 1$, and *split on the left* if there is a homomorphism $e\colon M \to A$ such that $e \circ f = 1$. These conditions turn out to be equivalent, and both are equivalent to the statement that M is essentially the direct sum of A and B. "Essentially" means that not only is M isomorphic to $A \oplus B$, but f can be identified with the injection of A into the direct sum, and g with the projection of the direct sum on B. We will see how to make this statement precise, but first we must turn to *diagram chasing*, which is a technique for proving assertions about commutative diagrams by sliding from one vertex to another. The best way to get accustomed to the method is to do examples. We will work one out in great detail in the text, and there will be more practice in the exercises, with solutions provided.

We will use the shorthand gf for $g \circ f$ and fm for $f(m)$.

4.7.2 The Five Lemma

Consider the following commutative diagram with exact rows.

$$\begin{array}{ccccccccc} D & \xrightarrow{e} & A & \xrightarrow{f} & M & \xrightarrow{g} & B & \xrightarrow{h} & C \\ \downarrow{\scriptstyle s} & & \downarrow{\scriptstyle t} & & \downarrow{\scriptstyle u} & & \downarrow{\scriptstyle v} & & \downarrow{\scriptstyle w} \\ D' & \xrightarrow[e']{} & A' & \xrightarrow[f']{} & M' & \xrightarrow[g']{} & B' & \xrightarrow[h']{} & C' \end{array}$$

If s, t, v and w are isomorphisms, so is u. (In fact, the hypotheses on s and w can be weakened to s surjective and w injective.)

Proof. The two parts of the proof are of interest in themselves, and are frequently called the "four lemma", since they apply to diagrams with four rather than five modules in each row.

(i) If t and v are surjective and w is injective, then u is surjective.

(ii) If s is surjective and t and v are injective, then u is injective.

[The pattern suggests a "duality" between injective and surjective maps. This idea will be explored in Chapter 10; see (10.1.4).] The five lemma follows from (i) and (ii). To prove (i), let $m' \in M'$. Then $g'm' \in B'$, and since v is surjective, we can write $g'm' = vb$ for some $b \in B$. By commutativity of the square on the right, $h'vb = whb$. But $h'vb = h'g'm' = 0$ by exactness of the bottom row at B', and we then have $whb = 0$. Thus $hb \in \ker w$, and since w is injective, we have $hb = 0$, so that $b \in \ker h = \operatorname{im} g$ by exactness of the top row at B. So we can write $b = gm$ for some $m \in M$. Now $g'm' = vb$ (see above) $= vgm = g'um$ by commutativity of the square $MBB'M'$. Therefore $m' - um \in \ker g' = \operatorname{im} f'$ by exactness of the bottom row at M'. Let $m' - um = f'a'$ for some $a' \in A'$. Since t is surjective, $a' = ta$ for some $a \in A$, and by commutativity of the square $AMM'A'$, $f'ta = ufa$, so $m' - um = ufa$, so $m' = u(m + fa)$. Consequently, m' belongs to the image of u, proving that u is surjective.

To prove (ii), suppose $m \in \ker u$. By commutativity, $g'um = vgm$, so $vgm = 0$. Since v is injective, $gm = 0$. Thus $m \in \ker g = \operatorname{im} f$ by exactness, say $m = fa$. Then

$0 = um = ufa = f'ta$ by commutativity. Thus $ta \in \ker f' = \operatorname{im} e'$ by exactness. If $ta = e'd'$, then since s is surjective, we can write $d' = sd$, so $ta = e'sd$. By commutativity, $e'sd = ted$, so $ta = ted$. By injectivity of t, $a = ed$. Therefore $m = fa = fed = 0$ by exactness. We conclude that u is injective. ♣

4.7.3 Corollary: The Short Five Lemma

Consider the following commutative diagram with exact rows. (Throughout this section, all maps in commutative diagrams and exact sequences are assumed to be R-module homomorphisms.)

$$\begin{array}{ccccccccc}
0 & \longrightarrow & A & \xrightarrow{f} & M & \xrightarrow{g} & B & \longrightarrow & 0 \\
 & & \downarrow{\scriptstyle t} & & \downarrow{\scriptstyle u} & & \downarrow{\scriptstyle v} & & \\
0 & \longrightarrow & A' & \xrightarrow[f']{} & M' & \xrightarrow[g']{} & B' & \longrightarrow & 0
\end{array}$$

If t and v are isomorphisms, so is u.

Proof. Apply the five lemma with $C = D = C' = D' = 0$, and s and w the identity maps. ♣

We can now deal with splitting of short exact sequences.

4.7.4 Proposition

Let

$$0 \longrightarrow A \xrightarrow{f} M \xrightarrow{g} B \longrightarrow 0$$

be a short exact sequence. The following conditions are equivalent, and define a *split exact sequence*.

(i) The sequence splits on the right.

(ii) The sequence splits on the left.

(iii) There is an isomorphism u of M and $A \oplus B$ such that the following diagram is commutative.

$$\begin{array}{ccccccccc}
0 & \longrightarrow & A & \xrightarrow{f} & M & \xrightarrow{g} & B & \longrightarrow & 0 \\
 & & \| & & \downarrow{\scriptstyle u} & & \| & & \\
0 & \longrightarrow & A & \xrightarrow[i]{} & A \oplus B & \xrightarrow[\pi]{} & B & \longrightarrow & 0
\end{array}$$

Thus M is isomorphic to the direct sum of A and B, and in addition, f can be identified with the injection i of A into $A \oplus B$, and g with the projection π of the direct sum onto B. (The double vertical bars indicate the identity map.)

Proof. It follows from our earlier discussion of diagram (3) that (iii) implies (i) and (ii). To show that (i) implies (iii), let h be a homomorphism of B into M such that $gh = 1$. We claim that

$$M = \ker g \oplus h(B).$$

First, suppose that $m \in M$. Write $m = (m - hgm) + hgm$; then $hgm \in h(B)$ and $g(m-hgm) = gm-ghgm = gm-1gm = gm-gm = 0$. Second, suppose $m \in \ker g \cap h(B)$, with $m = hb$. Then $0 = gm = ghb = 1b = b$, so $m = hb = h0 = 0$, proving the claim. Now since $\ker g = \operatorname{im} f$ by exactness, we may express any $m \in M$ in the form $m = fa + hb$. We take $um = a + b$, which makes sense because both f and h are injective and $f(A) \cap h(B) = 0$. This forces the diagram of (iii) to be commutative, and u is therefore an isomorphism by the short five lemma. Finally, we show that (ii) implies (iii). Let e be a homomorphism of M into A such that $ef = 1$. In this case, we claim that

$$M = f(A) \oplus \ker e.$$

If $m \in M$ then $m = fem + (m - fem)$ and $fem \in f(A)$, $e(m - fem) = em - efem = em - em = 0$. If $m \in f(A) \cap \ker e$, then, with $m = fa$, we have $0 = em = efa = a$, so $m = 0$, and the claim is verified. Now if $m \in M$ we have $m = fa + m'$ with $a \in A$ and $m' \in \ker e$. We take $u(m) = a + g(m') = a + gm$ since $gf = 0$. (The definition of u is unambiguous because f is injective and $f(A) \cap \ker e = 0$.) The choice of u forces the diagram to be commutative, and again u is an isomorphism by the short five lemma. ♣

4.7.5 Corollary

If the sequence

$$0 \longrightarrow A \xrightarrow{f} M \xrightarrow{g} B \longrightarrow 0$$

is split exact with splitting maps e and h as in (3), then the "backwards" sequence

$$0 \longleftarrow A \xleftarrow{e} M \xleftarrow{h} B \longleftarrow 0$$

is also split exact, with splitting maps g and f.

Proof. Simply note that $gh = 1$ and $ef = 1$. ♣

A device that I use to remember which way the splitting maps go (i.e., it's $ef = 1$, not $fe = 1$) is that the map that is applied first points inward toward the "center" M.

Problems For Section 4.7

Consider the following commutative diagram with exact rows:

$$\begin{array}{ccccccc} 0 & \longrightarrow & A & \xrightarrow{f} & B & \xrightarrow{g} & C \\ & & & & \downarrow{\scriptstyle v} & & \downarrow{\scriptstyle w} \\ 0 & \longrightarrow & A' & \xrightarrow[f']{} & B' & \xrightarrow[g']{} & C' \end{array}$$

Our objective in Problems 1–3 is to find a homomorphism $u\colon A \to A'$ such that the square $ABB'A'$, hence the entire diagram, is commutative.

1. Show that if u exists, it is unique.
2. If $a \in A$, show that $vfa \in \operatorname{im} f'$.
3. If $vfa = f'a'$, define ua appropriately.

Now consider another commutative diagram with exact rows:

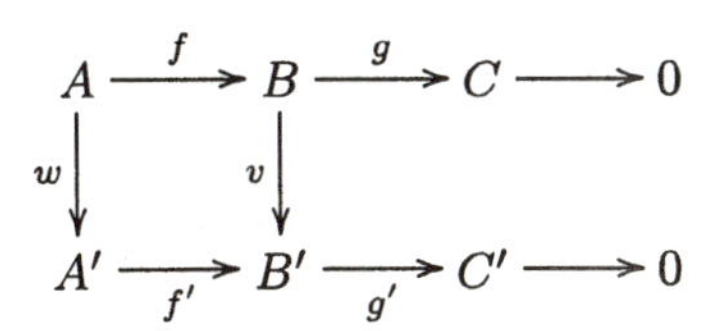

In Problems 4 and 5 we are to define $u\colon C \to C'$ so that the diagram will commute.

4. If $c \in C$, then since g is surjective, $c = gb$ for some $b \in B$. Write down the only possible definition of uc.
5. In Problem 4, b is not unique. Show that your definition of u does not depend on the particular b.

Problems 6–11 refer to the diagram of the short five lemma (4.7.3). Application of the four lemma is very efficient, but a direct attack is also good practice.

6. If t and v are injective, so is u.
7. If t and v are surjective, so is u.
8. If t is surjective and u is injective, then v is injective.
9. If u is surjective, so is v.

By Problems 8 and 9, if t and u are isomorphisms, so is v.

10. If u is injective, so is t.
11. If u is surjective and v is injective, then t is surjective.

Note that by Problems 10 and 11, if u and v are isomorphisms, so is t.

12. If you have not done so earlier, do Problem 8 directly, without appealing to the four lemma.
13. If you have not done so earlier, do Problem 11 directly, without appealing to the four lemma.

Enrichment

Chapters 1–4 form an idealized undergraduate course, written in the style of a graduate text. To help those seeing abstract algebra for the first time, I have prepared this section, which contains advice, explanations and additional examples for each section in the first four chapters.

Section 1.1

When we say that the rational numbers form a group under addition, we mean that rational numbers can be added and subtracted, and the result will inevitably be rational. Similarly for the integers, the real numbers, and the complex numbers. But the integers (even the nonzero integers) do not form a group under multiplication. If a is an integer other than ± 1, there is no integer b such that $ab = 1$. The nonzero rational numbers form a group under multiplication, as do the nonzero reals and the nonzero complex numbers. Not only can we add and subtract rationals, we can multiply and divide them (if the divisor is nonzero). The rational, reals and complex numbers are examples of *fields*, which will be studied systematically in Chapter 3.

Here is what the generalized associative law is saying. To compute the product of the elements a, b, c, d and e, one way is to first compute bc, then $(bc)d$, then $a((bc)d)$, and finally $[a((bc)d)e]$. Another way is (ab), then (cd), then $(ab)(cd)$, and finally $([(ab)(cd)]e)$. All procedures give the same result, which can therefore be written as $abcde$.

Notice that the solution to Problem 6 indicates how to construct a formal proof of 1.1.4.

Section 1.2

Groups whose descriptions differ may turn out to be isomorphic, and we already have an example from the groups discussed in this section. Consider the dihedral group D_6, with elements I, R, R^2, F, RF, R^2F. Let S_3 be the group of all permutations of $\{1, 2, 3\}$. We claim that D_6 and S_3 are isomorphic. This can be seen geometrically if we view D_6 as a group of permutations of the vertices of an equilateral triangle. Since D_6 has 6 elements and there are exactly 6 permutations of 3 symbols, we must conclude that D_6 and S_3 are essentially the same. To display an isomorphism explicitly, let R correspond to the permutation (1,2,3) and F to (2,3). Then

$$I = (1),\ R = (1,2,3),\ R^2 = (1,3,2),\ F = (2,3),\ RF = (1,2),\ R^2F = (1,3).$$

If G is a nonabelian group, then it must have an element of order 3 or more. (For example, S_3 has two elements of order 3.) In other words, if every element of G has order 1 or 2, then G is abelian. To prove this, let $a, b \in G$; we will show that $ab = ba$. We can assume with loss of generality that $a \neq 1$ and $b \neq 1$. But then $a^2 = 1$ and $b^2 = 1$, so that a is its own inverse, and similarly for b. If ab has order 1, then $ab = 1$, so a and b are inverses of each other. By uniqueness of inverses, $a = b$, hence $ab = ba$. If ab has order 2, then $abab = 1$, so $ab = b^{-1}a^{-1} = ba$.

Section 1.3

Here is another view of cosets that may be helpful. Suppose that a coded message is to be transmitted, and the message is to be represented by a code word x (an n-dimensional vector with components in some field). The allowable code words are solutions of $Ax = 0$, where A is an m by n matrix, hence the set H of code words is an abelian group under componentwise addition, a subgroup of the abelian group G of all n-dimensional vectors. (In fact, G and H are vector space, but let's ignore the additional structure.) Transmission is affected by noise, so that the received vector is of the form $z = x+y$, where y is another n-dimensional vector, called the *error vector* or *error pattern vector*. Upon receiving z, we calculate the *syndrome* $s = Az$. If s turns out to be the zero vector, we declare that no error has occurred, and the transmitted word is z. Of course our decision may be incorrect, but under suitable assumptions about the nature of the noise, our decision procedure will minimize the probability of making a mistake. Again, let's ignore this difficulty and focus on the algebraic aspects of the problem. We make the following claim:

Two vectors z_1 and z_2 have the same syndrome if and only if they lie in the same coset of H in G.

To prove this, observe that $Az_1 = Az_2$ iff $A(z_1 - z_2) = 0$ iff $z_1 - z_2 \in H$ iff $z_1 \in z_2 + H$. (We are working in an abelian group, so we use the additive notation $z_2 + H$ rather than the multiplicative notation z_2H.)

Now suppose that we agree that we are going to correct the error pattern y_1, in other words, if we receive $z = x_1 + y_1$, where x_1 is a code word, we will decode z as x_1. If we receive $z' = x_1' + y_1$, where x_1' is another code word, we decode z' as x_1'. Thus our procedure corrects y_1 regardless of the particular word transmitted. Here is a key algebraic observation:

If y_1 and y_2 are distinct vectors that lie in the same coset, it is impossible to correct both y_1 and y_2.

This holds because $y_1 = y_2 + x$ for some code word $x \neq 0$, hence $y_1 + 0 = y_2 + x$. Therefore we cannot distinguish between the following two possibilities:

1. The zero word is transmitted and the error pattern is y_1;
2. x is transmitted and the error pattern is y_2.

It follows that among all vectors in a given coset, equivalently among all vectors having the same syndrome, we can choose exactly one as a correctable error pattern. If the underlying field has only two elements 0 and 1, then (under suitable assumptions) it is best to choose to correct the pattern of minimum weight, that is, minimum number of 1's. In particular, if the coset is the subgroup H itself, then we choose the zero vector. This agrees with our earlier proposal: if the received vector z has zero syndrome, we decode z as z itself, thus "correcting" the zero pattern, in other words, declaring that there has been no error in transmission.

For further discussion and examples, see *Information Theory* by R. B. Ash, Dover 1991, Chapter 4.

Section 1.4

Here are some intuitive ideas that may help in visualizing the various isomorphism theorems. In topology, we can turn the real interval $[0, 1]$ into a circle by gluing the endpoints together, in other words identifying 0 and 1. Something similar is happening when we form the quotient group G/N where N is a normal subgroup of G. We have identified all the elements of N, and since the identity belongs to every subgroup, we can say that we have set everything in N equal to 1 (or 0 in the abelian case). Formally, (1.4.6) gives a correspondence between the subgroup of G/N consisting of the identity alone, and the subgroup N of G.

We have already seen an example of this identification process. In (1.2.4), we started with the free group G generated by the symbols R and F, and identified all sequences satisfying the relations $R^n = I$, $F^2 = I$, and $RF = FR^{-1}$ (equivalently $RFRF = I$). Here we would like to take N to be the subgroup of G generated by R^n, F^2, and $RFRF$, but N might not be normal. We will get around this technical difficulty when we discuss generators and relations in more detail in Section 5.8.

Section 1.5

Direct products provide a good illustration of the use of the first isomorphism theorem. Suppose that $G = H \times K$; what can we say about G/H? If $(h, k) \in G$, then $(h, k) = (h, 1_K)(1_H, k)$, and intuitively we have identified $(h, 1_K)$ with the identity $(1_H, 1_K)$. What we have left is $(1_H, k)$, and it appears that G/H should be isomorphic to K. To prove this formally, define $f\colon G \to K$ by $f(h, k) = k$. Then f is an epimorphism whose kernel is $\{(h, 1_K)\colon h \in H\}$, which can be identified with H. By the first isomorphism theorem, $G/H \cong K$.

Section 2.1

Here is an interesting ring that will come up in Section 9.5 in connection with group representation theory. Let $G = \{x_1, \dots x_m\}$ be a finite group, and let R be any ring. The *group ring* RG consists of all elements $r_1x_1 + \cdots + r_mx_m$. Addition of elements is componentwise, just as if the x_i were basis vectors of a vector space and the r_i were scalars in a field. Multiplication in RG is governed by the given multiplication in R, along with linearity. For example,

$$(r_1x_1 + r_2x_2)(s_1x_1 + s_2x_2) = r_1s_1x_1^2 + r_1s_2x_1x_2 + r_2s_1x_2x_1 + r_2s_2x_2^2.$$

The elements x_1^2, x_1x_2, x_2x_1, and x_2^2 belong to G, which need not be abelian. The elements r_1s_1, r_1s_2, r_2s_1, and r_2s_2 belong to R, which is not necessarily commutative. Thus it is essential to keep track of the order in which elements are written.

Section 2.2

Here is some additional practice with ideals in matrix rings. If I is an ideal of $M_n(R)$, we will show that I must have the form $M_n(I_0)$ for some unique ideal I_0 of R. [$M_n(I_0)$ is the set of all n by n matrices with entries in I_0.]

We note first that for any matrix A, we have $E_{ij}AE_{kl} = a_{jk}E_{il}$. This holds because $E_{ij}A$ puts row j of A in row i, and AE_{kl} puts column k of A in column l. Thus $E_{ij}AE_{kl}$ puts a_{jk} in the il position, with zeros elsewhere.

If I is an ideal of $M_n(R)$, let I_0 be the set of all entries a_{11}, where $A = (a_{ij})$ is a matrix in I. To verify that I_0 is an ideal, observe that $(A + B)_{11} = a_{11} + b_{11}$, $ca_{11} = (cE_{11}A)_{11}$, and $a_{11}c = (AE_{11}c)_{11}$. We will show that $I = M_n(I_0)$.

If $A \in I$, set $i = l = 1$ in the basic identity involving the elementary matrices E_{ij} (see the second paragraph above) to get $a_{jk}E_{11} \in I$. Thus $a_{jk} \in I_0$ for all j and k, so $A \in M_n(I_0)$.

Conversely, let $A \in M_n(I_0)$, so that $a_{il} \in I_0$ for all i, l. By definition of I_0, there is a matrix $B \in I$ such that $b_{11} = a_{il}$. Take $j = k = 1$ in the basic identity to get $E_{i1}BE_{1l} = b_{11}E_{il} = a_{il}E_{il}$. Consequently, $a_{il}E_{il} \in I$ for all i and l. But the sum of the matrices $a_{il}E_{il}$ over all i and l is simply A, and we conclude that $A \in I$.

To prove uniqueness, suppose that $M_n(I_0) = M_n(I_1)$. If $a \in I_0$, then $aE_{11} \in M_n(I_0) = M_n(I_1)$, so $a \in I_1$. A symmetrical argument completes the proof.

Section 2.3

If a and b are relatively prime integers, then a^i and b^j are relatively prime for all positive integers i and j. Here is an analogous result for ideals. Suppose that the ideals I_1 and I_2 of the ring R are relatively prime, so that $I_1 + I_2 = R$. Let us prove that I_1^2 and I_2 are relatively prime as well. By the definitions of the sum and product of ideals, we have

$$R = RR = (I_1 + I_2)(I_1 + I_2) = I_1^2 + I_1I_2 + I_2I_1 + I_2^2 \subseteq I_1^2 + I_2 \subseteq R$$

so $R = I_1^2 + I_2$, as asserted. Similarly, we can show that $R = I_1^3 + I_2$ by considering the product of $I_1^2 + I_2$ and $I_1 + I_2$. More generally, an induction argument shows that if

$$I_1 + \cdots + I_n = R,$$

then for all positive integers $m_1, \dots, m_n$ we have

$$I_1^{m_1} + \cdots + I_n^{m_n} = R.$$

Section 2.4

We have defined prime ideals only when the ring R is commutative, and it is natural to ask why this restriction is imposed. Suppose that we drop the hypothesis of commutativity, and define prime ideals as in (2.4.4). We can then prove that if P is a prime ideal, I and J are arbitrary ideals, and $P \supseteq IJ$, then either $P \supseteq I$ or $P \supseteq J$. [If the conclusion is false, there are elements $a \in I \setminus P$ and $b \in J \setminus P$. Then $ab \in IJ \subseteq P$, but $a \notin P$ and $b \notin P$, a contradiction.]

If we try to reverse the process and show that a proper ideal P such that $P \supseteq IJ$ implies $P \supseteq I$ or $P \supseteq J$ must be prime, we run into trouble. If ab belongs to P, then the principal ideal (ab) is contained in P. We would like to conclude that $(a)(b) \subseteq P$, so that $(a) \subseteq P$ or $(b) \subseteq P$, in other words, $a \in P$ or $b \in P$. But (ab) need not equal $(a)(b)$. For example, to express the product of the element $ar \in (a)$ and the element $sb \in (b)$ as a multiple of ab, we must invoke commutativity.

An explicit example: Let P be the zero ideal in the ring $M_n(R)$ of n by n matrices over a division ring R (see Section 2.2, exercises). Since $M_n(R)$ has no nontrivial two-sided ideals, $P \supseteq IJ$ implies $P \supseteq I$ or $P \supseteq J$. But $ab \in P$ does not imply $a \in P$ or $b \in P$, because the product of two nonzero matrices can be zero.

This example illustrates another source of difficulty. The zero ideal P is maximal, but $M_n(R)/P$ is not a division ring. Thus we cannot generalize (2.4.3) by dropping commutativity and replacing "field" by "division ring". [If R/M is a division ring, it does follow that M is a maximal ideal; the proof given in (2.4.3) works.]

Section 2.5

Let's have a brief look at polynomials in more than one variable; we will have much more to say in Chapter 8. For example, a polynomial $f(X, Y, Z)$ in 3 variables is a sum of monomials; a monomial is of the form $aX^iY^jZ^k$ where a belongs to the underlying ring R. The degree of such a monomial is $i+j+k$, and the degree of f is the maximum monomial degree. Formally, we can define $R[X, Y]$ as $(R[X])[Y]$, $R[X, Y, Z]$ as $(R[X, Y])[Z]$, etc.

Let f be a polynomial of degree n in $F[X, Y]$, where F is a field. There are many cases in which f has infinitely many roots in F. For example, consider $f(X, Y) = X + Y$ over the reals. The problem is that there is no direct extension of the division algorithm (2.5.1) to polynomials in several variables. The study of solutions to polynomial equations in more than one variable leads to algebraic geometry, which will be introduced in Chapter 8.

Section 2.6

We have shown in (2.6.8) that every principal ideal domain is a unique factorization domain. Here is an example of a UFD that is not a PID. Let $R = \mathbb{Z}[X]$, which will be shown to be a UFD in (2.9.6). Let I be the maximal ideal $< 2, X >$ (see (2.4.8)). If I is principal, then I consists of all multiples of a polynomial $f(X)$ with integer coefficients. Since $2 \in I$, we must be able to multiply $f(X) = a_0 + a_1X + \cdots + a_nX^n$ by a polynomial $g(X) = b_0 + b_1X + \cdots + b_mX^m$ and produce 2. There is no way to do this unless $f(X) = 1$ or 2. But if $f(X) = 1$ then $I = R$, a contradiction (a maximal ideal must be proper).

Thus $f(X) = 2$, but we must also be able to multiply $f(X)$ by some polynomial in $\mathbb{Z}[X]$ to produce X. This is impossible, and we conclude that I is not principal.

A faster proof that $\mathbb{Z}[X]$ is not a PID is as follows. In (2.4.8) we showed that $< 2 >$ and $< X >$ are prime ideals that are not maximal, and the result follows from (2.6.9). On the other hand, the first method produced an explicit example of an ideal that is not principal.

Section 2.7

It may be useful to look at the Gaussian integers in more detail, and identify the primes in $\mathbb{Z}[i]$. To avoid confusion, we will call a prime in $\mathbb{Z}[i]$ a *Gaussian prime* and a prime in $\mathbb{Z}$ a *rational prime.* Anticipating some terminology from algebraic number theory, we define the *norm* of the Gaussian integer $a + bi$ as $N(a + bi) = a^2 + b^2$. We will outline a sequence of results that determine exactly which Gaussian integers are prime. We use Greek letters for members of $\mathbb{Z}[i]$ and roman letters for ordinary integers.

1. α is a unit in $\mathbb{Z}[i]$ iff $N(\alpha) = 1$. Thus the only Gaussian units are ± 1 and $\pm i$.

[If $\alpha\beta = 1$, then $1 = N(1) = N(\alpha)N(\beta)$, so both $N(\alpha)$ and $N(\beta)$ must be 1.]

Let n be a positive integer.

2. If n is a Gaussian prime, then n is a rational prime not expressible as the sum of two squares.

[n is a rational prime because any factorization in $\mathbb{Z}$ is also a factorization in $\mathbb{Z}[i]$. If $n = x^2 + y^2 = (x + iy)(x - iy)$, then either $x + iy$ or $x - iy$ is a unit. By (1), $n = 1$, a contradiction.]

Now assume that n is a rational prime but not a Gaussian prime.

3. If $\alpha = a + bi$ is a nontrivial factor of n, then $gcd(a, b) = 1$.

[If the greatest common divisor d is greater than 1, then $d = n$. Thus α divides n and n divides α, so n and α are associates, a contradiction.]

4. n is a sum of two squares.

[Let $n = (a + bi)(c + di)$; since n is real we have $ad + bc = 0$, so a divides bc. By (3), a and b are relatively prime, so a divides c, say $c = ka$. Then $b(ka) = bc = -ad$, so $d = -bk$. Thus $n = ac - bd = ka^2 + kb^2 = k(a^2 + b^2)$. But $a + bi$ is a nontrivial factor of n, so $a^2 + b^2 = N(a + bi) > 1$. Since n is a rational prime, we must have $k = 1$ and $n = a^2 + b^2$.]

By the above results, we have:

5. If n is a positive integer, then n is a Gaussian prime if and only if n is a rational prime not expressible as the sum of two squares.

Now assume that $\alpha = a + bi$ is a Gaussian integer with both a and b nonzero. (The cases in which a or b is 0 are covered by (1) and (5).)

6. If $N(\alpha)$ is a rational prime, then α is a Gaussian prime.

[If $\alpha = \beta\gamma$ where β and γ are not units, then $N(\alpha) = N(\beta)N(\gamma)$, where $N(\beta)$ and $N(\gamma)$ are greater than 1, contradicting the hypothesis.]

Now assume that α is a Gaussian prime.

7. If $N(\alpha) = hk$ is a nontrivial factorization, so that $h > 1$ and $k > 1$, then α divides either h or k. If, say, α divides h, then so does its complex conjugate $\overline{\alpha}$.

[We have $N(\alpha) = a^2 + b^2 = (a + bi)(a - bi) = \alpha\overline{\alpha} = hk$. Since α divides the product hk, it must divide one of the factors. If $\alpha\beta = h$, take complex conjugates to conclude that $\overline{\alpha}\overline{\beta} = h$.]

8. $N(\alpha)$ is a rational prime.

[If not, then we can assume by (7) that α and $\overline{\alpha}$ divide h. If α and $\overline{\alpha}$ are not associates, then $N(\alpha) = \alpha\overline{\alpha}$ divides h, so hk divides h and therefore $k = 1$, a contradiction. If α and its conjugate are associates, then one is $\pm i$ times the other. The only way this can happen is if $\alpha = \gamma(1 \pm i)$ where γ is a unit. But then $N(\alpha) = N(\gamma)N(1 \pm i) = N(1 \pm i) = 2$, a rational prime.]

By the above results, we have:

9. If $\alpha = a + bi$ with $a \neq 0$, $b \neq 0$, then α is a Gaussian prime if and only if $N(\alpha)$ is a rational prime.

The assertions (5) and (9) give a complete description of the Gaussian primes, except that it would be nice to know when a rational prime p can be expressed as the sum of two squares. We have $2 = 1^2 + 1^2$, so 2 is not a Gaussian prime, in fact $2 = (1+i)(1-i)$. If p is an odd prime, then p is a sum of two squares iff $p \equiv 1 \bmod 4$, as we will prove at the beginning of Chapter 7. Thus we may restate (5) as follows:

10. If n is a positive integer, then n is a Gaussian prime iff n is a rational prime congruent to 3 mod 4.

[Note that a number congruent to 0 or 2 mod 4 must be even.]

Section 2.8

Suppose that R is an integral domain with quotient field F, and g is a ring homomorphism from R to an integral domain R'. We can then regard g as mapping R into the quotient field F' of R'. It is natural to try to extend g to a homomorphism $\overline{g}\colon F \to F'$. If $a, b \in R$ with $b \neq 0$, then $a = b(a/b)$, so we must have $g(a) = g(b)\overline{g}(a/b)$. Thus if an extension exists, it must be given by

$$\overline{g}(a/b) = g(a)[g(b)]^{-1}.$$

For this to make sense, we must have $g(b) \neq 0$ whenever $b \neq 0$, in other words, g is a monomorphism. [Note that if $x, y \in R$ and $g(x) = g(y)$, then $g(x-y) = 0$, hence $x-y = 0$, so $x = y$.] We will see in (3.1.2) that any homomorphism of fields is a monomorphism, so this condition is automatically satisfied. We can establish the existence of $\overline{g}$ by defining it as above and then showing that it is a well-defined ring homomorphism. This has already been done in Problem 8. We are in the general situation described in Problem 7, with S taken as the set of nonzero elements of R. We must check that $g(s)$ is a unit in F' for every $s \in S$, but this holds because $g(s)$ is a nonzero element of F'.

Section 2.9

Here is another useful result relating factorization over an integral domain to factorization over the quotient field. Suppose that f is a monic polynomial with integer coefficients, and that f can be factored as gh, where g and h are monic polynomials with rational coefficients. Then g and h must have integer coefficients. More generally, let D be a unique factorization domain with quotient field F, and let f be a monic polynomial in $D[X]$. If $f = gh$, with $g, h \in F[X]$, then g and h must belong to $D[X]$.

To prove this, we invoke the basic proposition (2.9.2) to produce a nonzero $\lambda \in F$ such that $\lambda g \in D[X]$ and $\lambda^{-1}h \in D[X]$. But g and h are monic, so $\lambda = 1$ and the result follows.

Let f be a cubic polynomial in $F[X]$. If f is reducible, it must have a linear factor and hence a root in F. We can check this easily if F is a finite field; just try all possibilities. A finite check also suffices when $F = \mathbb{Q}$, by the rational root test (Section 2.9, Problem 1). If g is a linear factor of f, then $f/g = h$ is quadratic. We can factor h as above, and in addition the quadratic formula is available if square roots can be extracted in F. In other words, if $a \in F$, then $b^2 = a$ for some $b \in F$.

Section 3.1

All results in this section are basic and should be studied carefully. You probably have some experience with polynomials over the rational numbers, so let's do an example with a rather different flavor. Let $F = \mathbb{F}_2$ be the field with two elements 0 and 1, and let $f \in F[X]$ be the polynomial $X^2 + X + 1$. Note that f is irreducible over F, because if f were factorable, it would have a linear factor and hence a root in F. This is impossible, as $f(0) = f(1) = 1 \neq 0$. If we adjoin a root α of f to produce an extension $F(\alpha)$, we know that f is the minimal polynomial of α over F, and that $F(\alpha)$ consists of all elements $b_0 + b_1\alpha$, with b_0 and b_1 in F. Since b_0 and b_1 take on values 0 and 1, we have constructed a field $F(\alpha)$ with 4 elements. Moreover, all nonzero elements of $F(\alpha)$ can be expressed as powers of α, as follows:

$\alpha^0 = 1$, $\alpha^1 = \alpha$, $\alpha^2 = -\alpha - 1 = 1 + \alpha$. (The last equality follows because $1 + 1 = 0$ in F.)

This is a typical computation involving finite fields, which will be studied in detail in Chapter 6.

Section 3.2

We found in Problem 3 that a splitting field for $X^4 - 2$ has degree 8 over $\mathbb{Q}$. If we make a seemingly small change and consider $f(X) = X^4 - 1$, the results are quite different. The roots of f are 1, i, -1 and $-i$. Thus $\mathbb{Q}(i)$ is the desired splitting field, and it has degree 2 over $\mathbb{Q}$ because the minimal polynomial of i over $\mathbb{Q}$ has degree 2.

A general problem suggested by this example is to describe a splitting field for $X^n - 1$ over $\mathbb{Q}$ for an arbitrary positive integer n. The splitting field is $\mathbb{Q}(\omega)$, where ω is a primitive n^{th} root of unity, for example, $\omega = e^{i2\pi/n}$. We will see in Section 6.5 that the degree of $\mathbb{Q}(\omega)$ over $\mathbb{Q}$ is $\varphi(n)$, where φ is the Euler phi function.

Section 3.3

In Problem 8 we used the existence of an algebraic closure of F to show that any set of nonconstant polynomials in $F[X]$ has a splitting field over F. Conversely, if we suppose that it is possible to find a splitting field K for an arbitrary family of polynomials over the field F, then the existence of an algebraic closure of F can be established quickly. Thus let K be a splitting field for the collection of all polynomials in $F[X]$, and let C be the algebraic closure of F in K (see (3.3.4)). Then by definition, C is an algebraic extension of F and every nonconstant polynomial in $F[X]$ splits over C. By (3.3.6), C is an algebraic closure of F.

Section 3.4

Let's have another look at Example 3.4.8 with $p = 2$ to get some additional practice with separability and inseparability. We have seen that $\sqrt{t}$ is not separable over F, in fact it is purely inseparable because its minimal polynomial $X^2 - t$ can be written as $(X - \sqrt{t})^2$. But if we adjoin a cube root of t, the resulting element $\sqrt[3]{t}$ *is* separable over F, because $X^3 - t$ has nonzero derivative, equivalently does not belong to $F[X^2]$ (see 3.4.3).

Notice also that adjoining $\sqrt{t}$ and $\sqrt[3]{t}$ is equivalent to adjoining $\sqrt[6]{t}$, in other words, $F(\sqrt{t}, \sqrt[3]{t}) = F(\sqrt[6]{t})$. To see this, first observe that if $\alpha = \sqrt[6]{t}$, then $\sqrt{t} = \alpha^3$ and $\sqrt[3]{t} = \alpha^2$. On the other hand, $(\sqrt{t}/\sqrt[3]{t})^6 = t$.

It is possible for an element α to be both separable and purely inseparable over F, but it happens if and only if α belongs to F. The minimal polynomial of α over F must have only one distinct root and no repeated roots, so $\min(\alpha, F) = X - \alpha$. But the minimal polynomial has coefficients in F (by definition), and the result follows.

Section 3.5

Suppose we wish to find the Galois group of the extension E/F, where $E = F(\alpha)$. Assume that α is algebraic over F with minimal polynomial f, and that f has n distinct roots $\alpha_1 = \alpha, \alpha_2, \dots, \alpha_n$ in some splitting field. If $\sigma \in \mathrm{Gal}(E/F)$, then σ permutes the roots of f by (3.5.1). Given any two roots α_i and α_j, $i \neq j$, we can find an F-isomorphism that carries α_i into α_j; see (3.2.3). Do not jump to the conclusion that all permutations are allowable, and therefore $\mathrm{Gal}(E/F)$ is isomorphic to S_n. For example, we may not be able to simultaneously carry α_1 into α_2 and α_3 into α_4. Another difficulty is that the F-isomorphism carrying α_i into α_j need not be an F-automorphism of E. This suggests that normality of the extension is a key property. If E/F is the non-normal extension of Example (3.5.10), the only allowable permutation is the identity.

Section 4.1

Finitely generated algebras over a commutative ring R frequently appear in applications to algebraic number theory and algebraic geometry. We say that A is a finitely generated R-algebra if there are finitely many elements $x_1, \dots, x_n$ in A such that every element of

A is a polynomial $f(x_1, \ldots, x_n)$ with coefficients in R. Equivalently, A is a homomorphic image of the polynomial ring $R[X_1, \ldots, X_n]$. The homomorphism is determined explicitly by mapping X_i to x_i, $i = 1, \ldots, n$. The polynomial $f(X_1, \ldots, X_n)$ is then mapped to $f(x_1, \ldots, x_n)$.

If every element is not just a polynomial in the x_i but a *linear combination* of the x_i with coefficients in R, then A is a finitely generated *module* over R. To see the difference clearly, look at the polynomial ring $R[X]$, which is a finitely generated R algebra. (In the above discussion we can take $n = 1$ and $x_1 = X$.) But if $f_1, \ldots, f_n$ are polynomials in $R[X]$ and the maximum degree of the f_i is m, there is no way to take linear combinations of the f_i and produce a polynomial of degree greater than m. Thus $R[X]$ is a not a finitely generated R-module.

Section 4.2

Here is some practice working with quotient modules. Let N be a submodule of the R-module M, and let π be the canonical map from M onto M/N, taking $x \in M$ to $x + N \in M/N$. Suppose that N_1 and N_2 are submodules of M satisfying

(a) $N_1 \leq N_2$;

(b) $N_1 \cap N = N_2 \cap N$;

(c) $\pi(N_1) = \pi(N_2)$.

Then $N_1 = N_2$.

To prove this, let $x \in N_2$. Hypothesis (c) says that $(N_1 + N)/N = (N_2 + N)/N$; we don't write N_i/N, $i = 1, 2$, because N is not necessarily a submodule of N_1 or N_2. Thus $x + N \in (N_2 + N)/N = (N_1 + N)/N$, so $x + N = y + N$ for some $y \in N_1$. By (a), $y \in N_2$, hence $x - y \in N_2 \cap N = N_1 \cap N$ by (b). Therefore $x - y$ and y both belong to N_1, and consequently so does x. We have shown that $N_2 \leq N_1$, and in view of hypothesis (a), we are finished.

Section 4.3

If M is a free R-module with basis $S = (x_i)$, then an arbitrary function f from S to an arbitrary R-module N has a unique extension to an R-homomorphism $\overline{f}\colon M \to N$; see (4.3.6).

This property characterizes free modules, in other words, if M is an R-module with a subset S satisfying the above property, then M is free with basis S. To see this, build a free module M' with basis $S' = (y_i)$ having the same cardinality as S. For example, we can take M' to be the direct sum of copies of R, as many copies as there are elements of S. Define $f\colon S \to S' \subseteq M'$ by $f(x_i) = y_i$, and let $\overline{f}$ be the unique extension of f to an R-homomorphism from M to M'. Similarly, define $g\colon S' \to S \subseteq M$ by $g(y_i) = x_i$, and let $\overline{g}$ be the unique extension of g to an R-homomorphism from M' to M. Note that $\overline{g}$ exists and is unique because M' is free. Now $\overline{g} \circ \overline{f}$ is the identity on S, so by uniqueness of extensions from S to M, $\overline{g} \circ \overline{f}$ is the identity on M. Similarly, $\overline{f} \circ \overline{g}$ is the identity on M'. Thus M and M' are not only isomorphic, but the isomorphism we have constructed carries S into S'. It follows that M is free with basis S.

This is an illustration of the characterization of an algebraic object by a *universal mapping property*. We will see other examples in Chapter 10.

Section 4.4

Here is some practice in decoding abstract presentations. An R-module can be defined as a representation of R in an endomorphism ring of an abelian group M. What does this mean?

First of all, for each $r \in R$, we have an endomorphism f_r of the abelian group M, given by $f_r(x) = rx$, $x \in M$. To say that f_r is an endomorphism is to say that $r(x+y) = rx+ry$, $x, y \in M$, $r \in R$.

Second, the mapping $r \to f_r$ is a ring homomorphism from R to $\text{End}_R(M)$. (Such a mapping is called a representation of R in $\text{End}_R(M)$.) This says that $f_{r+s}(x) = f_r(x) + f_s(x)$, $f_{rs}(x) = f_r(f_s(x))$, and $f_1(x) = x$. In other words, $(r+s)x = rx + sx$, $(rs)x = r(sx)$, and $1x = x$.

Thus we have found a fancy way to write the module axioms. If you are already comfortable with the informal view of a module as a "vector space over a ring", you are less likely to be thrown off stride by the abstraction.

Section 4.5

The technique given in Problems 1–3 for finding new bases and generators is worth emphasizing. We start with a matrix A to be reduced to Smith normal form. The equations $U = AX$ give the generators U of the submodule K in terms of the basis X of the free module M. The steps in the Smith calculation are of two types:

1. Premultiplication by an elementary row matrix R. This corresponds to changing generators via $V = RU$.
2. Postmultiplication by an elementary column matrix C. This corresponds to changing bases via $Y = C^{-1}X$.

Suppose that the elementary row matrices appearing in the calculation are $R_1, \dots, R_s$, in that order, and the elementary column matrices are $C_1, \dots, C_t$, in that order. Then the matrices Q and P are given by

$$Q = R_s \cdots R_2 R_1, \quad P^{-1} = C_1 C_2 \cdots C_t$$

hence $P = C_t^{-1} \cdots C_2^{-1} C_1^{-1}$. The final basis for M is $Y = PX$, and the final generating set for K is $V = QU = SY$, where $S = QAP^{-1}$ is the Smith normal form (see 4.4.2).

Section 4.6

Here is a result that is used in algebraic number theory. Let G be a free abelian group of rank n, and H a subgroup of G. By the simultaneous basis theorem, there is a basis

$y_1, \ldots y_n$ of G and there are positive integers $a_1, \ldots a_r$, $r \leq n$, such that a_i divides a_{i+1} for all i, and $a_1 y_1, \ldots, a_r y_r$ is a basis for H. We claim that the abelian group G/H is finite if and only if $r = n$, and in this case, the size of G/H is $|G/H| = a_1 a_2 \cdots a_r$.

To see this, look at the proof of (4.6.3) with R^n replaced by G and K by H. The argument shows that G/H is the direct sum of cyclic groups $\mathbb{Z}/\mathbb{Z}a_i$, $i = 1, \ldots, n$, with $a_i = 0$ for $r < i \leq n$. In other words, G/H is the direct sum of r finite cyclic groups (of order $a_1, \ldots, a_r$ respectively) and $n - r$ copies of $\mathbb{Z}$. The result follows.

Now assume that $r = n$, and let $x_1, \ldots, x_n$ and $z_1, \ldots, z_n$ be arbitrary bases for G and H respectively. Then each z_i is a linear combination of the x_i with integer coefficients; in matrix form, $z = Ax$. We claim that $|G/H|$ is the absolute value of the determinant of A. To verify this, first look at the special case $x_i = y_i$ and $z_i = a_i y_i$, $i = 1, \ldots, n$. Then A is a diagonal matrix with entries a_i, and the result follows. But the special case implies the general result, because any matrix corresponding to a change of basis of G or H is unimodular, in other words, has determinant ± 1. (See Section 4.4, Problem 1.)

Section 4.7

Here is some extra practice in diagram chasing. The diagram below is commutative with exact rows.

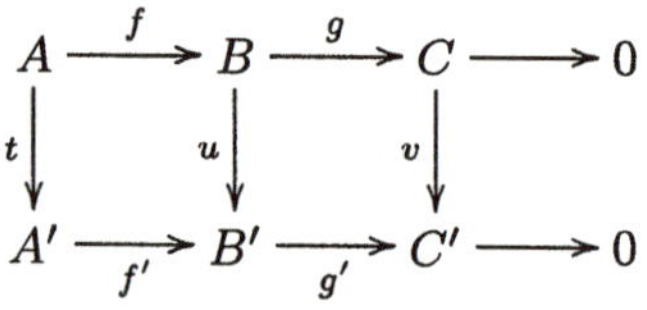

$$\begin{array}{ccccccc} A & \xrightarrow{f} & B & \xrightarrow{g} & C & \longrightarrow & 0 \\ \downarrow t & & \downarrow u & & \downarrow v & & \\ A' & \xrightarrow[f']{} & B' & \xrightarrow[g']{} & C' & \longrightarrow & 0 \end{array}$$

If t and u are isomorphisms, we will show that v is also an isomorphism. (The hypothesis on t can be weakened to surjectivity.)

Let $c' \in C'$; then $c' = g'b'$ for some $b' \in B'$. Since u is surjective, $g'b' = g'ub$ for some $b \in B$. By commutativity, $g'ub = vgb$, which proves that v is surjective.

Now assume $vc = 0$. Since g is surjective, $c = gb$ for some $b \in B$. By commutativity, $vgb = g'ub = 0$. Thus $ub \in \ker g' = \operatorname{im} f'$, so $ub = f'a'$ for some $a' \in A'$. Since t is surjective, $f'a' = f'ta$ for some $a \in A$. By commutativity, $f'ta = ufa$. We now have $ub = ufa$, so $b - fa \in \ker u$, hence $b = fa$ because u is injective. Consequently,

$$c = gb = gfa = 0$$

which proves that v is injective.

Chapter 5

Some Basic Techniques of Group Theory

5.1 Groups Acting on Sets

In this chapter we are going to analyze and classify groups, and, if possible, break down complicated groups into simpler components. To motivate the topic of this section, let's look at the following result.

5.1.1 Cayley's Theorem

Every group is isomorphic to a group of permutations.

Proof. The idea is that each element g in the group G corresponds to a permutation of the set G itself. If $x \in G$, then the permutation associated with g carries x into gx. If $gx = gy$, then premultiplying by g^{-1} gives $x = y$. Furthermore, given any $h \in G$, we can solve $gx = h$ for x. Thus the map $x \to gx$ is indeed a permutation of G. The map from g to its associated permutation is injective, because if $gx = hx$ for all $x \in G$, then (take $x = 1$) $g = h$. In fact the map is a homomorphism, since the permutation associated with hg is multiplication by hg, which is multiplication by g followed by multiplication by h, $h \circ g$ for short. Thus we have an embedding of G into the group of all permutations of the set G. ♣

In Cayley's theorem, a group acts on itself in the sense that each g yields a permutation of G. We can generalize to the notion of a group acting on an arbitrary set.

5.1.2 Definitions and Comments

The group G *acts on the set* X if for each $g \in G$ there is a mapping $x \to gx$ of X into itself, such that

(1) $h(gx) = (hg)x$ for every $g, h \in G$

(2) $1x = x$ for every $x \in X$.

As in (5.1.1), $x \to gx$ defines a permutation of X. The main point is that the action of g is a permutation because it has an inverse, namely the action of g^{-1}. (Explicitly, the inverse of $x \to gx$ is $y \to g^{-1}y$.) Again as in (5.1.1), the map from g to its associated permutation $\Phi(g)$ is a homomorphism of G into the group S_X of permutations of X. But we do not necessarily have an embedding. If $gx = hx$ for all x, then in (5.1.1) we were able to set $x = 1$, the identity element of G, but this resource is not available in general.

We have just seen that a group action induces a homomorphism from G to S_X, and there is a converse assertion. If Φ is a homomorphism of G to S_X, then there is a corresponding action, defined by $gx = \Phi(g)x, x \in X$. Condition (1) holds because Φ is a homomorphism, and (2) holds because $\Phi(1)$ must be the identity of S_X. The kernel of Φ is known as the *kernel of the action*; it is the set of all $g \in G$ such that $gx = x$ for all x, in other words, the set of g's that fix everything in X.

5.1.3 Examples

1. (*The regular action*) Every group acts on itself by multiplication on the left, as in (5.1.1). In this case, the homomorphism Φ is injective, and we say that the action is *faithful.*

[Similarly, we can define an action on the right by $(xg)h = x(gh)$, $x1 = x$, and then G acts on itself by right multiplication. The problem is that $\Phi(gh) = \Phi(h) \circ \Phi(g)$, an antihomomorphism. The damage can be repaired by writing function values as xf rather than $f(x)$, or by defining the action of g to be multiplication on the right by g^{-1}. We will avoid the difficulty by restricting to actions on the left.]

2. (*The trivial action*) We take $gx = x$ for all $g \in G$, $x \in X$. This action is highly unfaithful.

3. (*Conjugation on elements*) We use the notation $g \bullet x$ for the action of g on x, and we set $g \bullet x = gxg^{-1}$, called the *conjugate of* x *by* g, for g and x in the group G. Since $hgxg^{-1}h^{-1} = (hg)x(hg)^{-1}$ and $1x1^{-1} = x$, we have a legal action of G on itself. The kernel is

$$\{g \colon gxg^{-1} = x \text{ for all } x\}, \text{ that is, } \{g \colon gx = xg \text{ for all } x\}.$$

Thus the kernel is the set of elements that commute with everything in the group. This set is called the *center* of G, written $Z(G)$.

4. (*Conjugation on subgroups*) If H is a subgroup of G, we take $g \bullet H = gHg^{-1}$. Note that gHg^{-1} is a subgroup of G, called the *conjugate subgroup of* H *by* g, since $gh_1g^{-1}gh_2g^{-1} = g(h_1h_2)g^{-1}$ and $(ghg^{-1})^{-1} = gh^{-1}g^{-1}$. As in Example (3), we have a legal action of G on the set of subgroups of G.

5. (*Conjugation on subsets*) This is a variation of the previous example. In this case we let G act by conjugation on the collection of all subsets of G, not just subgroups. The verification that the action is legal is easier, because gHg^{-1} is certainly a subset of G.

6. (*Multiplication on left cosets*) Let G act on the set of left cosets of a fixed subgroup H by $g \bullet (xH) = (gx)H$. By definition of set multiplication, we have a legitimate action.

7. (*Multiplication on subsets*) Let G act on all subsets of G by $g \bullet S = gS = \{gx : x \in S\}$. Again the action is legal by definition of set multiplication.

Problems For Section 5.1

1. Let G act on left cosets of H by multiplication, as in Example 6. Show that the kernel of the action is a subgroup of H.
2. Suppose that H is a proper subgroup of G of index n, and that G is a *simple group*, that is, G has no normal subgroups except G itself and $\{1\}$. Show that$G$ can be embedded in S_n.
3. Suppose that G is an infinite simple group. Show that for every proper subgroup H of G, the index $[G : H]$ is infinite.
4. Let G act on left cosets of H by multiplication. Show that the kernel of the action is
$$N = \bigcap_{x \in G} xHx^{-1}.$$
5. Continuing Problem 4, if K is any normal subgroup of G contained in H, show that $K \leq N$. Thus N is the largest normal subgroup of G contained in H; N is called the *core* of H in G.
6. Here is some extra practice with left cosets of various subgroups. Let H and K be subgroups of G, and consider the map f which assigns to the coset $g(H \cap K)$ the pair of cosets (gH, gK). Show that f is well-defined and injective, and therefore
$$[G : H \cap K] \leq [G : H][G : K].$$
Thus (Poincaré) the intersection of finitely many subgroups of finite index also has finite index.
7. If $[G : H]$ and $[G : K]$ are finite and relatively prime, show that the inequality in the preceding problem is actually an equality.
8. Let H be a subgroup of G of finite index n, and let G act on left cosets xH by multiplication. Let N be the kernel of the action, so that $N \trianglelefteq H$ by Problem 1. Show that $[G : N]$ divides $n!$.
9. Let H be a subgroup of G of finite index $n > 1$. If $|G|$ does not divide $n!$, show that G is not simple.

5.2 The Orbit-Stabilizer Theorem

5.2.1 Definitions and Comments

Suppose that the group G acts on the set X. If we start with the element $x \in X$ and successively apply group elements in all possible ways, we get

$$B(x) = \{gx\colon g \in G\}$$

which is called the *orbit* of x under the action of G. The action is *transitive* (we also say that G *acts transitively* on X) if there is only one orbit, in other words, for any $x, y \in X$, there exists $g \in G$ such that $gx = y$. Note that the orbits partition X, because they are the equivalence classes of the equivalence relation given by $y \sim x$ iff $y = gx$ for some $g \in G$.

The *stabilizer* of an element $x \in X$ is

$$G(x) = \{g \in G\colon gx = x\},$$

the set of elements that leave x fixed. A direct verification shows that $G(x)$ is a subgroup. This is a useful observation because any set that appears as a stabilizer in a group action is guaranteed to be a subgroup; we need not bother to check each time.

Before proceeding to the main theorem, let's return to the examples considered in (5.1.3).

5.2.2 Examples

1. The regular action of G on G is transitive, and the stabilizer of x is the subgroup $\{1\}$.

2. The trivial action is not transitive (except in trivial cases), in fact, $B(x) = \{x\}$ for every x. The stabilizer of x is the entire group G.

3. Conjugation on elements is not transitive (see Problem 1). The orbit of x is the set of *conjugates* gxg^{-1} of x, that is,

$$B(x) = \{gxg^{-1}\colon g \in G\},$$

which is known as the *conjugacy class* of x. The stabilizer of x is

$$G(x) = \{g\colon gxg^{-1} = x\} = \{g\colon gx = xg\},$$

the set of group elements that commute with x. This set is called the *centralizer* of x, written $C_G(x)$. Similarly, the centralizer $C_G(S)$ of an arbitrary subset $S \subseteq G$ is defined as the set of elements of G that commute with everything in S. (Here, we do need to check that $C_G(S)$ is a subgroup, and this follows because $C_G(S) = \bigcap_{x\in S} C_G(x)$.)

4. Conjugation on subgroups is not transitive. The orbit of H is $\{gHg^{-1}\colon g \in G\}$, the collection of conjugate subgroups of H. The stabilizer of H is

$$\{g\colon gHg^{-1} = H\},$$

which is called the *normalizer* of H, written $N_G(H)$. If K is a subgroup of G containing H, we have

$$H \trianglelefteq K \text{ iff } gHg^{-1} = H \text{ for every } g \in K$$

and this holds iff K is a subgroup of $N_G(H)$. Thus $N_G(H)$ is the largest subgroup of G in which H is normal.

5. Conjugation on subsets is not transitive, and the orbit of the subset S is $\{gSg^{-1}\colon g \in G\}$. The stabilizer of S is the normalizer$N_G(S) = \{g\colon gSg^{-1} = S\}$.

6. Multiplication on left cosets is transitive; a solution of $g(xH) = yH$ for x is $x = g^{-1}y$. The stabilizer of xH is

$$\{g\colon gxH = xH\} = \{g\colon x^{-1}gx \in H\} = \{g\colon g \in xHx^{-1}\} = xHx^{-1},$$

the conjugate of H by x. Taking $x = 1$, we see that the stabilizer of H is H itself.

7. Multiplication on subsets is not transitive. The stabilizer of S is $\{g\colon gS = S\}$, the set of elements of G that permute the elements of S.

5.2.3 The Orbit-Stabilizer Theorem

Suppose that a group G acts on a set X. Let $B(x)$ be the orbit of $x \in X$, and let $G(x)$ be the stabilizer of x. Then the size of the orbit is the index of the stabilizer, that is,

$$|B(x)| = [G\colon G(x)].$$

Thus if G is finite, then $|B(x)| = |G|/|G(x)|$; in particular, the orbit size divides the order of the group.

Proof. If y belongs to the orbit of x, say $y = gx$. We take $f(y) = gH$, where $H = G(x)$ is the stabilizer of x. To check that f is a well-defined map of $B(x)$ to the set of left cosets of H, let $y = g_1x = g_2x$. Then $g_2^{-1}g_1x = x$, so $g_2^{-1}g_1 \in H$, i.e., $g_1H = g_2H$. Since g is an arbitrary element of G, f is surjective. If $g_1H = g_2H$, then $g_2^{-1}g_1 \in H$, so that $g_2^{-1}g_1x = x$, and consequently $g_1x = g_2x$. Thus if $y_1 = g_1x$, $y_2 = g_2x$, and $f(y_1) = f(y_2)$, then $y_1 = y_2$, proving f injective. ♣

Referring to (5.2.2), Example 3, we see that $B(x)$ is an orbit of size 1 iff x commutes with every $g \in G$, i.e., $x \in Z(G)$, the center of G. Thus if G is finite and we select one element x_i from each conjugacy class of size greater than 1, we get the *class equation*

$$|G| = |Z(G)| + \sum_i [G\colon C_G(x_i)].$$

We know that a group G acts on left cosets of a subgroup K by multiplication. To prepare for the next result, we look at the action of a *subgroup* H of G on left cosets of K. Since K is a left coset of K, it has an orbit given by $\{hK\colon h \in H\}$. The union of the sets hK is the set product HK. The stabilizer of K is not K itself, as in Example 6; it is $\{h \in H\colon hK = K\}$. But $hK = K (= 1K)$ if and only if $h \in K$, so the stabilizer is $H \cap K$.

5.2.4 Proposition

If H and K are subgroups of the finite group G, then

$$|HK| = \frac{|H||K|}{|H \cap K|}.$$

Proof. The cosets in the orbit of K are disjoint, and each has $|K|$ members. Since, as remarked above, the union of the cosets is HK, there must be exactly $|HK|/|K|$ cosets in the orbit. Since the index of the stabilizer of K is $|H/H \cap K|$, the result follows from the orbit-stabilizer theorem. ♣

Problems For Section 5.2

1. Let σ be the permutation $(1,2,3,4,5)$ and π the permutation $(1,2)(3,4)$. Then $\pi\sigma\pi^{-1}$, the conjugate of σ by π, can be obtained by applying π to the symbols of σ to get $(2,1,4,3,5)$. Reversing the process, if we are given $\tau = (1,2)(3,4)$ and we specify that $\mu\tau\mu^{-1} = (1,3)(2,5)$, we can take $\mu = \left[\begin{smallmatrix} 1 & 2 & 3 & 4 & 5 \\ 1 & 3 & 2 & 5 & 4 \end{smallmatrix}\right]$. This suggests that two permutations are conjugate if and only if they have the same cycle structure. Explain why this works.
2. Show that if S is any subset of G, then the centralizer of S is a normal subgroup of the normalizer of S. (Let the normalizer $N_G(S)$ act on S by conjugation on elements.)
3. Let $G(x)$ be the stabilizer of x under a group action. Show that stabilizers of elements in the orbit of x are conjugate subgroups. Explicitly, for every $g \in G$ and $x \in X$ we have

$$G(gx) = gG(x)g^{-1}.$$

4. Let G act on the set X. Show that for a given $x \in X$, $\Psi(gG(x)) = gx$ is a well-defined injective mapping of the set of left cosets of $G(x)$ into X, and is bijective if the action is transitive.
5. Continuing Problem 4, let G act transitively on X, and choose any $x \in X$. Show that the action of G on X is essentially the same as the action of G on the left cosets of the stabilizer subgroup $G(x)$. This is the meaning of the assertion that "any transitive G-set is isomorphic to a space of left cosets". Give an appropriate formal statement expressing this idea.
6. Suppose that G is a finite group, and for every $x, y \in G$ such that $x \neq 1$ and $y \neq 1$, x and y are conjugate. Show that the order of G must be 1 or 2.
7. First note that if r is a positive rational number and k a fixed positive integer, there are only finitely many positive integer solutions of the equation

$$\frac{1}{x_1} + \cdots + \frac{1}{x_k} = r.$$

Outline of proof: If x_k is the smallest x_i, the left side is at most k/x_k, so $1 \leq x_k \leq k/r$ and there are only finitely many choices for x_k. Repeat this argument for the equation $\frac{1}{x_1} + \cdots + \frac{1}{x_{k-1}} = r - \frac{1}{x_k}$.

Now set $r = 1$ and let $N(k)$ be an upper bound on all the x_i's in all possible solutions. If G is a finite group with exactly k conjugacy classes, show that the order of G is at most $N(k)$.

5.3 Application To Combinatorics

The theory of group actions can be used to solve a class of combinatorial problems. To set up a typical problem, consider the regular hexagon of Figure 5.3.1, and recall the dihedral group D_{12}, the group of symmetries of the hexagon (Section 1.2).

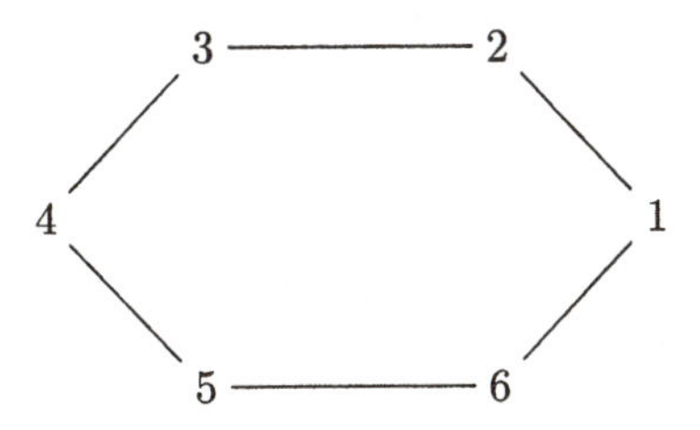

Figure 5.3.1

If R is rotation by 60 degrees and F is reflection about the horizontal line joining vertices 1 and 4, the 12 members of the group may be listed as follows.

$I =$ identity, $R = (1,2,3,4,5,6)$, $R^2 = (1,3,5)(2,4,6)$,
$R^3 = (1,4)(2,5)(3,6)$, $R^4 = (1,5,3)(2,6,4)$, $R^5 = (1,6,5,4,3,2)$
$F = (2,6)(3,5)$, $RF = (1,2)(3,6)(4,5)$, $R^2F = (1,3)(4,6)$
$R^3F = (1,4)(2,3)(5,6)$, $R^4F = (1,5)(2,4)$, $R^5F = (1,6)(2,5)(3,4)$.
(As before, RF means F followed by R.)

Suppose that we color the vertices of the hexagon, and we have n colors available (we are not required to use every color). How many distinct colorings are there? Since we may choose the color of any vertex in n ways, a logical answer is n^6. But this answer does not describe the physical situation accurately. To see what is happening, suppose we have two colors, yellow (Y) and blue (B). Then the coloring

$$C_1 = \begin{matrix} 1 & 2 & 3 & 4 & 5 & 6 \\ B & B & Y & Y & Y & B \end{matrix}$$

is mapped by RF to

$$C_2 = \begin{matrix} 1 & 2 & 3 & 4 & 5 & 6 \\ B & B & B & Y & Y & Y \end{matrix}$$

(For example, vertex 3 goes to where vertex 6 was previously, delivering the color yellow to vertex 6.) According to our counting scheme, C_2 is not the same as C_1. But imagine that we have two rigid necklaces in the form of a hexagon, one colored by C_1 and the other by C_2. If both necklaces were lying on a table, it would be difficult to argue that they are essentially different, since one can be converted to a copy of the other simply by flipping it over and then rotating it.

Let's try to make an appropriate mathematical model. Any group of permutations of a set X acts on X in the natural way: $g \bullet x = g(x)$. In particular, the dihedral group G acts on the vertices of the hexagon, and therefore on the set S of colorings of the vertices. The above discussion suggests that colorings in the same orbit should be regarded as equivalent, so the number of essentially different colorings is the number of orbits. The following result will help us do the counting.

5.3.1 Orbit-Counting Theorem

Let the finite group G act on the finite set X, and let $f(g)$ be the number of elements of X fixed by g, that is, the size of the set $\{x \in X\colon g(x) = x\}$. Then the number of orbits is

$$\frac{1}{|G|}\sum_{g \in G} f(g),$$

the average number of points left fixed by elements of G.

Proof. We use a standard combinatorial technique called "counting two ways". Let T be the set of all ordered pairs (g, x) such that $g \in G, x \in X$, and $gx = x$. For any $x \in X$, the number of g's such that $(g, x) \in T$ is the size of the stabilizer subgroup $G(x)$, hence

$$|T| = \sum_{x \in X} |G(x)|. \tag{1}$$

Now for any $g \in G$, the number of x's such that $(g, x) \in T$ is $f(g)$, the number of fixed points of g. Thus

$$|T| = \sum_{g \in G} f(g). \tag{2}$$

Divide (1) and (2) by the order of G to get

$$\sum_{x \in X} \frac{|G(x)|}{|G|} = \frac{1}{|G|}\sum_{g \in G} f(g). \tag{3}$$

But by the orbit-stabilizer theorem (5.2.3), $|G|/|G(x)|$ is $|B(x)|$,the size of the orbit of x. If, for example, an orbit has 5 members, then $1/5$ will appear 5 times in the sum on the left side of (3), for a total contribution of 1. Thus the left side of (3) is the total number of orbits. ♣

We can now proceed to the next step in the analysis.

5.3.2 Counting the Number of Colorings Fixed by a Given Permutation

Let $\pi = R^2 = (1, 3, 5)(2, 4, 6)$. Since $\pi(1) = 3$ and $\pi(3) = 5$, vertices 1,3 and 5 have the same color. Similarly, vertices 2,4 and 6 must have the same color. If there are n colors available, we can choose the color of each cycle in n ways, and the total number of choices is n^2. If $\pi = F = (2, 6)(3, 5)$, then as before we choose 1 color out of n for each cycle, but in this case we still have to color the vertices 1 and 4. Here is a general statement that covers both situations.

If π has c cycles, *counting cycles of length 1*, then the number of colorings fixed by π is n^c.

To emphasize the need to consider cycles of length 1, we can write F as (2,6)(3,5)(1)(4). From the cycle decompositions given at the beginning of the section, we have one permutation (the identity) with 6 cycles, three with 4 cycles, four with 3 cycles, two with 2 cycles, and two with 1 cycle. Thus the number of distinct colorings is

$$\frac{1}{12}(n^6 + 3n^4 + 4n^3 + 2n^2 + 2n).$$

5.3.3 A Variant

We now consider a slightly different question. How many distinct colorings of the vertices of a regular hexagon are there if we are forced to color exactly three vertices blue and three vertices yellow? The group G is the same as before, but the set S is different. Of the 64 possible colorings of the vertices, only $\binom{6}{3} = 20$ are legal, since 3 vertices out of 6 are chosen to be colored blue; the other vertices must be colored yellow. If π is a permutation of G, then within each cycle of π, all vertices have the same color, but in contrast to the previous example, we do not have a free choice of color for each cycle. To see this, consider $R^2 = (1,3,5)(2,4,6)$. The cycle $(1,3,5)$ can be colored blue and $(2,4,6)$ yellow, or vice versa, but it is not possible to color all six vertices blue, or to color all vertices yellow. Thus $f(R^2) = 2$. If $\pi = F = (2,6)(3,5)(1)(4)$, a fixed coloring is obtained by choosing one of the cycles of length 2 and one of the cycles of length 1 to be colored blue, thus producing 3 blue vertices. Consequently, $f(F) = 4$. To obtain $f(I)$, note that all legal colorings are fixed by I, so $f(I)$ is the number of colorings of 6 vertices with exactly 3 blue and 3 yellow vertices, namely, $\binom{6}{3} = 20$. From the cycle decompositions of the members of G, there are two permutations with $f = 2$, three with $f = 4$, and one with $f = 20$; the others have $f = 0$. Thus the number of distinct colorings is

$$\frac{1}{12}(2(2) + 3(4) + 20) = 3.$$

Problems For Section 5.3

1. Assume that two colorings of the vertices of a square are equivalent if one can be mapped into the other by a permutation in the dihedral group $G = D_8$. If n colors are available, find the number of distinct colorings.
2. In Problem 1, suppose that we color the sides of the square rather than the vertices. Do we get the same answer?
3. In Problem 1, assume that only two colors are available, white and green. There are 16 unrestricted colorings, but only 6 equivalence classes. List the equivalence classes explicitly.
4. Consider a rigid rod lying on the x-axis from $x = -1$ to $x = 1$, with three beads attached. The beads are located at the endpoints $(-1, 0)$ and $(1, 0)$, and at the center $(0, 0)$. The beads are to be painted using n colors, and two colorings are regarded as equivalent if one can be mapped into the other by a permutation in the group $G = \{I, \sigma\}$, where σ is the 180 degree rotation about the vertical axis. Find the number of distinct colorings.

5. In Problem 4, find the number of distinct colorings if the color of the central bead is always black.

6. Consider the group of rotations of the regular tetrahedron (see Figure 5.3.2); G consists of the following permutations.

   (i) The identity;

   (ii) Rotations by 120 degrees, clockwise or counterclockwise, about an axis through a vertex and the opposite face. There are 8 such rotations (choose 1 of 4 vertices, then choose a clockwise or counterclockwise direction);

   (iii) Rotations by 180 degrees about the line joining the midpoints of two nontouching edges. There are 3 such rotations.

   Argue geometrically to show that there are no other rotations in the group, and show that G is isomorphic to the alternating group A_4.

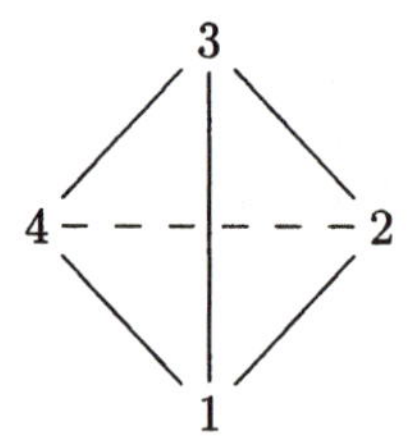

Figure 5.3.2

7. Given n colors, find the number of distinct colorings of the vertices of a regular tetrahedron, if colorings that can be rotated into each other are equivalent.

8. In Problem 7, assume that $n = 4$ and label the colors B,Y,W,G. Find the number of distinct colorings if exactly two vertices must be colored B.

9. The group G of rotations of a cube consists of the following permutations of the faces.

   (i) The identity.

   (ii) Rotations of ± 90 or 180 degrees about a line through the center of two opposite faces; there are $3 \times 3 = 9$ such rotations.

   (iii) Rotations of ± 120 degrees about a diagonal of the cube, i.e., a line joining two opposite vertices (vertices that are a maximal distance apart). There are 4 diagonals, so there are $4 \times 2 = 8$ such rotations.

   (iv) Rotations of 180 degrees about a line joining the midpoints of two opposite edges. There are 6 such rotations. (An axis of rotation is determined by selecting one of the four edges on the bottom of the cube, or one of the two vertical edges on the front face.)

   Argue geometrically to show that there are no other rotations in the group, and show that G is isomorphic to the symmetric group S_4.

10. If six colors are available and each face of a cube is painted a different color, find the number of distinct colorings.
11. Let G be the group of rotations of a regular p-gon, where p is an odd prime. If the vertices of the p-gon are to be painted using at most n colors, find the number of distinct colorings.
12. Use the result of Problem 11 to give an unusual proof of Fermat's little theorem.

5.4 The Sylow Theorems

Considerable information about the structure of a finite group G can be obtained by factoring the order of G. Suppose that $|G| = p^r m$ where p is prime, r is a positive integer, and p does not divide m. Then r is the highest power of p that divides the order of G. We will prove, among other things, that G must have a subgroup of order p^r, and any two such subgroups must be conjugate. We will need the following result about binomial coefficients.

5.4.1 Lemma

If $n = p^r m$ where p is prime, then $\binom{n}{p^r} \equiv m \bmod p$. Thus if p does not divide m, then it does not divide $\binom{p^r m}{p^r}$.

Proof. By the binomial expansion modulo p (see Section 3.4), which works for polynomials as well as for field elements, we have

$$(X+1)^{p^r} \equiv X^{p^r} + 1^{p^r} = X^{p^r} + 1 \bmod p.$$

Raise both sides to the power m to obtain

$$(X+1)^n \equiv (X^{p^r}+1)^m \bmod p.$$

On the left side, the coefficient of X^{p^r} is $\binom{n}{p^r}$, and on the right side, it is $\binom{m}{m-1} = m$. The result follows. ♣

5.4.2 Definitions and Comments

Let p be a prime number. The group G is said to be a *p-group* if the order of each element of G is a power of p. (The particular power depends on the element.) If G is a finite group, then G is a p-group iff the order of G is a power of p. [The "if" part follows from Lagrange's theorem, and the "only if" part is a corollary to the Sylow theorems; see (5.4.5).]

If $|G| = p^r m$, where p does not divide m, then a subgroup P of G of order p^r is called a *Sylow p-subgroup* of G. Thus P is a p-subgroup of G of maximum possible size.

5.4.3 The Sylow Theorems

Let G be a finite group of order $p^r m$, where p is prime, r is a positive integer, and p does not divide m. Then

(1) G has at least one Sylow p-subgroup, and every p-subgroup of G is contained in a Sylow p-subgroup.

(2) Let n_p be the number of Sylow p-subgroups of G. Then $n_p \equiv 1 \bmod p$ and n_p divides m.

(3) All Sylow p-subgroups are conjugate. Thus if we define an equivalence relation on subgroups by $H \sim K$ iff $H = gKg^{-1}$ for some $g \in G$, then the Sylow p-subgroups comprise a single equivalence class. [Note that the conjugate of a Sylow p-subgroup is also a Sylow p-subgroup, since it has the same number of elements p^r.]

Proof. (1) Let G act on subsets of G of size p^r by left multiplication. The number of such subsets is $\binom{p^r m}{p^r}$, which is not divisible by p by (5.4.1). Consequently, since orbits partition the set acted on by the group, there is at least one subset S whose orbit size is not divisible by p. If P is the stabilizer of S, then by (5.2.3), the size of the orbit is $[G : P] = |G|/|P| = p^r m/|P|$. For this to fail to be divisible by p, we must have $p^r \mid |P|$, and therefore $p^r \leq |P|$. But for any fixed $x \in S$, the map of P into S given by $g \to gx$ is injective. (The map is indeed into S because g belongs to the stabilizer of S, so that $gS = S$.) Thus $|P| \leq |S| = p^r$. We conclude that $|P| = p^r$, hence P is a Sylow p-subgroup.

So far, we have shown that a Sylow p-subgroup P exists, but not that every p-subgroup is contained in a Sylow p-subgroup. We will return to this in the course of proving (2) and (3).

(2) and (3) Let X be the set of all Sylow p-subgroups of G. Then $|X| = n_p$ and P acts on X by conjugation, i.e., $g \bullet Q = gQg^{-1}, g \in P$. By (5.2.3), the size of any orbit divides $|P| = p^r$, hence is a power of p. Suppose that there is an orbit of size 1, that is, a Sylow p-subgroup $Q \in X$ such that $gQg^{-1} = Q$, and therefore $gQ = Qg$, for every $g \in P$. (There is at least one such subgroup, namely P.) Then $PQ = QP$, so by (1.3.6), $PQ = \langle P, Q, \rangle$, the subgroup generated by P and Q. Since $|P| = |Q| = p^r$ it follows from (5.2.4) that $|PQ|$ is a power of p, say p^c. We must have $c \leq r$ because PQ is a subgroup of G (hence $|PQ|$ divides $|G|$). Thus

$$p^r = |P| \leq |PQ| \leq p^r, \text{ so } |P| = |PQ| = p^r.$$

But P is a subset of PQ, and since all sets are finite, we conclude that $P = PQ$, and therefore $Q \subseteq P$. Since both P and Q are of size p^r, we have $P = Q$. Thus there is only one orbit of size 1, namely $\{P\}$. Since by (5.2.3), all other orbit sizes are of the form p^c where $c \geq 1$, it follows that $n_p \equiv 1 \bmod p$.

Now let R be a p-subgroup of G, and let R act by multiplication on Y, the set of left cosets of P. Since $|Y| = [G : P] = |G|/|P| = p^r m/p^r = m$, p does not divide $|Y|$. Therefore some orbit size is not divisible by p. By (5.2.3), every orbit size divides $|R|$, hence is a power of p. (See (5.4.5) below. We are not going around in circles because (5.4.4) and (5.4.5) only depend on the existence of Sylow subgroups, which we have already established.) Thus there must be an orbit of size 1, say $\{gP\}$ with $g \in G$. If $h \in R$ then $hgP = gP$, that is, $g^{-1}hg \in P$, or equally well, $h \in gPg^{-1}$. Consequently, R is contained in a conjugate of P. If R is a Sylow p-subgroup to begin with, then R is a conjugate of P, completing the proof of (1) and (3).

To finish (2), we must show that n_p divides m. Let G act on subgroups by conjugation. The orbit of P has size n_p by (3), so by (5.2.3), n_p divides $|G| = p^r m$. But p cannot be a prime factor of n_p, since $n_p \equiv 1 \bmod p$. It follows that n_p must divide m. ♣

5.4.4 Corollary (Cauchy's Theorem)

If the prime p divides the order of G, then G has an element of order p.

Proof. Let P be a Sylow p-subgroup of G, and pick $x \in P$ with $x \neq 1$. The order of x is a power of p, say $|x| = p^k$. Then $x^{p^{k-1}}$ has order p. ♣

5.4.5 Corollary

The finite group G is a p-group if and only if the order of G is a power of p.

Proof. If the order of G is not a power of p, then it is divisible by some other prime q. But in this case, G has a Sylow q-subgroup, and therefore by (5.4.4), an element of order q. Thus G cannot be a p-group. The converse was done in (5.4.2). ♣

Problems For Section 5.4

1. Under the hypothesis of the Sylow theorems, show that G has a subgroup of index n_p.
2. Let P be a Sylow p-subgroup of the finite group G, and let Q be any p-subgroup. If Q is contained in the normalizer $N_G(P)$, show that PQ is a p-subgroup.
3. Continuing Problem 2, show that Q is contained in P.
4. Let P be a Sylow p-subgroup of the finite group G, and let H be a subgroup of G that contains the normalizer $N_G(P)$.

   (a) If $g \in N_G(H)$, show that P and gPg^{-1} are Sylow p-subgroups of H, hence they are conjugate in H.

   (b) Show that $N_G(H) = H$.

5. Let P be a Sylow p-subgroup of the finite group G, and let N be a normal subgroup of G. Assume that p divides $|N|$ and $|G/N|$, so that N and G/N have Sylow p-subgroups. Show that $[PN : P]$ and p are relatively prime, and then show that $P \cap N$ is a Sylow p-subgroup of N.
6. Continuing Problem 5, show that PN/N is a Sylow p-subgroup of G/N.
7. Suppose that P is the unique Sylow p-subgroup of G. [Equivalently, $P$is a normal Sylow p-subgroup of G; see (5.5.4).] Show that for each automorphism f of G, we have $f(P) = P$. [Thus P is a characteristic subgroup of G; see (5.7.1).]

8. The Sylow theorems are about subgroups whose order is a power of a prime p. Here is a result about subgroups of index p. Let H be a subgroup of the finite group G, and assume that $[G : H] = p$. Let N be a normal subgroup of G such that $N \le H$ and $[G : N]$ divides $p!$ (see Section 5.1, Problem 8). Show that $[H : N]$ divides $(p-1)!$.
9. Continuing Problem 8, let H be a subgroup of the finite group G, and assume that H has index p, where p is the smallest prime divisor of $|G|$. Show that $H \trianglelefteq G$.

5.5 Applications Of The Sylow Theorems

The Sylow theorems are of considerable assistance in the problem of classifying, up to isomorphism, all finite groups of a given order n. But in this area, proofs tend to involve intricate combinatorial arguments, best left to specialized texts in group theory. We will try to illustrate some of the basic ideas while keeping the presentation clean and crisp.

5.5.1 Definitions and Comments

A group G is *simple* if $G \neq \{1\}$ and the only normal subgroups of G are G itself and $\{1\}$. We will see later that simple groups can be regarded as building blocks for arbitrary finite groups. Abelian simple groups are already very familiar to us; they are the cyclic groups of prime order. For if $x \in G$, $x \neq 1$, then by simplicity (and the fact that all subgroups of an abelian group are normal), $G = \langle x \rangle$. If G is not of prime order, then G has a nontrivial proper subgroup by (1.1.4), so G cannot be simple.

The following results will be useful.

5.5.2 Lemma

If H and K are normal subgroups of G and the intersection of H and K is trivial (i.e., $\{1\}$), then $hk = kh$ for every $h \in H$ and $k \in K$.

Proof. We did this in connection with direct products; see the beginning of the proof of (1.5.2). ♣

5.5.3 Proposition

If P is a nontrivial finite p-group, then P has a nontrivial center.

Proof. Let P act on itself by conjugation; see (5.1.3) and (5.2.2), Example 3. The orbits are the conjugacy classes of P. The element x belongs to an orbit of size 1 iff x is in the center $Z(P)$, since $gxg^{-1} = x$ for all $g \in P$ iff $gx = xg$ for all $g \in P$ iff $x \in Z(P)$. By the orbit-stabilizer theorem, an orbit size that is greater than 1 must divide $|P|$, and therefore must be a positive power of p. If $Z(P) = \{1\}$, then we have one orbit of size 1, with all other orbit sizes $\equiv 0 \bmod p$. Thus $|P| \equiv 1 \bmod p$, contradicting the assumption that P is a nontrivial p-group. ♣

5.5.4 Lemma

P is a normal Sylow p-subgroup of G if and only if P is the unique Sylow p-subgroup of G.

Proof. By Sylow (3), the Sylow p-subgroups form a single equivalence class of conjugate subgroups. This equivalence class consists of a single element $\{P\}$ iff $gPg^{-1} = P$ for every $g \in G$, that is, iff $P \trianglelefteq G$. ♣

5.5.5 Proposition

Let G be a finite, nonabelian simple group. If the prime p divides the order of G, then the number n_p of Sylow p-subgroups of G is greater than 1.

Proof. If p is the only prime divisor of $|G|$, then G is a nontrivial p-group, hence $Z(G)$ is nontrivial by (5.5.3). Since $Z(G) \trianglelefteq G$ (see (5.1.3), Example 3), $Z(G) = G$, so that G is abelian, a contradiction. Thus $|G|$ is divisible by at least two distinct primes, so if P is a Sylow p-subgroup, then $\{1\} < P < G$. If $n_p = 1$, then there is a unique Sylow p-subgroup P, which is normal in G by (5.5.4). This contradicts the simplicity of G, so we must have $n_p > 1$. ♣

We can now derive some properties of groups whose order is the product of two distinct primes.

5.5.6 Proposition

Let G be a group of order pq, where p and q are distinct primes.

(i) If $q \not\equiv 1 \bmod p$, then G has a normal Sylow p-subgroup.

(ii) G is not simple.

(iii) If $p \not\equiv 1 \bmod q$ and $q \not\equiv 1 \bmod p$, then G is cyclic.

Proof. (i) By Sylow (2), $n_p \equiv 1 \bmod p$ and $n_p | q$, so $n_p = 1$. The result follows from (5.5.4).

(ii) We may assume without loss of generality that $p > q$. Then p cannot divide $q - 1$, so $q \not\equiv 1 \bmod p$. By (i), G has a normal Sylow p-subgroup, so G is not simple.

(iii) By (i), G has a normal Sylow p-subgroup P and a normal Sylow q-subgroup Q. Since P and Q are of prime order (p and q, respectively), they are cyclic. If x generates P and y generates Q, then $xy = yx$ by (5.5.2). [P and Q have trivial intersection because any member of the intersection has order dividing both p and q.] But then xy has order $pq = |G|$ (see Section 1.1, Problem 8). Thus $G = \langle xy \rangle$. ♣

We now look at the more complicated case $|G| = p^2q$. The combinatorial argument in the next proof is very interesting.

5.5.7 Proposition

Suppose that the order of the finite group G is p^2q, where p and q are distinct primes. Then G has either a normal Sylow p-subgroup or a normal Sylow q-subgroup. Thus G is not simple.

Proof. If the conclusion is false then n_p and n_q are both greater than 1. By Sylow (2), n_q divides p^2, so $n_q = p$ or p^2, and we will show that the second case leads to a contradiction. A Sylow q-subgroup Q is of order q and is therefore cyclic. Furthermore, every element of Q except the identity is a generator of Q. Conversely, any element of order q generates a Sylow q-subgroup. Since the only divisors of q are 1 and q, any two distinct Sylow q-subgroups have trivial intersection. Thus the number of elements of G of order q is exactly $n_q(q-1)$. If $n_q = p^2$, then the number of elements that are *not* of order q is

$$p^2q - p^2(q-1) = p^2.$$

Now let P be any Sylow p-subgroup of G. Then $|P| = p^2$, so no element of P can have order q (the orders must be 1, p or p^2). Since there are only p^2 elements of order unequal to q available, P takes care of all of them. Thus there cannot be another Sylow p-subgroup, so $n_p = 1$, a contradiction. We conclude that n_q must be p. Now by Sylow (2), $n_q \equiv 1 \bmod q$, hence $p \equiv 1 \bmod q$, so $p > q$. But n_p divides q, a prime, so $n_p = q$. Since $n_p \equiv 1 \bmod p$, we have $q \equiv 1 \bmod p$, and consequently $q > p$. Our original assumption that both n_p and n_q are greater than one has led inexorably to a contradiction. ♣

Problems For Section 5.5

1. Show that every group of order 15 is cyclic.
2. If $G/Z(G)$ is cyclic, show that $G = Z(G)$, and therefore G is abelian.
3. Show that for prime p, every group of order p^2 is abelian.
4. Let G be a group with $|G| = pqr$, where p, q and r are distinct primes and (without loss of generality) $p > q > r$. Show that $|G| \geq 1 + n_p(p-1) + n_q(q-1) + n_r(r-1)$.
5. Continuing Problem 4, if G is simple, show that n_p, n_q and n_r are all greater than 1. Then show that $n_p = qr$, $n_q \geq p$ and $n_r \geq q$.
6. Show that a group whose order is the product of three distinct primes is not simple.
7. Let G be a simple group of order $p^r m$, where $r \geq 1$, $m > 1$, and the prime p does not divide m. Let $n = n_p$ be the number of Sylow p-subgroups of G. If $H = N_G(P)$, where P is a Sylow p-subgroup of G, then $[G : H] = n$ (see Problem 1 of Section 5.4). Show that P cannot be normal in G (hence $n > 1$), and conclude that $|G|$ must divide $n!$.
8. If G is a group of order $250,000 = 2^4 5^6$, show that G is not simple.

5.6 Composition Series

5.6.1 Definitions and Comments

One way to break down a group into simpler components is via a *subnormal series*

$$1 = G_0 \trianglelefteq G_1 \trianglelefteq \cdots \trianglelefteq G_r = G.$$

"Subnormal" means that each subgroup G_i is normal in its successor G_{i+1}. In a *normal series*, the G_i are required to be normal subgroups of the entire group G. For convenience, the trivial subgroup $\{1\}$ will be written as 1.

Suppose that G_i is not a maximal normal subgroup of G_{i+1}, equivalently (by the correspondence theorem) G_{i+1}/G_i is not simple. Then the original subnormal series can be *refined* by inserting a group H such that $G_i \triangleleft H \triangleleft G_{i+1}$. We can continue refining in the hope that the process will terminate (it always will if G is finite). If all factors G_{i+1}/G_i are simple, we say that the group G has a *composition series*. [By convention, the trivial group has a composition series, namely $\{1\}$ itself.]

The Jordan-Hölder theorem asserts that if G has a composition series, the resulting *composition length* r and the *composition factors* G_{i+1}/G_i are unique (up to isomorphism and rearrangement). Thus all refinements lead to essentially the same result. Simple groups therefore give important information about arbitrary groups; if G_1 and G_2 have different composition factors, they cannot be isomorphic.

Here is an example of a composition series. Let S_4 be the group of all permutations of $\{1, 2, 3, 4\}$, and A_4 the subgroup of even permutations (normal in S_4 by Section 1.3, Problem 6). Let V be the four group (Section 1.2, Problem 6; normal in A_4, in fact in S_4, by direct verification). Let $\mathbb{Z}_2$ be any subgroup of V of order 2. Then

$$1 \triangleleft \mathbb{Z}_2 \triangleleft V \triangleleft A_4 \triangleleft S_4.$$

The proof of the Jordan-Hölder theorem requires some technical machinery.

5.6.2 Lemma

(i) If $K \trianglelefteq H \leq G$ and f is a homomorphism on G, then $f(K) \trianglelefteq f(H)$.

(ii) If $K \trianglelefteq H \leq G$ and $N \trianglelefteq G$, then $NK \trianglelefteq NH$.

(iii) If A, B, C and D are subgroups of G with $A \trianglelefteq B$ and $C \trianglelefteq D$, then $A(B \cap C) \trianglelefteq A(B \cap D)$, and by symmetry, $C(D \cap A) \trianglelefteq C(D \cap B)$.

(iv) In (iii), $A(B \cap C) \cap B \cap D = C(D \cap A) \cap D \cap B$.
Equivalently, $A(B \cap C) \cap D = C(D \cap A) \cap B$.

Proof. (i) For $h \in H, k \in K$, we have $f(h)f(k)f(h)^{-1} = f(hkh^{-1}) \in f(K)$.

(ii) Let f be the canonical map of G onto G/N. By (i) we have $NK/N \trianglelefteq NH/N$. The result follows from the correspondence theorem.

(iii) Apply (ii) with $G = B, N = A, K = B \cap C, H = B \cap D$.

(iv) The two versions are equivalent because $A(B \cap C) \leq B$ and $C(D \cap A) \leq D$. If x belongs to the set on the left, then $x = ac$ for some $a \in A, c \in B \cap C$, and x also belongs to D. But $x = c(c^{-1}ac) = ca^*$ for some $a^* \in A \trianglelefteq B$. Since $x \in D$ and $c \in C \leq D$, we have $a^* \in D$, hence $a^* \in D \cap A$. Thus $x = ca^* \in C(D \cap A)$, and since $x = ac$, with $a \in A \leq B$ and $c \in B \cap C \leq B$, $x \in C(D \cap A) \cap B$. Therefore the left side is a subset of the right side, and a symmetrical argument completes the proof. ♣

The diagram below is helpful in visualizing the next result.

$$\begin{array}{ccc} B & | & D \\ & | & \\ A & | & C \end{array}$$

To keep track of symmetry, take mirror images about the dotted line. Thus the group A will correspond to C, B to D, $A(B \cap C)$ to $C(D \cap A)$, and $A(B \cap D)$ to $C(D \cap B)$.

5.6.3 Zassenhaus Lemma

Let A, B, C and D be subgroups of G, with $A \trianglelefteq B$ and $C \trianglelefteq D$. Then

$$\frac{A(B \cap D)}{A(B \cap C)} \cong \frac{C(D \cap B)}{C(D \cap A)}.$$

Proof. By part (iii) of (5.6.2), the quotient groups are well-defined. An element of the group on the left is of the form $ayA(B \cap C), a \in A, y \in B \cap D$. But $ay = y(y^{-1}ay) = ya^*$, $a^* \in A$. Thus $ayA(B \cap C) = ya^*A(B \cap C) = yA(B \cap C)$. Similarly, an element of the right side is of the form $zC(D \cap A)$ with $z \in D \cap B = B \cap D$. Thus if $y, z \in B \cap D$, then

$$yA(B \cap C) = zA(B \cap C) \text{ iff } z^{-1}y \in A(B \cap C) \cap B \cap D$$

and by part (iv) of (5.6.2), this is equivalent to

$$z^{-1}y \in C(D \cap A) \cap D \cap B \text{ iff } yC(D \cap A) = zC(D \cap A).$$

Thus if h maps $yA(B \cap C)$ to $yC(D \cap A)$, then h is a well-defined bijection from the left to the right side of Zassenhaus' equation. By definition of multiplication in a quotient group, h is an isomorphism. ♣

5.6.4 Definitions and Comments

If a subnormal series is refined by inserting H between G_i and G_{i+1}, let us allow H to coincide with G_i or G_{i+1}. If all such insertions are strictly between the "endgroups", we will speak of a *proper refinement*. Two series are *equivalent* if they have the same length and their factor groups are the same, up to isomorphism and rearrangement.

5.6.5 Schreier Refinement Theorem

Let $1 = H_0 \trianglelefteq H_1 \trianglelefteq \cdots \trianglelefteq H_r = G$ and $1 = K_0 \trianglelefteq K_1 \trianglelefteq \cdots \trianglelefteq K_s = G$ be two subnormal series for the group G. Then the series have equivalent refinements.

Proof. Let $H_{ij} = H_i(H_{i+1} \cap K_j)$, $K_{ij} = K_j(K_{j+1} \cap H_i)$. By Zassenhaus we have

$$\frac{H_{i,j+1}}{H_{ij}} \cong \frac{K_{i+1,j}}{K_{ij}}.$$

(In (5.6.3) take $A = H_i$, $B = H_{i+1}$, $C = K_j$, $D = K_{j+1}$). We can now construct equivalent refinements; the easiest way to see this is to look at a typical concrete example. The first refinement will have r blocks of length s, and the second will have s blocks of length r. Thus the length will be rs in both cases. With $r = 2$ and $s = 3$, we have

$$1 = H_{00} \trianglelefteq H_{01} \trianglelefteq H_{02} \trianglelefteq H_{03} = H_1 = H_{10} \trianglelefteq H_{11} \trianglelefteq H_{12} \trianglelefteq H_{13} = H_2 = G,$$
$$1 = K_{00} \trianglelefteq K_{10} \trianglelefteq K_{20} = K_1 = K_{01} \trianglelefteq K_{11} \trianglelefteq K_{21} = K_2 = K_{02} \trianglelefteq K_{12} \trianglelefteq K_{22} = K_3 = G.$$

The corresponding factor groups are

$$H_{01}/H_{00} \cong K_{10}/K_{00}, \; H_{02}/H_{01} \cong K_{11}/K_{01}, \; H_{03}/H_{02} \cong K_{12}/K_{02}$$
$$H_{11}/H_{10} \cong K_{20}/K_{10}, \; H_{12}/H_{11} \cong K_{21}/K_{11}, \; H_{13}/H_{12} \cong K_{22}/K_{12}.$$

(Notice the pattern; in each isomorphism, the first subscript in the numerator is increased by 1 and the second subscript is decreased by 1 in going from left to right. The subscripts in the denominator are unchanged.) The factor groups of the second series are a reordering of the factor groups of the first series. ♣

The hard work is now accomplished, and we have everything we need to prove the main result.

5.6.6 Jordan-Hölder Theorem

If G has a composition series S (in particular if G is finite), then any subnormal series R without repetition can be refined to a composition series. Furthermore, any two composition series for G are equivalent.

Proof. By (5.6.5), R and S have equivalent refinements. Remove any repetitions from the refinements to produce equivalent refinements R_0 and S_0 without repetitions. But a composition series has no proper refinements, hence $S_0 = S$, proving the first assertion. If R is also a composition series, then $R_0 = R$ as well, and R is equivalent to S. ♣

Problems For Section 5.6

1. Show that if G has a composition series, so does every normal subgroup of G.
2. Give an example of a group that has no composition series.
3. Give an example of two nonisomorphic groups with the same composition factors, up to rearrangement.

Problems 4–9 will prove that the alternating group A_n is simple for all $n \geq 5$. (A_1 and A_2 are trivial and hence not simple; A_4 is not simple by the example given in (5.6.1); A_3 is cyclic of order 3 and is therefore simple.) In these problems, N stands for a normal subgroup of A_n.

4. Show that if $n \geq 3$, then A_n is generated by 3-cycles.
5. Show that if N contains a 3-cycle, then it contains all 3-cycles, so that $N = A_n$.
6. From now on, assume that N is a *proper* normal subgroup of A_n, and $n \geq 5$. Show that no permutation in N contains a cycle of length 4 or more.
7. Show that no permutation in N contains the product of two disjoint 3-cycles. Thus in view of Problems 4,5 and 6, every member of N is the product of an even number of disjoint transpositions.
8. In Problem 7, show that the number of transpositions in a nontrivial member of N must be at least 4.
9. Finally, show that the assumption that N contains a product of 4 or more disjoint transpositions leads to a contradiction, proving that $N = 1$, so that A_n is simple. It follows that a composition series for S_n is $1 \triangleleft A_n \triangleleft S_n$.
10. A *chief series* is a normal series without repetition that cannot be properly refined to another normal series. Show that if G has a chief series, then any normal series without repetition can be refined to a chief series. Furthermore, any two chief series of a given group are equivalent.
11. In a composition series, the factor groups G_{i+1}/G_i are required to be simple. What is the analogous condition for a chief series?

5.7 Solvable And Nilpotent Groups

Solvable groups are so named because of their connection with solvability of polynomial equations, a subject to be explored in the next chapter. To get started, we need a property of subgroups that is stronger than normality.

5.7.1 Definitions and Comments

A subgroup H of the group G is *characteristic* (in G) if for each automorphism f of G, $f(H) = H$. Thus f restricted to H is an automorphism of H. Consequently, if H is characteristic in G, then it is normal in G. If follows from the definition that if H is characteristic in K and K is characteristic in G, then H is characteristic in G. Another useful result is the following.

(1) If H is characteristic in K and K is normal in G, then H is normal in G.

To see this, observe that any inner automorphism of G maps K to itself, so restricts to an automorphism (not necessarily inner) of K. Further restriction to H results in an automorphism of H, and the result follows.

5.7.2 More Definitions and Comments

The *commutator subgroup* G' of a group G is the subgroup generated by all *commutators* $[x, y] = xyx^{-1}y^{-1}$. (Since $[x, y]^{-1} = [y, x]$, G' consists of all finite products of commutators.) Here are some basic properties.

(2) G' is characteristic in G.

This follows because any automorphism f maps a commutator to a commutator: $f[x,y] = [f(x), f(y)]$.

(3) G is abelian if and only if G' is trivial.

This holds because $[x,y] = 1$ iff $xy = yx$.

(4) G/G' is abelian. Thus forming the quotient of G by G', sometimes called *modding out by* G', in a sense "abelianizes" the group.

For $G'xG'y = G'yG'x$ iff $G'xy = G'yx$ iff $xy(yx)^{-1} \in G'$ iff $xyx^{-1}y^{-1} \in G'$, and this holds for all x and y by definition of G'.

(5) If $N \trianglelefteq G$, then G/N is abelian if and only if $G' \leq N$.

The proof of (4) with G' replaced by N shows that G/N is abelian iff all commutators belong to N, that is, iff $G' \leq N$.

The process of taking commutators can be iterated:

$$G^{(0)} = G, \ G^{(1)} = G', \ G^{(2)} = (G')',$$

and in general,

$$G^{(i+1)} = (G^{(i)})', \ i = 0, 1, 2, \ldots.$$

Since $G^{(i+1)}$ is characteristic in $G^{(i)}$, an induction argument shows that each $G^{(i)}$ is characteristic, hence normal, in G.

The group G is said to be *solvable* if $G^{(r)} = 1$ for some r. We then have a normal series

$$1 = G^{(r)} \trianglelefteq G^{(r-1)} \trianglelefteq \cdots \trianglelefteq G^{(0)} = G$$

called the *derived series* of G.

Every abelian group is solvable, by (3). Note that a group that is both simple and solvable must be cyclic of prime order. For the normal subgroup G' must be trivial; if it were G, then the derived series would never reach 1. By (3), G is abelian, and by (5.5.1), G must be cyclic of prime order.

A nonabelian simple group G (such as $A_n, n \geq 5$) cannot be solvable. For if G is nonabelian, then G' is not trivial. Thus $G' = G$, and as in the previous paragraph, the derived series will not reach 1.

There are several equivalent ways to describe solvability.

5.7.3 Proposition

The following conditions are equivalent.

(i) G is solvable.

(ii) G has a normal series with abelian factors.

(iii) G has a subnormal series with abelian factors.

Proof. Since (i) implies (ii) by (4) and (ii) implies (iii) by definition of normal and subnormal series, the only problem is (iii) implies (i). Suppose G has a subnormal series

$$1 = G_r \trianglelefteq G_{r-1} \trianglelefteq \cdots \trianglelefteq G_1 \trianglelefteq G_0 = G$$

with abelian factors. Since G/G_1 is abelian, we have $G' \leq G_1$ by (5), and an induction argument then shows that $G^{(i)} \leq G_i$ for all i. [The inductive step is $G^{(i+1)} = (G^{(i)})' \leq G_i' \leq G_{i+1}$ since G_i/G_{i+1} is abelian.] Thus $G^{(r)} \leq G_r = 1$. ♣

The next result gives some very useful properties of solvable groups.

5.7.4 Proposition

Subgroups and quotients of a solvable group are solvable. Conversely, if N is a normal subgroup of G and both N and G/N are solvable, then G is solvable.

Proof. If H is a subgroup of the solvable group G, then H is solvable because $H^{(i)} \leq G^{(i)}$ for all i. If N is a normal subgroup of the solvable group G, observe that commutators of G/N look like $xyx^{-1}y^{-1}N$, so $(G/N)' = G'N/N$. (Not G'/N, since N is not necessarily a subgroup of G'.) Inductively,

$$(G/N)^{(i)} = G^{(i)}N/N$$

and since N/N is trivial, G/N is solvable. Conversely, suppose that we have a subnormal series from $N_0 = 1$ to $N_r = N$, and a subnormal series from $G_0/N = 1$ (i.e., $G_0 = N$) to $G_s/N = G/N$ (i.e., $G_s = G$) with abelian factors in both cases. Then we splice the series of N_i's to the series of G_i's. The latter series is subnormal by the correspondence theorem, and the factors remain abelian by the third isomorphism theorem. ♣

5.7.5 Corollary

If G has a composition series, in particular if G is finite, then G is solvable if and only if the composition factors of G are cyclic of prime order.

Proof. Let G_{i+1}/G_i be a composition factor of the solvable group G. By (5.7.4), G_{i+1} is solvable, and again by (5.7.4), G_{i+1}/G_i is solvable. But a composition factor must be a simple group, so G_{i+1}/G_i is cyclic of prime order, as observed in (5.7.2). Conversely, if the composition factors of G are cyclic of prime order, then the composition series is a subnormal series with abelian factors. ♣

Nilpotent groups arise from a different type of normal series. We will get at this idea indirectly, and give an abbreviated treatment.

5.7.6 Proposition

If G is a finite group, the following conditions are equivalent, and define a *nilpotent group.* [Nilpotence of an arbitrary group will be defined in (5.7.8).]

(a) G is the direct product of its Sylow subgroups.

(b) Every Sylow subgroup of G is normal.

Proof. (a) implies (b): By (1.5.3), the factors of a direct product are normal subgroups.

(b) implies (a): By (5.5.4), there is a unique Sylow p_i-subgroup H_i for each prime divisor p_i of $|G|, i = 1, \dots, k$. By successive application of (5.2.4), we have $|H_1 \cdots H_k| = |H_1| \cdots |H_k|$, which is $|G|$ by definition of Sylow p-subgroup. Since all sets are finite, $G = H_1 \cdots H_k$. Furthermore, each $H_i \cap \prod_{j \neq i} H_j$ is trivial, because the orders of the H_i are powers of distinct primes. By (1.5.4), G is the direct product of the H_i. ♣

5.7.7 Corollary

Every finite abelian group and every finite p-group is nilpotent.

Proof. A finite abelian group must satisfy condition (b) of (5.7.6). If P is a finite p-group, then P has only one Sylow subgroup, P itself, so the conditions of (5.7.6) are automatically satisfied. ♣

We now connect this discussion with normal series. Suppose that we are trying to build a normal series for the group G, starting with $G_0 = 1$. We take G_1 to be $Z(G)$, the center of G; we have $G_1 \trianglelefteq G$ by (5.1.3), Example 3. We define G_2 by the correspondence theorem:

$$G_2/G_1 = Z(G/G_1)$$

and since $Z(G/G_1) \trianglelefteq G/G_1$, we have $G_2 \trianglelefteq G$. In general, we take

$$G_i/G_{i-1} = Z(G/G_{i-1}),$$

and by induction we have $G_i \trianglelefteq G$. The difficulty is that there is no guarantee that G_i will ever reach G. However, we will succeed if G is a finite p-group. The key point is that a nontrivial finite p-group has a nontrivial center, by (5.5.3). Thus by induction, G_i/G_{i-1} is nontrivial for every i, so $G_{i-1} < G_i$. Since G is finite, it must eventually be reached.

5.7.8 Definitions and Comments

A *central series* for G is a normal series $1 = G_0 \trianglelefteq G_1 \trianglelefteq \cdots \trianglelefteq G_r = G$ such that $G_i/G_{i-1} \subseteq Z(G/G_{i-1})$ for every $i = 1, \dots, r$. (The series just discussed is a special case called the *upper central series.*) An arbitrary group G is said to be *nilpotent* if it has a central series. Thus a finite p-group is nilpotent, and in particular, every Sylow p-subgroup is nilpotent. Now a direct product of a finite number of nilpotent groups is nilpotent. (If G_{ij} is the ith term of a central series of the jth factor H_j, with $G_{ij} = G$ if the series has already terminated at G, then $\prod_j G_{ij}$ will be the ith term of a central series for $\prod_j H_j$.) Thus a finite group that satisfies the conditions of (5.7.6) has a central series. Conversely, it can be shown that a finite group that has a central series satisfies (5.7.6), so the two definitions of nilpotence agree for finite groups.

Note that a nilpotent group is solvable. For if $G_i/G_{i-1} \subseteq Z(G/G_{i-1})$, then the elements of G_i/G_{i-1} commute with each other since they commute with everything in G/G_{i-1}; thus G_i/G_{i-1} is abelian. Consequently, a finite p-group is solvable.

Problems For Section 5.7

1. Give an example of a nonabelian solvable group.
2. Show that a solvable group that has a composition series must be finite.
3. Prove directly (without making use of nilpotence) that a finite p-group is solvable.
4. Give an example of a solvable group that is not nilpotent.
5. Show that if $n \geq 5$, then S_n is not solvable.
6. If P is a finite simple p-group, show that P has order p.
7. Let P be a nontrivial finite p-group. Show that P has a normal subgroup N whose index $[P : N]$ is p.
8. Let G be a finite group of order $p^r m$, where r is a positive integer and p does not divide m. Show that for any $k = 1, 2, \dots, r$, G has a subgroup of order p^k.
9. Give an example of a group G with a normal subgroup N such that N and G/N are abelian, but G is not abelian. (If "abelian" is replaced by "solvable", no such example is possible, by (5.7.4).)
10. If G is a solvable group, its *derived length*, $\mathrm{dl}(G)$, is the smallest nonnegative integer r such that $G^{(r)} = 1$. If N is a normal subgroup of the solvable group G, what can be said about the relation between $\mathrm{dl}(G)$, $\mathrm{dl}(N)$ and $\mathrm{dl}(G/N)$?

5.8 Generators And Relations

In (1.2.4) we gave an informal description of the dihedral group via generators and relations, and now we try to make the ideas more precise.

5.8.1 Definitions and Comments

The *free group* G on the set S (or the *free group with basis* S) consists of all *words* on S, that is, all finite sequences $x_1 \cdots x_n$, $n = 0, 1, \dots$, where each x_i is either an element of S or the inverse of an element of S. We regard the case $n = 0$ as the *empty word* λ. The group operation is concatenation, subject to the constraint that if s and s^{-1} occur in succession, they can be cancelled. The empty word is the identity, and inverses are calculated in the only reasonable way, for example, $(stu)^{-1} = u^{-1}t^{-1}s^{-1}$. We say that G is *free on* S.

Now suppose that G is free on S, and we attempt to construct a homomorphism f from G to an arbitrary group H. The key point is that f is completely determined by its values on S. If $f(s_1) = a$, $f(s_2) = b$, $f(s_3) = c$, then

$$f(s_1 s_2^{-1} s_3) = f(s_1)f(s_2)^{-1}f(s_3) = ab^{-1}c.$$

Here is the formal statement, followed by an informal proof.

5.8.2 Theorem

If G is free on S and g is an arbitrary function from S to a group H, then there is a unique homomorphism $f\colon G \to H$ such that $f = g$ on S.

Proof. The above discussion is a nice illustration of a concrete example with all the features of the general case. The analysis shows both existence and uniqueness of f. A formal proof must show that all aspects of the general case are covered. For example, if $u = s_1 s_2^{-1} s_3$ and $v = s_1 s_2^{-1} s_4^{-1} s_4 s_3$, then $f(u) = f(v)$, so that cancellation of $s_4^{-1} s_4$ causes no difficulty. Specific calculations of this type are rather convincing, and we will not pursue the formal details. (See, for example, Rotman, An Introduction to the Theory of Groups, pp. 343–345.) ♣

5.8.3 Corollary

Any group H is a homomorphic image of a free group.

Proof. Let S be a set of generators for H (if necessary, take $S = H$), and let G be free on S. Define $g(s) = s$ for all $s \in S$. If f is the unique extension of g to G, then since S generates H, f is an epimorphism. ♣

Returning to (1.2.4), we described a group H using generators R and F, and relations $R^n = I$, $F^2 = I$, $RF = FR^{-1}$. The last relation is equivalent to $RFRF = I$, since $F^2 = I$. The words R^n, F^2 and $RFRF$ are called *relators*, and the specification of generators and relations is called a *presentation*. We use the notation

$$H = \langle R, F \mid R^n, F^2, RFRF \rangle$$

or the long form

$$H = \langle R, F \mid R^n = I, F^2 = I, RF = FR^{-1} \rangle.$$

We must say precisely what it means to define a group by generators and relations, and show that the above presentation yields a group isomorphic to the dihedral group D_{2n}. We start with the free group on $\{R, F\}$ and set all relators equal to the identity. It is natural to mod out by the subgroup generated by the relators, but there is a technical difficulty; this subgroup is not necessarily normal.

5.8.4 Definition

Let G be free on the set S, and let K be a subset of G. We define the group $\langle S \mid K \rangle$ as $G/\overline{K}$, where $\overline{K}$ is the smallest normal subgroup of G containing K.

Unfortunately, it is a theorem of mathematical logic that there is no algorithm which when given a presentation, will find the order of the group. In fact, there is no algorithm to determine whether a given word of $\langle S \mid K \rangle$ coincides with the identity. Logicians say that the word problem for groups is unsolvable. But although there is no general solution, there are specific cases that can be analyzed, and the following result is very helpful.

5.8.5 Von Dyck's Theorem

Let $H = \langle S \mid K \rangle$ be a presentation, and let L be a group that is generated by the words in S. If L satisfies all the relations of K, then there is an epimorphism $\alpha\colon H \to L$. Consequently, $|H| \geq |L|$.

Proof. Let G be free on S, and let i be the identity map from S, regarded as a subset of G, to S, regarded as a subset of L. By (5.8.2), i has a unique extension to a homomorphism f of G into L, and in fact f is an epimorphism because S generates L. Now f maps any word of G to the same word in L, and since L satisfies all the relations, we have $K \subseteq \ker f$. But the kernel of f is a normal subgroup of G, hence $\overline{K} \subseteq \ker f$. The factor theorem provides an epimorphism $\alpha\colon G/\overline{K} \to L$. ♣

5.8.6 Justifying a presentation

If L is a finite group generated by the words of S, then in practice, the crucial step in identifying L with $H = \langle S \mid K \rangle$ is a proof that $|H| \leq |L|$. If we can accomplish this, then by (5.8.5), $|H| = |L|$. In this case, α is a surjective map of finite sets of the same size, so α is injective as well, hence is an isomorphism. For the dihedral group we have $H = \langle F, R \mid R^n, F^2, RFRF \rangle$ and $L = D_{2n}$. In (1.2.4) we showed that each word of H can be expressed as R^iF^j with $0 \leq i \leq n-1$ and $0 \leq j \leq 1$. Therefore $|H| \leq 2n = |D_{2n}| = |L|$. Thus the presentation H is a legitimate description of the dihedral group.

Problems For Section 5.8

1. Show that a presentation of the cyclic group of order n is $\langle a \mid a^n \rangle$.
2. Show that the quaternion group (see (2.1.3, Example 4)) has a presentation $\langle a, b \mid a^4 = 1, b^2 = a^2, ab = ba^{-1} \rangle$.
3. Show that $H = \langle a, b \mid a^3 = 1, b^2 = 1, ba = a^{-1}b \rangle$ is a presentation of S_3.
4. Is the presentation of a group unique?

In Problems 5–11, we examine a different way of assembling a group from subgroups, which generalizes the notion of a direct product. Let N be a normal subgroup of G, and H an arbitrary subgroup. We say that G is the *semidirect product* of N by H if $G = NH$ and $N \cap H = 1$. (If $H \trianglelefteq G$, we have the direct product.) For notational convenience, the letter n, possibly with subscripts, will always indicate a member of N, and similarly h will always belong to H. In Problems 5 and 6, we assume that G is the semidirect product of N by H.

5. If $n_1h_1 = n_2h_2$, show that $n_1 = n_2$ and $h_1 = h_2$.
6. If $i\colon N \to G$ is inclusion and $\pi\colon G \to H$ is projection ($\pi(nh) = h$), then the sequence

$$1 \rightarrow N \xrightarrow{i} G \xrightarrow{\pi} H \rightarrow 1$$

   is exact. Note that π is well-defined by Problem 5, and verify that π is a homomorphism. Show that the sequence *splits on the right*, i.e., there is a homomorphism $\psi\colon H \to G$ such that $\pi \circ \psi = 1$.
7. Conversely, suppose that the above exact sequence splits on the right. Since ψ is injective, we can regard H (and N as well) as subgroups of G, with ψ and i as inclusion maps. Show that G is the semidirect product of N by H.
8. Let N and H be arbitrary groups, and let f be a homomorphism of H into $\operatorname{Aut} N$, the group of automorphisms of N. Define a multiplication on $G = N \times H$ by

$$(n_1, h_1)(n_2, h_2) = (n_1 f(h_1)(n_2), h_1 h_2).$$

   [$f(h_1)(n_2)$ is the value of the automorphism $f(h_1)$ at the element n_2.] A lengthy but straightforward calculation shows that G is a group with identity $(1, 1)$ and inverses given by $(n, h)^{-1} = (f(h^{-1})(n^{-1}), h^{-1})$. Show that G is the semidirect product of $N \times \{1\}$ by $\{1\} \times H$.
9. Show that every semidirect product arises from the construction of Problem 8.
10. Show by example that it is possible for a short exact sequence of groups to split on the right but not on the left.

    [If $h\colon G \to N$ is a left-splitting map in the exact sequence of Problem 6, then h and π can be used to identify G with the direct product of N and H. Thus a left-splitting implies a right-splitting, but, unlike the result for modules in (4.7.4), not conversely.]
11. Give an example of a short exact sequence of groups that does not split on the right.
12. (The Frattini argument, frequently useful in a further study of group theory.) Let N be a normal subgroup of the finite group G, and let P be a Sylow p-subgroup of N. If $N_G(P)$ is the normalizer of P in G, show that $G = N_G(P)N$ ($= NN_G(P)$ by (1.4.3)).[If $g \in G$, look at the relation between P and gPg^{-1}.]
13. Let $N = \{1, a, a^2, \dots, a^{n-1}\}$ be a cyclic group of order n, and let $H = \{1, b\}$ be a cyclic group of order 2. Define $f\colon H \to \operatorname{Aut} N$ by taking $f(b)$ to be the automorphism that sends a to a^{-1}. Show that the dihedral group D_{2n} is the semidirect product of N by H. (See Problems 8 and 9 for the construction of the semidirect product.)
14. In Problem 13, replace N by an infinite cyclic group

$$\{\dots, a^{-2}, a^{-1}, 1, a, a^2, \dots\}.$$

    Give a presentation of the semidirect product of N by H. This group is called the *infinite dihedral group* D_∞.

Concluding Remarks

Suppose that the finite group G has a composition series

$$1 = G_0 \lhd G_1 \lhd \cdots \lhd G_r = G.$$

If $H_i = G_i/G_{i-1}$, then we say that G_i is an *extension of* G_{i-1} *by* H_i in the sense that $G_{i-1} \trianglelefteq G_i$ and $G_i/G_{i-1} \cong H_i$. If we were able to solve the extension problem (find all possible extensions of G_{i-1} by H_i) and we had a catalog of all finite simple groups, then we could build a catalog of all finite groups. This sharpens the statement made in (5.6.1) about the importance of simple groups.

Chapter 6

Galois Theory

6.1 Fixed Fields and Galois Groups

Galois theory is based on a remarkable correspondence between subgroups of the Galois group of an extension E/F and intermediate fields between E and F. In this section we will set up the machinery for the fundamental theorem. [A remark on notation: Throughout the chapter, the composition $\tau \circ \sigma$ of two automorphisms will be written as a product $\tau\sigma$.]

6.1.1 Definitions and Comments

Let $G = \text{Gal}(E/F)$ be the Galois group of the extension E/F. If H is a subgroup of G, the *fixed field* of H is the set of elements fixed by every automorphism in H, that is,

$$\mathcal{F}(H) = \{x \in E \colon \sigma(x) = x \text{ for every } \sigma \in H\}.$$

If K is an intermediate field, that is, $F \leq K \leq E$, define

$$\mathcal{G}(K) = \text{Gal}(E/K) = \{\sigma \in G \colon \sigma(x) = x \text{ for every } x \in K\}.$$

I like the term "*fixing group of* K" for $\mathcal{G}(K)$, since $\mathcal{G}(K)$ is the group of automorphisms of E that leave K fixed. Galois theory is about the relation between fixed fields and fixing groups. In particular, the next result suggests that the smallest subfield F corresponds to the largest subgroup G.

6.1.2 Proposition

Let E/F be a finite Galois extension with Galois group $G = \text{Gal}(E/F)$. Then

(i) The fixed field of G is F;

(ii) If H is a proper subgroup of G, then the fixed field of H properly contains F.

Proof. (i) Let F_0 be the fixed field of G. If σ is an F-automorphism of E, then by definition of F_0, σ fixes everything in F_0. Thus the F-automorphisms of G coincide with the F_0-automorphisms of G. Now by (3.4.7) and (3.5.8), E/F_0 is Galois. By (3.5.9), the size of the Galois group of a finite Galois extension is the degree of the extension. Thus $[E : F] = [E : F_0]$, so by (3.1.9), $F = F_0$.

(ii) Suppose that $F = \mathcal{F}(H)$. By the theorem of the primitive element (3.5.12), we have $E = F(\alpha)$ for some $\alpha \in E$. Define a polynomial $f(X) \in E[X]$ by

$$f(X) = \prod_{\sigma \in H} (X - \sigma(\alpha)).$$

If τ is any automorphism in H, then we may apply τ to f (that is, to the coefficients of f; we discussed this idea in the proof of (3.5.2)). The result is

$$(\tau f)(X) = \prod_{\sigma \in H} (X - (\tau\sigma)(\alpha)).$$

But as σ ranges over all of H, so does $\tau\sigma$, and consequently $\tau f = f$. Thus each coefficient of f is fixed by H, so $f \in F[X]$. Now α is a root of f, since $X - \sigma(\alpha)$ is 0 when $X = \alpha$ and σ is the identity. We can say two things about the degree of f:

(1) By definition of f, $\deg f = |H| < |G| = [E : F]$, and, since f is a multiple of the minimal polynomial of α over F,

(2) $\deg f \geq [F(\alpha) : F] = [E : F]$, and we have a contradiction. ♣

There is a converse to the first part of (6.1.2).

6.1.3 Proposition

Let E/F be a finite extension with Galois group G. If the fixed field of G is F, then E/F is Galois.

Proof. Let $G = \{\sigma_1, \dots, \sigma_n\}$, where σ_1 is the identity. To show that E/F is normal, we consider an irreducible polynomial $f \in F[X]$ with a root $\alpha \in E$. Apply each automorphism in G to α, and suppose that there are r distinct images $\alpha = \alpha_1 = \sigma_1(\alpha)$, $\alpha_2 = \sigma_2(\alpha), \dots, \alpha_r = \sigma_r(\alpha)$. If σ is any member of G, then σ will map each α_i to some α_j, and since σ is an injective map of the finite set $\{\alpha_1, \dots, \alpha_r\}$ to itself, it is surjective as well. To put it simply, σ permutes the α_i. Now we examine what σ does to the *elementary symmetric functions* of the α_i, which are given by

$$e_1 = \sum_{i=1}^{r} \alpha_i, \; e_2 = \sum_{i<j} \alpha_i \alpha_j, \; e_3 = \sum_{i<j<k} \alpha_i \alpha_j \alpha_k, \dots,$$

$$e_r = \prod_{i=1}^{r} \alpha_i.$$

Since σ permutes the α_i, it follows that $\sigma(e_i) = e_i$ for all i. Thus the e_i belong to the fixed field of G, which is F by hypothesis. Now we form a monic polynomial whose roots are the α_i:

$$g(X) = (X - \alpha_1) \cdots (X - \alpha_r) = X^r - e_1 X^{r-1} + e_2 X^{r-2} - \cdots + (-1)^r e_r.$$

Since the e_i belong to F, $g \in F[X]$, and since the α_i are in E, g splits over E. We claim that g is the minimal polynomial of α over F. To see this, let $h(X) = b_0+b_1X+\cdots+b_mX^m$ be any polynomial in $F[X]$ having α as a root. Applying σ_i to the equation

$$b_0 + b_1\alpha + \cdots b_m\alpha^m = 0$$

we have

$$b_0 + b_1\alpha_i + \cdots b_m\alpha_i^m = 0,$$

so that each α_i is a root of h, hence g divides h and therefore $g = \min(\alpha, F)$. But our original polynomial $f \in F[X]$ is irreducible and has α as a root, so it must be a constant multiple of g. Consequently, f splits over E, proving that E/F is normal. Since the α_i, $i = 1, \dots r$, are distinct, g has no repeated roots. Thus α is separable over F, which shows that the extension E/F is separable. ♣

It is profitable to examine elementary symmetric functions in more detail.

6.1.4 Theorem

Let f be a symmetric polynomial in the n variables $X_1, \dots, X_n$. [This means that if σ is any permutation in S_n and we replace X_i by $X_{\sigma(i)}$ for $i = 1, \dots, n$, then f is unchanged.] If $e_1, \dots, e_n$ are the elementary symmetric functions of the X_i, then f can be expressed as a polynomial in the e_i.

Proof. We give an algorithm. The polynomial f is a linear combination of monomials of the form $X_1^{r_1} \cdots X_n^{r_n}$, and we order the monomials lexicographically: $X_1^{r_1} \cdots X_n^{r_n} > X_1^{s_1} \cdots X_n^{s_n}$ iff the first disagreement between r_i and s_i results in $r_i > s_i$. Since f is symmetric, all terms generated by applying a permutation $\sigma \in S_n$ to the subscripts of $X_1^{r_1} \cdots X_n^{r_n}$ will also contribute to f. The idea is to cancel the leading terms (those associated with the monomial that is first in the ordering) by subtracting an expression of the form

$$e_1^{t_1} e_2^{t_2} \cdots e_n^{t_n} = (X_1 + \cdots + X_n)^{t_1} \cdots (X_1 \cdots X_n)^{t_n}$$

which has leading term

$$X_1^{t_1}(X_1X_2)^{t_2}(X_1X_2X_3)^{t_3} \cdots (X_1 \cdots X_n)^{t_n} \;=\; X_1^{t_1+\cdots+t_n} X_2^{t_2+\cdots+t_n} \cdots X_n^{t_n}.$$

This will be possible if we choose

$$t_1 = r_1 - r_2,\ t_2 = r_2 - r_3,\ \dots, t_{n-1} = r_{n-1} - r_n,\ t_n = r_n.$$

After subtraction, the resulting polynomial has a leading term that is below $X_1^{r_1} \cdots X_n^{r_n}$ in the lexicographical ordering. We can then repeat the procedure, which must terminate in a finite number of steps. ♣

6.1.5 Corollary

If g is a polynomial in $F[X]$ and $f(\alpha_1, \ldots, \alpha_n)$ is any symmetric polynomial in the roots $\alpha_1, \ldots, \alpha_n$ of g, then $f \in F[X]$.

Proof. We may assume without loss of generality that g is monic. Then in a splitting field of g we have

$$g(X) = (X - \alpha_1) \cdots (X - \alpha_n) = X^n - e_1 X^{n-1} + \cdots + (-1)^n e_n.$$

By (6.1.4), f is a polynomial in the e_i, and since the e_i are simply $\pm$ the coefficients of g, the coefficients of f are in F. ♣

6.1.6 Dedekind's Lemma

The result that the size of the Galois group of a finite Galois extension is the degree of the extension can be proved via Dedekind's lemma, which is of interest in its own right. Let G be a group and E a field. A *character* from G to E is a homomorphism from G to the multiplicative group E^* of nonzero elements of E. In particular, an automorphism of E defines a character with $G = E^*$, as does a monomorphism of E into a field L. Dedekind's lemma states that if $\sigma_1, \ldots, \sigma_n$ are distinct characters from G to E, then the σ_i are linearly independent over E. The proof is given in Problems 3 and 4.

Problems For Section 6.1

1. Express $X_1^2 X_2 X_3 + X_1 X_2^2 X_3 + X_1 X_2 X_3^2$ in terms of elementary symmetric functions.
2. Repeat Problem 1 for$X_1^2 X_2 + X_1^2 X_3 + X_1 X_2^2 + X_1 X_3^2 + X_2^2 X_3 + X_2 X_3^2 + 4X_1 X_2 X_3$.
3. To begin the proof of Dedekind's lemma, suppose that the σ_i are linearly dependent. By renumbering the σ_i if necessary, we have

$$a_1\sigma_1 + \cdots a_r\sigma_r = 0$$

where all a_i are nonzero and r is as small as possible. Show that for every h and $g \in G$, we have

$$\sum_{i=1}^{r} a_i \sigma_1(h) \sigma_i(g) = 0 \tag{1}$$

and

$$\sum_{i=1}^{r} a_i \sigma_i(h) \sigma_i(g) = 0. \tag{2}$$

[Equations (1) and (2) are not the same; in (1) we have $\sigma_1(h)$, not $\sigma_i(h)$.]

4. Continuing Problem 3, subtract (2) from (1) to get

$$\sum_{i=1}^{r} a_i (\sigma_1(h) - \sigma_i(h)) \sigma_i(g) = 0. \tag{3}$$

With g arbitrary, reach a contradiction by an appropriate choice of h.

5. If G is the Galois group of $\mathbb{Q}(\sqrt[3]{2})$ over $\mathbb{Q}$, what is the fixed field of G?
6. Find the Galois group of $\mathbb{C}/\mathbb{R}$.
7. Find the fixed field of the Galois group of Problem 6.

6.2 The Fundamental Theorem

With the preliminaries now taken care of, we can proceed directly to the main result.

6.2.1 Fundamental Theorem of Galois Theory

Let E/F be a finite Galois extension with Galois group G. If H is a subgroup of G, let $\mathcal{F}(H)$ be the fixed field of H, and if K is an intermediate field, let $\mathcal{G}(K)$ be $\text{Gal}(E/K)$, the fixing group of K (see (6.1.1)).

(1) $\mathcal{F}$ is a bijective map from subgroups to intermediate fields, with inverse $\mathcal{G}$. Both maps are inclusion-reversing, that is, if $H_1 \le H_2$ then $\mathcal{F}(H_1) \ge \mathcal{F}(H_2)$, and if $K_1 \le K_2$, then $\mathcal{G}(K_1) \ge \mathcal{G}(K_2)$.

(2) Suppose that the intermediate field K corresponds to the subgroup H under the Galois correspondence. Then

   (a) E/K is always normal (hence Galois);

   (b) K/F is normal if and only if H is a normal subgroup of G, and in this case,

   (c) the Galois group of K/F is isomorphic to the quotient group G/H. Moreover, whether or not K/F is normal,

   (d) $[K : F] = [G : H]$ and $[E : K] = |H|$.

(3) If the intermediate field K corresponds to the subgroup H and σ is any automorphism in G, then the field $\sigma K = \{\sigma(x) \colon x \in K\}$ corresponds to the conjugate subgroup $\sigma H \sigma^{-1}$. For this reason, σK is called a *conjugate subfield* of K.

The following diagram may aid the understanding.

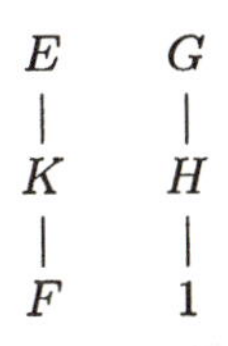

As we travel up the left side from smaller to larger fields, we move down the right side from larger to smaller groups. A statement about K/F, an extension at the bottom of the left side, corresponds to a statement about G/H, located at the top of the right side. Similarly, a statement about E/K corresponds to a statement about $H/1 = H$.

Proof. (1) First, consider the composite mapping $H \to \mathcal{F}(H) \to \mathcal{GF}(H)$. If $\sigma \in H$ then σ fixes $\mathcal{F}(H)$ by definition of fixed field, and therefore $\sigma \in \mathcal{GF}(H) = \text{Gal}(E/\mathcal{F}(H))$. Thus $H \subseteq \mathcal{GF}(H)$. If the inclusion is proper, then by (6.1.2) part (ii) with F replaced by $\mathcal{F}(H)$, we have $\mathcal{F}(H) > \mathcal{F}(H)$, a contradiction. [Note that E/K is a Galois extension for any intermediate field K, by (3.4.7) and (3.5.8).] Thus $\mathcal{GF}(H) = H$.

Now consider the mapping $K \to \mathcal{G}(K) \to \mathcal{FG}(K) = \mathcal{F}\,\mathrm{Gal}(E/K)$. By (6.1.2) part (i) with F replaced by K, we have $\mathcal{FG}(K) = K$. Since both $\mathcal{F}$ and $\mathcal{G}$ are inclusion-reversing by definition, the proof of (1) is complete.

(3) The fixed field of $\sigma H \sigma^{-1}$ is the set of all $x \in E$ such that $\sigma\tau\sigma^{-1}(x) = x$ for every $\tau \in H$. Thus

$$\mathcal{F}(\sigma H \sigma^{-1}) = \{x \in E \colon \sigma^{-1}(x) \in \mathcal{F}(H)\} = \sigma(\mathcal{F}(H)).$$

(2a) This was observed in the proof of (1).

(2b) If σ is an F-monomorphism of K into E, then by (3.5.2) and (3.5.6), σ extends to an F-monomorphism of E into itself, in other words (see (3.5.6)), an F-automorphism of E. Thus each such σ is the restriction to K of a member of G. Conversely, the restriction of an automorphism in G to K is an F-monomorphism of K into E. By (3.5.5) and (3.5.6), K/F is normal iff for every $\sigma \in G$ we have $\sigma(K) = K$. But by (3), $\sigma(K)$ corresponds to $\sigma H \sigma^{-1}$ and K to H. Thus K/F is normal iff $\sigma H \sigma^{-1} = H$ for every $\sigma \in G$, i.e., $H \trianglelefteq G$.

(2c) Consider the homomorphism of $G = \mathrm{Gal}(E/F)$ to $\mathrm{Gal}(K/F)$ given by $\sigma \to \sigma|_K$. The map is surjective by the argument just given in the proof of (2b). The kernel is the set of all automorphisms in G that restrict to the identity on K, that is, $\mathrm{Gal}(E/K) = H$. The result follows from the first isomorphism theorem.

(2d) By (3.1.9), $[E : F] = [E : K][K : F]$. The term on the left is $|G|$ by (3.5.9), and the first term on the right is $|\mathrm{Gal}(E/K)|$ by (2a), and this in turn is $|H|$ since $H = \mathcal{G}(K)$. Thus $|G| = |H|[K : F]$, and the result follows from Lagrange's theorem. [If K/F is normal, the proof is slightly faster. The first statement follows from (2c). To prove the second, note that by (3.1.9) and (3.5.9),

$$[E : K] = \frac{[E : F]}{[K : F]} = \frac{|G|}{|G/H|} = |H|.] \quad \clubsuit$$

The next result is reminiscent of the second isomorphism theorem, and is best visualized via the diamond diagram of Figure 6.2.1. In the diagram, EK is the *composite* of the two fields E and K, that is, the smallest field containing both E and K.

6.2.2 Theorem

Let E/F be a finite Galois extension and K/F an arbitrary extension. Assume that E and K are both contained in a common field, so that it is sensible to consider the composite EK. Then

(1) EK/K is a finite Galois extension;

(2) $\mathrm{Gal}(EK/K)$ is embedded in $\mathrm{Gal}(E/F)$, where the embedding is accomplished by restricting automorphisms in $\mathrm{Gal}(EK/K)$ to E;

(3) The embedding is an isomorphism if and only if $E \cap K = F$.

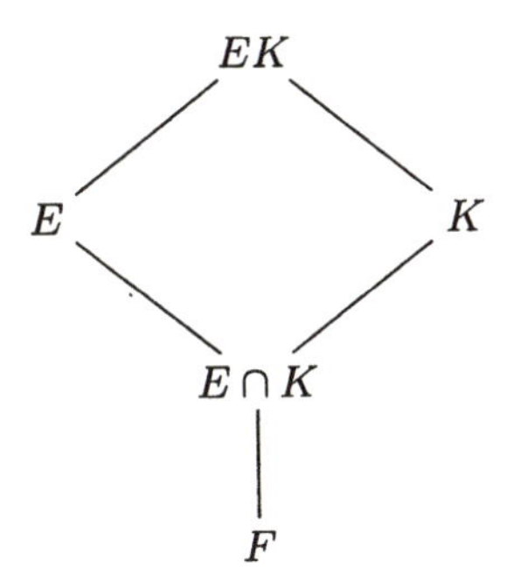

Figure 6.2.1

Proof. (1) By the theorem of the primitive element (3.5.12), we have $E = F[\alpha]$ for some $\alpha \in E$, so $EK = KF[\alpha] = K[\alpha]$. The extension $K[\alpha]/K$ is finite because α is algebraic over F, hence over K. Since α, regarded as an element of EK, is separable over F and hence over K, it follows that EK/K is separable. [To avoid breaking the main line of thought, this result will be developed in the exercises (see Problems 1 and 2).]

Now let f be the minimal polynomial of α over F, and g the minimal polynomial of α over K. Since $f \in K[X]$ and $f(\alpha) = 0$, we have $g \mid f$, and the roots of g must belong to $E \subseteq EK = K[\alpha]$ because E/F is normal. Therefore $K[\alpha]$ is a splitting field for g over K, so by (3.5.7), $K[\alpha]/K$ is normal.

(2) If σ is an automorphism in $\text{Gal}(EK/K)$, restrict σ to E, thus defining a homomorphism from $\text{Gal}(EK/K)$ to $\text{Gal}(E/F)$. (Note that $\sigma|_E$ is an automorphism of E because E/F is normal.) Now σ fixes K, and if σ belongs to the kernel of the homomorphism, then σ also fixes E, so σ fixes $EK = K[\alpha]$. Thus σ is the identity, and the kernel is trivial, proving that the homomorphism is actually an embedding.

(3) The embedding of (2) maps $\text{Gal}(EK/K)$ to a subgroup H of $\text{Gal}(E/F)$, and we will find the fixed field of H. By (6.1.2), the fixed field of $\text{Gal}(EK/K)$ is K, and since the embedding just restricts automorphisms to E, the fixed field of H must be $E \cap K$. By the fundamental theorem, $H = \text{Gal}(E/(E \cap K))$. Thus

$$H = \text{Gal}(E/F) \text{ iff } \text{Gal}(E/(E \cap K)) = \text{Gal}(E/F),$$

and by applying the fixed field operator $\mathcal{F}$, we see that this happens if and only if $E \cap K = F$. ♣

Problems For Section 6.2

1. Let $E = F(\alpha_1, \dots, \alpha_n)$, where each α_i is algebraic and separable over F. We are going to show that E is separable over F. Without loss of generality, we can assume that the characteristic of F is a prime p, and since F/F is separable, the result holds for $n = 0$. To carry out the inductive step, let $E_i = F(\alpha_1, \dots, \alpha_i)$, so that $E_{i+1} = E_i(\alpha_{i+1})$. Show that $E_{i+1} = E_i(E_{i+1}^p)$. (See Section 3.4, Problems 4–8, for the notation.)
2. Continuing Problem 1, show that E is separable over F.

3. Let $E = F(\alpha_1, \dots, \alpha_n)$, where each α_i is algebraic over F. If for each $i = 1, \dots, n$, all the conjugates of α_i (the roots of the minimal polynomial of α_i over F) belong to E, show that E/F is normal.

4. Suppose that $F = K_0 \le K_1 \le \cdots \le K_n = E$, where E/F is a finite Galois extension, and that the intermediate field K_i corresponds to the subgroup H_i under the Galois correspondence. Show that K_i/K_{i-1} is normal (hence Galois) if and only if $H_i \trianglelefteq H_{i-1}$, and in this case, $\mathrm{Gal}(K_i/K_{i-1})$ is isomorphic to H_{i-1}/H_i.

5. Let E and K be extensions of F, and assume that the composite EK is defined. If A is any set of generators for K over F (for example, $A = K$), show that $EK = E(A)$, the field formed from E by adjoining the elements of A.

6. Let E/F be a finite Galois extension with Galois group G, and let E'/F' be a finite Galois extension with Galois group G'. If τ is an isomorphism of E and E' with $\tau(F) = F'$, we expect intuitively that $G \cong G'$. Prove this formally.

7. Let K/F be a finite separable extension. Although K need not be a normal extension of F, we can form the normal closure N of K over F, as in (3.5.11). Then N/F is a Galois extension (see Problem 8 of Section 6.3); let G be its Galois group. Let $H = \mathrm{Gal}(N/K)$, so that the fixed field of H is K. If H' is a normal subgroup of G that is contained in H, show that the fixed field of H' is N.

8. Continuing Problem 7, show that H' is trivial, and conclude that

$$\bigcap_{g \in G} gHg^{-1} = \{1\}$$

where 1 is the identity automorphism.

6.3 Computing a Galois Group Directly

6.3.1 Definitions and Comments

Suppose that E is a splitting field of the separable polynomial f over F. The *Galois group of* f is the Galois group of the extension E/F. (The extension is indeed Galois; see Problem 8.) Given f, how can we determine its Galois group? It is not so easy, but later we will develop a systematic approach for polynomials of degree 4 or less. Some cases can be handled directly, and in this section we look at a typical situation. A useful observation is that the Galois group G of a finite Galois extension E/F acts *transitively* on the roots of any irreducible polynomial $h \in F[X]$ (assuming that one, hence every, root of h belongs to E). [Each $\sigma \in G$ permutes the roots by (3.5.1). If α and β are roots of h, then by (3.2.3) there is an F-isomorphism of $F(\alpha)$ and $F(\beta)$ carrying α to β. This isomorphism can be extended to an F-automorphism of E by (3.5.2), (3.5.5) and (3.5.6).]

6.3.2 Example

Let d be a positive integer that is not a perfect cube, and let θ be the positive cube root of d. Let $\omega = e^{i2\pi/3} = -\frac{1}{2} + i\frac{1}{2}\sqrt{3}$, so that $\omega^2 = e^{-i2\pi/3} = -\frac{1}{2} - i\frac{1}{2}\sqrt{3} = -(1+\omega)$. The minimal polynomial of θ over the rationals $\mathbb{Q}$ is $f(X) = X^3 - d$, because if f were reducible then it would have a linear factor and d would be a perfect cube. The minimal polynomial of ω over $\mathbb{Q}$ is $g(X) = X^2 + X + 1$. (If g were reducible, it would have a rational (hence real) root, so the discriminant would be nonnegative, a contradiction.) We will compute the Galois group G of the polynomial $f(X)g(X)$, which is the Galois group of $E = \mathbb{Q}(\theta, \omega)$ over $\mathbb{Q}$.

If the degree of $E/\mathbb{Q}$ is the product of the degrees of f and g, we will be able to make progress. We have $[\mathbb{Q}(\theta) : \mathbb{Q}] = 3$ and, since ω, a complex number, does not belong to $\mathbb{Q}(\theta)$, we have $[Q(\theta,\omega) : \mathbb{Q}(\theta)] = 2$. Thus $[\mathbb{Q}(\theta,\omega) : \mathbb{Q}] = 6$. But the degree of a finite Galois extension is the size of the Galois group by (3.5.9), so G has exactly 6 automorphisms. Now any $\sigma \in G$ must take θ to one of its conjugates, namely $\theta, \omega\theta$ or $\omega^2\theta$. Moreover, σ must take ω to a conjugate, namely ω or ω^2. Since σ is determined by its action on θ and ω, we have found all 6 members of G. The results can be displayed as follows.

$1\colon \theta \to \theta,\ \omega \to \omega$, order $= 1$
$\tau\colon \theta \to \theta,\ \omega \to \omega^2$, order $= 2$
$\sigma\colon \theta \to \omega\theta,\ \omega \to \omega$, order $= 3$
$\sigma\tau\colon \theta \to \omega\theta,\ \omega \to \omega^2$, order $= 2$
$\sigma^2\colon \theta \to \omega^2\theta,\ \omega \to \omega$, order $= 3$
$\tau\sigma\colon \theta \to \omega^2\theta,\ \omega \to \omega^2$, order $= 2$

Note that $\tau\sigma^2$ gives nothing new since $\tau\sigma^2 = \sigma\tau$. Similarly, $\sigma^2\tau = \tau\sigma$. Thus

$$\sigma^3 = \tau^2 = 1,\ \tau\sigma\tau^{-1} = \sigma^{-1}\ (= \sigma^2). \tag{1}$$

At this point we have determined the multiplication table of G, but much more insight is gained by observing that (1) gives a presentation of S_3 (Section 5.8, Problem 3). We conclude that $G \cong S_3$. The subgroups of G are

$$\{1\},\ G,\ \langle\sigma\rangle,\ \langle\tau\rangle,\ \langle\tau\sigma\rangle,\ \langle\tau\sigma^2\rangle$$

and the corresponding fixed fields are

$$E,\quad \mathbb{Q},\quad \mathbb{Q}(\omega),\quad \mathbb{Q}(\theta),\quad \mathbb{Q}(\omega\theta),\quad \mathbb{Q}(\omega^2\theta).$$

To show that the fixed field of $\langle\tau\sigma\rangle = \{1, \tau\sigma\}$ is $\mathbb{Q}(\omega\theta)$, note that $\langle\tau\sigma\rangle$ has index 3 in G, so by the fundamental theorem, the corresponding fixed field has degree 3 over $\mathbb{Q}$. Now $\tau\sigma$ takes $\omega\theta$ to $\omega^2\omega^2\theta = \omega\theta$ and $[\mathbb{Q}(\omega\theta) : \mathbb{Q}] = 3$ (because the minimal polynomial of $\omega\theta$ over $\mathbb{Q}$ is f). Thus $\mathbb{Q}(\omega\theta)$ is the entire fixed field. The other calculations are similar.

Problems For Section 6.3

1. Suppose that $E = F(\alpha)$ is a finite Galois extension of F, where α is a root of the irreducible polynomial $f \in F[X]$. Assume that the roots of f are $\alpha_1 = \alpha, \alpha_2, \dots, \alpha_n$. Describe, as best you can from the given information, the Galois group of E/F.
2. Let $E/\mathbb{Q}$ be a finite Galois extension, and let $x_1, \dots, x_n$ be a basis for E over $\mathbb{Q}$. Describe how you would find a primitive element, that is, an $\alpha \in E$ such that $E = \mathbb{Q}(\alpha)$. (Your procedure need not be efficient.)

3. Let G be the Galois group of a separable irreducible polynomial f of degree n. Show that G is isomorphic to a transitive subgroup H of S_n. [Transitivity means that if i and j belong to $\{1, 2, \dots, n\}$, then for some $\sigma \in H$ we have $\sigma(i) = j$. Equivalently, the *natural action* of H on $\{1, \dots, n\}$, given by $h \bullet x = h(x)$, is transitive.]
4. Use Problem 3 to determine the Galois group of an irreducible quadratic polynomial $aX^2 + bX + c \in F[X], a \neq 0$. Assume that the characteristic of F is not 2, so that the derivative of f is nonzero and f is separable.
5. Determine the Galois group of $(X^2 - 2)(X^2 - 3)$ over $\mathbb{Q}$.
6. In the Galois correspondence, suppose that K_i is the fixed field of the subgroup H_i, $i = 1, 2$. Identify the group corresponding to $K = K_1 \cap K_2$.
7. Continuing Problem 6, identify the fixed field of $H_1 \cap H_2$.
8. Suppose that E is a splitting field of a separable polynomial f over F. Show that E/F is separable. [Since the extension is finite by (3.2.2) and normal by (3.5.7), E/F is Galois.]
9. Let G be the Galois group of $f(X) = X^4 - 2$ over $\mathbb{Q}$. Thus if θ is the positive fourth root of 2, then G is the Galois group of $\mathbb{Q}(\theta, i)/\mathbb{Q}$. Describe all 8 automorphisms in G.
10. Show that G is isomorphic to the dihedral group D_8.
11. Define $\sigma(\theta) = i\theta$, $\sigma(i) = i$, $\tau(\theta) = \theta$, $\tau(i) = -i$, as in the solution to Problem 10. Find the fixed field of the normal subgroup $N = \{1, \sigma\tau, \sigma^2, \sigma^3\tau\}$ of G, and verify that the fixed field is a normal extension of $\mathbb{Q}$.

6.4 Finite Fields

Finite fields can be classified precisely. We will show that a finite field must have p^n elements, where p is a prime and n is a positive integer. In addition, there is (up to isomorphism) only one finite field with p^n elements. We sometimes use the notation $GF(p^n)$ for this field; GF stands for "Galois field". Also, the field with p elements will be denoted by $\mathbb{F}_p$ rather than $\mathbb{Z}_p$, to emphasize that we are working with fields.

6.4.1 Proposition

Let E be a finite field of characteristic p. Then $|E| = p^n$ for some positive integer n. Moreover, E is a splitting field for the separable polynomial $f(X) = X^{p^n} - X$ over $\mathbb{F}_p$, so that any finite field with p^n elements is isomorphic to E. Not only is E generated by the roots of f, but in fact E coincides with the set of roots of f.

Proof. Since E contains a copy of $\mathbb{F}_p$ (see (2.1.3), Example 2), we may view E as a vector space over $\mathbb{F}_p$. If the dimension of this vector space is n, then since each coefficient in a linear combination of basis vectors can be chosen in p ways, we have $|E| = p^n$.

Now let E^* be the multiplicative group of nonzero elements of E. If $\alpha \in E^*$, then $\alpha^{p^n - 1} = 1$ by Lagrange's theorem, so $\alpha^{p^n} = \alpha$ for every $\alpha \in E$, including $\alpha = 0$. Thus each element of E is a root of f, and f is separable by (3.4.5). Now f has at most p^n distinct roots, and as we have already identified the p^n elements of E as roots of f, in fact f has p^n distinct roots and every root of f must belong to E. ♣

6.4.2 Corollary

If E is a finite field of characteristic p, then $E/\mathbb{F}_p$ is a Galois extension. The Galois group is cyclic and is generated by the Frobenius automorphism $\sigma(x) = x^p$, $x \in E$.

Proof. E is a splitting field for a separable polynomial over $\mathbb{F}_p$, so $E/\mathbb{F}_p$ is Galois; see (6.3.1). Since $x^p = x$ for each $x \in \mathbb{F}_p$, $\mathbb{F}_p$ is contained in the fixed field $\mathcal{F}(\langle\sigma\rangle)$. But each element of the fixed field is a root of $X^p - X$, so $\mathcal{F}(\langle\sigma\rangle)$ has at most p elements. Consequently, $\mathcal{F}(\langle\sigma\rangle) = \mathbb{F}_p$. Now $\mathbb{F}_p = \mathcal{F}(\text{Gal}(E/\mathbb{F}_p))$ by (6.1.2), so by the fundamental theorem, $\text{Gal}(E/\mathbb{F}_p) = \langle\sigma\rangle$. ♣

6.4.3 Corollary

Let E/F be a finite extension of a finite field, with $|E| = p^n$, $|F| = p^m$. Then E/F is a Galois extension. Moreover, m divides n, and $\text{Gal}(E/F)$ is cyclic and is generated by the automorphism $\tau(x) = x^{p^m}$, $x \in E$. Furthermore, F is the only subfield of E of size p^m.

Proof. If the degree of E/F is d, then as in (6.4.1), $(p^m)^d = p^n$, so $d = n/m$ and $m \mid n$. We may then reproduce the proof of (6.4.2) with $\mathbb{F}_p$ replaced by F, σ by τ, x^p by x^{p^m}, and X^p by X^{p^m}. Uniqueness of F as a subfield of E with p^m elements follows because there is only one splitting field over $\mathbb{F}_p$ for $X^{p^m} - X$ inside E; see (3.2.1). ♣

How do we know that finite fields (other than the $\mathbb{F}_p$) exist? There is no problem. Given any prime p and positive integer n, we can construct $E = GF(p^n)$ as a splitting field for $X^{p^n} - X$ over $\mathbb{F}_p$. We have just seen that if E contains a subfield F of size p^m, then m is a divisor of n. The converse is also true, as a consequence of the following basic result.

6.4.4 Theorem

The multiplicative group of a finite field is cyclic. More generally, if G is a finite subgroup of the multiplicative group of an arbitrary field, then G is cyclic.

Proof. G is a finite abelian group, hence contains an element g whose order r is the *exponent* of G, that is, the least common multiple of the orders of all elements of G; see Section 1.1, Problem 9. Thus if $x \in G$ then the order of x divides r, so $x^r = 1$. Therefore each element of G is a root of $X^r - 1$, so $|G| \le r$. But $|G|$ is a multiple of the order of every element, so $|G|$ is at least as big as the least common multiple, so $|G| \ge r$. We conclude that the order and the exponent are the same. But then g has order $|G|$, so $G = \langle g\rangle$ and G is cyclic. ♣

6.4.5 Proposition

$GF(p^m)$ is a subfield of $E = GF(p^n)$ if and only if m is a divisor of n.

Proof. The "only if" part follows from (6.4.3), so assume that m divides n. If t is any positive integer greater than 1, then $m \mid n$ iff $(t^m - 1) \mid (t^n - 1)$. (A formal proof is not difficult, but I prefer to do an ordinary long division of $t^n - 1$ by $t^m - 1$. The successive quotients are $t^{n-m}, t^{n-2m}, t^{n-3m}, \dots$, so the division will be successful iff $n - rm = 0$ for some positive integer r.) Taking $t = p$, we see that $p^m - 1$ divides $|E^*|$, so by (6.4.4) and (1.1.4), E^* has a subgroup H of order $p^m - 1$. By Lagrange's theorem, each $x \in H \cup \{0\}$ satisfies $x^{p^m} = x$. As in the proof of (6.4.1), $H \cup \{0\}$ coincides with the set of roots of $X^{p^m} - X$. Thus we may construct entirely inside $GF(p^n)$ a splitting field for $X^{p^m} - X$ over $\mathbb{F}_p$. But this splitting field is a copy of $GF(p^m)$. ♣

In practice, finite fields are constructed by adjoining roots of carefully selected irreducible polynomials over $\mathbb{F}_p$. The following result is very helpful.

6.4.6 Theorem

Let p be a prime and n a positive integer. Then $X^{p^n} - X$ is the product of all monic irreducible polynomials over $\mathbb{F}_p$ whose degree divides n.

Proof. Let us do all calculations inside $E = GF(p^n) =$ the set of roots of $f(X) = X^{p^n} - X$. If $g(X)$ is any monic irreducible factor of $f(X)$, and deg $g = m$, then all roots of g lie in E. If α is any root of g, then $\mathbb{F}_p(\alpha)$ is a finite field with p^m elements, so m divides n by (6.4.5) or (6.4.3). Conversely, let $g(X)$ be a monic irreducible polynomial over $\mathbb{F}_p$ whose degree m is a divisor of n. Then by (6.4.5), E contains a subfield with p^m elements, and this subfield must be isomorphic to $\mathbb{F}_p(\alpha)$. If $\beta \in E$ corresponds to α under this isomorphism, then $g(\beta) = 0$ (because $g(\alpha) = 0$) and $f(\beta) = 0$ (because $\beta \in E$). Since g is the minimal polynomial of β over $\mathbb{F}_p$, it follows that $g(X)$ divides $f(X)$. By (6.4.1), the roots of f are distinct, so no irreducible factor can appear more than once. The theorem is proved. ♣

6.4.7 The Explicit Construction of a Finite Field

By (6.4.4), the multiplicative group E^* of a finite field $E = GF(p^n)$ is cyclic, so E^* can be generated by a single element α. Thus $E = \mathbb{F}_p(\alpha) = \mathbb{F}_p[\alpha]$, so that α is a primitive element of E. The minimal polynomial of α over $\mathbb{F}_p$ is called a *primitive polynomial.* The key point is that the nonzero elements of E are not simply the nonzero polynomials of degree at most $n-1$ in α, they are the *powers of* α. This is significant in applications to coding theory. Let's do an example over $\mathbb{F}_2$.

The polynomial $g(X) = X^4 + X + 1$ is irreducible over $\mathbb{F}_2$. One way to verify this is to factor $X^{16} - X = X^{16} + X$ over $\mathbb{F}_2$; the factors are the (necessarily monic) irreducible polynomials of degrees 1,2 and 4. To show that g is primitive, we compute powers of α:

$\alpha^0 = 1$, $\alpha^1 = \alpha$, $\alpha^2 = \alpha^2$, $\alpha^3 = \alpha^3$, $\alpha^4 = 1 + \alpha$ (since $g(\alpha) = 0$),

$\alpha^5 = \alpha + \alpha^2$, $\alpha^6 = \alpha^2 + \alpha^3$, $\alpha^7 = \alpha^3 + \alpha^4 = 1 + \alpha + \alpha^3$, $\alpha^8 = \alpha + \alpha^2 + \alpha^4 = 1 + \alpha^2$ (since 1+1=0 in $\mathbb{F}_2$),

$\alpha^9 = \alpha + \alpha^3$, $\alpha^{10} = 1 + \alpha + \alpha^2$, $\alpha^{11} = \alpha + \alpha^2 + \alpha^3$, $\alpha^{12} = 1 + \alpha + \alpha^2 + \alpha^3$, $\alpha^{13} = 1 + \alpha^2 + \alpha^3$, $\alpha^{14} = 1 + \alpha^3$,

and at this point we have all $2^4 - 1 = 15$ nonzero elements of $GF(16)$. The pattern now repeats, beginning with $\alpha^{15} = \alpha + \alpha^4 = 1$.

For an example of a non-primitive polynomial, see Problem 1.

Problems For Section 6.4

1. Verify that the irreducible polynomial $X^4+X^3+X^2+X+1 \in \mathbb{F}_2[X]$ is not primitive.
2. Let F be a finite field and d a positive integer. Show that there exists an irreducible polynomial of degree d in $F[X]$.
3. In (6.4.5) we showed that $m \mid n$ iff $(t^m - 1) \mid (t^n - 1)$ $(t = 2, 3, \dots)$. Show that an equivalent condition is $(X^m - 1)$ divides $(X^n - 1)$.

   If E is a finite extension of a finite field, or more generally a finite separable extension of a field F, then by the theorem of the primitive element, $E = F(\alpha)$ for some $\alpha \in E$. We now develop a condition equivalent to the existence of a primitive element.
4. Let E/F be a finite extension, with $E = F(\alpha)$ and $F \leq L \leq E$. Suppose that the minimal polynomial of α over L is $g(X) = \sum_{i=0}^{r-1} b_i X^i + X^r$, and let $K = F(b_0, \dots, b_{r-1})$. If h is the minimal polynomial of α over K, show that $g = h$, and conclude that $L = K$.
5. Continuing Problem 4, show that there are only finitely many intermediate fields L between E and F.
6. Conversely, let $E = F(\alpha_1, \dots, \alpha_n)$ be a finite extension with only finitely many intermediate fields between E and F. We are going to show by induction that E/F has a primitive element. If $n = 1$ there is nothing to prove, so assume the result holds for all integers less than n. If $L = F(\alpha_1, \dots, \alpha_{n-1})$, show that $E = F(\beta, \alpha_n)$ for some $\beta \in L$.
7. Now assume (without loss of generality) that F is infinite. Show that there are distinct elements $c, d \in F$ such that $F(c\beta + \alpha_n) = F(d\beta + \alpha_n)$.
8. Continuing Problem 7, show that $E = F(c\beta + \alpha_n)$. Thus a finite extension has a primitive element iff there are only finitely many intermediate fields.
9. Let α be an element of the finite field $GF(p^n)$. Show that α and α^p have the same minimal polynomial over F_p.
10. Suppose that α is an element of order 13 in the multiplicative group of nonzero elements in $GF(3^n)$. Partition the integers $\{0, 1, \dots, 12\}$ into disjoint subsets such that if i and j belong to the same subset, then α^i and α^j have the same minimal polynomial. Repeat for α an element of order 15 in $GF(2^n)$. [Note that elements of the specified orders exist, because 13 divides $26 = 3^3 - 1$ and $15 = 2^4 - 1$.]

6.5 Cyclotomic Fields

6.5.1 Definitions and Comments

Cyclotomic extensions of a field F are formed by adjoining n^{th} roots of unity. Formally, a *cyclotomic extension* of F is a splitting field E for $f(X) = X^n - 1$ over F. The roots of f are called *n^{th} roots of unity*, and they form a multiplicative subgroup of the group E^* of nonzero elements of E. This subgroup must be cyclic by (6.4.4). A *primitive n^{th} root of unity* is one whose order in E^* is n.

It is tempting to say "obviously, primitive n^{th} roots of unity must exist, just take a generator of the cyclic subgroup". But suppose that F has characteristic p and p divides n, say $n = mp$. If ω is an nth root of unity, then

$$0 = \omega^n - 1 = (\omega^m - 1)^p$$

so the order of ω must be less than n. To avoid this difficulty, we assume that the characteristic of F does not divide n. Then $f'(X) = nX^{n-1} \neq 0$, so the greatest common divisor of f and f' is constant. By (3.4.2), f is separable, and consequently E/F is Galois. Since there are n distinct n^{th} roots of unity, there must be a primitive n^{th} root of unity ω, and for any such ω, we have $E = F(\omega)$.

If σ is any automorphism in the Galois group $\text{Gal}(E/F)$, then σ must take a primitive root of unity ω to another primitive root of unity ω^r, where r and n are relatively prime. (See (1.1.5).) We can identify σ with r, and this shows that $\text{Gal}(E/F)$ is isomorphic to a subgroup of U_n, the group of units mod n. Consequently, the Galois group is abelian.

Finally, by the fundamental theorem (or (3.5.9)), $[E : F] = |\text{Gal}(E/F)|$, which is a divisor of $|U_n| = \varphi(n)$.

Cyclotomic fields are of greatest interest when the underlying field F is $\mathbb{Q}$, the rational numbers, and from now on we specialize to that case. The primitive n^{th} roots of unity are $e^{i2\pi r/n}$ where r and n are relatively prime. Thus there are $\varphi(n)$ primitive n^{th} roots of unity. Finding the minimal polynomial of a primitive n^{th} root of unity requires some rather formidable equipment.

6.5.2 Definition

The nth *cyclotomic polynomial* is defined by

$$\Psi_n(X) = \prod_i (X - \omega_i)$$

where the ω_i are the primitive n^{th} roots of unity in the field $\mathbb{C}$ of complex numbers. Thus the degree of $\Psi_n(X)$ is $\varphi(n)$.

From the definition, we have $\Psi_1(X) = X - 1$ and $\Psi_2(X) = X + 1$. In general, the cyclotomic polynomials can be calculated by the following recursion formula, in which d runs through all positive divisors of n.

6.5.3 Proposition

$$X^n - 1 = \prod_{d|n} \Psi_d(X).$$

In particular, if p is prime, then

$$\Psi_p(X) = \frac{X^p - 1}{X - 1} = X^{p-1} + X^{p-2} + \cdots + X + 1.$$

Proof. If ω is an nth root of unity, then its order in $\mathbb{C}^*$ is a divisor d of n, and in this case, ω is a primitive d^{th} root of unity, hence a root of $\Psi_d(X)$. Conversely, if $d \mid n$, then any root of $\Psi_d(X)$ is a dth, hence an nth, root of unity. ♣

¿From (6.5.3) we have
$\Psi_3(X) = X^2 + X + 1$,
$\Psi_4(X) = X^2 + 1$, $\Psi_5(X) = X^4 + X^3 + X^2 + X + 1$,
$\Psi_6(X) = \frac{X^6-1}{(X-1)(X+1)(X^2+X+1)} = \frac{X^6-1}{(X^3-1)(X+1)} = \frac{X^3+1}{X+1} = X^2 - X + 1$.
It is a natural conjecture that all coefficients of the cyclotomic polynomials are integers, and this turns out to be correct.

6.5.4 Proposition

$\Psi_n(X) \in \mathbb{Z}[X]$.

Proof. By (6.5.3), we have

$$X^n - 1 = [\prod_{d|n, d<n} \Psi_d(X)]\Psi_n(X).$$

By definition, the cyclotomic polynomials are monic, and by induction hypothesis, the expression in brackets is a monic polynomial in $\mathbb{Z}[X]$. Thus $\Psi_n(X)$ is the quotient of two monic polynomials with integer coefficients. At this point, all we know for sure is that the coefficients of $\Psi_n(X)$ are complex numbers. But if we apply ordinary long division, even in $\mathbb{C}$, we know that the process will terminate, and this forces the quotient $\Psi_n(X)$ to be in $\mathbb{Z}[X]$. ♣

We now show that the nth cyclotomic polynomial is the minimal polynomial of each primitive n^{th} root of unity.

6.5.5 Theorem

$\Psi_n(X)$ is irreducible over $\mathbb{Q}$.

Proof. Let ω be a primitive n^{th} root of unity, with minimal polynomial f over $\mathbb{Q}$. Since ω is a root of $X^n - 1$, we have $X^n - 1 = f(X)g(X)$ for some $g \in \mathbb{Q}[X]$. Now it follows from (2.9.2) that if a monic polynomial over $\mathbb{Z}$ is the product of two monic polynomials f and g over $\mathbb{Q}$, then in fact the coefficients of f and g are integers.

If p is a prime that does not divide n, we will show that ω^p is a root of f. If not, then it is a root of g. But $g(\omega^p) = 0$ implies that ω is a root of $g(X^p)$, so $f(X)$ divides $g(X^p)$, say $g(X^p) = f(X)h(X)$. As above, $h \in \mathbb{Z}[X]$. But by the binomial expansion modulo p, $g(X)^p \equiv g(X^p) = f(X)h(X) \bmod p$. Reducing the coefficients of a polynomial $k(X)$ mod p is equivalent to viewing it as an element $\overline{k} \in \mathbb{F}_p[X]$, so we may write $\overline{g}(X)^p = \overline{f}(X)\overline{h}(X)$. Then any irreducible factor of $\overline{f}$ must divide $\overline{g}$, so $\overline{f}$ and $\overline{g}$ have a common factor. But then $X^n - \overline{1}$ has a multiple root, contradicting (3.4.2). [This is where we use the fact that p does not divide n.]

Now we claim that every primitive n^{th} root of unity is a root of f, so that deg $f \geq \varphi(n) =$deg Ψ_n, and therefore $f = \Psi_n$ by minimality of f. The best way to visualize this is via a concrete example with all the features of the general case. If ω is a primitive n^{th} root of unity where $n = 175$, then ω^{72} is a primitive n^{th} root of unity because 72 and 175 are relatively prime. Moreover, since $72 = 2^3 \times 3^2$, we have

$$\omega^{72} = (((((\omega)^2)^2)^2)^3)^3$$

and the result follows. ♣

6.5.6 Corollary

The Galois group G of the nth cyclotomic extension $\mathbb{Q}(\omega)/\mathbb{Q}$ is isomorphic to the group U_n of units mod n.

Proof. By the fundamental theorem, $|G| = [\mathbb{Q}(\omega) : \mathbb{Q}] = \deg \Psi_n = \varphi(n) = |U_n|$. Thus the monomorphism of G and a subgroup of U_n (see (6.5.1)) is surjective. ♣

Problems For Section 6.5

1. If p is prime and p divides n, show that $\Psi_{pn}(X) = \Psi_n(X^p)$. (This formula is sometimes useful in computing the cyclotomic polynomials.)
2. Show that the group of automorphisms of a cyclic group of order n is isomorphic to the group U_n of units mod n. (This can be done directly, but it is easier to make use of the results of this section.)

   We now do a detailed analysis of subgroups and intermediate fields associated with the cyclotomic extension $\mathbb{Q}_7 = \mathbb{Q}(\omega)/\mathbb{Q}$ where $\omega = e^{i2\pi/7}$ is a primitive 7th root of unity. The Galois group G consists of automorphisms σ_i, $i = 1, 2, 3, 4, 5, 6$, where $\sigma_i(\omega) = \omega^i$.
3. Show that σ_3 generates the cyclic group G.
4. Show that the subgroups of G are $\langle 1 \rangle$ (order 1), $\langle \sigma_6 \rangle$ (order 2), $\langle \sigma_2 \rangle$ (order 3), and $G = \langle \sigma_3 \rangle$ (order 6).
5. The fixed field of $\langle 1 \rangle$ is $\mathbb{Q}_7$ and the fixed field of G is $\mathbb{Q}$. Let K be the fixed field of $\langle \sigma_6 \rangle$. Show that $\omega + \omega^{-1} \in K$, and deduce that $K = \mathbb{Q}(\omega + \omega^{-1}) = \mathbb{Q}(\cos 2\pi/7)$.
6. Let L be the fixed field of $\langle \sigma_2 \rangle$. Show that $\omega + \omega^2 + \omega^4$ belongs to L but not to $\mathbb{Q}$.
7. Show that $L = \mathbb{Q}(\omega + \omega^2 + \omega^4)$.
8. If $q = p^r$, p prime, $r > 0$, show that

$$\Psi_q(X) = t^{p-1} + t^{p-2} + \cdots + 1$$

   where $t = X^{p^{r-1}}$.
9. Assuming that the first 6 cyclotomic polynomials are available [see after (6.5.3)], calculate $\Psi_{18}(X)$ in an effortless manner.

6.6 The Galois Group of a Cubic

Let f be a polynomial over F, with distinct roots $x_1, \ldots, x_n$ in a splitting field E over F. The Galois group G of f permutes the x_i, but which permutations belong to G? When f is a quadratic, the analysis is straightforward, and is considered in Section 6.3, Problem 4. In this section we look at cubics (and some other manageable cases), and the appendix to Chapter 6 deals with the quartic.

6.6.1 Definitions and Comments

Let f be a polynomial with roots $x_1, \ldots, x_n$ in a splitting field. Define

$$\Delta(f) = \prod_{i<j}(x_i - x_j).$$

The *discriminant* of f is defined by

$$D(f) = \Delta^2 = \prod_{i<j}(x_i - x_j)^2.$$

Let's look at a quadratic polynomial $f(X) = X^2 + bX + c$, with roots $\frac{1}{2}(-b \pm \sqrt{b^2 - 4c})$. In order to divide by 2, we had better assume that the characteristic of F is not 2, and this assumption is usually made before defining the discriminant. In this case we have $(x_1 - x_2)^2 = b^2 - 4c$, a familiar formula. Here are some basic properties of the discriminant.

6.6.2 Proposition

Let E be a splitting field of the separable polynomial f over F, so that E/F is Galois.

(a) $D(f)$ belongs to the base field F.

(b) Let σ be an automorphism in the Galois group G of f. Then σ is an even permutation (of the roots of f) iff $\sigma(\Delta) = \Delta$, and σ is odd iff $\sigma(\Delta) = -\Delta$.

(c) $G \subseteq A_n$, that is, G consists entirely of even permutations, iff $D(f)$ is the square of an element of F (for short, $D \in F^2$).

Proof. Let us examine the effect of a transposition $\sigma = (i, j)$ on Δ. Once again it is useful to consider a concrete example with all the features of the general case. Say $n = 15, i = 7, j = 10$. Then

$$x_3 - x_7 \to x_3 - x_{10},\ x_3 - x_{10} \to x_3 - x_7$$

$$x_{10} - x_{12} \to x_7 - x_{12},\ x_7 - x_{12} \to x_{10} - x_{12}$$

$$x_7 - x_8 \to x_{10} - x_8,\ x_8 - x_{10} \to x_8 - x_7$$

$$x_7 - x_{10} \to x_{10} - x_7.$$

The point of the computation is that the net effect of (i, j) on Δ is to take $x_i - x_j$ to its negative. Thus $\sigma(\Delta) = -\Delta$ when σ is a transposition. Thus if σ is any permutation, we have $\sigma(\Delta) = \Delta$ if Δ is even, and $\sigma(\Delta) = -\Delta$ if σ is odd. Consequently, $\sigma(\Delta^2) =$

$(\sigma(\Delta))^2 = \Delta^2$, so D belongs to the fixed field of G, which is F. This proves (a), and (b) follows because $\Delta \neq -\Delta$ (remember that the characteristic of F is not 2). Finally $G \subseteq A_n$ iff $\sigma(\Delta) = \Delta$ for every $\sigma \in G$ iff $\Delta \in \mathcal{F}(G) = F$. ♣

6.6.3 The Galois Group of a Cubic

In the appendix to Chapter 6, it is shown that the discriminant of the abbreviated cubic $X^3 + pX + q$ is $-4p^3 - 27q^2$, and the discriminant of the general cubic $X^3 + aX^2 + bX + c$ is

$$a^2(b^2 - 4ac) - 4b^3 - 27c^2 + 18abc.$$

Alternatively, the change of variable $Y = X + \frac{a}{3}$ eliminates the quadratic term without changing the discriminant.

We now assume that the cubic polynomial f is irreducible as well as separable. Then the Galois group G is isomorphic to a transitive subgroup of S_3 (see Section 6.3, Problem 3). By direct enumeration, G must be A_3 or S_3, and by (6.6.2(c)), $G = A_3$ iff the discriminant D is a square in F.

If $G = A_3$, which is cyclic of order 3, there are no proper subgroups except $\{1\}$, so there are no intermediate fields strictly between E and F. However, if $G = S_3$, then the proper subgroups are

$$\{1, (2,3)\}, \ \{1, (1,3)\}, \ \{1, (1,2)\}, \ A_3 = \{1, (1,2,3), (1,3,2)\}.$$

If the roots of f are α_1, α_2 and α_3, then the corresponding fixed fields are

$$F(\alpha_1), \ F(\alpha_2), \ F(\alpha_3), \ F(\Delta)$$

where A_3 corresponds to $F(\Delta)$ because only even permutations fix Δ.

6.6.4 Example

Let $f(X) = X^3 - 31X + 62$ over $\mathbb{Q}$. An application of the rational root test (Section 2.9, Problem 1) shows that f is irreducible. The discriminant is $-4(-31)^3 - 27(62)^2 = 119164 - 103788 = 15376 = (124)^2$, which is a square in $\mathbb{Q}$. Thus the Galois group of f is A_3.

We now develop a result that can be applied to certain cubics, but which has wider applicability as well. The preliminary steps are also of interest.

6.6.5 Some Generating Sets of S_n

(i) S_n is generated by the transpositions $(1,2), (1,3), \dots, (1,n)$.
[An arbitrary transposition (i,j) can be written as $(1,i)(1,j)(1,i)$.]

(ii) S_n is generated by transpositions of adjacent digits, i.e., $(1,2), (2,3), \dots, (n-1,n)$.
[Since $(1, j-1)(j-1, j)(1, j-1) = (1, j)$, we have

$$(1,2)(2,3)(1,2) = (1,3), \quad (1,3)(3,4)(1,3) = (1,4), \text{ etc.},$$

and the result follows from (i).]

(iii) S_n is generated by the two permutations $\sigma_1 = (1,2)$ and $\tau = (1,2,\dots,n)$.
[If $\sigma_2 = \tau\sigma_1\tau^{-1}$, then σ_2 is obtained by applying τ to the symbols of σ_1 (see Section 5.2, Problem 1). Thus $\sigma_2 = (2,3)$. Similarly,

$$\sigma_3 = \tau\sigma_2\tau^{-1} = (3,4), \dots, \sigma_{n-1} = \tau\sigma_{n-2}\tau^{-1} = (n-1,n),$$

and the result follows from (ii).]

(iv) S_n is generated by $(1,2)$ and $(2,3,\dots,n)$.
[$(1,2)(2,3,\dots,n) = (1,2,3,\dots,n)$, and (iii) applies.]

6.6.6 Lemma

If f is an irreducible separable polynomial over F of degree n, and G is the Galois group of f, then n divides $|G|$. If n is a prime number p, then G contains a p-cycle.

Proof. If α is any root of f, then $[F(\alpha) : F] = n$, so by the fundamental theorem, G contains a subgroup whose index is n. By Lagrange's theorem, n divides $|G|$. If $n = p$, then by Cauchy's theorem, G contains an element σ of order p. We can express σ as a product of disjoint cycles, and the length of each cycle must divide the order of σ. Since p is prime, σ must consist of disjoint p-cycles. But a single p-cycle already uses up all the symbols to be permuted, so σ is a p-cycle. ♣

6.6.7 Proposition

If f is irreducible over $\mathbb{Q}$ and of prime degree p, and f has exactly two nonreal roots in the complex field $\mathbb{C}$, then the Galois group G of f is S_p.

Proof. By (6.6.6), G contains a p-cycle σ. Now one of the elements of G must be complex conjugation τ, which is an automorphism of $\mathbb{C}$ that fixes $\mathbb{R}$ (hence $\mathbb{Q}$). Thus τ permutes the two nonreal roots and leaves the $p-2$ real roots fixed, so τ is a transposition. Since p is prime, σ^k is a p-cycle for $k = 1, \dots, p-1$. It follows that by renumbering symbols if necessary, we can assume that $(1,2)$ and $(1,2,\dots,p)$ belong to G. By (6.6.5) part (iii), $G = S_p$. ♣

Problems For Section 6.6

In Problems 1–4, all polynomials are over the rational field $\mathbb{Q}$, and in each case, you are asked to find the Galois group G.

1. $f(X) = X^3 - 2$ (do it two ways)
2. $f(X) = X^3 - 3X + 1$
3. $f(X) = X^5 - 10X^4 + 2$
4. $f(X) = X^3 + 3X^2 - 2X + 1$ (calculate the discriminant in two ways)
5. If f is a separable cubic, not necessarily irreducible, then there are other possibilities for the Galois group G of f besides S_3 and A_3. What are they?

6. Let f be an irreducible cubic over $\mathbb{Q}$ with exactly one real root. Show that $D(f) < 0$, and conclude that the Galois group of f is S_3.

7. Let f be an irreducible cubic over $\mathbb{Q}$ with 3 distinct real roots. Show that $D(f) > 0$, so that the Galois group is A_3 or S_3 according as $\sqrt{D} \in \mathbb{Q}$ or $\sqrt{D} \notin \mathbb{Q}$

6.7 Cyclic and Kummer Extensions

The problem of solving a polynomial equation by radicals is thousands of years old, but it can be given a modern flavor. We are looking for roots of $f \in F[X]$, and we are only allowed to use algorithms that do ordinary arithmetic plus the extraction of nth roots. The idea is to identify those polynomials whose roots can be found in this way. Now if $a \in F$ and our algorithm computes $\theta = \sqrt[n]{a}$ in some extension field of F, then θ is a root of $X^n - a$, so it is natural to study splitting fields of $X^n - a$.

6.7.1 Assumptions, Comments and a Definition

Assume

(i) E is a splitting field for $f(X) = X^n - a$ over F, where $a \neq 0$.

(ii) F contains a primitive n^{th} root of unity ω.

These are natural assumption if we want to allow the computation of nth roots. If θ is any root of f in E, then the roots of f are $\theta, \omega\theta, \dots, \omega^{n-1}\theta$. (The roots must be distinct because a, hence θ, is nonzero.) Therefore $E = F(\theta)$. Since f is separable, the extension E/F is Galois (see (6.3.1)). If $G = \text{Gal}(E/F)$, then $|G| = [E : F]$ by the fundamental theorem (or by (3.5.9)).

In general, a *cyclic extension* is a Galois extension whose Galois group is cyclic.

6.7.2 Theorem

Under the assumptions of (6.7.1), E/F is a cyclic extension and the order of the Galois group G is a divisor of n. We have $|G| = n$ if and only if $f(X)$ is irreducible over F.

Proof. Let $\sigma \in G$; since σ permutes the roots of f by (3.5.1), we have $\sigma(\theta) = \omega^{u(\sigma)}\theta$. [Note that σ fixes ω by (ii).] We identify integers $u(\sigma)$ with the same residue mod n. If $\sigma_i(\theta) = \omega^{u(\sigma_i)}\theta$, $i = 1, 2$, then

$$\sigma_1(\sigma_2(\theta)) = \omega^{u(\sigma_1)+u(\sigma_2)}\theta,$$

so

$$u(\sigma_1\sigma_2) = u(\sigma_1) + u(\sigma_2)$$

and u is a group homomorphism from G to $\mathbb{Z}_n$. If $u(\sigma)$ is 0 mod n, then $\sigma(\theta) = \theta$, so σ is the identity and the homomorphism is injective. Thus G is isomorphic to a subgroup of $\mathbb{Z}_n$, so G is cyclic and $|G|$ divides n.

If f is irreducible over F, then $|G| = [E : F] = [F(\theta) : F] = \deg f = n$. If f is not irreducible over F, let g be a proper irreducible factor. If β is a root of g in E, then β is also a root of f, so $E = F(\beta)$ and $|G| = [E : F] = [F(\beta) : F] = \deg g < n$. ♣

Thus splitting fields of $X^n - a$ give rise to cyclic extensions. Conversely, we can prove that a cyclic extension comes from such a splitting field.

6.7.3 Theorem

Let E/F be a cyclic extension of degree n, where F contains a primitive n^{th} root of unity ω. Then for some nonzero $a \in F$, $f(X) = X^n - a$ is irreducible over F and E is a splitting field for f over F.

Proof. Let σ be a generator of the Galois group of the extension. By Dedekind's lemma (6.1.6), the distinct automorphisms $1, \sigma, \sigma^2, \dots, \sigma^{n-1}$ are linearly independent over E. Thus $1 + \omega\sigma + \omega^2\sigma^2 + \cdots + \omega^{n-1}\sigma^{n-1}$ is not identically 0, so for some $\beta \in E$ we have

$$\theta = \beta + \omega\sigma(\beta) + \cdots + \omega^{n-1}\sigma^{n-1}(\beta) \neq 0.$$

Now

$$\sigma(\theta) = \sigma(\beta) + \omega\sigma^2(\beta) + \cdots + \omega^{n-2}\sigma^{n-1}(\beta) + \omega^{n-1}\sigma^n(\beta) = \omega^{-1}\theta$$

since $\sigma^n(\beta) = \beta$. We take $a = \theta^n$. To prove that $a \in F$, note that

$$\sigma(\theta^n) = (\sigma(\theta))^n = (\omega^{-1}\theta)^n = \theta^n$$

and therefore σ fixes θ^n. Since σ generates G, all other members of G fix θ^n, hence a belongs to the fixed field of $\mathrm{Gal}(E/F)$, which is F.

Now by definition of a, θ is a root of $f(X) = X^n - a$, so the roots of $X^n - a$ are $\theta, \omega\theta, \dots, \omega^{n-1}\theta$. Therefore $F(\theta)$ is a splitting field for f over F. Since $\sigma(\theta) = \omega^{-1}\theta$, the distinct automorphisms $1, \sigma, \dots, \sigma^{n-1}$ can be restricted to distinct automorphisms of $F(\theta)$. Consequently,

$$n \leq |\,\mathrm{Gal}(F(\theta)/F)| = [F(\theta) : F] \leq \deg f = n$$

so $[F(\theta) : F] = n$. It follows that $E = F(\theta)$ and (since f must be the minimal polynomial of θ over F) f is irreducible over F. ♣

A finite abelian group is a direct product of cyclic groups (or direct sum, in additive notation; see (4.6.4)). It is reasonable to expect that our analysis of cyclic Galois groups will help us to understand abelian Galois groups.

6.7.4 Definition

A *Kummer extension* is a finite Galois extension with an abelian Galois group.

6.7.5 Theorem

Let E/F be a finite extension, and assume that F contains a primitive n^{th} root of unity ω. Then E/F is a Kummer extension whose Galois group G has an exponent dividing n if and only if there are nonzero elements $a_1, \dots, a_r \in F$ such that E is a splitting field of $(X^n - a_1) \cdots (X^n - a_r)$ over F. [For short, $E = F(\sqrt[n]{a_1}, \dots, \sqrt[n]{a_r})$.]

Proof. We do the "if" part first. As in (6.7.1), we have $E = F(\theta_1, \dots, \theta_r)$ where θ_i is a root of $X^n - a_i$. If $\sigma \in \mathrm{Gal}(E/F)$, then σ maps θ_i to another root of $X^n - a_i$, so

$$\sigma(\theta_i) = \omega^{u_i(\sigma)}\theta_i.$$

Thus if σ and τ are any two automorphisms in the Galois group G, then $\sigma\tau = \tau\sigma$ and G is abelian. [The u_i are integers, so $u_i(\sigma) + u_i(\tau) = u_i(\tau) + u_i(\sigma)$.] Now restrict attention to the extension $F(\theta_i)$. By (6.7.2), the Galois group of $F(\theta_i)/F$ has order dividing n, so $\sigma^n(\theta_i) = \theta_i$ for all $i = 1, \dots, r$. Thus σ^n is the identity, and the exponent of G is a divisor of n.

For the "only if" part, observe that since G is a finite abelian group, it is a direct product of cyclic groups $C_1, \dots, C_r$. For each $i = 1, \dots, r$, let H_i be the product of the C_j for $j \neq i$; by (1.5.3), $H_i \trianglelefteq G$. We have $G/H_i \cong C_i$ by the first isomorphism theorem. (Consider the projection mapping $x_1 \cdots x_r \to x_i \in C_i$.) Let K_i be the fixed field of H_i. By the fundamental theorem, K_i/F is a Galois extension and its Galois group is isomorphic to G/H_i, hence isomorphic to C_i. Thus K_i/F is a cyclic extension of degree $d_i = |C_i|$, and d_i is a divisor of n. (Since G is the direct product of the C_i, some element of G has order d_i, so d_i divides the exponent of G and therefore divides n.) We want to apply (6.7.3) with n replaced by d_i, and this is possible because F contains a primitive d_i^{th} root of unity, namely ω^{n/d_i}. We conclude that $K_i = F(\theta_i)$, where $\theta_i^{d_i}$ is a nonzero element $b_i \in F$. But $\theta_i^n = \theta_i^{d_i(n/d_i)} = b_i^{n/d_i} = a_i \in F$.

Finally, in the Galois correspondence, the intersection of the H_i is paired with the composite of the K_i, which is $F(\theta_1, \dots, \theta_r)$; see Section 6.3, Problem 7. But $\bigcap_{i=1}^r H_i = 1$, so $E = F(\theta_1, \dots, \theta_r)$, and the result follows. ♣

Problems For Section 6.7

1. Find the Galois group of the extension $\mathbb{Q}(\sqrt{2}, \sqrt{3}, \sqrt{5}, \sqrt{7})$ [the splitting field of $(X^2 - 2)(X^2 - 3)(X^2 - 5)(X^2 - 7)$] over $\mathbb{Q}$.
2. Suppose that E is a splitting field for $f(X) = X^n - a$ over F, $a \neq 0$, but we drop the second assumption in (6.7.1) that F contains a primitive n^{th} root of unity. Is it possible for the Galois group of E/F to be cyclic?
3. Let E be a splitting field for $X^n - a$ over F, where $a \neq 0$, and assume that the characteristic of F does not divide n. Show that E contains a primitive n^{th} root of unity.

We now assume that E is a splitting field for $f(X) = X^p - c$ over F, where $c \neq 0$, p is prime and the characteristic of F is not p. Let ω be a primitive p^{th} root of unity in E (see Problem 3). Assume that f is not irreducible over F, and let g be an irreducible factor of f of degree d, where $1 \leq d < p$. Let θ be a root of g in E.

4. Let g_0 be the product of the roots of g. (Since g_0 is $\pm$ the constant term of g, $g_0 \in F$.) Show that $g_0^p = \theta^{dp} = c^d$.
5. Since d and p are relatively prime, there are integers a and b such that $ad + bp = 1$. Use this to show that if $X^p - c$ is not irreducible over F, then it must have a root in F.

6. Continuing Problem 5, show that if $X^p - c$ is not irreducible over F, then $E = F(\omega)$.
7. Continuing Problem 6, show that if $X^p - c$ is not irreducible over F, then $X^p - c$ splits over F if and only if F contains a primitive p^{th} root of unity.

Let E/F be a cyclic Galois extension of prime degree p, where p is the characteristic of F. Let σ be a generator of $G = \text{Gal}(E/F)$. It is a consequence of Hilbert's Theorem 90 (see the Problems for Section 7.3) that there is an element $\theta \in E$ such that $\sigma(\theta) = \theta + 1$. Prove the *Artin-Schreier theorem*:

8. $E = F(\theta)$.
9. θ is a root of $f(X) = X^p - X - a$ for some $a \in F$.
10. f is irreducible over F (hence $a \neq 0$).

Conversely, Let F be a field of prime characteristic p, and let E be a splitting field for $f(X) = X^p - X - a$, where a is a nonzero element of F.

11. If θ is any root of f in E, show that $E = F(\theta)$ and that f is separable.
12. Show that every irreducible factor of f has the same degree d, where $d = 1$ or p. Thus if $d = 1$, then $E = F$, and if $d = p$, then f is irreducible over F.
13. If f is irreducible over F, show that the Galois group of f is cyclic of order p.

6.8 Solvability By Radicals

6.8.1 Definitions and Comments

We wish to solve the polynomial equation $f(X) = 0$, $f \in F[X]$, under the restriction that we are only allowed to perform ordinary arithmetic operations (addition, subtraction, multiplication and division) on the coefficients, along with extraction of nth roots (for any $n = 2, 3, \dots$). A sequence of operations of this type gives rise to a sequence of extensions

$$F \le F(\alpha_1) \le F(\alpha_1, \alpha_2) \le \cdots \le F(\alpha_1, \dots, \alpha_r) = E$$

where $\alpha_1^{n_1} \in F$ and $\alpha_i^{n_i} \in F(\alpha_1, \dots, \alpha_{i-1}), i = 2, \dots, r$. Equivalently, we have

$$F = F_0 \le F_1 \le \cdots \le F_r = E$$

where $F_i = F_{i-1}(\alpha_i)$ and $\alpha_i^{n_i} \in F_{i-1}, i = 1, \dots, r$. We say that E is a *radical extension* of F. It is convenient (and legal) to assume that $n_1 = \cdots = n_r = n$. (Replace each n_i by the product of all the n_i. To justify this, observe that if α^j belongs to a field L, then $\alpha^{mj} \in L, m = 2, 3, \dots$.) Unless otherwise specified, we will make this assumption in all hypotheses, conclusions and proofs.

We have already seen three explicit classes of radical extensions: cyclotomic, cyclic and Kummer. (In the latter two cases, we assume that the base field contains a primitive n^{th} root of unity.)

We say that the polynomial $f \in F[X]$ is *solvable by radicals* if the roots of f lie in some radical extension of F, in other words, there is a radical extension E of F such that f splits over E.

Since radical extensions are formed by successively adjoining n^{th} roots, it follows that the transitivity property holds: If E is a radical extension of F and L is a radical extension of E, then L is a radical extension of F.

A radical extension is always finite, but it need not be normal or separable. We will soon specialize to characteristic 0, which will force separability, and we can achieve normality by taking the normal closure (see (3.5.11)).

6.8.2 Proposition

Let E/F be a radical extension, and let N be the normal closure of E over F. Then N/F is also a radical extension.

Proof. E is obtained from F by successively adjoining $\alpha_1, \dots, \alpha_r$, where α_i is the nth root of an element in F_{i-1}. On the other hand, N is obtained from F by adjoining not only the α_i, but their conjugates $\alpha_{i1}, \dots, \alpha_{im(i)}$. For any fixed i and j, there is an automorphism $\sigma \in \mathrm{Gal}(N/F)$ such that $\sigma(\alpha_i) = \alpha_{ij}$ (see (3.2.3), (3.5.5) and (3.5.6)). Thus

$$\alpha_{ij}^n = \sigma(\alpha_i)^n = \sigma(\alpha_i^n)$$

and since α_i^n belongs to $F(\alpha_1, \dots, \alpha_{i-1})$, it follows from (3.5.1) that $\sigma(\alpha_i^n)$ belongs to the splitting field K_i of $\prod_{j=1}^{i-1} \min(\alpha_j, F)$ over F. [Take $K_1 = F$, and note that since $\alpha_1^n = b_1 \in F$, we have $\sigma(\alpha_1^n) = \sigma(b_1) = b_1 \in F$. Alternatively, observe that by (3.5.1), σ must take a root of $X^n - b_1$ to another root of this polynomial.] Thus we can display N as a radical extension of F by successively adjoining

$$\alpha_{11}, \dots, \alpha_{1m(1)}, \dots, \alpha_{r1}, \dots, \alpha_{rm(r)}. \quad ♣$$

6.8.3 Preparation for the Main Theorem

If F has characteristic 0, then a primitive n^{th} root of unity ω can be adjoined to F to reach an extension $F(\omega)$; see (6.5.1). If E is a radical extension of F and $F = F_0 \le F_1 \le \cdots \le F_r = E$, we can replace F_i by $F_i(\omega), i = 1, \dots, r$, and $E(\omega)$ will be a radical extension of F. By (6.8.2), we can pass from $E(\omega)$ to its normal closure over F. Here is the statement we are driving at:

Let $f \in F[X]$, where F has characteristic 0. If f is solvable by radicals, then there is a Galois radical extension $N = F_r \ge \cdots \ge F_1 \ge F_0 = F$ containing a splitting field K for f over F, such that each intermediate field $F_i, i = 1, \dots, r$, contains a primitive n^{th} root of unity ω. We can assume that $F_1 = F(\omega)$ and for $i > 1$, F_i is a splitting field for $X^n - b_i$ over F_{i-1}. [Look at the end of the proof of (6.8.2).] By (6.5.1), F_1/F is a cyclotomic (Galois) extension, and by (6.7.2), each $F_i/F_{i-1}, i = 2, \dots, r$ is a cyclic (Galois) extension.

We now do some further preparation. Suppose that K is a splitting field for f over F, and that the Galois group of K/F is solvable, with

$$\operatorname{Gal}(K/F) = H_0 \trianglerighteq H_1 \trianglerighteq \cdots \trianglerighteq H_r = 1$$

with each H_{i-1}/H_i abelian. By the fundamental theorem (and Section 6.2, Problem 4), we have the corresponding sequence of fixed fields

$$F = K_0 \leq K_1 \leq \cdots \leq K_r = K$$

with K_i/K_{i-1} Galois and $\operatorname{Gal}(K_i/K_{i-1})$ isomorphic to H_{i-1}/H_i. Let us adjoin a primitive n^{th} root of unity ω to each K_i, so that we have fields $F_i = K_i(\omega)$ with

$$F \leq F_0 \leq F_1 \leq \cdots \leq F_r.$$

We take $n = |\operatorname{Gal}(K/F)|$. Since F_i can be obtained from F_{i-1} by adjoining everything in $K_i \setminus K_{i-1}$, we have

$$F_i = F_{i-1}K_i = K_iF_{i-1}$$

the composite of F_{i-1} and $K_i, i = 1, \ldots, r$. We may now apply Theorem 6.2.2. In the diamond diagram of Figure 6.2.1, at the top of the diamond we have F_i, on the left K_i, on the right F_{i-1}, and on the bottom $K_i \cap F_{i-1} \supseteq K_{i-1}$ (see Figure 6.8.1). We conclude that F_i/F_{i-1} is Galois, with a Galois group isomorphic to a subgroup of $\operatorname{Gal}(K_i/K_{i-1})$. Since $\operatorname{Gal}(K_i/K_{i-1}) \cong H_{i-1}/H_i$, it follows that $\operatorname{Gal}(F_i/F_{i-1})$ is abelian. Moreover, the exponent of this Galois group divides the order of H_0, which coincides with the size of $\operatorname{Gal}(K/F)$. (This explains our choice of n.)

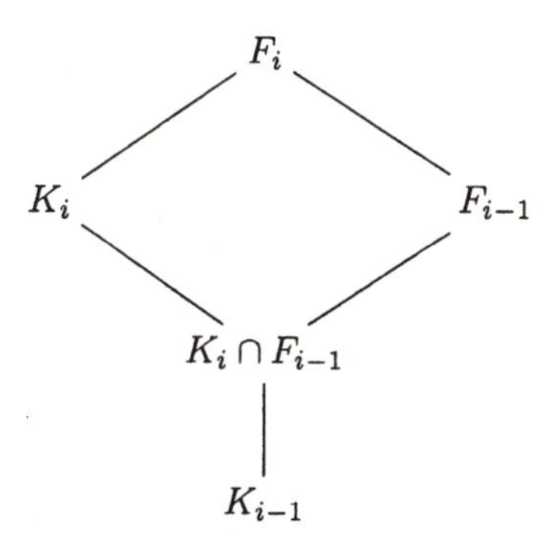

Figure 6.8.1

6.8.4 Galois' Solvability Theorem

Let K be a splitting field for f over F, where F has characteristic 0. Then f is solvable by radicals if and only if the Galois group of K/F is solvable.

Proof. If f is solvable by radicals, then as in (6.8.3), we have

$$F = F_0 \le F_1 \le \cdots \le F_r = N$$

where N/F is Galois, N contains a splitting field K for f over F, and each F_i/F_{i-1} is Galois with an abelian Galois group. By the fundamental theorem (and Section 6.2, Problem 4), the corresponding sequence of subgroups is

$$1 = H_r \trianglelefteq H_{r-1} \trianglelefteq \cdots \trianglelefteq H_0 = G = \text{Gal}(N/F)$$

with each H_{i-1}/H_i abelian. Thus G is solvable, and since

$$\text{Gal}(K/F) \cong \text{Gal}(N/F)/\text{Gal}(N/K)$$

[map $\text{Gal}(N/F) \to \text{Gal}(K/F)$ by restriction; the kernel is $\text{Gal}(N/K)$], $\text{Gal}(K/F)$ is solvable by (5.7.4).

Conversely, assume that $\text{Gal}(K/F)$ is solvable. Again as in (6.8.3), we have

$$F \le F_0 \le F_1 \le \cdots \le F_r$$

where $K \le F_r$, each F_i contains a primitive n^{th} root of unity, with $n = |\text{Gal}(K/F)|$, and $\text{Gal}(F_i/F_{i-1})$ is abelian with exponent dividing n for all $i = 1, \ldots, r$. Thus each F_i/F_{i-1} is a Kummer extension whose Galois group has an exponent dividing n. By (6.7.5) (or (6.5.1) for the case $i = 1$), each F_i/F_{i-1} is a radical extension. By transitivity (see (6.8.1)), F_r is a radical extension of F. Since $K \subseteq F_r$, f is solvable by radicals. ♣

6.8.5 Example

Let $f(X) = X^5 - 10X^4 + 2$ over the rationals. The Galois group of f is S_5, which is not solvable. (See Section 6.6, Problem 3 and Section 5.7, Problem 5.) Thus f is not solvable by radicals.

There is a fundamental idea that needs to be emphasized. The significance of Galois' solvability theorem is not simply that there are some examples of bad polynomials. The key point is there is no *general* method for solving a polynomial equation over the rationals by radicals, if the degree of the polynomial is 5 or more. If there were such a method, then in particular it would work on Example (6.8.5), a contradiction.

Problems For Section 6.8

In the exercises, we will sketch another classical problem, that of constructions with ruler and compass. In Euclidean geometry, we start with two points $(0,0)$ and $(1,0)$, and we are allowed the following constructions.

(i) Given two points P and Q, we can draw a line joining them;

(ii) Given a point P and a line L, we can draw a line through P parallel to L;

(iii) Given a point P and a line L, we can draw a line through P perpendicular to L;

(iv) Given two points P and Q, we can draw a circle with center at P passing through Q;

(v) Let A, and similarly B, be a line or a circle. We can generate new points, called *constructible points*, by forming the intersection of A and B. If $(c, 0)$ (equivalently $(0, c)$) is a constructible point, we call c a *constructible number*. It follows from (ii) and (iii) that (a, b) is a constructible point iff a and b are constructible numbers. It can be shown that every rational number is constructible, and that the constructible numbers form a field. Now in (v), the intersection of A and B can be found by ordinary arithmetic plus at worst the extraction of a square root. Conversely, the square roof of any nonnegative constructible number can be constructed. Therefore c is constructible iff there are real fields $\mathbb{Q} = F_0 \leq F_1 \cdots \leq F_r$ such that $c \in F_r$ and each $[F_i : F_{i-1}]$ is 1 or 2. Thus if c is constructible, then c is algebraic over $\mathbb{Q}$ and $[\mathbb{Q}(c) : \mathbb{Q}]$ is a power of 2.

1. (Trisecting the angle) If it is possible to trisect any angle with ruler and compass, then in particular a 60 degree angle can be trisected, so that $\alpha = \cos 20°$ is constructible. Using the identity

$$e^{i3\theta} = \cos 3\theta + i \sin 3\theta = (\cos\theta + i \sin\theta)^3,$$

reach a contradiction.

2. (Duplicating the cube) Show that it is impossible to construct, with ruler and compass, a cube whose volume is exactly 2. (The side of such a cube would be $\sqrt[3]{2}$.)

3. (Squaring the circle) Show that if it were possible to construct a square with area π, then π would be algebraic over $\mathbb{Q}$. (It is known that π is transcendental over $\mathbb{Q}$.)

To construct a regular n-gon, that is, a regular polygon with n sides, $n \geq 3$,we must be able to construct an angle of $2\pi/n$; equivalently, $\cos 2\pi/n$ must be a constructible number. Let $\omega = e^{i2\pi/n}$, a primitive n^{th} root of unity.

4. Show that $[\mathbb{Q}(\omega) : \mathbb{Q}(\cos 2\pi/n)] = 2$.

5. Show that if a regular n-gon is constructible, then the Euler phi function $\varphi(n)$ is a power of 2.

Conversely, assume that $\varphi(n)$ is a power of 2.

6. Show that $\mathrm{Gal}(\mathbb{Q}(\cos 2\pi/n)/\mathbb{Q})$ is a 2-group, that is, a p-group with $p = 2$.

7. By Section 5.7, Problem 7, every nontrivial finite p-group has a subnormal series in which every factor has order p. Use this (with $p = 2$) to show that a regular n-gon is constructible.

8. ¿From the preceding, a regular n-gon is constructible if and only if $\varphi(n)$ is a power of 2. Show that an equivalent condition is that $n = 2^s q_1 \cdots q_t, s, t = 0, 1, \dots,$ where the q_i are distinct *Fermat primes*, that is, primes of the form $2^m + 1$ for some positive integer m.

9. Show that if $2^m + 1$ is prime, then m must be a power of 2. The only known Fermat primes have $m = 2^a$, where $a = 0, 1, 2, 3, 4$ ($2^{32} + 1$ is divisible by 641). [The key point is that if a is odd, then $X + 1$ divides $X^a + 1$ in $\mathbb{Z}[X]$; the quotient is $X^{a-1} - X^{a-2} + \cdots - X + 1$ (since $a - 1$ is even).]

Let F be the field of rational functions in n variables $e_1, \dots, e_n$ over a field K with characteristic 0, and let $f(X) = X^n - e_1 X^{n-1} + e_2 X^{n-2} - \cdots + (-1)^n e_n \in F[X]$. If

$\alpha_1, \ldots, \alpha_n$ are the roots of f in a splitting field over F, then the e_i are the elementary symmetric functions of the α_i. Let $E = F(\alpha_1, \ldots, \alpha_n)$, so that E/F is a Galois extension and $G = \text{Gal}(E/F)$ is the Galois group of f.

10. Show that $G \cong S_n$.

11. What can you conclude from Problem 10 about solvability of equations?

6.9 Transcendental Extensions

6.9.1 Definitions and Comments

An extension E/F such that at least one $\alpha \in E$ is not algebraic over F is said to be *transcendental.* An idea analogous to that of a basis of an arbitrary vector space V turns out to be profitable in studying transcendental extensions. A basis for V is a subset of V that is linearly independent and spans V. A key result, whose proof involves the Steinitz exchange, is that if $\{x_1, \ldots, x_m\}$ spans V and S is a linearly independent subset of V, then $|S| \leq m$. We are going to replace linear independence by algebraic independence and spanning by algebraic spanning. We will find that every transcendental extension has a transcendence basis, and that any two transcendence bases for a given extension have the same cardinality. All these terms will be defined shortly. The presentation in the text will be quite informal; I believe that this style best highlights the strong connection between linear and algebraic independence. An indication of how to formalize the development is given in a sequence of exercises. See also Morandi, "Fields and Galois Theory", pp. 173–182.

Let E/F be an extension. The elements $t_1, \ldots, t_n \in E$ are *algebraically dependent over F* (or the set $\{t_1, \ldots, t_n\}$ is algebraically dependent over F) if there is a nonzero polynomial $f \in F[X_1, \ldots, X_n]$ such that $f(t_1, \ldots, t_n) = 0$; otherwise the t_i are *algebraically independent over F*. Algebraic independence of an infinite set means algebraic independence of every finite subset.

Now if a set T spans a vector space V, then each x in V is a linear combination of elements of T, so that x depends on T in a linear fashion. Replacing "linear" by "algebraic", we say that the element $t \in E$ *depends algebraically on T* over F if t is algebraic over $F(T)$, the field generated by T over F (see Section 3.1, Problem 1). We say that T *spans E algebraically over F* if each t in E depends algebraically on T over F, that is, E is an algebraic extension of $F(T)$. A *transcendence basis* for E/F is a subset of E that is algebraically independent over F and spans E algebraically over F. (From now on, we will frequently regard F as fixed and drop the phrase "over F".)

6.9.2 Lemma

If S is a subset of E, the following conditions are equivalent.

(i) S is a transcendence basis for E/F;

(ii) S is a maximal algebraically independent set;

(iii) S is a minimal algebraically spanning set.

Thus by (ii), S is a transcendence basis for E/F iff S is algebraically independent and E is algebraic over $F(S)$.

Proof. (i) implies (ii): If $S \subset T$ where T is algebraically independent, let $u \in T \setminus S$. Then u cannot depend on S algebraically (by algebraic independence of T), so S cannot span E algebraically.

(ii) implies (i): If S does not span E algebraically, then there exists $u \in E$ such that u does not depend algebraically on S. But then $S \cup \{u\}$ is algebraically independent, contradicting maximality of S.

(i) implies (iii): If $T \subset S$ and T spans E algebraically, let $u \in S \setminus T$. Then u depends algebraically on T, so $T \cup \{u\}$, hence S, is algebraically dependent, a contradiction.

(iii) implies (i): If S is algebraically dependent, then some $u \in S$ depends algebraically on $T = S \setminus \{u\}$. But then T spans E algebraically, a contradiction. ♣

6.9.3 Proposition

Every transcendental extension has a transcendence basis.

Proof. The standard argument via Zorn's lemma that an arbitrary vector space has a maximal linearly independent set (hence a basis) shows that an arbitrary transcendental extension has a maximal algebraically independent set, which is a transcendence basis by (6.9.2). ♣

For completeness, if E/F is an algebraic extension, we can regard $\emptyset$ as a transcendence basis.

6.9.4 The Steinitz Exchange

If $\{x_1, \dots, x_m\}$ spans E algebraically and $S \subseteq E$ is algebraically independent, then $|S| \leq m$.

Proof. Suppose that S has at least $m+1$ elements $y_1, \dots, y_{m+1}$. Since the x_i span E algebraically, y_1 depends algebraically on $x_1, \dots, x_m$. The algebraic dependence relation must involve at least one x_i, say x_1. (Otherwise, S would be algebraically dependent.) Then x_1 depends algebraically on $y_1, x_2, \dots, x_m$, so $\{y_1, x_2, \dots, x_m\}$ spans E algebraically. We claim that for every $i = 1, \dots, m$, $\{y_1, \dots, y_i, x_{i+1}, \dots, x_m\}$ spans E algebraically. We have just proved the case $i = 1$. If the result holds for i, then y_{i+1} depends algebraically on $\{y_1, \dots, y_i, x_{i+1}, \dots, x_m\}$, and the dependence relation must involve at least one x_j, say x_{i+1} for convenience. (Otherwise, S would be algebraically dependent.) Then x_{i+1} depends algebraically on $y_1, \dots, y_{i+1}, x_{i+2}, \dots, x_m$, so $\{y_1, \dots, y_{i+1}, x_{i+2}, \dots, x_m\}$ spans E algebraically, completing the induction.

Since there are more y's than x's, eventually the x's disappear, and $y_1, \dots, y_m$ span E algebraically. But then y_{m+1} depends algebraically on $y_1, \dots, y_m$, contradicting the algebraic independence of S. ♣

6.9.5 Corollary

Let S and T be transcendence bases of E. Then either S and T are both finite or they are both infinite; in the former case, $|S| = |T|$.

Proof. Assume that one of the transcendence bases, say T, is finite. By (6.9.4), $|S| \le |T|$, so S is finite also. By a symmetrical argument, $|T| \le |S|$, so $|S| = |T|$. ♣

6.9.6 Proposition

If S and T are arbitrary transcendence bases for E, then $|S| = |T|$. [The common value is called the *transcendence degree* of E/F.]

Proof. By (6.9.5), we may assume that S and T are both infinite. Let $T = \{y_i : i \in I\}$. If $x \in S$, then x depends algebraically on finitely many elements $y_{i_1}, \dots, y_{i_r}$ in T. Define $I(x)$ to be the set of indices $\{i_1, \dots, i_r\}$. It follows that $I = \cup\{I(x) : x \in S\}$. For if j belongs to none of the $I(x)$, then we can remove y_j from T and the resulting set will still span E algebraically, contradicting (6.9.2) part (iii). Now an element of $\cup\{I(x) : x \in S\}$ is determined by selecting an element $x \in S$ and then choosing an index in $I(x)$. Since $I(x)$ is finite, we have $|I(x)| \le \aleph_0$. Thus

$$|I| = |\bigcup\{I(x) : x \in S\}| \le |S|\aleph_0 = |S|$$

since S is infinite. Thus $|T| \le |S|$. By symmetry, $|S| = |T|$. ♣

6.9.7 Example

Let $E = F(X_1, \dots, X_n)$ be the field of rational functions in the variables $X_1, \dots, X_n$ with coefficients in F. If $f(X_1, \dots, X_n) = 0$, then f is the zero polynomial, so $S = \{X_1, \dots, X_n\}$ is an algebraically independent set. Since $E = F(S)$, E is algebraic over $F(S)$ and therefore S spans E algebraically. Thus S is a transcendence basis.

Now let $T = \{X_1^{u_1}, \dots, X_n^{u_n}\}$, where $u_1, \dots, u_n$ are arbitrary positive integers. We claim that T is also a transcendence basis. As above, T is algebraically independent. Moreover, each X_i is algebraic over $F(T)$. To see what is going on, look at a concrete example, say $T = \{X_1^5, X_2^3, X_3^4\}$. If $f(Z) = Z^3 - X_2^3 \in F(T)[Z]$, then X_2 is a root of f, so X_2, and similarly each X_i, is algebraic over $F(T)$. By (3.3.3), E is algebraic over $F(T)$, so T is a transcendence basis.

Problems For Section 6.9

1. If S is an algebraically independent subset of E over F, T spans E algebraically over F, and $S \subseteq T$, show that there is a transcendence basis B such that $S \subseteq B \subseteq T$.
2. Show that every algebraically independent set can be extended to a transcendence basis, and that every algebraically spanning set contains a transcendence basis.
3. Prove carefully, for an extension E/F and a subset $T = \{t_1, \dots, t_n\} \subseteq E$, that the following conditions are equivalent.

(i) T is algebraically independent over F;

(ii) For every $i = 1, \dots, n$, t_i is transcendental over $F(T \setminus \{t_i\})$;

(iii) For every $i = 1, \dots, n$, t_i is transcendental over $F(t_1, \dots, t_{i-1})$ (where the statement for $i = 1$ is that t_1 is transcendental over F).

4. Let S be a subset of E that is algebraically independent over F. Show that if $t \in E \setminus S$, then t is transcendental over $F(S)$ if and only if $S \cup \{t\}$ is algebraically independent over F.

[Problems 3 and 4 suggest the reasoning that is involved in formalizing the results of this section.]

5. Let $F \leq K \leq E$, with S a subset of K that is algebraically independent over F, and T a subset of E that is algebraically independent over K. Show that $S \cup T$ is algebraically independent over F, and $S \cap T = \emptyset$.
6. Let $F \leq K \leq E$, with S a transcendence basis for K/F and T a transcendence basis for E/K. Show that $S \cup T$ is a transcendence basis for E/F. Thus if tr deg abbreviates transcendence degree, then by Problem 5,

$$\operatorname{tr}\deg(E/F) = \operatorname{tr}\deg(K/F) + \operatorname{tr}\deg(E/K).$$

7. Let E be an extension of F, and $T = \{t_1, \dots, t_n\}$ a finite subset of E. Show that $F(T)$ is F-isomorphic to the rational function field $F(X_1, \dots, X_n)$ if and only if T is algebraically independent over F.
8. An *algebraic function field* F in one variable over K is a field F/K such that there exists $x \in F$ transcendental over K with $[F : K(x)] < \infty$. If $z \in F$, show that z is transcendental over K iff $[F : K(z)] < \infty$.
9. Find the transcendence degree of the complex field over the rationals.

Appendix To Chapter 6

We will develop a method for calculating the discriminant of a polynomial and apply the result to a cubic. We then calculate the Galois group of an arbitrary quartic.

A6.1 Definition

If $x_1, \dots, x_n$ $(n \geq 2)$ are arbitrary elements of a field, the *Vandermonde determinant* of the x_i is

$$\det V = \begin{vmatrix} 1 & 1 & \cdots & 1 \\ x_1 & x_2 & \cdots & x_n \\ & \vdots & & \\ x_1^{n-1} & x_2^{n-1} & \cdots & x_n^{n-1} \end{vmatrix}$$

A6.2 Proposition

$$\det V = \prod_{i<j}(x_j - x_i).$$

Proof. $\det V$ is a polynomial h of degree $1 + 2 + \cdots + (n-1) = \binom{n}{2}$ in the variables $x_1, \dots, x_n$, as is $g = \prod_{i<j}(x_j - x_i)$. If $x_i = x_j$ for $i < j$, then the determinant is 0, so by the remainder theorem (2.5.2), each factor of g, hence g itself, divides h. Since h and g have the same degree, $h = cg$ for some constant c. Now look at the leading terms of h and g, i.e., those terms in which x_n appears to as high a power as possible, and subject to this constraint, x_{n-1} appears to as high a power as possible, etc. In both cases, the leading term is $x_2 x_3^2 \cdots x_n^{n-1}$, and therefore c must be 1. (For this step it is profitable to regard the x_i as abstract variables in a polynomial ring. Then monomials $x_1^{r_1} \cdots x_n^{r_n}$ with different sequences $(r_1, \dots, r_n)$ of exponents are linearly independent.) ♣

A6.3 Corollary

If f is a polynomial in $F[X]$ with roots $x_1, \dots, x_n$ in some splitting field over F, then the discriminant of f is $(\det V)^2$.

Proof. By definition of the discriminant D of f (see 6.6.1), we have $D = \Delta^2$ where $\Delta = \pm \det V$. ♣

A6.4 Computation of the Discriminant

The square of the determinant of V is $\det(VV^t)$, which is the determinant of

$$\begin{bmatrix} 1 & 1 & \cdots & 1 \\ x_1 & x_2 & \cdots & x_n \\ & \vdots & & \\ x_1^{n-1} & x_2^{n-1} & \cdots & x_n^{n-1} \end{bmatrix} \begin{bmatrix} 1 & x_1 & \cdots & x_1^{n-1} \\ 1 & x_2 & \cdots & x_2^{n-1} \\ & \vdots & & \\ 1 & x_n & \cdots & x_n^{n-1} \end{bmatrix}$$

and this in turn is

$$\begin{vmatrix} t_0 & t_1 & \cdots & t_{n-1} \\ t_1 & t_2 & \cdots & t_n \\ & \vdots & & \\ t_{n-1} & t_n & \cdots & t_{2n-2} \end{vmatrix}$$

where the *power sums* t_r are given by

$$t_0 = n, \ t_r = \sum_{i=1}^{n} x_i^r, \ r \geq 1.$$

We must express the power sums in terms of the coefficients of the polynomial f. This will involve, improbably, an exercise in differential calculus. We have

$$F(z) = \prod_{i=1}^{n}(1 - x_i z) = \sum_{i=0}^{n} c_i z^i \text{ with } c_0 = 1;$$

the variable z ranges over real numbers. Take the logarithmic derivative of F to obtain

$$\frac{F'(z)}{F(z)} = \frac{d}{dz} \log F(z) = \sum_{i=1}^{n} \frac{-x_i}{1 - x_i z} = -\sum_{i=1}^{n} \sum_{j=0}^{\infty} x_i^{j+1} z^j = -\sum_{j=0}^{\infty} t_{j+1} z^j.$$

Thus

$$F'(z) + F(z) \sum_{j=0}^{\infty} t_{j+1} z^j = 0,$$

that is,

$$\sum_{i=1}^{n} i c_i z^{i-1} + \sum_{i=0}^{n} c_i z^i \sum_{j=1}^{\infty} t_j z^{j-1} = 0.$$

Equating powers of z^{r-1}, we have, assuming that $n \geq r$,

$$rc_r + c_0 t_r + c_1 t_{r-1} + \cdots + c_{r-1} t_1 = 0; \tag{1}$$

if $r > n$, the first summation does not contribute, and we get

$$t_r + c_1 t_{r-1} + \cdots + c_n t_{r-n} = 0. \tag{2}$$

Our situation is a bit awkward here because the roots of $F(z)$ are the reciprocals of the x_i. The x_i are the roots of $\sum_{i=0}^{n} a_i z^i$ where $a_i = c_{n-i}$ (so that $a_n = c_0 = 1$). The results can be expressed as follows.

A6.5 Newton's Identities

If $f(X) = \sum_{i=0}^{n} a_i X^i$ (with $a_n = 1$) is a polynomial with roots $x_1, \ldots, x_n$, then the power sums t_i satisfy

$$t_r + a_{n-1} t_{r-1} + \cdots + a_{n-r+1} t_1 + r a_{n-r} = 0, \ r \leq n \tag{3}$$

and

$$t_r + a_{n-1} t_{r-1} + \cdots + a_0 t_{r-n} = 0, \ r > n. \tag{4}$$

A6.6 The Discriminant of a Cubic

First consider the case where the X^2 term is missing, so that $f(X) = X^3 + pX + q$. Then $n = t_0 = 3, a_0 = q, a_1 = p, a_2 = 0$ $(a_3 = 1)$. Newton's identities yield

$t_1 + a_2 = 0, t_1 = 0; \quad t_2 + a_2 t_1 + 2a_1 = 0, t_2 = -2p;$
$t_3 + a_2 t_2 + a_1 t_1 + 3a_0 = 0, t_3 = -3a_0 = -3q;$
$t_4 + a_2 t_3 + a_1 t_2 + a_0 t_1 = 0, t_4 = -p(-2p) = 2p^2$

$$D = \begin{vmatrix} 3 & 0 & -2p \\ 0 & -2p & -3q \\ -2p & -3q & 2p^2 \end{vmatrix} = -4p^3 - 27q^2.$$

We now go to the general case $f(X) = X^3 + aX^2 + bX + c$. The quadratic term can be eliminated by the substitution $Y = X + \frac{a}{3}$. Then

$$f(X) = g(Y) = (Y - \frac{a}{3})^3 + a(Y - \frac{a}{3})^2 + b(Y - \frac{a}{3}) + c$$

$$= Y^3 + pY + q \text{ where } \quad p = b - \frac{a^2}{3}, q = \frac{2a^3}{27} - \frac{ba}{3} + c.$$

Since the roots of f are translations of the roots of g by the same constant, the two polynomials have the same discriminant. Thus $D = -4p^3 - 27q^2$, which simplifies to

$$D = a^2(b^2 - 4ac) - 4b^3 - 27c^2 + 18abc.$$

We now consider the Galois group of a quartic $X^4 + aX^3 + bX^2 + cX + d$, assumed irreducible and separable over a field F. As above, the translation $Y = X + \frac{a}{4}$ eliminates the cubic term without changing the Galois group, so we may assume that $f(X) = X^4 + qX^2 + rX + s$. Let the roots of f be x_1, x_2, x_3, x_4 (distinct by separability), and let V be the four group, realized as the subgroup of S_4 containing the permutations $(1,2)(3,4)$, $(1,3)(2,4)$ and $(1,4)(2,3)$, along with the identity. By direct verification (i.e., brute force), $V \trianglelefteq S_4$. If G is the Galois group of f (regarded as a group of permutations of the roots), then $V \cap G \trianglelefteq G$ by the second isomorphism theorem.

A6.7 Lemma

$\mathcal{F}(V \cap G) = F(u, v, w)$, where

$$u = (x_1 + x_2)(x_3 + x_4), \quad v = (x_1 + x_3)(x_2 + x_4), \quad w = (x_1 + x_4)(x_2 + x_3).$$

Proof. Any permutation in V fixes u, v and w, so $\mathcal{G}F(u, v, w) \supseteq V \cap G$. If $\sigma \in G$ but $\sigma \notin V \cap G$ then (again by direct verification) σ moves at least one of u, v, w. For example, (1,2,3) sends u to w, and (1,2) sends v to w. Thus $\sigma \notin \mathcal{G}F(u, v, w)$. Therefore $\mathcal{G}F(u, v, w) = V \cap G$, and an application of the fixed field operator $\mathcal{F}$ completes the proof. ♣

A6.8 Definition

The *resolvent cubic* of $f(X) = X^4 + qX^2 + rX + s$ is $g(X) = (X - u)(X - v)(X - w)$.

To compute g, we must express its coefficients in terms of q, r and s. First note that $u - v = -(x_1 - x_4)(x_2 - x_3)$, $u - w = -(x_1 - x_3)(x_2 - x_4)$, $v - w = -(x_1 - x_2)(x_3 - x_4)$. Thus f and g have the same discriminant. Now

$$X^4 + qX^2 + rX + s = (X^2 + kX + l)(X^2 - kX + m)$$

where the appearance of k and $-k$ is explained by the missing cubic term. Equating coefficients gives $l + m - k^2 = q$, $k(m - l) = r$, $lm = s$. Solving the first two equations for m and adding, we have $2m = k^2 + q + r/k$, and solving the first two equations for l and adding, we get $2l = k^2 + q - r/k$. Multiply the last two equations and use $lm = s$ to get a cubic in k^2, namely

$$k^6 + 2qk^4 + (q^2 - 4s)k^2 - r^2 = 0.$$

(This gives a method for actually finding the roots of a quartic.) To summarize,

$$f(X) = (X^2 + kX + l)(X^2 - kX + m)$$

where k^2 is a root of

$$h(X) = X^3 + 2qX^2 + (q^2 - 4s)X - r^2.$$

We claim that the roots of h are simply $-u, -v, -w$. For if we arrange the roots of f so that x_1 and x_2 are the roots of $X^2 + kX + l$, and x_3 and x_4 are the roots of $X^2 - kX + m$, then $k = -(x_1 + x_2), -k = -(x_3 + x_4)$, so $-u = k^2$. The argument for $-v$ and $-w$ is similar. Therefore to get g from h, we simply change the sign of the quadratic and constant terms, and leave the linear term alone.

A6.9 An Explicit Formula For The Resolvent Cubic:

$$g(X) = X^3 - 2qX^2 + (q^2 - 4s)X + r^2.$$

We need some results concerning subgroups of S_n, $n \geq 3$.

A6.10 Lemma

(i) A_n is generated by 3-cycles, and every 3-cycle is a commutator.

(ii) The only subgroup of S_n with index 2 is A_n.

Proof. For the first assertion of (i), see Section 5.6, Problem 4. For the second assertion of (i), note that

$$(a, b)(a, c)(a, b)^{-1}(a, c)^{-1} = (a, b)(a, c)(a, b)(a, c) = (a, b, c).$$

To prove (ii), let H be a subgroup of S_n with index 2; H is normal by Section 1.3, Problem 6. Thus S_n/H has order 2, hence is abelian. But then by (5.7.2), part 5, $S'_n \leq H$, and since A_n also has index 2, the same argument gives $S'_n \leq A_n$. By (i), $A_n \leq S'_n$, so $A_n = S'_n \leq H$. Since A_n and H have the same finite number of elements $n!/2$, it follows that $H = A_n$. ♣

A6.11 Proposition

Let G be a subgroup of S_4 whose order is a multiple of 4, and let V be the four group (see the discussion preceding A6.7). Let m be the order of the quotient group $G/(G \cap V)$. Then

(a) If $m = 6$, then $G = S_4$;

(b) If $m = 3$, then $G = A_4$;

(c) If $m = 1$, then $G = V$;

(d) If $m = 2$, then $G = D_8$ or $\mathbb{Z}_4$ or V;

(e) If G acts transitively on $\{1, 2, 3, 4\}$, then the case $G = V$ is excluded in (d). [In all cases, equality is up to isomorphism.]

Proof. If $m = 6$ or 3, then since $|G| = m|G \cap V|$, 3 is a divisor of $|G|$. By hypothesis, 4 is also a divisor, so $|G|$ is a multiple of 12. By A6.10 part (ii), G must be S_4 or A_4. But

$$|S_4/(S_4 \cap V)| = |S_4/V| = 24/4 = 6$$

and

$$|A_4/(A_4 \cap V)| = |A_4/V| = 12/4 = 3$$

proving both (a) and (b). If $m = 1$, then $G = G \cap V$, so $G \le V$, and since $|G|$ is a multiple of 4 and $|V| = 4$, we have $G = V$, proving (c).

If $m = 2$, then $|G| = 2|G \cap V|$, and since $|V| = 4$, $|G \cap V|$ is 1, 2 or 4. If it is 1, then $|G| = 2 \times 1 = 2$, contradicting the hypothesis. If it is 2, then $|G| = 2 \times 2 = 4$, and $G = \mathbb{Z}_4$ or V (the only groups of order 4). Finally, assume $|G \cap V| = 4$, so $|G| = 8$. But a subgroup of S_4 of order 8 is a Sylow 2-subgroup, and all such subgroups are conjugate and therefore isomorphic. One of these subgroups is D_8, since the dihedral group of order 8 is a group of permutations of the 4 vertices of a square. This proves (d).

If $m = 2$, G acts transitively on $\{1, 2, 3, 4\}$ and $|G| = 4$, then by the orbit-stabilizer theorem, each stabilizer subgroup $G(x)$ is trivial (since there is only one orbit, and its size is 4). Thus every permutation in G except the identity moves every integer $1, 2, 3, 4$. Since $|G \cap V| = 2$, G consists of the identity, one other element of V, and two elements not in V, which must be 4-cycles. But a 4-cycle has order 4, so G must be cyclic, proving (e). ♣

A6.12 Theorem

Let f be an irreducible separable quartic, with Galois group G. Let m be the order of the Galois group of the resolvent cubic. Then:

(a) If $m = 6$, then $G = S_4$;

(b) If $m = 3$, then $G = A_4$;

(c) If $m = 1$, then $G = V$;

(d) If $m = 2$ and f is irreducible over $L = F(u, v, w)$, where u, v and w are the roots of the resolvent cubic, then $G = D_8$;

(e) If $m = 2$ and f is reducible over L, then $G = \mathbb{Z}_4$.

Proof. By A6.7 and the fundamental theorem, $[G : G \cap V] = [L : F]$. Now the roots of the resolvent cubic g are distinct, since f and g have the same discriminant. Thus L is a splitting field of a separable polynomial, so L/F is Galois. Consequently, $[L : F] = m$ by (3.5.9). To apply (A6.11), we must verify that $|G|$ is a multiple of 4. But this follows from the orbit-stabilizer theorem: since G acts transitively on the roots of f, there is only one orbit, of size $4 = |G|/|G(x)|$. Now (A6.11) yields (a), (b) and (c), and if $m = 2$, then $G = D_8$ or $\mathbb{Z}_4$.

To complete the proof, assume that $m = 2$ and $G = D_8$. Thinking of D_8 as the group of symmetries of a square with vertices 1,2,3,4, we can take D_8 to be generated by $(1, 2, 3, 4)$ and $(2, 4)$, with $V = \{1, (1, 2)(3, 4), (1, 3)(2, 4), (1, 4)(2, 3)\}$. The elements of V are symmetries of the square, hence belong to D_8; thus $V = G \cap V = \text{Gal}(E/L)$ by (A6.7). [E is a splitting field for f over F.] Since V is transitive, for each $i, j = 1, 2, 3, 4$, $i \neq j$, there is an L-automorphism τ of E such that $\tau(x_i) = x_j$. Applying τ to the equation $h(x_i) = 0$, where h is the minimal polynomial of x_i over L, we see that each x_j is a root of h, and therefore $f \mid h$. But $h \mid f$ by minimality of h, so $h = f$, proving that f is irreducible over L.

Finally, assume $m = 2$ and $G = \mathbb{Z}_4$, which we take as $\{1, (1, 2, 3, 4), (1, 3)(2, 4), (1, 4, 3, 2)\}$. Then $G \cap V = \{1, (1, 3)(2, 4)\}$, which is not transitive. Thus for some $i \neq j$, x_i and x_j are not roots of the same irreducible polynomial over L. In particular, f is reducible over L. ♣

A6.13 Example

Let $f(X) = X^4 + 3X^2 + 2X + 1$ over $\mathbb{Q}$, with $q = 3, r = 2, s = 1$. The resolvent cubic is, by (A6.9), $g(X) = X^3 - 6X^2 + 5X + 4$. To calculate the discriminant of g, we can use the general formula in (A6.6), or compute $g(X + 2) = (X + 2)^3 - 6(X + 2)^2 + 5(X + 2) + 4 = X^3 - 7X - 2$. [The rational root test gives irreducibility of g and restricts a factorization of f to $(X^2 + aX \pm 1)(X^2 - aX \pm 1)$, $a \in \mathbb{Z}$, which is impossible. Thus f is irreducible as well.] We have $D(g) = -4(-7)^3 - 27(-2)^2 = 1264$, which is not a square in $\mathbb{Q}$. Thus $m = 6$, so the Galois group of f is S_4.

Chapter 7

Introducing Algebraic Number Theory

(Commutative Algebra 1)

The general theory of commutative rings is known as *commutative algebra*. The main applications of this discipline are to algebraic number theory, to be discussed in this chapter, and algebraic geometry, to be introduced in Chapter 8.

Techniques of abstract algebra have been applied to problems in number theory for a long time, notably in the effort to prove Fermat's Last Theorem. As an introductory example, we will sketch a problem for which an algebraic approach works very well. If p is an odd prime and $p \equiv 1 \bmod 4$, we will prove that p is the sum of two squares, that is, p can be expressed as $x^2 + y^2$ where x and y are integers. Since $\frac{p-1}{2}$ is even, it follows that -1 is a quadratic residue (that is, a square) mod p. [Pair each of the numbers 2,3, $\dots, p-2$ with its multiplicative inverse mod p and pair 1 with $p-1 \equiv -1 \bmod p$. The product of the numbers 1 through $p-1$ is, mod p,

$$1 \times 2 \times \cdots \times \frac{p-1}{2} \times -1 \times -2 \times \cdots \times -\frac{p-1}{2}$$

and therefore $\left[\left(\frac{p-1}{2}\right)!\right]^2 \equiv -1 \bmod p$.]

If $-1 \equiv x^2 \bmod p$, then p divides x^2+1. Now we enter the ring of Gaussian integers and factor x^2+1 as $(x+i)(x-i)$. Since p can divide neither factor, it follows that p is not prime in $\mathbb{Z}[i]$, so we can write $p = \alpha\beta$ where neither α nor β is a unit.

Define the *norm* of $\gamma = a + bi$ as $N(\gamma) = a^2 + b^2$. Then $N(\gamma) = 1$ iff $\gamma = \pm 1$ or $\pm i$ iff γ is a unit. (See Section 2.1, Problem 5.) Thus

$$p^2 = N(p) = N(\alpha)N(\beta) \text{ with } N(\alpha) > 1 \text{ and } N(\beta) > 1,$$

so $N(\alpha) = N(\beta) = p$. If $\alpha = x + iy$, then $p = x^2 + y^2$.

Conversely, if p is an odd prime and $p = x^2 + y^2$, then p is congruent to 1 mod 4. (If x is even, then $x^2 \equiv 0$ mod 4, and if x is odd, then $x^2 \equiv 1$ mod 4. We cannot have x and y both even or both odd, since p is odd.)

It is natural to conjecture that we can identify those primes that can be represented as $x^2 + |d|y^2$, where d is a negative integer, by working in the ring $\mathbb{Z}[\sqrt{d}]$. But the Gaussian integers ($d = -1$) form a Euclidean domain, in particular a unique factorization domain. On the other hand, unique factorization fails for $d \leq -3$ (Section 2.7, Problem 7), so the above argument collapses. [Recall from (2.6.4) that in a UFD, an element p that is not prime must be reducible.] Difficulties of this sort led Kummer to invent "ideal numbers", which later became ideals at the hands of Dedekind. We will see that although a ring of algebraic integers need not be a UFD, unique factorization of ideals will always hold.

7.1 Integral Extensions

If E/F is a field extension and $\alpha \in E$, then α is algebraic over F iff α is a root of a polynomial with coefficients in F. We can assume if we like that the polynomial is monic, and this turns out to be crucial in generalizing the idea to ring extensions.

7.1.1 Definitions and Comments

In this chapter, unless otherwise specified, *all rings are assumed commutative.* Let A be a subring of the ring R, and let $x \in R$. We say that x is *integral over* A if x is a root of a monic polynomial f with coefficients in A. The equation $f(X) = 0$ is called an *equation of integral dependence* for x over A. If x is a real or complex number that is integral over $\mathbb{Z}$, then x is called an *algebraic integer*. Thus for every integer d, $\sqrt{d}$ is an algebraic integer, as is any n^{th} root of unity. (The monic polynomials are, respectively, $X^2 - d$ and $X^n - 1$.) In preparation for the next result on conditions equivalent to integrality, note that $A[x]$, the set of polynomials in x with coefficients in A, is an A-module. (The sum of two polynomials is a polynomial, and multiplying a polynomial by a member of A produces another polynomial over A.)

7.1.2 Proposition

Let A be a subring of R, with $x \in R$. The following conditions are equivalent:

(i) x is integral over A;

(ii) The A-module $A[x]$ is finitely generated;

(iii) x belongs to a subring B of R such that $A \subseteq B$ and B is a finitely generated A-module.

Proof. (i) implies (ii). If x is a root of a monic polynomial over A of degree n, then x^n and all higher powers of x can be expressed as linear combinations of lower powers of x. Thus $1, x, x^2, , \dots, x^{n-1}$ generate $A[x]$ over A.

(ii) implies (iii). Take $B = A[x]$.

(iii) implies (i). If $\beta_1, \dots, \beta_n$ generate B over A, then $x\beta_i$ is a linear combination of the β_j, say $x\beta_i = \sum_{j=1}^n c_{ij}\beta_j$. Thus if β is a column vector whose components are the β_i, I is an n by n identity matrix, and $C = [c_{ij}]$, then

$$(xI - C)\beta = 0,$$

and if we premultiply by the adjoint matrix of $xI - C$ (as in Cramer's rule), we get

$$[\det(xI - C)]I\beta = 0,$$

hence $[\det(xI - C)]b = 0$ for every $b \in B$. Since B is a ring we may set $b = 1$ and conclude that x is a root of the monic polynomial $\det(XI - C)$ in $A[X]$. ♣

For other equivalent conditions, see Problems 1 and 2.

We are going to prove a transitivity property for integral extensions (analogous to (3.3.5)), and the following result will be helpful.

7.1.3 Lemma

Let A be a subring of R, with $x_1, \dots, x_n \in R$. If x_1 is integral over A, x_2 is integral over $A[x_1]$, ... , and x_n is integral over $A[x_1, \dots, x_{n-1}]$, then $A[x_1, \dots, x_n]$ is a finitely generated A-module.

Proof. The $n = 1$ case follows from (7.1.2), part (ii). Going from $n - 1$ to n amounts to proving that if A, B and C are rings, with C a finitely generated B-module and B a finitely generated A-module, then C is a finitely generated A-module. This follows by a brief computation:

$$C = \sum_{j=1}^{r} By_j, \quad B = \sum_{k=1}^{s} Az_k \quad \text{so} \quad C = \sum_{j=1}^{r}\sum_{k=1}^{s} Ay_jz_k. \quad ♣$$

7.1.4 Transitivity of Integral Extensions

Let A, B and C be subrings of R. If C is integral over B, that is, each element of C is integral over B, and B is integral over A, then C is integral over A.

Proof. Let $x \in C$, with $x^n + b_{n-1}x^{n-1} + \cdots + b_1x + b_0 = 0$, $b_i \in B$. Then x is integral over $A[b_0, \dots, b_{n-1}]$. Each b_i is integral over A, hence over $A[b_0, \dots, b_{i-1}]$. By (7.1.3), $A[b_0, \dots, b_{n-1}, x]$ is a finitely generated A-module. By (7.1.2), part (iii), x is integral over A. ♣

7.1.5 Definitions and Comments

If A is a subring of R, the *integral closure of A in R* is the set A_c of elements of R that are integral over A. Note that $A \subseteq A_c$ because each $a \in A$ is a root of $X - a$. We say that A is *integrally closed* in R if $A_c = A$. If we simply say that A is *integrally closed* without reference to R, we assume that A is an integral domain with quotient field K, and A is integrally closed in K.

If x and y are integral over A, then just as in the proof of (7.1.4), it follows from (7.1.3) that $A[x, y]$ is a finitely generated A-module. Since $x + y, x - y$ and xy belong to this module, they are integral over A by (7.1.2) part (iii). The important conclusion is that

$$A_c \text{ is a subring of } R \text{ containing } A.$$

If we take the integral closure of the integral closure, we get nothing new.

7.1.6 Proposition

The integral closure A_c of A in R is integrally closed in R.

Proof. By definition, A_c is integral over A. If x is integral over A_c, then as in the proof of (7.1.4), x is integral over A, so that $x \in A_c$. ♣

We can identify a large class of integrally closed rings.

7.1.7 Proposition

If A is a UFD, then A is integrally closed.

Proof. If x belongs to the quotient field K, then we can write $x = a/b$ where $a, b \in A$, with a and b relatively prime. If x is integral over A, then there is an equation of the form

$$(a/b)^n + a_{n-1}(a/b)^{n-1} + \cdots + a_1(a/b) + a_0 = 0$$

with $a_i \in A$. Multiplying by b^n, we have $a^n + bc = 0$, with $c \in A$. Thus b divides a^n, which cannot happen for relatively prime a and b unless b has no prime factors at all, in other words, b is a unit. But then $x = ab^{-1} \in A$. ♣

We can now discuss one of the standard setups for doing algebraic number theory.

7.1.8 Definitions and Comments

A *number field* is a subfield L of the complex numbers $\mathbb{C}$ such that L is a finite extension of the rationals $\mathbb{Q}$. Thus the elements of L are algebraic numbers. The integral closure of $\mathbb{Z}$ in L is called the ring of *algebraic integers* (or simply *integers*) of L. In the next section, we will find the algebraic integers explicitly when L is a quadratic extension.

Problems For Section 7.1

1. Show that in (7.1.2) another equivalent condition is the following:

   (iv) There is a subring B of R such that B is a finitely generated A-module and $xB \subseteq B$.

   If R is a field, show that the assumption that B is a subring can be dropped (as long as $B \neq 0$).

2. A module is said to be *faithful* if its annihilator is 0. Show that in (7.1.2) the following is another equivalent condition:
   (v) There is a faithful $A[x]$-module B that is finitely generated as an A-module.

Let A be a subring of the integral domain B, with B integral over A. In Problems 3–5 we are going to show that A is a field if and only if B is a field.

3. Assume that B is a field, and let a be a nonzero element of A. Then since $a^{-1} \in B$, there is an equation of the form

$$(a^{-1})^n + c_{n-1}(a^{-1})^{n-1} + \cdots + c_1 a^{-1} + c_0 = 0$$

   with $c_i \in A$. Show that $a^{-1} \in A$, proving that A is a field.
4. Now assume that A is a field, and let b be a nonzero element of B. By (7.1.2) part (ii), $A[b]$ is a finite-dimensional vector space over A. Let f be the A-linear transformation on this vector space given by multiplication by b, in other words, $f(z) = bz, z \in A[b]$. Show that f is injective.
5. Show that f is surjective as well, and conclude that B is a field.

In Problems 6–8, let A be a subring of B, with B integral over A. Let Q be a prime ideal of B and let $P = Q \cap A$.

6. Show that P is a prime ideal of A, and that A/P can be regarded as a subring of B/Q.
7. Show that B/Q is integral over A/P.
8. Show that P is a maximal ideal of A if and only if Q is a maximal ideal of B.

7.2 Quadratic Extensions of the Rationals

We will determine the algebraic integers of $L = \mathbb{Q}(\sqrt{d})$, where d is a square-free integer (a product of distinct primes). The restriction on d involves no loss of generality; for example, $\mathbb{Q}(\sqrt{12}) = \mathbb{Q}(\sqrt{3})$. The minimal polynomial of $\sqrt{d}$ over $\mathbb{Q}$ is $X^2 - d$, which has roots $\pm\sqrt{d}$. The extension $L/\mathbb{Q}$ is Galois, and the Galois group consists of the identity and the automorphism $\sigma(a + b\sqrt{d}) = a - b\sqrt{d},\ a, b \in \mathbb{Q}$.

A remark on notation: To make sure that there is no confusion between algebraic integers and ordinary integers, we will use the term *rational integer* for a member of $\mathbb{Z}$.

7.2.1 Lemma

If a and b are rational numbers, then $a + b\sqrt{d}$ is an algebraic integer if and only if $2a$ and $a^2 - db^2$ belong to $\mathbb{Z}$. In this case, $2b$ is also in $\mathbb{Z}$.

Proof. Let $x = a + b\sqrt{d}$, so that $\sigma(x) = a - b\sqrt{d}$. Then $x + \sigma(x) = 2a \in \mathbb{Q}$ and $x\sigma(x) = a^2 - db^2 \in \mathbb{Q}$. Now if x is an algebraic integer, then x is a root of a monic polynomial $f \in \mathbb{Z}[X]$. But $f(\sigma(x)) = \sigma(f(x))$ since σ is an automorphism, so $\sigma(x)$ is also a root of f and hence an algebraic integer. By (7.1.5), $2a$ and $a^2 - db^2$ are also algebraic integers, as well as rational numbers. By (7.1.7), $\mathbb{Z}$ is integrally closed, so $2a$

and $a^2 - db^2$ belong to $\mathbb{Z}$. The converse holds because $a+b\sqrt{d}$ is a root of $(X-a)^2 = db^2$, i.e., $X^2 - 2aX + a^2 - db^2 = 0$.

Now if $2a$ and $a^2 - db^2$ are rational integers, then $(2a)^2 - d(2b)^2 = 4(a^2 - db^2) \in \mathbb{Z}$, so $d(2b)^2 \in \mathbb{Z}$. If $2b \notin \mathbb{Z}$, then its denominator would include a prime factor p , which would appear as p^2 in the denominator of $(2b)^2$. Multiplication of $(2b)^2$ by d cannot cancel the p^2 because d is square-free, and the result follows. ♣

7.2.2 Corollary

The set B of algebraic integers of $\mathbb{Q}(\sqrt{d})$, d square-free, can be described as follows.

(i) If $d \not\equiv 1 \mod 4$, then B consists of all $a + b\sqrt{d}$, $a, b \in \mathbb{Z}$;

(ii) If $d \equiv 1 \mod 4$, then B consists of all $\frac{u}{2} + \frac{v}{2}\sqrt{d}$, $u, v \in Z$, where u and v have the same parity (both even or both odd).

[Note that since d is square-free, it is not divisible by 4, so the condition in (i) is $d \equiv 2$ or 3 mod 4.]

Proof. By (7.2.1), the algebraic integers are of the form $\frac{u}{2} + \frac{v}{2}\sqrt{d}$ where $u, v \in \mathbb{Z}$ and $\frac{u^2}{4} - \frac{dv^2}{4} \in \mathbb{Z}$, i.e., $u^2 - dv^2 \equiv 0 \mod 4$. It follows that u and v have the same parity. (The square of an even number is congruent to 0 and the square of an odd number to 1 mod 4.) Moreover, the "both odd" case can only occur when $d \equiv 1 \mod 4$. The "both even" case is equivalent to $\frac{u}{2}, \frac{v}{2} \in \mathbb{Z}$, and the result follows. ♣

We can express these results in a more convenient form. We will show in (7.4.10) that the set B of algebraic integers in any number field L is a free $\mathbb{Z}$-module of rank $n = [L : \mathbb{Q}]$. A basis for this module is called an *integral basis* or *$\mathbb{Z}$-basis* for B.

7.2.3 Theorem

Let B be the algebraic integers of $\mathbb{Q}(\sqrt{d})$, d square-free.

(i) If $d \not\equiv 1 \mod 4$, then 1 and $\sqrt{d}$ form an integral basis of B;

(ii) If $d \equiv 1 \mod 4$, then 1 and $\frac{1}{2}(1 + \sqrt{d})$ form an integral basis.

Proof. (i) By (7.2.2), 1 and $\sqrt{d}$ span B over $\mathbb{Z}$, and they are linearly independent because $\sqrt{d}$ is irrational.

(ii) By (7.2.2), 1 and $\frac{1}{2}(1 + \sqrt{d})$ are algebraic integers. To show that they span B, consider $\frac{1}{2}(u + v\sqrt{d})$, where u and v have the same parity. Then

$$\frac{1}{2}(u + v\sqrt{d}) = \left(\frac{u-v}{2}\right)(1) + v\left[\frac{1}{2}(1+\sqrt{d})\right]$$

with $\frac{u-v}{2}$ and v in $\mathbb{Z}$. Finally, to show linear independence, assume that $a, b \in \mathbb{Z}$ and

$$a + b\left[\frac{1}{2}(1+\sqrt{d})\right] = 0.$$

Then $2a + b + b\sqrt{d} = 0$, which forces $a = b = 0$. ♣

Problems For Section 7.2

1. Let $L = \mathbb{Q}(\alpha)$, where α is a root of the irreducible quadratic $X^2 + bX + c$, with $b, c \in \mathbb{Q}$. Show that $L = \mathbb{Q}(\sqrt{d})$ for some square-free integer d. Thus the analysis of this section covers all possible quadratic extensions of $\mathbb{Q}$.
2. Show that the quadratic extensions $\mathbb{Q}(\sqrt{d})$, d square-free, are all distinct.
3. Continuing Problem 2, show that in fact no two distinct quadratic extensions of $\mathbb{Q}$ are $\mathbb{Q}$-isomorphic.

Cyclotomic fields do not exhibit the same behavior. Let $\omega_n = e^{i2\pi/n}$, a primitive n^{th} root of unity. By a direct computation, we have $\omega_{2n}^2 = \omega_n$, and

$$-\omega_{2n}^{n+1} = -e^{i\pi(n+1)/n} = e^{i\pi}e^{i\pi}e^{i\pi/n} = \omega_{2n}.$$

4. Show that if n is odd, then $\mathbb{Q}(\omega_n) = \mathbb{Q}(\omega_{2n})$.
5. If x is an algebraic integer, show that the minimal polynomial of x over $\mathbb{Q}$ has coefficients in $\mathbb{Z}$. (This will be a consequence of the general theory to be developed in this chapter, but it is accessible now without heavy machinery.) Consequently, an algebraic integer that belongs to $\mathbb{Q}$ in fact belongs to $\mathbb{Z}$. (The minimal polynomial of $r \in \mathbb{Q}$ over $\mathbb{Q}$ is $X - r$.)
6. Give an example of a quadratic extension of $\mathbb{Q}$ that is also a cyclotomic extension.
7. Show that an integral basis for the ring of algebraic integers of a number field L is, in particular, a basis for L over $\mathbb{Q}$.

7.3 Norms and Traces

7.3.1 Definitions and Comments

If E/F is a field extension of finite degree n, then in particular, E is an n-dimensional vector space over F, and the machinery of basic linear algebra becomes available. If x is any element of E, we can study the F-linear transformation $m(x)$ given by multiplication by x, that is, $m(x)y = xy$. We define the *norm* and the *trace* of x, relative to the extension E/F, as

$$N[E/F](x) = \det m(x) \text{ and } T[E/F](x) = \text{ trace } m(x).$$

We will write $N(x)$ and $T(x)$ if E/F is understood. If the matrix $A(x) = [a_{ij}(x)]$ represents $m(x)$ with respect to some basis for E over F, then the norm of x is the determinant of $A(x)$ and the trace of x is the trace of $A(x)$, that is, the sum of the main diagonal entries. The *characteristic polynomial* of x is defined as the characteristic polynomial of the matrix $A(x)$, that is,

$$\text{char}[E/F](x) = \det[XI - A(x)]$$

where I is an n by n identity matrix. If E/F is understood, we will refer to the *characteristic polynomial of* x, written $\text{char}(x)$.

7.3.2 Example

Let $E = \mathbb{C}$ and $F = \mathbb{R}$. A basis for $\mathbb{C}$ over $\mathbb{R}$ is $\{1, i\}$ and, with $x = a + bi$, we have

$$(a+bi)(1) = a(1) + b(i) \text{ and } (a+bi)(i) = -b(1) + a(i).$$

Thus

$$A(a+bi) = \begin{bmatrix} a & -b \\ b & a \end{bmatrix}.$$

The norm, trace and characteristic polynomial of $a + bi$ are

$$N(a+bi) = a^2 + b^2,\ T(a+bi) = 2a,\ \text{char}(a+bi) = X^2 - 2aX + a^2 + b^2.$$

The computation is exactly the same if $E = \mathbb{Q}(i)$ and $F = \mathbb{Q}$. Notice that the coefficient of X is minus the trace and the constant term is the norm. In general, it follows from the definition of characteristic polynomial that

$$\text{char}(x) = X^n - T(x)X^{n-1} + \cdots + (-1)^n N(x).$$

[The only terms multiplying X^{n-1} in the expansion of the determinant are $-a_{ii}(x)$, $i = 1, \dots, n$. Set $X = 0$ to show that the constant term of char(x) is $(-1)^n \det A(x)$.]

7.3.3 Lemma

If E is an extension of F and $x \in E$, then $N(x)$, $T(x)$ and the coefficients of char(x) belong to F. If $a \in F$, then

$$N(a) = a^n,\ T(a) = na, \text{ and } \text{char}(a) = (X - a)^n.$$

Proof. The first assertion follows because the entries of the matrix $A(x)$ are in F. The second statement holds because if $a \in F$, the matrix representing multiplication by a is aI. ♣

It is natural to look for a connection between the characteristic polynomial of x and the minimal polynomial of x over F.

7.3.4 Proposition

$$\text{char}[E/F](x) = [\min(x, F)]^r$$

where $r = [E : F(x)]$.

Proof. First assume that $r = 1$, so that $E = F(x)$. By the Cayley-Hamilton theorem, the linear transformation $m(x)$ satisfies char(x), and since $m(x)$ is multiplication by x, x itself is a root of char(x). Thus $\min(x, F)$ divides char(x). But both polynomials have degree n, and the result follows. In the general case, let $y_1, \dots, y_s$ be a basis for $F(x)$ over F, and let $z_1, \dots, z_r$ be a basis for E over $F(x)$. Then the $y_i z_j$ form a basis for E over F. Let $A = A(x)$ be the matrix representing multiplication by x in the extension $F(x)/F$, so that $xy_i = \sum_k a_{ki}y_k$, and $x(y_iz_j) = \sum_k a_{ki}(y_kz_j)$. Order the basis for E/F as $y_1z_1, y_2z_1, \dots, y_sz_1; y_1z_2, y_2z_2, \dots, y_sz_2; \dots; y_1z_r, y_2z_r, \dots, y_sz_r$. Then $m(x)$ is represented in E/F as

$$\begin{bmatrix} A & 0 & \dots & 0 \\ 0 & A & \dots & 0 \\ \vdots & \vdots & & \vdots \\ 0 & 0 & \dots & A \end{bmatrix}$$

Thus $\text{char}[E/F](x) = [\det(XI - A)]^r$, which by the $r = 1$ case coincides with $[\min(x, F)]^r$. ♣

7.3.5 Corollary

Let $[E : F] = n$, and $[F(x) : F] = d$. Let $x_1, \dots, x_d$ be the roots of $\min(x, F)$ in a splitting field (counting multiplicity). Then

$$N(x) = \left(\prod_{i=1}^{d} x_i\right)^{n/d}, \quad T(x) = \frac{n}{d}\sum_{i=1}^{d} x_i$$

and

$$\text{char}(x) = \left[\prod_{i=1}^{d}(X - x_i)\right]^{n/d}.$$

Proof. The formula for the characteristic polynomial follows from (7.3.4). The norm is $(-1)^n$ times the constant term of $\text{char}(x)$ (see (7.3.2)), hence is

$$(-1)^n(-1)^n\left(\prod_{i=1}^{d} x_i\right)^{n/d}.$$

Finally, if $\min(x, F) = X^d + a_{d-1}X^{d-1} + \dots + a_1X + a_0$, then the coefficient of X^{n-1} in $[\min(x, F)]^{n/d}$ is $\frac{n}{d}a_{d-1} = -\frac{n}{d}\sum_{i=1}^{d} x_i$. Since the trace is the negative of this coefficient [see (7.3.2)], the result follows. ♣

If E is a separable extension of F, there are very useful alternative expressions for the trace and norm.

7.3.6 Proposition

Let E/F be a separable extension of degree n, and let $\sigma_1, \dots, \sigma_n$ be the distinct F-monomorphisms of E into an algebraic closure of E, or equally well into a normal extension L of F containing E. Then

$$T[E/F](x) = \sum_{i=1}^{n} \sigma_i(x) \text{ and } N[E/F](x) = \prod_{i=1}^{n} \sigma_i(x).$$

Consequently, $T(ax+by) = aT(x)+bT(y)$ and $N(xy) = N(x)N(y)$ for $x, y \in E, a, b \in F$.

Proof. Each of the d distinct F-embeddings τ_i of $F(x)$ into L takes x into a unique conjugate x_i, and extends to exactly $\frac{n}{d} = [E : F(x)]$ F-embeddings of E into L, all of which also take x to x_i[see (3.5.1), (3.2.3) and (3.5.2)]. Thus

$$\sum_{i=1}^{n} \sigma_i(x) = \frac{n}{d} \sum_{i=1}^{d} \tau_i(x) = T(x)$$

and

$$\prod_{i=1}^{n} \sigma_i(x) = \left[\prod_{i=1}^{d} \tau_i(x)\right]^{n/d} = N(x)$$

by (7.3.5). ♣

The linearity of T and the multiplicativity of N hold without any assumption of separability, since in (7.3.1) we have $m(ax+by) = am(x) + bm(y)$ and $m(xy) = m(x) \circ m(y)$.

7.3.7 Corollary (Transitivity of Trace and Norm)

If $F \leq K \leq E$, where E/F is finite and separable, then

$$T[E/F] = T[K/F] \circ T[E/K] \text{ and } N[E/F] = N[K/F] \circ N[E/K].$$

Proof. Let $\sigma_1, \dots, \sigma_n$ be the distinct F-embeddings of K into L, and let $\tau_1, \dots, \tau_m$ be the distinct K-embeddings of E into L, where L is the normal closure of E over F. By (6.3.1) and (3.5.11), L/F is Galois, and by (3.5.2), (3.5.5) and (3.5.6), each mapping σ_i and τ_j extends to an automorphism of L. Therefore it makes sense to allow the mappings to be composed. By (7.3.6),

$$T[K/F](T[E/K])(x) = \sum_{i=1}^{n} \sigma_i\Big(\sum_{j=1}^{m} \tau_j(x)\Big) = \sum_{i=1}^{n}\sum_{j=1}^{m} \sigma_i\tau_j(x).$$

Now each $\sigma_i\tau_j$ is an F-embedding of E into L, and the number of mappings is $mn = [E : K][K : F] = [E : F]$. Furthermore, the $\sigma_i\tau_j$ are distinct when restricted to E. For if $\sigma_i\tau_j = \sigma_k\tau_l$ on E, hence on K, then $\sigma_i = \sigma_k$ on K (because $\tau_j = \tau_l =$ the identity on K). Thus $i = k$, so that $\tau_j = \tau_l$ on E. But then $j = l$. By (7.3.6), $T[K/F](T[E/K])(x) = T[E/F](x)$. The norm is handled the same way, with sums replaced by products. ♣

7.3.8 Corollary

If E/F is a finite separable extension, then $T[E/F](x)$ cannot be 0 for all $x \in E$.

Proof. If $T(x) = 0$ for all x, then by (7.3.6), $\sum_{i=1}^{n} \sigma_i(x) = 0$ for all x. This contradicts Dedekind's lemma (6.1.6). ♣

A statement equivalent to (7.3.8) is that if E/F is finite and separable, then the "trace form" [the bilinear form $(x, y) \to T[E/F](xy)$] is nondegenerate, i.e., if $T(xy) = 0$ for all y, then $x = 0$. For if $x \neq 0, T(x_0) \neq 0$, and $T(xy) = 0$ for all y, choose y so that $xy = x_0$ to reach a contradiction.

7.3.9 The Basic Setup For Algebraic Number Theory

Let A be an integral domain with quotient field K, and let L be a finite separable extension of K. Let B be the set of elements of L that are integral over A, that is, B is the integral closure of A in L. The diagram below summarizes all the information.

$$\begin{array}{ccc} L & \text{—} & B \\ | & & | \\ K & \text{—} & A \end{array}$$

In the most important special case, $A = \mathbb{Z}$, $K = \mathbb{Q}$, L is a number field, and B is the ring of algebraic integers of L. From now on, we will refer to (7.3.9) as the *AKLB setup.*

7.3.10 Proposition

If $x \in B$, then the coefficients of $\text{char}[L/K](x)$ and $\min(x, K)$ are integral over A. In particular, $T[L/K](x)$ and $N[L/K](x)$ are integral over A, by (7.3.2). If A is integrally closed, then by (7.3.3), the coefficients belong to A.

Proof. The coefficients of $\min(x, K)$ are sums of products of the roots x_i, so by (7.1.5) and (7.3.4), it suffices to show that the x_i are integral over A. Each x_i is a conjugate of x over K, so by (3.2.3) there is a K-isomorphism $\tau_i \colon K(x) \to K(x_i)$ such that $\tau_i(x) = x_i$. If we apply τ_i to an equation of integral dependence for x over A, we get an equation of integral dependence for x_i over A. ♣

Problems For Section 7.3

1. If $E = \mathbb{Q}(\sqrt{d})$ and $x = a + b\sqrt{d} \in E$, find the norm and trace of x.
2. If $E = \mathbb{Q}(\theta)$ where θ is a root of the irreducible cubic $X^3 - 3X + 1$, find the norm and trace of θ^2.
3. Find the trace of the primitive 6th root of unity ω in the cyclotomic extension $\mathbb{Q}_6$.

We will now prove *Hilbert's Theorem 90*: If E/F is a cyclic extension with $[E : F] = n$ and Galois group $G = \{1, \sigma, \dots, \sigma^{n-1}\}$ generated by σ, and $x \in E$, then

(i) $N(x) = 1$ if and only if there exists $y \in E$ such that $x = y/\sigma(y)$;

(ii) $T(x) = 0$ if and only if there exists $z \in E$ such that $x = z - \sigma(z)$.

4. Prove the "if" parts of (i) and (ii) by direct computation.

By Dedekind's lemma, $1, \sigma, \sigma^2, \dots, \sigma^{n-1}$ are linearly independent over E, so

$$1 + x\sigma + x\sigma(x)\sigma^2 + \cdots + x\sigma(x)\cdots\sigma^{n-2}(x)\sigma^{n-1}$$

is not identically 0 on E.

5. Use this to prove the "only if" part of (i).

   By (7.3.8), there is an element $u \in E$ whose trace is not 0. Let

$$w = x\sigma(u) + (x + \sigma(x))\sigma^2(u) + \cdots + (x + \sigma(x) + \cdots + \sigma^{n-2}(x))\sigma^{n-1}(u)$$

hence

$$\sigma(w) = \sigma(x)\sigma^2(u) + (\sigma(x) + \sigma^2(x))\sigma^3(u) + \cdots + (\sigma(x) + \sigma^2(x) + \cdots + \sigma^{n-1}(x))\sigma^n(u)$$

6. If $T(x) = 0$, show that $w - \sigma(w) = xT(u)$.
7. If $z = w/T(u)$, show that $z - \sigma(z) = x$, proving the "only if" part of (ii).
8. In Hilbert's Theorem 90, are the elements y and z unique?
9. Let θ be a root of $X^4 - 2$ over $\mathbb{Q}$. Find the trace over $\mathbb{Q}$ of θ, θ^2, θ^3 and $\sqrt{3}\theta$.
10. Continuing Problem 9, show that $\sqrt{3}$ cannot belong to $\mathbb{Q}[\theta]$.

7.4 The Discriminant

We have met the discriminant of a polynomial in connection with Galois theory (Section 6.6). There is also a discriminant in algebraic number theory. The two concepts are unrelated at first glance, but there is a connection between them. We assume the basic $AKLB$ setup of (7.3.9), with $n = [L : K]$.

7.4.1 Definition

The *discriminant* of the n-tuple $x = (x_1, \dots, x_n)$ of elements of L is

$$D(x) = \det(T[L/K](x_i x_j)).$$

Thus we form a matrix whose ij element is the trace of $x_i x_j$, and take the determinant of the matrix. By (7.3.3), $D(x)$ belongs to K. If the x_i are in B, then by (7.3.10), $D(x)$ is integral over A, hence belongs to A if A is integrally closed.

The discriminant behaves quite reasonably under linear transformation:

7.4.2 Lemma

If $y = Cx$, where C is an n by n matrix over K and x and y are n-tuples written as column vectors, then $D(y) = (\det C)^2 D(x)$.

Proof. The trace of $y_r y_s$ is

$$T\Big(\sum_{i,j} c_{ri} c_{sj} x_i x_j\Big) = \sum_{i,j} c_{ri} T(x_i x_j) c_{sj}$$

hence

$$(T(y_r y_s)) = C(T(x_i x_j))C'$$

where C' is the transpose of C. The result follows upon taking determinants. ♣

Here is an alternative expression for the discriminant.

7.4.3 Lemma

Let $\sigma_1, \dots, \sigma_n$ be the K-embeddings of L into an algebraic closure of L, as in (7.3.6). Then $D(x) = [\det(\sigma_i(x_j))]^2$.

Thus we form the matrix whose ij element is $\sigma_i(x_j)$, take the determinant and square the result.

Proof. By (7.3.6),

$$T(x_i x_j) = \sum_k \sigma_k(x_i x_j) = \sum_k \sigma_k(x_i)\sigma_k(x_j)$$

so if C is the matrix whose ij entry is $\sigma_i(x_j)$, then

$$(T(x_i x_j)) = C'C$$

and again the result follows upon taking determinants. ♣

The discriminant "discriminates" between bases and non-bases, as follows.

7.4.4 Proposition

If $x = (x_1, \dots, x_n)$, then the x_i form a basis for L over K if and only if $D(x) \neq 0$.

Proof. If $\sum_j c_j x_j = 0$, with the $c_j \in K$ and not all 0, then $\sum_j c_j \sigma_i(x_j) = 0$ for all i, so the columns of the matrix $B = (\sigma_i(x_j))$ are linearly dependent. Thus linear dependence of the x_i implies that $D = 0$. Conversely, assume that the x_i are linearly independent (and therefore a basis since $n = [L : K]$). If $D = 0$, then the rows of B are linearly dependent, so for some $c_i \in K$, not all 0, we have $\sum_i c_i \sigma_i(x_j) = 0$ for all j. Since the x_j form a basis, we have $\sum_i c_i \sigma_i(u) = 0$ for all $u \in L$, so the monomorphisms σ_i are linearly dependent. This contradicts Dedekind's lemma. ♣

We now make the connection between the discriminant defined above and the discriminant of a polynomial defined previously.

7.4.5 Proposition

Assume that $L = K(x)$, and let f be the minimal polynomial of x over K. Let D be the discriminant of the basis $1, x, x^2, \dots, x^{n-1}$ for L over K. Then D is the discriminant of the polynomial f.

Proof. Let $x_1, \dots, x_n$ be the roots of f in a splitting field, with $x_1 = x$. Let σ_i be the K-embedding that takes x to x_i, $i = 1, \dots, n$. Then $\sigma_i(x^j) = x_i^j$, so by (7.4.3), D is the square of the determinant of the matrix

$$\begin{bmatrix} 1 & x_1 & x_1^2 & \cdots & x_1^{n-1} \\ 1 & x_2 & x_2^2 & \cdots & x_2^{n-1} \\ \vdots & \vdots & & \vdots & \\ 1 & x_n & x_n^2 & \cdots & x_n^{n-1} \end{bmatrix}$$

and the result follows from the formula for a Vandermonde determinant; see (A6.2). ♣

7.4.6 Corollary

Under the hypothesis of (7.4.5),

$$D = (-1)^{\binom{n}{2}} N[L/K](f'(x))$$

where f' is the derivative of f.

Proof. Let $a = (-1)^{\binom{n}{2}}$. By (7.4.5),

$$D = \prod_{i<j}(x_i - x_j)^2 = a\prod_{i\neq j}(x_i - x_j) = a\prod_i \prod_{j\neq i}(x_i - x_j).$$

But $f(X) = (X - x_1)\cdots(X - x_n)$, so

$$f'(x_i) = \sum_k \prod_{j\neq k}(X - x_j)$$

with X replaced by x_i. When X is replaced by x_i, only the $k = i$ term is nonzero, hence

$$f'(x_i) = \prod_{j\neq i}(x_i - x_j).$$

Consequently,

$$D = a\prod_{i=1}^{n} f'(x_i).$$

But

$$f'(x_i) = f'(\sigma_i(x)) = \sigma_i(f'(x))$$

so by (7.3.6),

$$D = aN[L/K](f'(x)). \quad ♣$$

The discriminant of an integral basis for a number field has special properties. We will get at these results by considering the general $AKLB$ setup, adding some additional conditions as we go along.

7.4.7 Lemma

There is a basis for L/K consisting entirely of elements of B.

Proof. Let $x_1, \dots, x_n$ be a basis for L over K. Each x_i is algebraic over K, and therefore satisfies a polynomial equation of the form

$$a_m x_i^m + \cdots + a_1 x_i + a_0 = 0$$

with $a_m \neq 0$ and the $a_i \in A$. (Initially, we only have $a_i \in K$, but then a_i is the ratio of two elements in A, and we can form a common denominator.) Multiply the equation by a_m^{m-1} to obtain an equation of integral dependence for $y_i = a_m x_i$ over A. The y_i form the desired basis. ♣

7.4.8 Theorem

Suppose we have a nondegenerate symmetric bilinear form on an n-dimensional vector space V, written for convenience using inner product notation (x, y). If $x_1, \dots, x_n$ is any basis for V, then there is a basis $y_1, \dots, y_n$ for V, called the *dual basis referred to* V, such that

$$(x_i, y_j) = \delta_{ij} = \begin{cases} 1, & i = j; \\ 0, & i \neq j. \end{cases}$$

This is a standard (and quite instructive) result in linear algebra, and it will be developed in the exercises.

7.4.9 Theorem

If A is a principal ideal domain, then B is a free A-module of rank n.

Proof. By (7.3.8), the trace is a nondegenerate symmetric bilinear form on the n-dimensional vector space L over K. By (7.1.7), A is integrally closed, so by (7.3.10), the trace of any element of L belongs to A. Now let $x_1, \dots, x_n$ be any basis for L over K consisting of algebraic integers, and let $y_1, \dots, y_n$ be the dual basis referred to L (see (7.4.8)). If $z \in B$, then we can write $z = \sum_{j=1}^n a_j y_j$ with $a_j \in K$. We know that the trace of $x_i z$ belongs to A, and we also have

$$T(x_i z) = T\Big(\sum_{j=1}^n a_j x_i y_j\Big) = \sum_{j=1}^n a_j T(x_i y_j) = \sum_{j=1}^n a_j \delta_{ij} = a_i.$$

Thus each a_i belongs to A, so that B is an A-submodule of the free A-module $\oplus_{j=1}^n A y_j$. By (4.6.2), B is a free A-module of rank at most n. But by (7.4.7), B contains a basis for L over K, and if we wish, we can assume that this basis is $x_1, \dots, x_n$. Then B contains the free A-module $\oplus_{j=1}^n A x_j$, so the rank of B as an A-module is at least n, and hence exactly n. ♣

7.4.10 Corollary

The set B of algebraic integers in any number field L is a free $\mathbb{Z}$-module of rank $n = [L : \mathbb{Q}]$. Therefore B has an integral basis. The discriminant is the same for every integral basis; it is known as the *field discriminant.*

Proof. Take $A = \mathbb{Z}$ in (7.4.9) to show that B has an integral basis. The transformation matrix C between two integral bases (see (7.4.2)) is invertible, and both C and C^{-1} have rational integer coefficients. Take determinants in the equation $CC^{-1} = I$ to conclude that $\det C$ is a unit in $\mathbb{Z}$. Therefore $\det C = \pm 1$, so by (7.4.2), all integral bases have the same discriminant. ♣

Problems For Section 7.4

Let $x_1, \dots, x_n$ be a basis for the vector space V, and let (x, y) be a nondegenerate symmetric bilinear form on V. We now supply the details of the proof of (7.4.8).

1. For any $y \in V$, the mapping $x \to (x, y)$ is a linear form $l(y)$, i.e., a linear map from V to the field of scalars. Show that the linear transformation $y \to l(y)$ from V to V^*, the dual space of V (i.e., the space of all linear forms on V), is injective.
2. Show that any linear form on V is $l(y)$ for some $y \in V$.
3. Let $f_1, \dots, f_n$ be the dual basis corresponding to $x_1, \dots, x_n$. Thus each f_j belongs to V^* (*not* V) and $f_j(x_i) = \delta_{ij}$. If $f_j = l(y_j)$, show that $y_1, \dots, y_n$ is the required dual basis referred to V.
4. Show that $x_i = \sum_{j=1}^n (x_i, x_j)y_j$. Thus in order to compute the dual basis referred to V in terms of the original basis, we must invert the matrix $((x_i, x_j))$.
5. A matrix C with coefficients in $\mathbb{Z}$ is said to be *unimodular* if its determinant is ± 1. Show that C is unimodular if and only if C is invertible and its inverse has coefficients in $\mathbb{Z}$.
6. Show that the field discriminant of the quadratic extension $\mathbb{Q}(\sqrt{d})$, d square-free, is
$$D = \begin{cases} 4d & \text{if } d \not\equiv 1 \bmod 4; \\ d & \text{if } d \equiv 1 \bmod 4. \end{cases}$$
7. Let $x_1, \dots, x_n$ be arbitrary algebraic integers in a number field, and consider the determinant of the matrix $(\sigma_i(x_j))$, as in (7.4.3). The direct expansion of the determinant has $n!$ terms. Let P be the sum of those terms in the expansion that have plus signs in front of them, and N the sum of those terms prefixed by minus signs. Thus the discriminant D of $(x_1, \dots, x_n)$ is $(P - N)^2$. Show that $P + N$ and PN are fixed by each σ_i, and deduce that $P + N$ and PN are rational numbers.
8. Continuing Problem 7, show that $P + N$ and PN are rational integers.
9. Continuing Problem 8, prove *Stickelberger's theorem:* $D \equiv 0$ or $1 \bmod 4$.
10. Let L be a number field of degree n over $\mathbb{Q}$, and let $y_1, \dots, y_n$ be a basis for L over $\mathbb{Q}$ consisting of algebraic integers. Let $x_1, \dots, x_n$ be an integral basis. Show that if the discriminant $D(y_1, \dots, y_n)$ is square-free, then each x_i can be expressed as a linear combination of the y_j with integer coefficients.

11. Continuing Problem 10, show that if $D(y_1, \dots, y_n)$ is square-free, then $y_1, \dots, y_n$ is an integral basis.
12. Is the converse of the result of Problem 11 true?
13. In the standard $AKLB$ setup (see (7.3.9)), show that L is the quotient field of B.

7.5 Noetherian and Artinian Modules and Rings

7.5.1 Definitions and Comments

In this section, rings are *not* assumed commutative. Let M be an R-module, and suppose that we have an increasing sequence of submodules $M_1 \le M_2 \le M_3 \le \dots$, or a decreasing sequence $M_1 \ge M_2 \ge M_3 \ge \dots$. We say that the sequence *stabilizes* if for some t, $M_t = M_{t+1} = M_{t+2} = \dots$. The question of stabilization of sequences of submodules appears in a fundamental way in many areas of abstract algebra and its applications. We have already made contact with the idea; see (2.6.6) and the introductory remarks in Section 4.6.

The module M is said to satisfy the *ascending chain condition (acc)* if every increasing sequence of submodules stabilizes; M satisfies the *descending chain condition (dcc)* if every decreasing sequence of submodules stabilizes.

7.5.2 Proposition

The following conditions on an R-module M are equivalent, and define a *Noetherian module*:

(1) M satisfies the acc;

(2) Every nonempty collection of submodules of M has a maximal element (with respect to inclusion).

The following conditions on M are equivalent, and define an *Artinian module*:

(1') M satisfies the dcc;

(2') Every nonempty collection of submodules of M has a minimal element.

Proof. Assume (1), and let $\mathcal{S}$ be a nonempty collection of submodules. Choose $M_1 \in \mathcal{S}$. If M_1 is maximal, we are finished; otherwise we have $M_1 < M_2$ for some $M_2 \in \mathcal{S}$. If we continue inductively, the process must terminate at a maximal element; otherwise the acc would be violated.

Conversely, assume (2), and let $M_1 \le M_2 \le \dots$. The sequence must stabilize; otherwise $\{M_1, M_2, \dots\}$ would be a nonempty collection of submodules with no maximal element. The proof is exactly the same in the Artinian case, with all inequalities reversed. ♣

There is another equivalent condition in the Noetherian case.

7.5.3 Proposition

M is Noetherian iff every submodule of M is finitely generated.

Proof. If the sequence $M_1 \le M_2 \le \dots$ does not stabilize, let $N = \cup_{r=1}^{\infty} M_r$. Then N is a submodule of M, and it cannot be finitely generated. For if $x_1, \dots, x_s$ generate N, then for sufficiently large t, all the x_i belong to M_t. But then $N \subseteq M_t \subseteq M_{t+1} \subseteq \dots \subseteq N$, so $M_t = M_{t+1} = \dots$. Conversely, assume that the acc holds, and let $N \le M$. If $N \ne 0$, choose $x_1 \in N$. If $Rx_1 = N$, then N is finitely generated. Otherwise, there exists $x_2 \notin Rx_1$. If x_1 and x_2 generate N, we are finished. Otherwise, there exists $x_3 \notin Rx_1 + Rx_2$. The acc forces the process to terminate at some stage t, in which case $x_1, \dots, x_t$ generate N. ♣

The analogous equivalent condition in the Artinian case (see Problem 8) is that every quotient module M/N is *finitely cogenerated*, that is, if the intersection of a collection of submodules of M/N is 0, then there is a finite subcollection whose intersection is 0.

7.5.4 Definitions and Comments

A ring R is *Noetherian* [resp. *Artinian*] if it is Noetherian [resp. Artinian] as a module over itself. If we need to distinguish between R as a left, as opposed to right, R-module, we will refer to a *left Noetherian* and a *right Noetherian* ring, and similarly for Artinian rings. This problem will not arise until Chapter 9.

7.5.5 Examples

1. Every PID is Noetherian.

   This follows from (7.5.3), since every ideal is generated by a single element.

2. $\mathbb{Z}$ is Noetherian (a special case of Example 1) but not Artinian. There are many descending chains of ideals that do not stabilize, e.g.,

$$\mathbb{Z} \supset (2) \supset (4) \supset (8) \supset \dots.$$

   We will prove in Chapter 9 that an Artinian ring must also be Noetherian.

3. If F is a field, then the polynomial ring $F[X]$ is Noetherian (another special case of Example 1) but not Artinian. A descending chain of ideals that does not stabilize is

$$(X) \supset (X^2) \supset (X^3) \supset \dots.$$

4. The ring $F[X_1, X_2, \dots]$ of polynomials over F in infinitely many variables is neither Artinian nor Noetherian. A descending chain of ideals that does not stabilize is constructed as in Example 3, and an ascending chain of ideals that does not stabilize is

$$(X_1) \subset (X_1, X_2) \subset (X_1, X_2, X_3) \subset \dots.$$

7.5.6 Remark

The following observations will be useful in deriving properties of Noetherian and Artinian modules. If $N \le M$, then a submodule L of M that contains N can always be written in the form $K+N$ for some submodule K. ($K = L$ is one possibility.) By the correspondence theorem,

$$(K_1 + N)/N = (K_2 + N)/N \text{ implies } K_1 + N = K_2 + N \text{ and}$$
$$(K_1 + N)/N \le (K_2 + N)/N \text{ implies } K_1 + N \le K_2 + N.$$

7.5.7 Proposition

If N is a submodule of M, then M is Noetherian [resp. Artinian] if and only if N and M/N are Noetherian [resp. Artinian].

Proof. Assume M is Noetherian. Then N is Noetherian by (2) of (7.5.2), since a submodule of N must also be a submodule of M. By (7.5.6), an ascending chain of submodules of M/N looks like $(M_1 + N)/N \le (M_2 + N)/N \le \cdots$. But then the $M_i + N$ form an ascending sequence of submodules of M, which must stabilize. Consequently, the sequence $(M_i + N)/N, i = 1, 2, \ldots,$ must stabilize.

Conversely, assume that N and M/N are Noetherian, and let $M_1 \le M_2 \le \cdots$ be an increasing sequence of submodules of M. Take i large enough so that both sequences $\{M_i \cap N\}$ and $\{M_i + N\}$ have stabilized. If $x \in M_{i+1}$, then $x + N \in M_{i+1} + N = M_i + N$, so $x = y + z$ where $y \in M_i$ and $z \in N$. Thus $x - y \in M_{i+1} \cap N = M_i \cap N$, and since $y \in M_i$ we have $x \in M_i$ as well. Consequently, $M_i = M_{i+1}$ and the sequence of M_i's has stabilized. The Artinian case is handled by reversing inequalities (and interchanging indices i and $i+1$ in the second half of the proof). ♣

7.5.8 Corollary

If $M_1, \ldots, M_n$ are Noetherian [resp. Artinian] R-modules, then so is $M_1 \oplus M_2 \oplus \cdots \oplus M_n$.

Proof. It suffices to consider $n = 2$ (induction will take care of higher values of n). The submodule $N = M_1$ of $M = M_1 \oplus M_2$ is Noetherian by hypothesis, and $M/N \cong M_2$ is also Noetherian (apply the first isomorphism theorem to the natural projection of M onto M_2). By (7.5.7), M is Noetherian. The Artinian case is done the same way. ♣

7.5.9 Corollary

If M is a finitely generated module over the Noetherian [resp. Artinian] ring R, then M is Noetherian [resp. Artinian]

Proof. By (4.3.6), M is a quotient of a free module L of finite rank. Since L is the direct sum of a finite number of copies of R, the result follows from (7.5.8) and (7.5.7). ♣

Ascending and descending chains of submodules are reminiscent of normal and subnormal series in group theory, and in fact we can make a precise connection.

7.5.10 Definitions

A *series* of *length* n for a module M is a sequence of the form

$$M = M_0 \geq M_1 \geq \cdots \geq M_n = 0.$$

The series is called a *composition series* if each factor module M_i/M_{i+1} is *simple.* [A module is simple if it is nonzero and has no submodules except itself and 0. We will study simple modules in detail in Chapter 9.] Thus we are requiring the series to have no proper refinement. Two series are *equivalent* if they have the same length and the same factor modules, up to isomorphism and rearrangement.

By convention, the zero module has a composition series, namely $\{0\}$ itself.

7.5.11 Jordan-Hölder Theorem For Modules

If M has a composition series, then any two composition series for M are equivalent. Furthermore, any strictly decreasing sequence of submodules can be refined to a composition series.

Proof. The development of the Jordan-Hölder theorem for groups can be taken over verbatim if we change multiplicative to additive notation. In particular, we can reproduce the preliminary lemma (5.6.2), the Zassenhaus lemma (5.6.3), the Schreier refinement theorem (5.6.5), and the Jordan-Hölder Theorem (5.6.6). We need not worry about normality of subgroups because in an abelian group, all subgroups are normal. As an example of the change in notation, the Zassenhaus lemma becomes

$$\frac{A + (B \cap D)}{A + (B \cap C)} \cong \frac{C + (D \cap B)}{C + (D \cap A)}.$$

This type of proof can be irritating, because it forces readers to look at the earlier development and make sure that everything does carry over. A possible question is "Why can't a composition series $\mathcal{S}$ of length n coexist with an *infinite* ascending or descending chain?" But if such a situation occurs, we can form a series $\mathcal{T}$ for M of length $n+1$. By Schreier, $\mathcal{S}$ and $\mathcal{T}$ have equivalent refinements. Since $\mathcal{S}$ has no proper refinements, and equivalent refinement have the same length, we have $n \geq n+1$, a contradiction. ♣

We can now relate the ascending and descending chain conditions to composition series.

7.5.12 Theorem

The R-module M has a composition series if and only if M is both Noetherian and Artinian.

Proof. The "only if" part was just done at the end of the proof of (7.5.11). Thus assume that M is Noetherian and Artinian. Assuming (without loss of generality) that $M \neq 0$, it follows from (2) of (7.5.2) that $M_0 = M$ has a maximal proper submodule M_1. Now M_1 is Noetherian by (7.5.7), so if $M_1 \neq 0$, then M_1 has a maximal proper submodule M_2. Continuing inductively, we must reach 0 at some point because M is Artinian. By construction, each M_i/M_{i+1} is simple, and we have a composition series for M. ♣

Here is a connection with algebraic number theory.

7.5.13 Proposition

In the basic $AKLB$ setup of (7.3.9), assume that A is integrally closed. If A is a Noetherian ring, then so is B. In particular, the ring of algebraic integers in a number field is Noetherian.

Proof. By the proof of (7.4.9), B is a submodule of a free A-module M of finite rank. (The assumption that A is a PID in (7.4.9) is used to show that A is integrally closed, and we have this by hypothesis. The PID assumption is also used to show that B is a free A-module, but we do not need this in the present argument.) By (7.5.8), M is Noetherian, so by (7.5.7), B is a Noetherian A-module. An ideal of B is, in particular, an A-submodule of B, hence is finitely generated over A and therefore over B. Thus B is a Noetherian ring. ♣

Problems For Section 7.5

1. Let p be a fixed prime, and let A be the abelian group of all rational numbers a/p^n, $n = 0, 1, \ldots, a \in \mathbb{Z}$, where all calculations are modulo 1, in other words, A is a subgroup of $\mathbb{Q}/\mathbb{Z}$. Let A_n be the subgroup $\{0, 1/p^n, 2/p^n, \ldots, (p^n - 1)/p^n\}$. Show that A is not a Noetherian $\mathbb{Z}$-module.
2. Continuing Problem 1, if B is a proper subgroup of A, show that B must be one of the A_n. Thus A is an Artinian $\mathbb{Z}$-module. [This situation cannot arise for *rings*, where Artinian implies Noetherian.]
3. If V is a vector space, show that V is finite-dimensional iff V is Noetherian iff V is Artinian iff V has a composition series.
4. Define the *length* of a module M [notation $l(M)$] as the length of a composition series for M. (If M has no composition series, take $l(M) = \infty$.) Suppose that we have a short exact sequence

$$0 \longrightarrow N \xrightarrow{f} M \xrightarrow{g} M/N \longrightarrow 0$$

   Show that $l(M)$ is finite if and only if $l(N)$ and $l(M/N)$ are both finite.
5. Show that l is *additive*, that is,

$$l(M) = l(N) + l(M/N).$$

6. Let S be a subring of the ring R, and assume that S is a Noetherian ring. If R is finitely generated as a module over S, show that R is also a Noetherian ring.
7. Let R be a ring, and assume that the polynomial ring $R[X]$ is Noetherian. Does it follow that R is Noetherian?
8. Show that a module M is Artinian if and only if every quotient module M/N is finitely cogenerated.

7.6 Fractional Ideals

Our goal is to establish unique factorization of ideals in a Dedekind domain, and to do this we will need to generalize the notion of ideal. First, some preliminaries.

7.6.1 Definition

If $I_1, \ldots, I_n$ are ideals, the *product* $I_1 \cdots I_n$ is the set of all finite sums $\sum_i a_{1i}a_{2i} \cdots a_{ni}$, where $a_{ki} \in I_k$, $k = 1, \ldots, n$. It follows from the definition that the product is an ideal contained in each I_j.

7.6.2 Lemma

If P is a prime ideal that contains a product $I_1 \cdots I_n$ of ideals, then P contains I_j for some j.

Proof. If not, let $a_j \in I_j \setminus P$, $j = 1, \ldots, n$. Then $a_1 \cdots a_n$ belongs to $I_1 \cdots I_n \subseteq P$, and since P is prime, some a_j belongs to P, a contradiction. ♣

7.6.3 Proposition

If I is a nonzero ideal of the Noetherian integral domain R, then I contains a product of nonzero prime ideals.

Proof. Assume the contrary. If $\mathcal{S}$ is the collection of all nonzero ideals that do not contain a product of nonzero prime ideals, then since R is Noetherian, $\mathcal{S}$ has a maximal element J, and J cannot be prime because it belongs to $\mathcal{S}$. Thus there are elements $a, b \in R$ with $a \notin J$, $b \notin J$, and $ab \in J$. By maximality of J, the ideals $J + Ra$ and $J + Rb$ each contain a product of nonzero prime ideals, hence so does $(J + Ra)(J + Rb) \subseteq J + Rab = J$. This is a contradiction. [Notice that we must use the fact that a product of nonzero ideals is nonzero, and this is where the hypothesis that R is an integral domain comes in.] ♣

7.6.4 Corollary

If I is an ideal of the Noetherian ring R (not necessarily an integral domain), then I contains a product of prime ideals.

Proof. Repeat the proof of (7.6.3) with the word "nonzero" deleted. ♣

Ideals in the ring of integers are of the form $n\mathbb{Z}$, the set of multiples of n. A set of the form $\frac{3}{2}\mathbb{Z}$ is not an ideal because it is not a subset of $\mathbb{Z}$, yet it behaves in a similar manner. The set is closed under addition and multiplication by an integer, and it becomes an ideal of $\mathbb{Z}$ if we simply multiply all the elements by 2. It will be profitable to study sets of this type.

7.6.5 Definitions

Let R be an integral domain, with K its quotient field, and let I be an R-submodule of K. We say that I is a *fractional ideal* of R if $rI \subseteq R$ for some nonzero $r \in R$. We will call r a *denominator* of I. An ordinary ideal of R is a fractional ideal (take $r = 1$), and will often be referred to an as *integral ideal.*

7.6.6 Lemma

(i) If I is a finitely generated R-submodule of K, then I is a fractional ideal.

(ii) If R is Noetherian and I is a fractional ideal of R, then I is a finitely generated R-submodule of K.

(iii) If I and J are fractional ideals with denominators r and s respectively, then $I \cap J$, $I + J$ and IJ are fractional ideals with respective denominators r (or s), rs and rs. [The product of fractional ideals is defined exactly as in (7.6.1).]

Proof. (i) If $x_1 = a_1/b_1, \dots, x_n = a_n/b_n$ generate I and $b = b_1 \cdots b_n$, then $bI \subseteq R$.

(ii) If $rI \subseteq R$, then $I \subseteq r^{-1}R$. As an R-module, $r^{-1}R$ is isomorphic to R and is therefore Noetherian. Consequently, I is finitely generated.

(iii) It follows from the definition (7.6.5) that the intersection, sum and product of fractional ideals are fractional ideals. The assertions about denominators are proved by noting that $r(I \cap J) \subseteq rI \subseteq R$, $rs(I + J) \subseteq rI + sJ \subseteq R$, and $rsIJ = (rI)(sJ) \subseteq R$. ♣

The product of two nonzero fractional ideals is a nonzero fractional ideal, and the multiplication is associative (since multiplication in R is associative). There is an identity element, namely R, since $RI \subseteq I = 1I \subseteq RI$. We will show that if R is a Dedekind domain, then every nonzero fractional ideal has a multiplicative inverse, so the nonzero fractional ideals form a group.

7.6.7 Definitions and Comments

A *Dedekind domain* is an integral domain R such that

(1) R is Noetherian,

(2) R is integrally closed, and

(3) Every nonzero prime ideal of R is maximal.

Every PID is a Dedekind domain, by (7.5.5), (7.1.7), (2.6.8) and (2.6.9). We will prove that the algebraic integers of a number field form a Dedekind domain. But as we know, the ring of algebraic integers need not be a PID, or even a UFD (see the discussion at the beginning of this chapter, and the exercises in Section 7.7).

7.6.8 Lemma

Let I be a nonzero prime ideal of the Dedekind domain R, and let $J = \{x \in K : xI \subseteq R\}$. Then $R \subset J$.

Proof. Since $RI \subseteq R$, it follows that R is a subset of J. Pick a nonzero element $a \in I$, so that I contains the principal ideal Ra. Let n be the smallest positive integer such that Ra contains a product $P_1 \cdots P_n$ of n nonzero prime ideals. Since R is Noetherian, there is such an n by (7.6.3), and by (7.6.2), I contains one of the P_i, say P_1. But in a Dedekind domain, every nonzero prime ideal is maximal, so $I = P_1$. Assuming $n \geq 2$, set $I_1 = P_2 \cdots P_n$, so that $Ra \not\supseteq I_1$ by minimality of n. Choose $b \in I_1$ with $b \notin Ra$. Now $II_1 \subseteq Ra$, in particular, $Ib \subseteq Ra$, hence $Iba^{-1} \subseteq R$. (Note that a has an inverse in K, but not necessarily in R.) Thus $ba^{-1} \in J$, but $ba^{-1} \notin R$, for if so, $b \in Ra$, contradicting the choice of b.

The case $n = 1$ must be handled separately. In this case, $P_1 = I \supseteq Ra \supseteq P_1$, so $I = Ra$. Thus Ra is a proper ideal, and we can choose $b \in R$ with $b \notin Ra$. Then $ba^{-1} \notin R$, but $ba^{-1}I = ba^{-1}Ra = bR \subseteq R$, so $ba^{-1} \in J$. ♣

We now prove that in (7.6.8), J is the inverse of I.

7.6.9 Proposition

Let I be a nonzero prime ideal of the Dedekind domain R, and let $J = \{x \in K : xI \subseteq R\}$. Then J is a fractional ideal and $IJ = R$.

Proof. By definition, J is an R-submodule of K. If r is a nonzero element of I and $x \in J$, then $rx \in R$, so $rJ \subseteq R$ and J is a fractional ideal. Now $IJ \subseteq R$ by definition of J, so IJ is an integral ideal. Since (using (7.6.8)) $I = IR \subseteq IJ \subseteq R$, maximality of I implies that either $IJ = I$ or $IJ = R$. In the latter case, we are finished, so assume $IJ = I$.

If $x \in J$, then $xI \subseteq IJ = I$, and by induction, $x^n I \subseteq I$ for all $n = 1, 2, \ldots$. Let r be any nonzero element of I. Then $rx^n \in x^n I \subseteq I \subseteq R$, so $R[x]$ is a fractional ideal. Since R is Noetherian, part (ii) of (7.6.6) implies that $R[x]$ is a finitely generated R-submodule of K. By (7.1.2), x is integral over R. But R, a Dedekind domain, is integrally closed, so $x \in R$. Therefore $J \subseteq R$, contradicting (7.6.8). ♣

Problems For Section 7.6

1. Show that a proper ideal P is prime if and only if for all ideals A and B, $P \supseteq AB$ implies that $P \supseteq A$ or $P \supseteq B$.

We are going to show that if an ideal I is contained in the union of the prime ideals $P_1, \ldots, P_n$, then I is contained in some P_i. Equivalently, if for all $i = 1, \ldots, n$, we have $I \not\subseteq P_i$, then $I \not\subseteq \cup_{i=1}^n P_i$. There is no problem when $n = 1$, so assume the result holds for $n - 1$ prime ideals. By the induction hypothesis, for each i there exists $x_i \in I$ with $x_i \notin P_j$, $j \neq i$.

2. Show that we can assume without loss of generality that $x_i \in P_i$ for all i.
3. Continuing Problem 2, let $x = \sum_{i=1}^n x_1 \cdots x_{i-1} x_{i+1} \cdots x_n$. Show that $x \in I$ but $x \notin \cup_{i=1}^n P_i$, completing the proof.
4. If I and J are relatively prime ideals ($I + J = R$), show that $IJ = I \cap J$. More generally, if $I_1, \ldots, I_n$ are relatively prime in pairs (see (2.3.7)), show that $I_1 \cdots I_n = \cap_{i=1}^n I_i$.

5. Show that if a Dedekind domain R is a UFD, then R is a PID.
6. Suppose that in (7.6.9), we would like to invert every maximal ideal of R, rather than the nonzero prime ideals. What is a reasonable hypothesis to add about R?
7. Let R be an integral domain with quotient field K. If K is a fractional ideal of R, show that $R = K$.
8. Let P_1 and P_2 be relatively prime ideals in the ring R. Show that P_1^r and P_2^s are relatively prime for arbitrary positive integers r and s.

7.7 Unique Factorization of Ideals in a Dedekind Domain

In the previous section, we inverted nonzero prime ideals in a Dedekind domain. We must now extend this result to nonzero fractional ideals.

7.7.1 Theorem

If I is a nonzero fractional ideal of the Dedekind domain R, then I can be factored uniquely as $P_1^{n_1} P_2^{n_2} \cdots P_r^{n_r}$ where the P_i are prime ideals and the n_i are integers. Consequently, the nonzero fractional ideals form a group under multiplication.

Proof. First consider the existence of such a factorization. Without loss of generality, we can restrict to integral ideals. [Note that if $r \neq 0$ and $rI \subseteq R$, then $I = (Rr)^{-1}(rI)$.] By convention, we regard R as the product of the empty collection of prime ideals, so let $\mathcal{S}$ be the set of all nonzero proper ideals of R that cannot be factored in the given form, with all n_i *positive* integers. [This trick will yield the useful result that the factorization of integral ideals only involves positive exponents.] Since R is Noetherian, $\mathcal{S}$, if nonempty, has a maximal element I_0, which is contained in a maximal ideal I. By (7.6.9), I has an inverse fractional ideal J. Thus by (7.6.8) and (7.6.9),

$$I_0 = I_0 R \subseteq I_0 J \subseteq IJ = R.$$

Therefore $I_0 J$ is an integral ideal, and we claim that $I_0 \subset I_0 J$. For if $I_0 = I_0 J$, the last paragraph of the proof of (7.6.9) can be reproduced with I replaced by I_0 to reach a contradiction. By maximality of I_0, $I_0 J$ is a product of prime ideals, say $I_0 J = P_1 \cdots P_r$ (with repetition allowed). Multiply both sides by the prime ideal I to conclude that I_0 is a product of prime ideals, contradicting $I_0 \in \mathcal{S}$. Thus $\mathcal{S}$ must be empty, and the existence of the desired factorization is established.

To prove uniqueness, suppose that we have two prime factorizations

$$P_1^{n_1} \cdots P_r^{n_r} = Q_1^{t_1} \cdots Q_s^{t_s}$$

where again we may assume without loss of generality that all exponents are positive. [If P^{-n} appears, multiply both sides by P^n.] Now P_1 contains the product of the $P_i^{n_i}$, so by (7.6.2), P_1 contains Q_j for some j. By maximality of Q_j, $P_1 = Q_j$, and we may renumber so that $P_1 = Q_1$. Multiply by the inverse of P_1 (a fractional ideal, but there is no problem) to cancel P_1 and Q_1, and continue inductively to complete the proof. ♣

7.7.2 Corollary

A nonzero fractional ideal I is an integral ideal if and only if all exponents in the prime factorization of I are nonnegative.

Proof. The "only if" part was noted in the proof of (7.7.1). The "if" part follows because a power of an integral ideal is still an integral ideal. ♣

7.7.3 Corollary

Denote by $n_P(I)$ the exponent of the prime ideal P in the factorization of I. (If P does not appear, take $n_P(I) = 0$.) If I_1 and I_2 are nonzero fractional ideals, then $I_1 \supseteq I_2$ if and only if for every prime ideal P of R, $n_P(I_1) \leq n_P(I_2)$.

Proof. We have $I_2 \subseteq I_1$ iff $I_2 I_1^{-1} \subseteq R$, and by (7.7.2), this happens iff for every P, $n_P(I_2) - n_P(I_1) \geq 0$. ♣

7.7.4 Definition

Let I_1 and I_2 be nonzero integral ideals. We say that I_1 *divides* I_2 if $I_2 = JI_1$ for some integral ideal J. Just as with integers, an equivalent statement is that each prime factor of I_1 is a factor of I_2.

7.7.5 Corollary

If I_1 and I_2 are nonzero integral ideals, then I_1 divides I_2 if and only if $I_1 \supseteq I_2$. In other words, for these ideals,

DIVIDES MEANS CONTAINS.

Proof. By (7.7.4), I_1 divides I_2 iff $n_P(I_1) \leq n_P(I_2)$ for every prime ideal P. By (7.7.3), this is equivalent to $I_1 \supseteq I_2$. ♣

The next result explains why Dedekind domains are important in algebraic number theory.

7.7.6 Theorem

In the basic $AKLB$ setup of (7.3.9), if A is a Dedekind domain, then so is B. In particular, the ring of algebraic integers in a number field is a Dedekind domain. In addition, B is a finitely generated A-module and the quotient field of B is L.

Proof. By (7.1.6), B is integrally closed in L. The proof of (7.4.7), with x_i replaced by an arbitrary element of L, shows that L is the quotient field of B. Therefore B is integrally closed. By (7.5.13), B is a Noetherian ring, and the proof of (7.5.13) shows that B is a Noetherian, hence finitely generated, A-module.

It remains to prove that every nonzero prime ideal Q of B is maximal. Choose any nonzero element x of Q. Since $x \in B$, x satisfies a polynomial equation

$$x^n + a_{n-1}x^{n-1} + \cdots + a_1 x + a_0 = 0$$

with the $a_i \in A$. If we take the positive integer n as small as possible, then $a_0 \neq 0$ by minimality of n. Solving for a_0, we see that $a_0 \in Bx \cap A \subseteq Q \cap A$, so $P = Q \cap A \neq 0$. But P is the preimage of the prime ideal Q under the inclusion map of A into B. Therefore P is a nonzero prime, hence maximal, ideal of the Dedekind domain A. Consequently, A/P is a field.

Now A/P can be identified with a subring of the integral domain B/Q via $y + P \to y + Q$. Moreover, B/Q is integral over A/P. [B is integral over A, and we can simply use the same equation of integral dependence.] It follows from Section 7.1, Problem 5, that B/Q is a field, so Q is a maximal ideal. ♣

Problems For Section 7.7

By (7.2.3), the ring B of algebraic integers in $\mathbb{Q}(\sqrt{-5})$ is $\mathbb{Z}[\sqrt{-5}]$. We will show that $\mathbb{Z}[\sqrt{-5}]$ is not a unique factorization domain. (For a different approach, see Section 2.7, Problems 5–7.) Consider the factorization

$$(1 + \sqrt{-5})(1 - \sqrt{-5}) = (2)(3).$$

1. By computing norms, verify that all four of the above factors are irreducible.
2. Show that the only units of B are ± 1.
3. Show that no factor on one side of the above equation is an associate of a factor on the other side, so unique factorization fails.
4. We can use the prime factorization of ideals in a Dedekind domain to compute the greatest common divisor and the least common multiple of two nonzero ideals I and J, exactly as with integers. Show that the greatest common divisor of I and J is $I + J$ and the least common multiple is $I \cap J$.
5. A Dedekind domain R comes close to being a PID in the following sense. (All ideals are assumed nonzero.) If I is an integral ideal, in fact if I is a fractional ideal, show that there is an integral ideal J such that IJ is a principal ideal of R.
6. Show that the ring of algebraic integers in $\mathbb{Q}(\sqrt{-17})$ is not a unique factorization domain.
7. In Problem 6, the only algebraic integers of norm 1 are ± 1. Show that this property does not hold for the algebraic integers in $\mathbb{Q}(\sqrt{-3})$.

7.8 Some Arithmetic in Dedekind Domains

Unique factorization of ideals in a Dedekind domain permits calculations that are analogous to familiar manipulations involving ordinary integers. In this section, we illustrate some of the ideas.

Let $P_1, \ldots, P_n$ be distinct nonzero prime ideals of the Dedekind domain R, and let $J = P_1 \cdots P_n$. Let Q_i be the product of the P_j with P_i omitted, that is,

$$Q_i = P_1 \cdots P_{i-1} P_{i+1} \cdots P_n.$$

(If $n = 1$, we take $Q_i = R$.) If I is any nonzero ideal of R, then by unique factorization, $IQ_i \supset IJ$. For each $i = 1, \ldots, n$, choose an element a_i belonging to IQ_i but not to IJ, and let $a = \sum_{i=1}^{n} a_i$.

7.8.1 Lemma

$a \in I$ but for each i, $a \notin IP_i$. (In particular, $a \neq 0$.)

Proof. Since each a_i belongs to $IQ_i \subseteq I$, we have $a \in I$. Now a_i cannot belong to IP_i, for if so, $a_i \in IP_i \cap IQ_i$, which is the least common multiple of IP_i and IQ_i (see Section 7.7, Problem 4). But by definition of Q_i, the least common multiple is simply IJ, and this contradicts the choice of a_i. We break up the sum defining a as follows:

$$a = (a_1 + \cdots + a_{i-1}) + a_i + (a_{i+1} + \cdots + a_n). \tag{1}$$

If $j \neq i$, then $a_j \in IQ_j \subseteq IP_i$, so the first and third terms of (1) belong to IP_i. Since $a_i \notin IP_i$, as found above, we have $a \notin IP_i$. ♣

In Section 7.7, Problem 5, we found that any nonzero ideal is a factor of a principal ideal. We can sharpen this result as follows.

7.8.2 Proposition

Let I be a nonzero ideal of the Dedekind domain R. Then there is a nonzero ideal I' such that II' is a principal ideal (a). Moreover, if J is an arbitrary nonzero ideal of R, I' can be chosen to be relatively prime to J.

Proof. Let $P_1, \ldots, P_n$ be the distinct prime divisors of J, and choose a as in (7.8.1). Then $a \in I$, so $(a) \subseteq I$. Since divides means contains (see (7.7.5)), I divides (a), so $(a) = II'$ for some nonzero ideal I'. If I' is divisible by P_i, then $I' = P_i I_0$ for some nonzero ideal I_0, and $(a) = IP_i I_0$. Consequently, $a \in IP_i$, contradicting (7.8.1). ♣

7.8.3 Corollary

A Dedekind domain with only finitely many prime ideals is a PID.

Proof. Let J be the product of all the nonzero prime ideals. If I is any nonzero ideal, then by (7.8.2) there is a nonzero ideal I' such that II' is a principal ideal (a), with I' relatively prime to J. But then the set of prime factors of I' is empty, so that $I' = R$. Thus $(a) = II' = IR = I$. ♣

The next result shows that a Dedekind domain is not too far away from a principal ideal domain.

7.8.4 Corollary

Let I be a nonzero ideal of the Dedekind domain R, and let a be any nonzero element of I. Then I can be generated by two elements, one of which is a.

Proof. Since $a \in I$, we have $(a) \subseteq I$, so I divides (a), say $(a) = IJ$. By (7.8.2) there is a nonzero ideal I' such that II' is a principal ideal (b) and I' is relatively prime to J. If gcd stands for greatest common divisor, then the ideal generated by a and b is

$$\gcd((a),(b)) = \gcd(IJ, II') = I$$

since $\gcd(J, I') = (1)$. ♣

Problems For Section 7.8

1. Let $I(R)$ be the group of nonzero fractional ideals of a Dedekind domain R. If $P(R)$ is the subset of $I(R)$ consisting of all nonzero *principal fractional ideals* Rx, $x \in K$, show that $P(R)$ is a subgroup of $I(R)$. The quotient group $C(R) = I(R)/P(R)$ is called the *ideal class group* of R. Since R is commutative, $C(R)$ is abelian, and it can be shown that in the number field case, $C(R)$ is finite.
2. Continuing Problem 1, show that $C(R)$ is trivial iff R is a PID.

We will now go through the factorization of an ideal in a number field. The necessary background is developed in a course in algebraic number theory, but some of the manipulations are accessible to us now. By (7.2.3), the ring B of algebraic integers of the number field $\mathbb{Q}(\sqrt{-5})$ is $\mathbb{Z}[\sqrt{-5}]$. (Note that $-5 \equiv 3 \bmod 4$.) If we wish to factor the ideal $(2) = 2B$ in B, the idea is to factor $x^2 + 5 \bmod 2$, and the result is $x^2 + 5 \equiv (x+1)^2 \bmod 2$. Identifying x with $\sqrt{-5}$, we form the ideal $P_2 = (2, 1 + \sqrt{-5})$, which turns out to be prime. The desired factorization is $(2) = P_2^2$. This technique works if $B = \mathbb{Z}[\alpha]$, where the number field L is $\mathbb{Q}(\alpha)$.

3. Show that $1 - \sqrt{-5} \in P_2$, and conclude that $6 \in P_2^2$.
4. Show that $2 \in P_2^2$, hence $(2) \subseteq P_2^2$.
5. Expand $P_2^2 = (2, 1+\sqrt{-5})(2, 1+\sqrt{-5})$, and conclude that $P_2^2 \subseteq (2)$.
6. Following the technique suggested in the above problems, factor $x^2 + 5 \bmod 3$, and conjecture that the prime factorization of (3) in the ring of integers of $\mathbb{Q}(\sqrt{-5})$ is $(3) = P_3 P_3'$ for appropriate P_3 and P_3'.
7. With P_3 and P_3' as found in Problem 6, verify that $(3) = P_3 P_3'$.

7.9 p-adic Numbers

We will give a very informal introduction to this basic area of number theory. (For further details, see Gouvea, "p-adic Numbers".) Throughout the discussion, p is a fixed prime.

7.9.1 Definitions and Comments

A *p-adic integer* can be described in several ways. One representation is via a series

$$x = a_0 + a_1 p + a_2 p^2 + \cdots, \quad a_i \in \mathbb{Z}. \tag{1}$$

(Let's ignore the problem of convergence for now.) The partial sums are $x_n = a_0 + a_1 p + \cdots + a_n p^n$, so that $x_n - x_{n-1} = a_n p^n$. A p-adic integer can also be defined as a sequence of integers $x = \{x_0, x_1, \ldots,\}$ satisfying

$$x_n \equiv x_{n-1} \mod p^n, \ n = 1, 2, \ldots. \tag{2}$$

Given a sequence satisfying (2), we can recover the coefficients of the series (1) by

$$a_0 = x_0, \ a_1 = \frac{x_1 - x_0}{p}, \ a_2 = \frac{x_2 - x_1}{p^2}, \ldots.$$

The sequences x and y are regarded as defining the same p-adic integer iff $x_n \equiv y_n \mod p^{n+1}$, $n = 0, 1, \ldots$. Replacing each x_n by the smallest nonnegative integer congruent to it mod p^{n+1} is equivalent to restricting the a_i in (1) to $\{0, 1, \ldots, p-1\}$. [We call this the *standard representation.*] Thus (1) is the limiting case in some sense of an expansion in base p.

Sums and products of p-adic integers can be defined by polynomial multiplication if (1) is used. With the representation (2), we take

$$x + y = \{x_n + y_n\}, \quad xy = \{x_n y_n\}.$$

With addition and multiplication defined in this way, we get the *ring of p-adic integers*, denoted by θ_p. (A more common notation is $\mathbb{Z}_p$, with the ring of integers modulo p written as $\mathbb{Z}/p\mathbb{Z}$. Since the integers mod p occur much more frequently in this text than the p-adic integers, and $\mathbb{Z}_p$ is a bit simpler than $\mathbb{Z}/p\mathbb{Z}$, I elected to use $\mathbb{Z}_p$ for the integers mod p.) The rational integers $\mathbb{Z}$ form a subring of θ_p via $x = \{x, x, x, \ldots\}$.

We now identify the units of θ_p.

7.9.2 Proposition

The p-adic integer $x = \{x_n\}$ is a unit of θ_p (also called a *p-adic unit*) if and only if $x_0 \not\equiv 0 \mod p$. In particular, a rational integer a is a p-adic unit if and only if $a \not\equiv 0 \mod p$.

Proof. If $(a_0 + a_1 p + \cdots)(b_0 + b_1 p + \cdots) = 1$, then $a_0 b_0 = 1$, so $a_0 \not\equiv 0 \mod p$, proving the "only if" part. Thus assume that $x_0 \not\equiv 0 \mod p$. By (2), $x_n \equiv x_{n-1} \equiv \cdots \equiv x_0 \mod p$, so $x_n \not\equiv 0 \mod p$. Therefore x_n and p^{n+1} are relatively prime, so there exists y_n such that $x_n y_n \equiv 1 \mod p^{n+1}$, hence mod p^n. Now by (2), $x_n \equiv x_{n-1} \mod p^n$, so $x_n y_{n-1} \equiv x_{n-1} y_{n-1} \equiv 1 \mod p^n$. Thus $x_n y_n \equiv x_n y_{n-1} \mod p^n$, so $y_n \equiv y_{n-1} \mod p^n$. The sequence $y = \{y_n\}$ is therefore a p-adic integer, and by construction, $xy = 1$. ♣

7.9.3 Corollary

Every nonzero p-adic integer has the form $x = p^n u$ where $n \geq 0$ and u is a p-adic unit. Consequently, θ_p is an integral domain. Furthermore, θ_p has only one prime element p, and every $x \in \theta_p$ is a power of p, up to multiplication by a unit.

Proof. The series representation for x has a nonzero term $a_n p^n$ of lowest degree n, where a_n can be taken between 1 and $p-1$. Factor out p^n to obtain $x = p^n u$, where u is a unit by (7.9.2). ♣

7.9.4 Definitions and Comments

The quotient field $\mathbb{Q}_p$ of θ_p is called the field of *p-adic numbers*. By (7.9.3), each $\alpha \in \mathbb{Q}_p$ has the form $p^m u$, where m is an integer (possibly negative) and u is a unit in θ_p. Thus α has a "Laurent expansion"

$$\frac{a_{-r}}{p^r} + \cdots + \frac{a_{-1}}{p} + a_0 + a_1 p + \cdots .$$

Another representation is $\alpha = x/p^r$, where x is a p-adic integer and $r \geq 0$. This version is convenient for doing addition and multiplication in $\mathbb{Q}_p$.

The rationals $\mathbb{Q}$ are a subfield of $\mathbb{Q}_p$. To see this, let a/b be a rational number in lowest terms (a and b relatively prime). If p does not divide b, then by (7.9.2), b is a unit of θ_p. Since $a \in \mathbb{Z} \subseteq \theta_p$, we have $a/b \in \theta_p$. If $b = p^t b'$ where p does not divide b', we can factor out p^t and reduce to the previous case. Thus a/b always belongs to $\mathbb{Q}_p$, and $a/b \in \theta_p$ iff p does not divide b. Rational numbers belonging to θ_p are sometimes called *p-integers*.

We now outline a procedure for constructing the p-adic numbers formally.

7.9.5 Definitions and Comments

The *p-adic valuation* on $\mathbb{Q}_p$ is defined by

$$v_p(p^m u) = m.$$

In general, a *valuation* v on a field F is a real-valued function on $F \setminus \{0\}$ satisfying:

(a) $v(xy) = v(x) + v(y)$;

(b) $v(x+y) \geq \min(v(x), v(y))$.

By convention, we take $v(0) = +\infty$.

The representation $x = p^m u$ shows that v_p is indeed a valuation on $\mathbb{Q}_p$. If c is any real number greater than 1, then the valuation v induces an *absolute value* on F, namely,

$$|x| = c^{-v(x)}.$$

When $v = v_p$, the constant c is usually taken to be p, and we obtain the *p-adic absolute value*

$$|x|_p = p^{-v_p(x)}.$$

Thus the p-adic absolute value of p^n is p^{-n}, which approaches 0 exponentially as n approaches infinity.

In general, an *absolute value* on a field F is a real-valued function on F such that:

(i) $|x| \geq 0$, with equality if and only if $x = 0$;

(ii) $|xy| = |x||y|$;

(iii) $|x + y| \leq |x| + |y|$.

By (b), an absolute value induced by a valuation satisfies a property that is stronger than (iii):

(iv) $|x + y| \leq \max(|x|, |y|)$.

An absolute value satisfying (iv) is said to be *nonarchimedian.*

7.9.6 Proposition

Let F be the quotient field of an integral domain R. The absolute value $|\ |$ on F is nonarchimedian if and only if $|n| \leq 1$ for every integer $n = 1 \pm \cdots \pm 1 \in R$.

Proof. Assume a nonarchimedian absolute value. By property (ii) of (7.9.5), $|\pm 1| = 1$. If $|n| \leq 1$, then by property (iv), $|n \pm 1| \leq 1$, and the desired conclusion follows by induction. Conversely, assume that the absolute value of every integer is at most 1. To prove (iv), it suffices to show that $|x + 1| \leq \max(|x|, 1)$ for every $x \in F$. [If $y \neq 0$ in (iv), divide by $|y|$.] By the binomial theorem,

$$|x + 1|^n = |\sum_{r=0}^{n} \binom{n}{r} x^r| \leq \sum_{r=0}^{n} |\binom{n}{r}||x|^r.$$

By hypothesis, the integer $\binom{n}{r}$ has absolute value at most 1. If $|x| > 1$, then $|x|^r \leq |x|^n$ for all $r = 0, 1, \ldots, n$. If $|x| \leq 1$, then $|x|^r \leq 1$. Consequently,

$$|x + 1|^n \leq (n + 1) \max(|x|^n, 1).$$

Take n^{th} roots and let $n \to \infty$ to get $|x + 1| \leq \max(|x|, 1)$. ♣

The next result may seem innocuous, but it leads to a remarkable property of nonarchimedian absolute values.

7.9.7 Proposition

If $|\ |$ is a nonarchimedian absolute value, then

$$|x| \neq |y| \text{ implies } |x + y| = \max(|x|, |y|).$$

Proof. First note that $|-y| = |(-1)y| = |-1||y| = |y|$. We can assume without loss of generality that $|x| > |y|$. Using property (iv) of (7.9.5), we have

$$|x| = |x + y - y| \leq \max(|x + y|, |y|) = |x + y|.$$

[Otherwise, $\max(|x + y|, |y|) = |y|$, hence $|x| \leq |y| < |x|$, a contradiction.] Since $|x + y| \leq \max(|x|, |y|) = |x|$, the result follows. ♣

Any absolute value determines a metric via $d(x, y) = |x - y|$. This distance function can be used to measure the length of the sides of a triangle.

7.9.8 Corollary

With respect to the metric induced by a nonarchimedian absolute value, all triangles are isosceles.

Proof. Let the vertices of the triangle be x, y and z. Then

$$|x - y| = |(x - z) + (z - y)|.$$

If $|x - z| = |z - y|$, then two side lengths are equal. If $|x - z| \neq ||z - y|$, then by (7.9.7), $|x - y| = \max(|x - z|, |z - y|)$, and again two side lengths are equal. ♣

We now look at the p-adic numbers from the viewpoint of valuation theory.

7.9.9 Definitions and Comments

Let $|\ |$ be a nonarchimedian absolute value on the field F. The *valuation ring* of $|\ |$ is

$$\theta = \{x \in F : |x| \leq 1\}.$$

In the p-adic case, $\theta = \{x \in \mathbb{Q}_p : v_p(x) \geq 0\} = \theta_p$. By properties (ii) and (iv) of (7.9.5), θ is a subring of F.

The *valuation ideal* of $|\ |$ is

$$\beta = \{x \in F : |x| < 1\}.$$

In the p-adic case, $\beta = \{x \in \mathbb{Q}_p : v_p(x) \geq 1\} = p\theta_p$, those p-adic integers whose series representation has no constant term. To verify that β is an ideal of θ, note that if $x, y \in \beta$ and $r \in \theta$, then $|rx| = |r||x| \leq |x| < 1$ and $|x + y| \leq \max(|x|, |y|) < 1$.

Now if $x \in \theta \setminus \beta$, then $|x| = 1$, hence $|x^{-1}| = 1/|x| = 1$, so $x^{-1} \in \theta$ and x is a unit of θ. On the other hand, if $x \in \beta$, then x cannot be a unit of θ. [If $xy = 1$, then $1 = |x||y| \leq |x| < 1$, a contradiction.] Thus the ideal β is the set of all nonunits of θ. No proper ideal I of θ can contain a unit, so $I \subseteq \beta$. It follows that β is the unique maximal ideal of θ. A ring with a unique maximal ideal is called a *local ring*. We will meet such rings again when we examine the localization process in Section 8.5.

To construct the p-adic numbers, we start with the p-adic valuation on the integers, and extend it to the rationals in the natural way: $v_p(a/b) = v_p(a) - v_p(b)$. The p-adic valuation then determines the p-adic absolute value, which induces a metric d on $\mathbb{Q}$.

[Because d comes from a nonarchimedian absolute value, it will satisfy the *ultrametric inequality* $d(x, y) \leq \max(d(x, z), d(z, y))$, which is stronger than the triangle inequality.] The process of constructing the real numbers by completing the rationals using equivalence classes of Cauchy sequences is familiar. The same process can be carried out using the p-adic absolute value rather than the usual absolute value on $\mathbb{Q}$. The result is a complete metric space, the field of p-adic numbers, in which $\mathbb{Q}$ is dense.

Ostrowski's theorem says that the usual absolute value $|\ |_\infty$ on $\mathbb{Q}$, along with the p-adic absolute values $|\ |_p$ for all primes p, and the *trivial* absolute value ($|0| = 0; |x| = 1$ for $x \neq 0$), essentially exhaust all possibilities. To be more precise, two absolute values on a field F are *equivalent* if the corresponding metrics on F induce the same topology. Any nontrivial absolute value on $\mathbb{Q}$ is equivalent to $|\ |_\infty$ or to one of the $|\ |_p$.

Problems For Section 7.9

1. Take $p = 3$, and compute the standard representation of $(2 + p + p^2)(2 + p^2)$ in two ways, using (1) and (2) of (7.9.1). Check the result by computing the product using ordinary multiplication of two integers, and then expanding in base $p = 3$.
2. Express the p-adic integer -1 as an infinite series of the form (1), using the standard representation.
3. Show that every absolute value on a finite field is trivial.
4. Show that an absolute value is archimedian iff the set $S = \{|n| : n \in \mathbb{Z}\}$ is unbounded.
5. Show that a field that has an archimedian absolute value must have characteristic 0.
6. Show that an infinite series $\sum z_n$ of p-adic numbers converges if and only if $z_n \to 0$ as $n \to \infty$.
7. Show that the sequence $a_n = n!$ of p-adic integers converges to 0.
8. Does the sequence $a_n = n$ converge in $\mathbb{Q}_p$?

Chapter 8

Introducing Algebraic Geometry

(Commutative Algebra 2)

We will develop enough geometry to allow an appreciation of the Hilbert Nullstellensatz, and look at some techniques of commutative algebra that have geometric significance. As in Chapter 7, unless otherwise specified, all rings will be assumed commutative.

8.1 Varieties

8.1.1 Definitions and Comments

We will be working in $k[X_1, \dots, X_n]$, the ring of polynomials in n variables over the field k. (Any application of the Nullstellensatz requires that k be algebraically closed, but we will not make this assumption until it becomes necessary.) The set $A^n = A^n(k)$ of all n-tuples with components in k is called *affine n-space.* If S is a set of polynomials in $k[X_1, \dots, X_n]$, then the zero-set of S, that is, the set $V = V(S)$ of all $x \in A^n$ such that $f(x) = 0$ for every $f \in S$, is called a *variety.* (The term "affine variety" is more precise, but we will use the short form because we will not be discussing projective varieties.) Thus a variety is the solution set of simultaneous polynomial equations.

If I is the ideal generated by S, then I consists of all finite linear combinations $\sum g_i f_i$ with $g_i \in k[X_1, \dots, X_n]$ and $f_i \in S$. It follows that $V(S) = V(I)$, so every variety is the variety of some ideal. We now prove that we can make A^n into a topological space by taking varieties as the closed sets.

8.1.2 Proposition

(1) If $V_\alpha = V(I_\alpha)$ for all $\alpha \in T$, then $\bigcap V_\alpha = V(\bigcup I_\alpha)$. Thus an arbitrary intersection of varieties is a variety.

(2) If $V_j = V(I_j)$, $j = 1, \dots, r$, then $\bigcup_{j=1}^r V_j = V(\{f_1 \cdots f_r \colon f_j \in I_j, 1 \le j \le r\})$. Thus a finite union of varieties is a variety.

(3) $A^n = V(0)$ and $\emptyset = V(1)$, so the entire space and the empty set are varieties.

Consequently, there is a topology on A^n, called the *Zariski topology*, such that the closed sets and the varieties coincide.

Proof. (1) If $x \in A^n$, then $x \in \bigcap V_\alpha$ iff every polynomial in every I_α vanishes at x iff $x \in V(\bigcup I_\alpha)$.

(2) $x \in \bigcup_{j=1}^r V_j$ iff for some j, every $f_j \in I_j$ vanishes at x iff $x \in V(\{f_1 \cdots f_r \colon f_j \in I_j$ for all $j\})$.

(3) The zero polynomial vanishes everywhere and a nonzero constant polynomial vanishes nowhere. ♣

Note that condition (2) can also be expressed as

$$\cup_{j=1}^r V_j = V\left(\prod_{j=1}^r I_j\right) = V\left(\cap_{j=1}^r I_j\right).$$

[See (7.6.1) for the definition of a product of ideals.]

We have seen that every subset of $k[X_1, \dots, X_n]$, in particular every ideal, determines a variety. We can reverse this process as follows.

8.1.3 Definitions and Comments

If X is an arbitrary subset of A^n, we define the *ideal of* X as $I(X) = \{f \in k[X_1, \dots, X_n] \colon f$ vanishes on $X\}$. By definition we have:

(4) If $X \subseteq Y$ then $I(X) \supseteq I(Y)$; if $S \subseteq T$ then $V(S) \supseteq V(T)$.

Now if S is any set of polynomials, define $IV(S)$ as $I(V(S))$, the ideal of the zero-set of S; we are simply omitting parentheses for convenience. Similarly, if X is any subset of A^n, we can define $VI(X)$, $IVI(X)$, $VIV(S)$, and so on. From the definitions we have:

(5) $IV(S) \supseteq S$; $VI(X) \supseteq X$.

[If $f \in S$ then f vanishes on $V(S)$, hence $f \in IV(S)$. If $x \in X$ then every polynomial in $I(X)$ vanishes at x, so x belongs to the zero-set of $I(X)$.]

If we keep applying V's and I's alternately, the sequence stabilizes very quickly:

(6) $VIV(S) = V(S)$; $IVI(X) = I(X)$.

[In each case, apply (4) and (5) to show that the left side is a subset of the right side. If $x \in V(S)$ and $f \in IV(S)$ then $f(x) = 0$, so $x \in VIV(S)$. If $f \in I(X)$ and $x \in VI(X)$ then x belongs to the zero-set of $I(X)$, so $f(x) = 0$. Thus f vanishes on $VI(X)$, so $f \in IVI(X)$.]

Since every polynomial vanishes on the empty set (vacuously), we have:

(7) $I(\emptyset) = k[X_1, \dots, X_n]$.

The next two properties require a bit more effort.

(8) If k is an infinite field, then $I(A^n) = \{0\}$;

(9) If $x = (a_1, \dots, a_n) \in A^n$, then $I(\{x\}) = (X_1 - a_1, \dots, X_n - a_n)$.

Property (8) holds for $n = 1$ since a nonconstant polynomial in one variable has only finitely many zeros. Thus $f \neq 0$ implies that $f \notin I(A^1)$. If $n > 1$, let $f = a_r X_1^r + \cdots + a_1 X_1 + a_0$ where the a_i are polynomials in $X_2, \dots, X_n$ and $a_r \neq 0$. By the induction hypothesis, there is a point $(x_2, \dots, x_n)$ at which a_r does not vanish. Fixing this point, we can regard f as a polynomial in X_1, which cannot possibly vanish at all $x_1 \in k$. Thus $f \notin I(A^n)$.

To prove (9), note that the right side is contained in the left side because $X_i - a_i$ is 0 when $X_i = a_i$. Also, the result holds for $n = 1$ by the remainder theorem (2.5.2). Thus assume $n > 1$ and let $f = b_r X_1^r + \cdots + b_1 X_1 + b_0 \in I(\{x\})$, where the b_i are polynomials in $X_2, \dots, X_n$ and $b_r \neq 0$. By the division algorithm (2.5.1), we have

$$f = (X_1 - a_1)g(X_1, \dots, X_n) + h(X_2, \dots, X_n)$$

and h must vanish at $(a_2, \dots, a_n)$. By the induction hypothesis, $h \in (X_2 - a_2, \dots, X_n - a_n)$, hence $f \in (X_1 - a_1, X_2 - a_2, \dots, X_n - a_n)$.

Problems For Section 8.1

A variety is said to be *reducible* if it can be expressed as the union of two proper subvarieties; otherwise the variety is *irreducible*. In Problems 1–4, we are going to show that a variety V is irreducible if and only if $I(V)$ is a prime ideal.

1. Assume that $I(V)$ is not prime, and let $f_1 f_2 \in I(V)$ with $f_1, f_2 \notin I(V)$. If $x \in V$, show that $x \notin V(f_1)$ implies $x \in V(f_2)$ (and similarly, $x \notin V(f_2)$ implies $x \in V(f_1)$).
2. Show that V is reducible.
3. Show that if V and W are varieties with $V \subset W$, then $I(V) \supset I(W)$.
4. Now assume that $V = V_1 \bigcup V_2$, with $V_1, V_2 \subset V$. By Problem 3, we can choose $f_i \in I(V_i)$ with $f_i \notin I(V)$. Show that $f_1 f_2 \in I(V)$, so $I(V)$ is not a prime ideal.
5. Show that any variety is the union of finitely many irreducible subvarieties.
6. Show that the decomposition of Problem 5 is unique, assuming that we discard any subvariety that is contained in another one.
7. Assume that k is algebraically closed. Suppose that A^n is covered by open sets $A^n \setminus V(I_i)$ in the Zariski topology. Let I is the ideal generated by the I_i, so that $I = \sum I_i$, the set of all finite sums $x_{i_1} + \cdots x_{i_r}$ with $x_{i_j} \in I_{i_j}$. Show that $1 \in I$. (You may appeal to the weak Nullstellensatz, to be proved in Section 8.4.)
8. Show that A^n is compact in the Zariski topology.

8.2 The Hilbert Basis Theorem

If S is a set of polynomials in $k[X_1, \dots, X_n]$, we have defined the variety $V(S)$ as the zero-set of S, and we know that $V(S) = V(I)$, where I is the ideal generated by S. Thus any set of simultaneous polynomial equations defines a variety. In general, infinitely many equations may be involved, but as Hilbert proved, an infinite collection of equations can always be replaced by a finite collection. The reason is that every ideal of $k[X_1, \dots, X_n]$ has a finite set of generators, in other words, $k[X_1, \dots, X_n]$ is a Noetherian ring. The field k is, in particular, a PID, so k is Noetherian. The key step is to show that if R is a Noetherian ring, then the polynomial ring in n variables over R is also Noetherian.

8.2.1 Hilbert Basis Theorem

If R is a Noetherian ring, then $R[X_1, \dots, X_n]$ is also Noetherian.

Proof. By induction, we can assume $n = 1$. Let I be an ideal of $R[X]$, and let J be the ideal of all leading coefficients of polynomials in I. (The leading coefficient of $5X^2 - 3X + 17$ is 5; the leading coefficient of the zero polynomial is 0.) By hypothesis, J is finitely generated, say by $a_1, \dots, a_n$. Let f_i be a polynomial in I whose leading coefficient is a_i, and let d_i be the degree of f_i. Let I^* consist of all polynomials in I of degree at most $d = \max\{d_i \colon 1 \leq i \leq n\}$. Then I^* is an R-submodule of the free R-module M of all polynomials $b_0 + b_1X + \cdots + b_dX^d, b_i \in R$. Now a finitely generated free R-module is a finite direct sum of copies of R, hence M, and therefore I^*, is Noetherian. Thus I^* can be generated by finitely many polynomials $g_1, \dots, g_m$. Take I_0 to be the ideal of $R[X]$ generated by $f_1, \dots, f_n, g_1, \dots, g_m$. We will show that $I_0 = I$, proving that I is finitely generated.

First observe that $f_i \in I$ and $g_j \in I^* \subseteq I$, so $I_0 \subseteq I$. Thus we must show that each $h \in I$ belongs to I_0.

Case 1: $\deg h \leq d$

Then $h \in I^*$, so h is a linear combination of the g_j (with coefficients in $R \subseteq R[X]$), so $h \in I_0$.

Case 2: $\deg h = r > d$

Let a be the leading coefficient of h. Since $a \in J$, we have $a = \sum_{i=1}^{n} c_i a_i$ with the $c_i \in R$. Let

$$q = h - \sum_{i=1}^{n} c_i X^{r-d_i} f_i \in I.$$

The coefficient of X^r in q is

$$a - \sum_{i=1}^{n} c_i a_i = 0$$

so that $\deg q < r$. We can iterate this degree-reduction process until the resulting polynomial has degree d or less, and therefore belongs to I_0. But then h is a finite linear combination of the f_i and g_j. ♣

8.2.2 Corollary

Every variety is the intersection of finitely many *hypersurfaces* (zero-sets of single polynomials).

Proof. Let $V = V(I)$ be a variety. By (8.2.1), I has finitely many generators $f_1, \dots, f_r$. But then $V = \bigcap_{i=1}^r V(f_i)$. ♣

8.2.3 Formal Power Series

The argument used to prove the Hilbert basis theorem can be adapted to show that if R is Noetherian, then the ring $R[[X]]$ of formal power series is Noetherian. We cannot simply reproduce the proof because an infinite series has no term of highest degree, but we can look at the *lowest* degree term. If $f = a_r X^r + a_{r+1} X^{r+1} + \cdots$, where r is a nonnegative integer and $a_r \neq 0$, let us say that f has degree r and leading coefficient a_r. (If $f = 0$, take the degree to be infinite and the leading coefficient to be 0.)

If I is an ideal of $R[[X]]$, we must show that I is finitely generated. We will inductively construct a sequence of elements $f_i \in R[[X]]$ as follows. Let f_1 have minimal degree among elements of I. Suppose that we have chosen $f_1, \dots, f_i$, where f_i has degree d_i and leading coefficient a_i. We then select f_{i+1} satisfying the following three requirements:

1. f_{i+1} belongs to I;
2. a_{i+1} does not belong to $(a_1, \dots, a_i)$, the ideal of R generated by the a_j, $j = 1, \dots, i$;
3. Among all elements satisfying the first two conditions, f_{i+1} has minimal degree.

The second condition forces the procedure to terminate in a finite number of steps; otherwise there would be an infinite ascending chain $(a_1) \subset (a_1, a_2) \subset (a_1, a_2, a_3) \subset \cdots$. If stabilization occurs at step k, we will show that I is generated by $f_1, \dots, f_k$.

Let $g = aX^d + \cdots$ be an element of I of degree d and leading coefficient a. Then $a \in (a_1, \dots, a_k)$ (Problem 1).

Case 1: $d \geq d_k$. Since $d_i \leq d_{i+1}$ for all i (Problem 2), we have $d \geq d_i$ for $i = 1, \dots, k$. Now $a = \sum_{i=1}^k c_{i0} a_i$ with the $c_{i0} \in R$. Define

$$g_0 = \sum_{i=1}^k c_{i0} X^{d-d_i} f_i$$

so that g_0 has degree d and leading coefficient a, and consequently $g - g_0$ has degree greater than d. Having defined $g_0, \dots, g_r \in (f_1, \dots, f_k)$ such that $g - \sum_{i=0}^r g_i$ has degree greater than $d + r$, say

$$g - \sum_{i=0}^r g_i = bX^{d+r+1} + \dots.$$

(The argument is the same if the degree is greater than $d + r + 1$.) Now $b \in (a_1, \dots, a_k)$ (Problem 1 again), so

$$b = \sum_{i=1}^k c_{i,r+1} a_i$$

with $c_{i,r+1} \in R$. We define

$$g_{r+1} = \sum_{i=1}^{k} c_{i,r+1} X^{d+r+1-d_i} f_i$$

so that $g - \sum_{i=0}^{r+1} g_i$ has degree greater than $d + r + 1$. Thus

$$g = \sum_{r=0}^{\infty} g_r = \sum_{r=0}^{\infty} \sum_{i=1}^{k} c_{ir} X^{d+r-d_i} f_i$$

and it follows upon reversing the order of summation that $g \in (f_1, \dots, f_k)$. (The reversal is legal because the inner summation is finite. For a given nonnegative integer j, there are only finitely many terms of the form bX^j.)

Case 2: $d < d_k$. As above, $a \in (a_1, \dots, a_k)$, so there is a smallest m between 1 and k such that $a \in (a_1, \dots, a_m)$. It follows that $d \geq d_m$ (Problem 3). As in case 1 we have $a = \sum_{i=1}^{m} c_i a_i$ with $c_i \in R$. Define

$$h = \sum_{i=1}^{m} c_i X^{d-d_i} f_i \in (f_1, \dots, f_k) \subseteq I.$$

The leading coefficient of h is a, so the degree of $g - h$ is greater than d. We replace g by $g - h$ and repeat the procedure. After at most $d_k - d$ iterations, we produce an element $g - \sum h_i$ in I of degree at least d_k, with all $h_i \in (f_1, \dots, f_k)$. By the analysis of case 1, $g \in (f_1, \dots, f_k)$.

Problems For Section 8.2

1. Justify the step $a \in (a_1, \dots, a_k)$ in (8.2.3).
2. Justify the step $d_i \leq d_{i+1}$ in (8.2.3).
3. Justify the step $d \geq d_m$ in (8.2.3).
4. Let R be a subring of the ring S, and assume that S is finitely generated as an algebra over R. In other words, there are finitely many elements $x_1, \dots, x_n \in S$ such that the smallest subring of S containing the x_i and all elements of R is S itself. Show that S is a homomorhic image of the polynomial ring $R[X_1, \dots, X_n]$.
5. Continuing Problem 4, show that if R is Noetherian, then S is also Noetherian.

8.3 The Nullstellensatz: Preliminaries

We have observed that every variety V defines an ideal $I(V)$ and every ideal I defines a variety $V(I)$. Moreover, if $I(V_1) = I(V_2)$, then $V_1 = V_2$ by (6) of (8.1.3). But it is entirely possible for many ideals to define the same variety. For example, the ideals (f) and (f^m) need not coincide, but their zero-sets are identical. Appearances to the contrary, the two statements in part (6) of (8.1.3) are not symmetrical. A variety V is, by definition, always expressible as $V(S)$ for some collection S of polynomials, but an ideal I need not be of the special form $I(X)$. Hilbert's Nullstellensatz says that if two ideals define the same variety, then, informally, the ideals are the same "up to powers". More precisely, if g belongs to one of the ideals, then g^r belongs to the other ideal for some positive integer r. Thus the only factor preventing a one-to-one correspondence between ideals and varieties is that a polynomial can be raised to a power without affecting its zero-set. In this section we collect some results needed for the proof of the Nullstellensatz. We begin by showing that each point of A^n determines a maximal ideal.

8.3.1 Lemma

If $a = (a_1, \dots, a_n) \in A^n$, then $I = (X_1 - a_1, \dots, X_n - a_n)$ is a maximal ideal.

Proof. Suppose that I is properly contained in the ideal J, with $f \in J \setminus I$. Apply the division algorithm n times to get

$$f = A_1(X_1 - a_1) + A_2(X_2 - a_2) + \cdots + A_n(X_n - a_n) + b$$

where $A_1 \in k[X_1, \dots, X_n]$, $A_2 \in k[X_2, \dots, X_n]$, $\dots$, $A_n \in k[X_n]$, $b \in k$. Note that b cannot be 0 since $f \notin I$. But $f \in J$, so by solving the above equation for b we have $b \in J$, hence $1 = (1/b)b \in J$. Consequently, $J = k[X_1, \dots, X_n]$. ♣

The following definition will allow a precise statement of the Nullstellensatz.

8.3.2 Definition

The *radical* of an ideal I (in any commutative ring R) is the set of all elements $f \in R$ such that $f^r \in I$ for some positive integer r.

A popular notation for the radical of I is $\sqrt{I}$. If f^r and g^s belong to I, then by the binomial theorem, $(f + g)^{r+s-1} \in I$, and it follows that $\sqrt{I}$ is an ideal.

8.3.3 Lemma

If I is any ideal of $k[X_1, \dots, X_n]$, then $\sqrt{I} \subseteq IV(I)$.

Proof. If $f \in \sqrt{I}$, then $f^r \in I$ for some positive integer r. But then f^r vanishes on $V(I)$, hence so does f. Therefore $f \in IV(I)$. ♣

The Nullstellensatz states that $IV(I) = \sqrt{I}$, and the hard part is to prove that $IV(I) \subseteq \sqrt{I}$. The technique is known as the "Rabinowitsch trick", and it is indeed very clever. Assume that $f \in IV(I)$. We introduce a new variable Y and work in $k[X_1, \dots, X_n, Y]$. If I is an ideal of $k[X_1, \dots, X_n]$, then by the Hilbert basis theorem, I is finitely generated, say by $f_1, \dots, f_m$. Let I^* be the ideal of $k[X_1, \dots, X_n, Y]$ generated by $f_1, \dots, f_m, 1 - Yf$. [There is a slight ambiguity: by $f_i(X_1, \dots, X_n, Y)$ we mean $f_i(X_1, \dots, X_n)$, and similarly for f.] At an appropriate moment we will essentially set Y equal to $1/f$ and come back to the original problem.

8.3.4 Lemma

If $(a_1, \ldots, a_n, a_{n+1})$ is any point in A^{n+1} and $(a_1, \ldots, a_n) \in V(I)$ (in other words, the f_i, $i = 1, \ldots, m$, vanish at $(a_1, \ldots, a_n)$), then $(a_1, \ldots, a_n, a_{n+1}) \notin V(I^*)$.

Proof. We are assuming that $f \in IV(I)$, so that f vanishes on the zero-set of $\{f_1, \ldots, f_m\}$. In particular, $f(a_1, \ldots, a_n) = 0$. The value of $1 - Yf$ at $(a_1, \ldots, a_n, a_{n+1})$ is therefore $1 - a_{n+1}f(a_1, \ldots, a_n) = 1 - a_{n+1}(0) = 1 \neq 0$. But $1 - Yf \in I^*$, so $(a_1, \ldots, a_n, a_{n+1})$ does not belong to the zero-set of I^*. ♣

8.3.5 Lemma

If $(a_1, \ldots, a_n, a_{n+1})$ is any point in A^{n+1} and $(a_1, \ldots, a_n) \notin V(I)$, then $(a_1, \ldots, a_n, a_{n+1}) \notin V(I^*)$. Consequently, by (8.3.4), $V(I^*) = \emptyset$.

Proof. By hypothesis, $f_i(a_1, \ldots, a_n, a_{n+1}) \neq 0$ for some i, and since $f_i \in I^*$, $(a_1, \ldots, a_{n+1})$ cannot belong to the zero-set of I^*. ♣

At this point we are going to assume what is called the weak Nullstellensatz, namely that if I is a proper ideal of $k[X_1, \ldots, X_n]$, then $V(I)$ is not empty.

8.3.6 Lemma

There are polynomials $g_1, \ldots, g_m, h \in k[X_1, \ldots, X_n, Y]$ such that

$$1 = \sum_{i=1}^{m} g_i f_i + h(1 - Yf). \tag{1}$$

This equation also holds in the rational function field $k(X_1, \ldots, X_n, Y)$ consisting of quotients of polynomials in $k[X_1, \ldots, X_n, Y]$.

Proof. By (8.3.4) and (8.3.5), $V(I^*) = \emptyset$, so by the weak Nullstellensatz, $I^* = k[X_1, \ldots, X_n, Y]$. In particular, $1 \in I^*$, and since I^* is generated by $f_1, \ldots, f_m, 1 - Yf$, there is an equation of the specified form. The equation holds in the rational function field because a polynomial is a rational function. ♣

8.3.7 The Rabinowitsch Trick

The idea is to set $Y = 1/f$, so that (1) becomes

$$1 = \sum_{i=1}^{m} g_i(X_1, \ldots, X_n, 1/f(X_1, \ldots, X_n)) f_i(X_1, \ldots, X_n). \tag{2}$$

Is this legal? First of all, if f is the zero polynomial, then certainly $f \in \sqrt{I}$, so we can assume $f \neq 0$. To justify replacing Y by $1/f$, consider the ring homomorphism from $k[X_1, \ldots, X_n, Y]$ to $k(X_1, \ldots, X_n)$ determined by $X_i \to X_i$, $i = 1, \ldots, n$, $Y \to 1/f(X_1, \ldots, X_n)$. Applying this mapping to (1), we get (2). Now the right side of (2) is a sum of rational functions whose denominators are various powers of f. If f^r is the highest power that appears, we can absorb all denominators by multiplying (2) by f^r. The result is an equation of the form

$$f^r = \sum_{i=1}^{m} h_i(X_1,\dots,X_n)f_i(X_1,\dots,X_n)$$

where the h_i are *polynomials* in $k[X_1,\dots,X_n]$. Consequently, $f^r \in I$. ♣

The final ingredient is a major result in its own right.

8.3.8 Noether Normalization Lemma

Let A be a finitely generated k-algebra, where k is a field. In other words, there are finitely many elements $x_1,\dots,x_n$ in A that generate A over k in the sense that every element of A is a polynomial in the x_i. Equivalently, A is a homomorphic image of the polynomial ring $k[X_1,\dots,X_n]$ via the map determined by $X_i \to x_i$, $i = 1,\dots,n$.

There exists a subset $\{y_1,\dots,y_r\}$ of A such that the y_i are algebraically independent over k and A is integral over $k[y_1,\dots,y_r]$.

Proof. Let $\{x_1,\dots,x_r\}$ be a maximal algebraically independent subset of $\{x_1,\dots,x_n\}$. If $n = r$ we are finished, since we can take $y_i = x_i$ for all i. Thus assume $n > r$, in which case $x_1,\dots,x_n$ are algebraically dependent over k. Thus there is a nonzero polynomial $f \in k[X_1,\dots,X_n]$ such that $f(x_1,\dots,x_n) = 0$. We can assume $n > 1$, for if $n = 1$ and $r = 0$, then $A = k[x_1]$ and we can take $\{y_1,\dots,y_r\}$ to be the empty set.

We first assume that k is infinite and give a proof by induction on n. (It is possible to go directly to the general case, but the argument is not as intricate for an infinite field.) Decompose f into its homogeneous components (sums of monomials of the same degree). Say that g is the homogeneous component of maximum degree d. Then, regarding g as a polynomial in X_n whose coefficients are polynomials in the other X_i, we have, relabeling variables if necessary, $g(X_1,\dots,X_{n-1},1) \neq 0$. Since k is infinite, it follows from (8.1.3) part (8) that there are elements $a_1,\dots,a_{n-1} \in k$ such that $g(a_1,\dots,a_{n-1},1) \neq 0$. Set $z_i = x_i - a_i x_n$, $i = 1,\dots,n-1$, and plug into $f(x_1,\dots,x_n) = 0$ to get an equation of the form

$$g(a_1,\dots,a_{n-1},1)x_n^d + \text{terms of degree less than } d \text{ in } x_n = 0.$$

A concrete example may clarify the idea. If $f(x_1,x_2) = g(x_1,x_2) = x_1^2x_2^3$ and $x_1 = z_1 + a_1x_2$, then the substitution yields

$$(z_1^2 + 2a_1z_1x_2 + a_1^2x_2^2)x_2^3$$

which indeed is $g(a_1,1)x_2^5$ plus terms of degree less than 5 in x_2. Divide by $g(a_1,\dots,a_{n-1},1) \neq 0$ to conclude that x_n is integral over $B = k[z_1,\dots,z_{n-1}]$. By the induction hypothesis, there are elements $y_1,\dots,y_r$ algebraically independent over k such that B is integral over $k[y_1,\dots,y_r]$. But the $x_i, i < n$, are integral over B since $x_i = z_i + a_ix_n$. By transitivity (see (7.1.4)), $x_1,\dots,x_n$ are integral over $k[y_1,\dots,y_r]$. Thus (see (7.1.5)) A is integral over $k[y_1,\dots,y_r]$.

Now we consider arbitrary k. As before, we produce a nonzero polynomial f such that $f(x_1,\dots,x_n) = 0$. We assign a weight $w_i = s^{n-i}$ to the variable X_i, where s is a large positive integer. (It suffices to take s greater than the total degree of f, that is, the sum of the degrees of all monomials in f.) If $h = \lambda X_1^{a_1} \cdots X_n^{a_n}$ is a monomial of f, we define the weight of h as $w(h) = \sum_{i=1}^n a_i w_i$. The point is that if $h' = \mu X_1^{b_1} \cdots X_n^{b_n}$, then $w(h) > w(h')$ iff $h > h'$ in the lexicographic ordering, that is, for some m we have $a_i = b_i$ for $i \leq m$, and $a_{m+1} > b_{m+1}$. We take h to be the monomial of maximum weight. (If two monomials differ in the lexicographic ordering, they must have different weights.) Set $z_i = x_i - x_n^{w_i}$, $1 \leq i \leq n-1$, and plug into $f(x_1, \dots, x_n) = 0$ to get

$$cx_n^{w(h)} + \text{terms of lower degree in } x_n = 0.$$

For example, if $f(x_1, x_2) = h(x_1, x_2) = x_1^3 x_2^2$, then $x_1 = z_1 + x_2^{w_1}$ gives

$$(z_1^3 + 3z_1^2 x_2^{w_1} + 3z_1 x_2^{2w_1} + x_2^{3w_1})x_2^2$$

and $w(h) = 3w_1 + 2w_2 = 3w_1 + 2$ since $s^{n-2} = s^0 = 1$. Thus x_n is integral over $B = k[z_1, \dots, z_{n-1}]$, and an induction argument finishes the proof as in the first case. ♣

8.3.9 Corollary

Let B be a finitely generated k-algebra, where k is a field. If I is a maximal ideal of B, then B/I is a finite extension of k.

Proof. The field k can be embedded in B/I via $c \to c + I$, $c \in k$. [If $c \in I$, $c \neq 0$, then $c^{-1}c = 1 \in I$, a contradiction.] Since $A = B/I$ is also a finitely generated k-algebra, it follows from (8.3.8) that there is a subset $\{y_1, \dots, y_r\}$ of A with the y_i algebraically independent over k and A integral over $k[y_1, \dots, y_r]$. Now A is a field (because I is a maximal ideal), and therefore so is $k[y_1, \dots, y_r]$ (see the Problems in Section 7.1). But this will lead to a contradiction if $r \geq 1$, because $1/y_1 \notin k[y_1, \dots, y_r]$. (If $1/y_1 = g(y_1, \dots, y_r) \in k[y_1, \dots, y_r]$, then $y_1 g(y_1, \dots, y_r) = 1$, contradicting algebraic independence.) Thus r must be 0, so A is integral, hence algebraic, over the field k. Therefore A is generated over k by finitely many algebraic elements, so by (3.3.3), A is a finite extension of k. ♣

8.3.10 Corollary

Let A be a finitely generated k-algebra, where k is a field. If A is itself a field, then A is a finite extension of k.

Proof. As in (8.3.9), with B/I replaced by A. ♣

Problems For Section 8.3

1. Let S be a multiplicative subset of the ring R (see (2.8.1)). If I is an ideal that is disjoint from S, then by Zorn's lemma, there is an ideal J that is maximal among ideals disjoint from S. Show that J must be prime.

2. Show that the radical of the ideal I is the intersection of all prime ideals containing I. [If $f^r \in I \subseteq P$, P prime, then $f \in P$. Conversely, assume $f \notin \sqrt{I}$. With a clever choice of multiplicative set S, show that for some prime ideal P containing I, we have $f \notin P$.]
3. An *algebraic curve* is a variety defined by a nonconstant polynomial in two variables. Show (using the Nullstellensatz) that the polynomials f and g define the same algebraic curve iff f divides some power of g and g divides some power of f. Equivalently, f and g have the same irreducible factors.
4. Show that the variety V defined over the complex numbers by the two polynomials $Y^2 - XZ$ and $Z^2 - X^2Y$ is the union of the line L given by $Y = Z = 0$, X arbitrary, and the set W of all (t^3, t^4, t^5), $t \in \mathbb{C}$.
5. The *twisted cubic* is the variety V defined over the complex numbers by $Y - X^2$ and $Z - X^3$. In parametric form, $V = \{(t, t^2, t^3) : t \in \mathbb{C}\}$. Show that V is irreducible. [The same argument works for any variety that can be parametrized over an infinite field.]
6. Find parametrizations of the following algebraic curves over the complex numbers. (It is permissible for your parametrizations to fail to cover finitely many points of the curve.)
   (a) The unit circle $x^2 + y^2 = 1$;
   (b) The cuspidal cubic $y^2 = x^3$;
   (c) The nodal cubic $y^2 = x^2 + x^3$.
7. Let f be an irreducible polynomial, and g an arbitrary polynomial, in $k[x, y]$. If f does not divide g, show that the system of simultaneous equations $f(x, y) = g(x, y) = 0$ has only finitely many solutions.

8.4 The Nullstellensatz: Equivalent Versions And Proof

We are now in position to establish the equivalence of several versions of the Nullstellensatz.

8.4.1 Theorem

For any field k and any positive integer n, the following statements are equivalent.

(1) **Maximal Ideal Theorem** The maximal ideals of $k[X_1, \ldots, X_n]$ are the ideals of the form $(X_1 - a_1, \ldots, X_n - a_n)$, $a_1, \ldots, a_n \in k$. Thus maximal ideals correspond to points.

(2) **Weak Nullstellensatz** If I is an ideal of $k[X_1, \ldots, X_n]$ and $V(I) = \emptyset$, then $I = k[X_1, \ldots, X_n]$. Equivalently, if I is a proper ideal, then $V(I)$ is not empty.

(3) **Nullstellensatz** If I is an ideal of $k[X_1, \ldots, X_n]$, then

$$IV(I) = \sqrt{I}.$$

(4) k is algebraically closed.

Proof. (1) implies (2). Let I be a proper ideal, and let J be a maximal ideal containing I. By (8.1.3), part (4), $V(J) \subseteq V(I)$, so it suffices to show that $V(J)$ is not empty. By (1), J has the form $(X_1 - a_1, \dots, X_n - a_n)$. But then $a = (a_1, \dots, a_n) \in V(J)$. [In fact $V(J) = \{a\}$.]

(2) implies (3). This was done in Section 8.3.

(3) implies (2). We use the fact that the radical of an ideal I is the intersection of all prime ideals containing I; see Section 8.3, Problem 2. Let I be a proper ideal of $k[X_1, \dots, X_n]$. Then I is contained in a maximal, hence prime, ideal P. By the result just quoted, $\sqrt{I}$ is also contained in P, hence $\sqrt{I}$ is a proper ideal. By (3), $IV(I)$ is a proper ideal. But if $V(I) = \emptyset$, then by (8.1.3) part (7), $IV(I) = k[X_1, \dots, X_n]$, a contradiction.

(2) implies (1). If I is a maximal ideal, then by (2) there is a point $a = (a_1, \dots, a_n) \in V(I)$. Thus every $f \in I$ vanishes at a, in other words, $I \subseteq I(\{a\})$. But $(X_1 - a_1, \dots, X_n - a_n) = I(\{a\})$; to see this, decompose $f \in I(\{a\})$ as in the proof of (8.3.1). Therefore the maximal ideal I is contained in the maximal ideal $(X_1 - a_1, \dots, X_n - a_n)$, and it follows that $I = (X_1 - a_1, \dots, X_n - a_n)$.

(4) implies (1). Let I be a maximal ideal of $k[X_1, \dots, X_n]$, and let $K = k[X_1, \dots, X_n]/I$, a field containing an isomorphic copy of k via $c \to c + I$, $c \in k$. By (8.3.9), K is a finite extension of k, so by (4), $K = k$. But then $X_i + I = a_i + I$ for some $a_i \in k$, $i = 1, \dots, n$. Therefore $X_i - a_i$ is zero in $k[X_1, \dots, X_n]/I$, in other words, $X_i - a_i \in I$. Consequently, $I \supseteq (X_1 - a_1, \dots, X_n - a_n)$, and we must have equality by (8.3.1).

(1) implies (4). Let f be a nonconstant polynomial in $k[X_1]$ with no root in k. We can regard f is a polynomial in n variables with no root in A^n. Let I be a maximal ideal containing the proper ideal (f). By (1), I is of the form $(X_1 - a_1, \dots, X_n - a_n) = I(\{a\})$ for some $a = (a_1, \dots, a_n) \in A^n$. Therefore f vanishes at a, a contradiction. ♣

8.4.2 Corollary

If the ideals I and J define the same variety and a polynomial g belongs to one of the ideals, then some power of g belongs to the other ideal.

Proof. If $V(I) = V(J)$, then by the Nullstellensatz, $\sqrt{I} = \sqrt{J}$. If $g \in I \subseteq \sqrt{I}$, then $g^r \in J$ for some positive integer r. ♣

8.4.3 Corollary

The maps $V \to I(V)$ and $I \to V(I)$ set up a one-to-one correspondence between varieties and *radical ideals* (defined by $I = \sqrt{I}$).

Proof. By (8.1.3) part 6, $VI(V) = V$. By the Nullstellensatz, $IV(I) = \sqrt{I} = I$ for radical ideals. It remains to prove that for any variety V, $I(V)$ is a radical ideal. If $f^r \in I(V)$, then f^r, hence f, vanishes on V, so $f \in I(V)$. ♣

8.4.4 Corollary

Let $f_1, \dots, f_r, g \in k[X_1, \dots, X_n]$, and assume that g vanishes wherever the f_i all vanish. Then there are polynomials $h_1, \dots, h_r \in k[X_1, \dots, X_n]$ and a positive integer s such that $g^s = h_1 f_1 + \cdots + h_r f_r$.

Proof. Let I be the ideal generated by $f_1, \dots, f_r$. Then $V(I)$ is the set of points at which all f_i vanish, so that $IV(I)$ is the set of polynomials that vanish wherever all f_i vanish. Thus g belongs to $IV(I)$, which is $\sqrt{I}$ by the Nullstellensatz. Consequently, for some positive integer s, we have $g^s \in I$, and the result follows. ♣

Problems For Section 8.4

1. Let f be a polynomial in $k[X_1, \dots, X_n]$, and assume that the factorization of f into irreducibles is $f = f_1^{n_1} \cdots f_r^{n_r}$. Show that the decomposition of the variety $V(f)$ into irreducible subvarieties (Section 8.1, Problems 5 and 6) is given by $V(f) = \cup_{i=1}^r V(f_i)$.
2. Under the hypothesis of Problem 1, show that $IV(f) = (f_1 \cdots f_r)$.
3. Show that there is a one-to-one correspondence between irreducible polynomials in $k[X_1, \dots, X_n]$ and irreducible hypersurfaces (see (8.2.2))in $A^n(k)$, if polynomials that differ by a nonzero multiplicative constant are identified.
4. For any collection of subsets X_i of A^n, show that $I(\cup_i X_i) = \cap_i I(X_i)$.
5. Show that every radical ideal I of $k[X_1, \dots, X_n]$ is the intersection of finitely many prime ideals.
6. In Problem 5, show that the decomposition is unique, subject to the condition that the prime ideals P are *minimal*, that is, there is no prime ideal Q with $I \subseteq Q \subset P$.
7. Suppose that X is a variety in A^2, defined by equations $f_1(x, y) = \cdots = f_m(x, y) = 0$, $m \geq 2$. Let g be the greatest common divisor of the f_i. If g is constant, show that X is a finite set (possibly empty).
8. Show that every variety in A^2 except for A^2 itself is the union of a finite set and an algebraic curve.
9. Give an example of two distinct irreducible polynomials in $k[X, Y]$ with the same zero-set, and explain why this cannot happen if k is algebraically closed.
10. Give an explicit example of the failure of a version of the Nullstellensatz in a non-algebraically closed field.

8.5 Localization

8.5.1 Geometric Motivation

Suppose that V is an irreducible variety, so that $I(V)$ is a prime ideal. A polynomial g will belong to $I(V)$ if and only if it vanishes on V. If we are studying rational functions f/g in the neighborhood of a point $x \in V$, we must have $g(x) \neq 0$. It is very convenient to have every polynomial $g \notin I(V)$ available as a legal object, even though g may vanish at some points of V. The technical device that makes this possible is the construction of the ring of fractions $S^{-1}R$, the localization of R by S, where $R = k[X_1, \dots, X_n]$ and S is the multiplicative set $R \setminus I(V)$. We will now study the localization process in general.

8.5.2 Notation

Recalling the setup of Section 2.8, let S be a multiplicative subset of the ring R, and $S^{-1}R$ the ring of fractions of R by S. Let h be the natural homomorphism of R into $S^{-1}R$, given by $h(a) = a/1$. If X is any subset of R, define $S^{-1}X = \{x/s\colon x \in X, s \in S\}$. We will be especially interested in such a set when X is an ideal.

If I and J are ideals of R, the *product* of I and J, denoted by IJ, is defined (as in (7.6.1)) as the set of all finite sums $\sum_i x_i y_i$, $x_i \in I$, $y_i \in J$. It follows from the definition that IJ is an ideal. The *sum* of two ideals has already been defined in (2.2.8).

8.5.3 Lemma

If I is an ideal of R, then $S^{-1}I$ is an ideal of $S^{-1}R$. If J is another ideal of R, the

(i) $S^{-1}(I + J) = S^{-1}I + S^{-1}J$;

(ii) $S^{-1}(IJ) = (S^{-1}I)(S^{-1}J)$;

(iii) $S^{-1}(I \cap J) = S^{-1}I \cap S^{-1}J$;

(iv) $S^{-1}I$ is a proper ideal iff $S \cap I = \emptyset$.

Proof. The definition of addition and multiplication in $S^{-1}R$ implies that $S^{-1}I$ is an ideal, and that in (i), (ii) and (iii), the left side is contained in the right side. The reverse inclusions in (i) and (ii) follow from

$$\frac{a}{s} + \frac{b}{t} = \frac{at + bs}{st}, \qquad \frac{a}{s}\frac{b}{t} = \frac{ab}{st}.$$

To prove (iii), let $a/s = b/t$ where $a \in I$, $b \in J$, $s, t \in S$. There exists $u \in S$ such that $u(at - bs) = 0$. Then $a/s = uat/ust = ubs/ust \in S^{-1}(I \cap J)$.

Finally, if $s \in S \cap I$ then $1/1 = s/s \in S^{-1}I$, so $S^{-1}I = S^{-1}R$. Conversely, if $S^{-1}I = S^{-1}R$, then $1/1 = a/s$ for some $a \in I$, $s \in S$. There exists $t \in S$ such that $t(s - a) = 0$, so $at = st \in S \cap I$. ♣

Ideals in $S^{-1}R$ must be of a special form.

8.5.4 Lemma

If J is an ideal of $S^{-1}R$ and $I = h^{-1}(J)$, then I is an ideal of R and $S^{-1}I = J$.

Proof. I is an ideal by the basic properties of preimages of sets. Let $a/s \in S^{-1}I$, with $a \in I$ and $s \in S$. Then $a/1 \in J$, so $a/s = (a/1)(1/s) \in J$. Conversely, let $a/s \in J$, with $a \in R$, $s \in S$. Then $h(a) = a/1 = (a/s)(s/1) \in J$, so $a \in I$ and $a/s \in S^{-1}I$. ♣

Prime ideals yield sharper results.

8.5.5 Lemma

If I is any ideal of R, then $I \subseteq h^{-1}(S^{-1}I)$, with equality if I is prime and disjoint from S.

Proof. If $a \in I$, then $h(a) = a/1 \in S^{-1}I$. Thus assume that I is prime and disjoint from S, and let $a \in h^{-1}(S^{-1}I)$. Then $h(a) = a/1 \in S^{-1}I$, so $a/1 = b/s$ for some $b \in I$, $s \in S$. There exists $t \in S$ such that $t(as - b) = 0$. Thus $ast = bt \in I$, with $st \notin I$ since $S \cap I = \emptyset$. Since I is prime, we have $a \in I$. ♣

8.5.6 Lemma

If I is a prime ideal of R disjoint from S, then $S^{-1}I$ is a prime ideal of $S^{-1}R$.

Proof. By (8.5.3), part (iv), $S^{-1}I$ is a proper ideal. Let $(a/s)(b/t) = ab/st \in S^{-1}I$, with $a, b \in R$, $s, t \in S$. Then $ab/st = c/u$ for some $c \in I$, $u \in S$. There exists $v \in S$ such that $v(abu - cst) = 0$. Thus $abuv = cstv \in I$, and $uv \notin I$ because $S \cap I = \emptyset$. Since I is prime, $ab \in I$, hence $a \in I$ or $b \in I$. Therefore either a/s or b/t belongs to $S^{-1}I$. ♣

The sequence of lemmas can be assembled to give a precise conclusion.

8.5.7 Theorem

There is a one-to-one correspondence between prime ideals P of R that are disjoint from S and prime ideals Q of $S^{-1}R$, given by

$$P \to S^{-1}P \text{ and } Q \to h^{-1}(Q).$$

Proof. By (8.5.4), $S^{-1}(h^{-1}(Q)) = Q$, and by (8.5.5), $h^{-1}(S^{-1}P) = P$. By (8.5.6), $S^{-1}P$ is a prime ideal, and $h^{-1}(Q)$ is a prime ideal by the basic properties of preimages of sets. If $h^{-1}(Q)$ meets S, then by (8.5.3) part (iv), $Q = S^{-1}(h^{-1}(Q)) = S^{-1}R$, a contradiction. Thus the maps $P \to S^{-1}P$ and $Q \to h^{-1}(Q)$ are inverses of each other, and the result follows. ♣

8.5.8 Definitions and Comments

If P is a prime ideal of R, then $S = R \setminus P$ is a multiplicative set. In this case, we write $R(P)$ for $S^{-1}R$, and call it the *localization of* R *at* P. (The usual notation is R_P, but it's easier to read without subscripts.) If I is an ideal of R, we write $I(P)$ for $S^{-1}I$. We are going to show that $R(P)$ is a *local ring*, that is, a ring with a unique maximal ideal. First we give some conditions equivalent to the definition of a local ring.

8.5.9 Proposition

For a ring R, the following conditions are equivalent.

(i) R is a local ring;

(ii) There is a proper ideal I of R that contains all nonunits of R;

(iii) The set of nonunits of R is an ideal.

Proof. (i) implies (ii). If a is a nonunit, then (a) is a proper ideal, hence is contained in the unique maximal ideal I.

(ii) implies (iii). If a and b are nonunits, so are $a+b$ and ra. If not, then I contains a unit, so $I = R$, a contradiction.

(iii) implies (i). If I is the ideal of nonunits, then I is maximal, because any larger ideal J would have to contain a unit, so that $J = R$. If H is any proper ideal, then H cannot contain a unit, so $H \subseteq I$. Therefore I is the unique maximal ideal. ♣

8.5.10 Theorem

$R(P)$ is a local ring.

Proof. Let Q be a maximal ideal of $R(P)$. Then Q is prime, so by (8.5.7), $Q = I(P)$ for some prime ideal I of R that is disjoint from S, in other words, contained in P. Thus $Q = I(P) \subseteq P(P)$. If $P(P) = R(P)$, then by (8.5.3) part (iv), P is not disjoint from $S = R \backslash P$, which is impossible. Therefore $P(P)$ is a proper ideal containing every maximal ideal, so it must be the unique maximal ideal. ♣

If R is an integral domain and S is the set of all nonzero elements of R, then $S^{-1}R$ is the quotient field of R. In this case, $S^{-1}R$ is a local ring, because any field is a local ring. ($\{0\}$ is the unique maximal ideal.) Alternatively, we can appeal to (8.5.10) with $P = \{0\}$.

8.5.11 Localization of Modules

If M is an R-module and S a multiplicative subset of R, we can essentially repeat the construction of Section 2.8 to form the localization $S^{-1}M$ of M by S, and thereby divide elements of M by elements of S. If $x, y \in M$ and $s, t \in S$, we call (x, s) and (y, t) equivalent if for some $u \in S$, $u(tx - sy) = 0$. The equivalence class of (x, s) is denoted by x/s, and addition is defined by

$$\frac{x}{s} + \frac{y}{t} = \frac{tx + sy}{st}.$$

If $a/s \in S^{-1}R$ and $x/t \in S^{-1}M$, we define

$$\frac{a}{s}\frac{x}{t} = \frac{ax}{st}.$$

In this way, $S^{-1}M$ becomes an $S^{-1}R$-module. Exactly as in (8.5.3), if M and N are submodules of a module L, then

$$S^{-1}(M+N) = S^{-1}M + S^{-1}N \text{ and } S^{-1}(M \cap N) = S^{-1}M \cap S^{-1}N.$$

Further properties will be given in the exercises.

Problems For Section 8.5

1. Let M be a maximal ideal of R, and assume that for every $x \in M$, $1+x$ is a unit. Show that R is a local ring (with maximal ideal M). [Show that if $x \notin M$, then x is a unit, and apply (8.5.9).]
2. Show that if p is prime and n is a positive integer, then $\mathbb{Z}_{p^n}$ is a local ring with maximal ideal (p).
3. Let R be the ring of all n by n matrices with coefficients in a field F. If A is a nonzero element of R and 1 is the identity matrix, is $\{1, A, A^2, \dots\}$ always a multiplicative set?

Let S be a multiplicative subset of the ring R. We are going to construct a mapping from R-modules to $S^{-1}R$-modules, and another mapping from R-module homomorphisms to $S^{-1}R$-module homomorphisms, as follows. If M is an R-module, we let $M \to S^{-1}M$. If $f\colon M \to N$ is an R-module homomorphism, we define $S^{-1}f\colon S^{-1}M \to S^{-1}N$ by

$$\frac{x}{s} \to \frac{f(x)}{s}.$$

Since f is a homomorphism, so is $S^{-1} f$.

4. If $g\colon N \to L$ and composition of functions is written as a product, show that $S^{-1}(gf) = S^{-1}(g)S^{-1}(f)$, and if 1_M is the identity mapping on M, then $S^{-1}(1_M) = 1_{S^{-1}M}$. We say that S^{-1} is a *functor* from the *category* of R-modules to the category of $S^{-1}R$-modules. This terminology will be explained in great detail in Chapter 10.
5. If

$$M \overset{f}{\rightarrow} N \overset{g}{\rightarrow} L$$

is an exact sequence, show that

$$S^{-1}M \overset{S^{-1}f}{\longrightarrow} S^{-1}N \overset{S^{-1}g}{\longrightarrow} S^{-1}L$$

is exact. We say that S^{-1} is an *exact functor*. Again, we will study this idea in Chapter 10.

6. Let R be the ring of rational functions f/g with $f, g \in k[X_1, \dots, X_n]$ and $g(a) \neq 0$, where $a = (a_1, \dots, a_n)$ is a fixed point in A^n. Show that R is a local ring, and identify the unique maximal ideal.
7. If M is an R-module and S is a multiplicative subset of R, denote $S^{-1}M$ by M_S. If N is a submodule of M, show that $(M/N)_S \cong M_S/N_S$.

8.6 Primary Decomposition

We have seen that every radical ideal in $k[X_1, \dots, X_n]$ can be expressed as an intersection of finitely many prime ideals (Section 8.4, Problem 5). A natural question is whether a similar result holds for arbitrary ideals. The answer is yes if we generalize from prime to primary ideals.

8.6.1 Definitions and Comments

The ideal Q in the ring R is *primary* if Q is proper and whenever a product ab belongs to Q, either $a \in Q$ or $b^n \in Q$ for some positive integer n. [The condition on b is equivalent to $b \in \sqrt{Q}$.] An equivalent statement is that $R/Q \neq 0$ and whenever $(a+Q)(b+Q) = 0$ in R/Q, either $a + Q = 0$ or $(b+Q)^n = 0$ for some positive integer n. This says that if $b + Q$ is a zero-divisor in R/Q, then it is *nilpotent*, that is, some power of $b + Q$ is 0.

It follows from the definition that every prime ideal is primary. Also, if Q is primary, then $\sqrt{Q}$ is the smallest prime ideal containing Q. [Since $\sqrt{Q}$ is the intersection of all prime ideals containing Q (Section 8.3, Problem 2), it suffices to show that $\sqrt{Q}$ is prime. But if $a^n b^n \in Q$, then $a^n \in Q$ or $b^{nm} \in Q$ for some m, so either a or b must belong to $\sqrt{Q}$. Note also that since Q is proper, it is contained in a maximal, hence prime, ideal, so $\sqrt{Q}$ is also proper.]

If Q is primary and $\sqrt{Q} = P$, we say that Q is *P-primary.*

8.6.2 Examples

1. In $\mathbb{Z}$, the primary ideals are $\{0\}$ and (p^r), where p is prime. In $\mathbb{Z}_6$, 2 and 3 are zero-divisors that are not nilpotent, and a similar situation will occur in $\mathbb{Z}_m$ whenever more than one prime appears in the factorization of m.

2. Let $R = k[X,Y]$ where k is any field, and take $Q = (X, Y^3)$, the ideal generated by X and Y^3. This is a nice example of analysis in quotient rings. We are essentially setting X and Y^3 equal to zero, and this collapses the ring R down to polynomials $a_0 + a_1 Y + a_2 Y^2$, with the $a_i \in k$ and arithmetic mod Y^3. Formally, R/Q is isomorphic to $k[Y]/(Y^3)$. The zero-divisors in R/Q are of the form $cY + dY^2$, $c \in k$, and they are nilpotent. Thus Q is primary. If $f \in R$, then the only way for f not to belong to the radical of Q is for the constant term of f to be nonzero. Thus $\sqrt{Q} = (X, Y)$, a maximal ideal by (8.3.1).

Now we claim that Q cannot be a power of a prime ideal; this will be a consequence of the next result.

8.6.3 Lemma

If P is a prime ideal, then for every positive integer n, $\sqrt{P^n} = P$.

Proof. Since P is a prime ideal containing P^n, $\sqrt{P^n} \subseteq P$. If $x \in P$, then $x^n \in P^n$, so $x \in \sqrt{P^n}$. ♣

Returning to Example 2 of (8.6.2), if $Q = (X, Y^3)$ is a prime power P^n, then its radical is P, so P must be (X, Y). But $X \in Q$ and $X \notin P^n, n \geq 2$; since Y belongs to P but not Q, we have reached a contradiction.

After a preliminary definition, we will give a convenient sufficient condition for an ideal to be primary.

8.6.4 Definition

The *nilradical* $\mathcal{N}(R)$ of a ring R is the set of nilpotent elements of R, that is, $\{x \in R\colon x^n = 0 \text{ for some positive integer } n\}$. Thus $\mathcal{N}(R)$ is the radical of the zero ideal, which is the intersection of all prime ideals of R.

8.6.5 Proposition

If the radical of the ideal Q is maximal, then Q is primary.

Proof. Since $\sqrt{Q}$ is maximal, it must be the only prime ideal containing Q. By the correspondence theorem and the fact that the preimage of a prime ideal is a prime ideal (cf. (8.5.7)), R/Q has exactly one prime ideal, which must coincide with $\mathcal{N}(R/Q)$. Any element of R/Q that is not a unit generates a proper ideal, which is contained in a maximal ideal, which again must be $\mathcal{N}(R/Q)$. Thus every element of R/Q is either a unit or nilpotent. Since a zero-divisor cannot be a unit, every zero-divisor of R/Q is nilpotent, so Q is primary. ♣

8.6.6 Corollary

If M is a maximal ideal, then M^n is M-primary for all $n = 1, 2, \dots$.

Proof. By (8.6.3), the radical of M^n is M, and the result follows from (8.6.5). ♣

Here is another useful property.

8.6.7 Proposition

If Q is a finite intersection of P-primary ideals Q_i, $i = 1, \dots, n$, then Q is P-primary.

Proof. First note that the radical of a finite intersection of ideals is the intersection of the radicals (see Problem 1). It follows that the radical of Q is P, and it remains to show that Q is primary. If $ab \in Q$ but $a \notin Q$, then for some i we have $a \notin Q_i$. Since Q_i is P-primary, b belongs to $P = \sqrt{Q_i}$. But then some power of b belongs to Q. ♣

We are going to show that in a Noetherian ring, every proper ideal I has a primary decomposition, that is, I can be expressed as a finite intersection of primary ideals.

8.6.8 Lemma

Call an ideal I *irreducible* if for any ideals J and K, $I = J \cap K$ implies that $I = J$ or $I = K$. If R is Noetherian, then every ideal of R is a finite intersection of irreducible ideals.

Proof. Suppose that the collection $\mathcal{S}$ of all ideals that cannot be so expressed is nonempty. Since R is Noetherian, $\mathcal{S}$ has a maximal element I, necessarily reducible. Let $I = J \cap K$, where I is properly contained in both J and K. By maximality of I, the ideals J and K are finite intersections of irreducible ideals, and consequently so is I, contradicting $I \in \mathcal{S}$. ♣

If we can show that every irreducible proper ideal is primary, we then have the desired primary decomposition. Let us focus on the chain of reasoning we will follow. If I is an irreducible proper ideal of R, then by the correspondence theorem, 0 is an irreducible ideal of the Noetherian ring R/I. If we can show that 0 is primary in R/I, then again by the correspondence theorem, I is primary in R.

8.6.9 Primary Decomposition Theorem

Every proper ideal in a Noetherian ring R has a primary decomposition. (We can drop the word "proper" if we regard R as the intersection of the empty collection of primary ideals.)

Proof. By the above discussion, it suffices to show that if 0 is an irreducible ideal of R, then it is primary. Let $ab = 0$ with $a \neq 0$. Since R is Noetherian, the sequence of annihilators

$$\operatorname{ann} b \subseteq \operatorname{ann} b^2 \subseteq \operatorname{ann} b^3 \subseteq \cdots$$

stabilizes, so $\operatorname{ann} b^n = \operatorname{ann} b^{n+1}$ for some n. If we can show that

$$(a) \cap (b^n) = 0$$

we are finished, because $a \neq 0$ and the zero ideal is irreducible (by hypothesis). Thus let $x = ca = db^n$ for some $c, d \in R$. Then $bx = cab = db^{n+1} = 0$ (because $ab = 0$), so d annihilates b^{n+1}, hence d annihilates b^n. Thus $x = db^n = 0$. ♣

Problems For Section 8.6

1. If $I_1, \dots, I_n$ are arbitrary ideals, show that

$$\sqrt{\bigcap_{i=1}^{n} I_i} = \bigcap_{i=1}^{n} \sqrt{I_i}.$$

2. Let I be the ideal $(XY - Z^2)$ in $k[X, Y, Z]$, where k is any field, and let $R = k[X, Y, Z]/I$. If P is the ideal $(X + I, Z + I)$, show that P is prime.
3. Continuing Problem 2, show that P^2, whose radical is prime by (8.6.3) and which is a power of a prime, is nevertheless not primary.
4. Let $R = k[X, Y]$, and let $P_1 = (X)$, $P_2 = (X, Y)$, $Q = (X^2, Y)$. Show that P_1 is prime and P_2^2 and Q are P_2-primary.
5. Continuing Problem 4, let $I = (X^2, XY)$. Show that $P_1 \cap P_2^2$ and $P_1 \cap Q$ are both primary decompositions of I.

Notice that the radicals of the components of the primary decomposition (referred to as the primes *associated* with I) are P_1 and P_2 in both cases. [P_1 is prime, so $\sqrt{P_1} = P_1$; $P_2 \subseteq \sqrt{Q}$ and P_2 is maximal, so $P_2 = \sqrt{Q}$;] Uniqueness questions involving primary decompositions are treated in detail in textbooks on commutative algebra.

6. We have seen in Problem 5 of Section 8.4 that every radical ideal in $R = k[X_1, \dots, X_n]$ is the intersection of finitely many prime ideals. Show that this result holds in an arbitrary Noetherian ring R.

7. Let $R = k[X, Y]$ and let I_n be the ideal (X^3, XY, Y^n). Show that for every positive integer n, I_n is a primary ideal of R.

8.7 Tensor Product of Modules Over a Commutative Ring

8.7.1 Motivation

In many areas of algebra and its applications, it is useful to multiply, in a sensible way, an element x of an R-module M by an element y of an R-module N. In group representation theory, M and N are free modules, in fact finite-dimensional vector spaces, with bases $\{x_i\}$ and $\{y_j\}$. Thus if we specify that multiplication is linear in each variable, then we need only specify products of x_i and y_j. We require that the these products, to be denoted by $x_i \otimes y_j$, form a basis for a new R-module T.

If $f\colon R \to S$ is a ring homomorphism and M is an S-module, then M becomes an R-module via $rx = f(r)x$, $r \in R$, $x \in M$. This is known as restriction of scalars. In algebraic topology and algebraic number theory, it is often desirable to reverse this process. If M is an R-module, we want to extend the given multiplication rx, $r \in R$, $x \in M$, to multiplication of an arbitrary $s \in S$ by $x \in M$. This is known as extension of scalars, and it becomes possible with the aid of the tensor product construction.

The tensor product arises in algebraic geometry in the following way. Let M be the coordinate ring of a variety V in affine space A^m, in other words, M is the set of all polynomial functions from V to the base field k. Let N be the coordinate ring of the variety W in A^n. Then the cartesian product $V \times W$ is a variety in A^{m+n}, and its coordinate ring turns out to be the tensor product of M and N.

Let's return to the first example above, where M and N are free modules with bases $\{x_i\}$ and $\{y_j\}$. Suppose that f is a bilinear map from $M \times N$ to an R-module P. (In other words, f is R-linear in each variable.) Information about f can be completely encoded into a function g of one variable, where g is an R-module homomorphism from T to P. We take $g(x_i \otimes y_j) = f(x_i, y_j)$ and extend by linearity. Thus f is the composition of the bilinear map h from $M \times N$ to T specified by $(x_i, y_j) \to x_i \otimes y_j$, followed by g. To summarize:

Every bilinear mapping on $M \times N$ can be factored through T.

The R-module T is called the tensor product of M and N, and we write $T = M \otimes_R N$. We are going to construct a tensor product of arbitrary modules over a commutative ring, and sketch the generalization to noncommutative rings.

8.7.2 Definitions and Comments

Let M and N be arbitrary R-modules, and let F be a free R-module with basis $M \times N$. Let G be the submodule of F generated by the "relations"

$$(x+x',y)-(x,y)-(x',y); \quad (x,y+y')-(x,y)-(x,y');$$
$$(rx,y)-r(x,y); \quad (x,ry)-r(x,y)$$

where $x, x' \in M$, $y, y' \in N$, $r \in R$. Define the *tensor product* of M and N (over R) as

$$T = M \otimes_R N = F/G$$

and denote the element $(x,y)+G$ of T by $x \otimes y$. Thus the general element of T is a finite sum of the form

$$t = \sum_i x_i \otimes y_i \tag{1}$$

with $x_i \in M$ and $y_i \in N$. It is important to note that the representation (1) is not necessarily unique.

The relations force $x \otimes y$ to be linear in each variable, so that

$$x \otimes (y+y') = x \otimes y + x \otimes y', \ (x+x') \otimes y = x \otimes y + x' \otimes y, \tag{2}$$
$$r(x \otimes y) = rx \otimes y = x \otimes ry. \tag{3}$$

See Problem 1 for details. Now if f is a bilinear mapping from $M \times N$ to the R-module P, then f extends uniquely to a homomorphism from F to P, also called f. Bilinearity means that the kernel of f contains G, so by the factor theorem, there is a unique R-homomorphism $g\colon T \to P$ such that $g(x \otimes y) = f(x,y)$ for all $x \in M$, $y \in N$. As in (8.7.1), if we compose the bilinear map $h\colon (x,y) \to x \otimes y$ with g, the result is f. Again, we say that

Every bilinear mapping on $M \times N$ can be factored through T.

We have emphasized this sentence, known as a *universal mapping property* (abbreviated UMP), because along with equations (1), (2) and (3), it indicates how the tensor product is applied in practice. The detailed construction we have just gone through can now be forgotten. In fact any two R-modules that satisfy the universal mapping property are isomorphic. The precise statement and proof of this result will be developed in the exercises.

In a similar fashion, using multilinear rather than bilinear maps, we can define the tensor product of any finite number of R-modules. [In physics and differential geometry, a tensor is a multilinear map on a product $M_1 \times \cdots \times M_r$, where each M_i is either a finite-dimensional vector space V or its dual space V^*. This suggests where the terminology "tensor product" comes from.]

In the discussion to follow, M, N and P are R-modules. The ring R is assumed fixed, and we will usually write $\otimes$ rather than $\otimes_R$.

8.7.3 Proposition

$M \otimes N \cong N \otimes M$.

Proof. Define a bilinear mapping $f\colon M \times N \to N \otimes M$ by $f(x, y) = y \otimes x$. By the UMP, there is a homomorphism $g\colon M \otimes N \to N \otimes M$ such that $g(x \otimes y) = y \otimes x$. Similarly, there is a homomorphism $g'\colon N \otimes M \to M \otimes N$ with $g'(y \otimes x) = x \otimes y$. Thus g is an isomorphism (with inverse g'). ♣

8.7.4 Proposition

$M \otimes (N \otimes P) \cong (M \otimes N) \otimes P$.

Proof. Define $f\colon M \times N \times P \to (M \otimes N) \otimes P$ by $f(x, y, z) = (x \otimes y) \otimes z$. The UMP produces $g\colon M \times (N \otimes P) \to (M \otimes N) \otimes P$ with $g((x, (y \otimes z))) = (x \otimes y) \otimes z$. [We are applying the UMP for each fixed $x \in M$, and assembling the maps to produce g.] Since g is bilinear (by Equations (2) and (3)), the UMP yields $h\colon M \otimes (N \otimes P) \to (M \otimes N) \otimes P$ with $h(x \otimes (y \otimes z)) = (x \otimes y) \otimes z$. Exactly as in (8.7.3), we can construct the inverse of h, so h is the desired isomorphism. ♣

8.7.5 Proposition

$M \otimes (N \oplus P) \cong (M \otimes N) \oplus (M \otimes P)$.

Proof. Let f be an arbitrary bilinear mapping from $M \times (N \oplus P)$ to Q. If $x \in M$, $y \in N$, $z \in P$, then $f(x, y + z) = f(x, y) + f(x, z)$. The UMP gives homomorphisms $g_1\colon M \otimes N \to Q$ and $g_2\colon M \otimes P \to Q$ such that $g_1(x \otimes y) = f(x, y)$ and $g_2(x \otimes z) = f(x, z)$. The maps g_1 and g_2 combine to give $g\colon (M \otimes N) \oplus (M \otimes P) \to Q$ such that

$$g((x \otimes y) + (x' \otimes z)) = g_1(x \otimes y) + g_2(x' \otimes z).$$

In particular, with $x' = x$,

$$g((x \otimes y) + (x \otimes z)) = f(x, y + z),$$

so if $h\colon M \times (N \oplus P) \to M \otimes (N \oplus P)$ is defined by

$$h(x, y + z) = (x \otimes y) + (x \otimes z),$$

then $f = gh$. Thus $(M \otimes N) \oplus (M \otimes P)$ satisfies the universal mapping property, hence must be isomorphic to the tensor product. ♣

8.7.6 Proposition

Regarding R as a module over itself, $R \otimes_R M \cong M$.

Proof. The map $(r, x) \to rx$ of $R \times M \to M$ is bilinear, so there is a homomorphism $g\colon R \otimes M \to M$ such that $g(r \otimes x) = rx$. Define $h\colon M \to R \otimes M$ by $h(x) = 1 \otimes x$. Then $h(rx) = 1 \otimes rx = r1 \otimes x = r \otimes x$. Thus g is an isomorphism (with inverse h). ♣

8.7.7 Corollary

Let R^m be the direct sum of m copies of R, and M^m the direct sum of m copies of M. Then $R^m \otimes M \cong M^m$.

Proof. By (8.7.5), $R^m \otimes M$ is isomorphic to the direct sum of m copies of $R \otimes M$, which is isomorphic to M^m by (8.7.6). ♣

8.7.8 Proposition

$R^m \otimes R^n \cong R^{mn}$. Moreover, if $\{x_1, \dots, x_m\}$ is a basis for R^m and $\{y_1, \dots, y_n\}$ is a basis for R^n, then $\{x_i \otimes y_j, i = 1, \dots, m, j = 1, \dots, n\}$ is a basis for R^{mn}.

Proof. This follows from the discussion in (8.7.1). [The first assertion can also be proved by taking $M = R^n$ in (8.7.7).] ♣

8.7.9 Tensor Product of Homomorphisms

Let $f_1 \colon M_1 \to N_1$ and $f_2 \colon M_2 \to N_2$ be R-module homomorphisms. The map $(x_1, x_2) \to f_1(x_1) \otimes f_2(x_2)$ of $M_1 \times M_2$ into $N_1 \otimes N_2$ is bilinear, and induces a unique $f \colon M_1 \otimes M_2 \to N_1 \otimes N_2$ such that

$$f(x_1 \otimes x_2) = f_1(x_1) \otimes f_2(x_2), \ x_1 \in M_1, x_2 \in M_2.$$

We write $f = f_1 \otimes f_2$, and call it the *tensor product* of f_1 and f_2. Similarly, if $g_1 \colon N_1 \to P_1$ and $g_2 \colon N_2 \to P_2$, then we can compose $g_1 \otimes g_2$ with $f_1 \otimes f_2$, and

$$(g_1 \otimes g_2)(f_1 \otimes f_2)(x_1 \otimes x_2) = g_1 f_1(x_1) \otimes g_2 f_2(x_2),$$

hence

$$(g_1 \otimes g_2) \circ (f_1 \otimes f_2) = (g_1 \circ f_1) \otimes (g_2 \circ f_2).$$

When $M_1 = N_1 = V$, a free R-module of rank m, and $M_2 = N_2 = W$, a free R-module of rank n, there is a very concrete interpretation of the tensor product of the endomorphisms $f \colon V \to V$ and $g \colon W \to W$. If f is represented by the matrix A and g by the matrix B, then the action of f and g on basis elements is given by

$$f(v_j) = \sum_i a_{ij} v_i, \quad g(w_l) = \sum_k b_{kl} w_k$$

where i and j range from 1 to m, and k and l range from 1 to n. Thus

$$(f \otimes g)(v_j \otimes w_l) = f(v_j) \otimes g(w_l) = \sum_{i,k} a_{ij} b_{kl} (v_i \otimes w_k).$$

The mn by mn matrix representing the endomorphism $f \otimes g \colon V \otimes W \to V \otimes W$ is denoted by $A \otimes B$ and called the *tensor product* or *Kronecker product* of A and B. It is given by

$$A \otimes B = \begin{bmatrix} a_{11}B & a_{12}B & \cdots & a_{1m}B \\ \vdots & & & \vdots \\ a_{m1}B & a_{m2}B & \cdots & a_{mm}B \end{bmatrix}.$$

The ordering of the basis of $V \otimes W$ is

$$v_1 \otimes w_1, \ldots, v_1 \otimes w_n, \ldots, v_m \otimes w_1, \ldots, v_m \otimes w_n.$$

To determine the column of $A \otimes B$ corresponding to $v_j \otimes w_l$, locate the $a_{ij}B$ block ($i = 1, \ldots, m$; j fixed) and proceed to column l of B. As we move down this column, the indices i and k vary according to the above ordering of basis elements. If this road map is not clear, perhaps writing out the entire matrix for $m = 2$ and $n = 3$ will help.

Problems For Section 8.7

1. Verify Equations (2) and (3) of (8.7.2).
2. If m and n are relatively prime, show that $\mathbb{Z}_m \otimes_{\mathbb{Z}} \mathbb{Z}_n = 0$.
3. If A is a finite abelian group and $\mathbb{Q}$ is the additive group of rationals, show that $A \otimes_{\mathbb{Z}} \mathbb{Q} = 0$. Generalize to a wider class of abelian groups A.
4. The definition of $M \otimes_R N$ via a universal mapping property is as follows. The tensor product is an R-module T along with a bilinear map $h \colon M \times N \to T$ such that given any bilinear map $f \colon M \times N \to P$, there is a unique R-homomorphism $g \colon T \to P$ such that $f = gh$. See the diagram below.

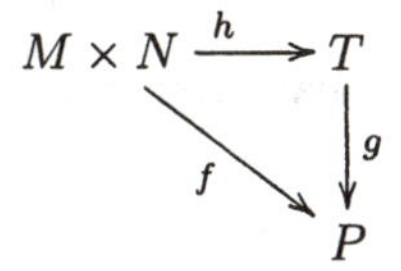

   Now suppose that another R-module T', along with a bilinear mapping $h' \colon M \times N \to T'$, satisfies the universal mapping property. Using the above diagram with $P = T'$ and f replaced by h', we get a unique homomorphism $g \colon T \to T'$ such that $h' = gh$. Reversing the roles of T and T', we get $g' \colon T' \to T$ such that $h = g'h'$.

   Show that T and T' are isomorphic.
5. Consider the element $n \otimes x$ in $\mathbb{Z} \otimes \mathbb{Z}_n$, where x is any element of $\mathbb{Z}_n$ and we are tensoring over $\mathbb{Z}$, i.e., $R = \mathbb{Z}$. Show that $n \otimes x = 0$.
6. Continuing Problem 5, take $x \neq 0$ and regard $n \otimes x$ as an element of $n\mathbb{Z} \otimes \mathbb{Z}_n$ rather than $\mathbb{Z} \otimes \mathbb{Z}_n$. Show that $n \otimes x \neq 0$.
7. Let M, N, M', N' be arbitrary R-modules, where R is a commutative ring. Show that the tensor product of homomorphisms induces a linear map from $\operatorname{Hom}_R(M, M') \otimes_R \operatorname{Hom}_R(N, N')$ to $\operatorname{Hom}_R(M \otimes_R N, M' \otimes_R N')$.
8. Let M be a free R-module of rank m, and N a free R-module of rank n. Show that there is an R-module isomorphism of $\operatorname{End}_R(M) \otimes_R \operatorname{End}_R(N)$ and $\operatorname{End}_R(M \otimes N)$.

8.8 General Tensor Products

We now consider tensor products of modules over noncommutative rings. A natural question is "Why not simply repeat the construction of (8.7.2) for an arbitrary ring R?".

But this construction forces

$$rx \otimes sy = r(x \otimes sy) = rs(x \otimes y)$$

and

$$rx \otimes sy = s(rx \otimes y) = sr(x \otimes y)$$

which cannot hold in general if R is noncommutative. A solution is to modify the construction so that the tensor product T is only an abelian group. Later we can investigate conditions under which T has a module structure as well.

8.8.1 Definitions and Comments

Let M be a right R-module and N a left R-module. (We often abbreviate this as M_R and ${}_RN$.) Let $f\colon M \times N \to P$, where P is an abelian group. The map f is *biadditive* if it is additive in each variable, that is, $f(x+x', y) = f(x, y)+f(x', y)$ and $f(x, y+y') = f(x, y)+f(x, y')$ for all $x, x' \in M$, $y, y' \in N$. The map f is *R-balanced* if $f(xr, y) = f(x, ry)$ for all $x \in M$, $y \in N$, $r \in R$. As before, the key idea is the *universal mapping property*: Every biadditive, R-balanced map can be factored through the tensor product.

8.8.2 Construction of the General Tensor Product

If M_R and ${}_RN$, let F be the free abelian group with basis $M \times N$. Let G be the subgroup of R generated by the relations

$$\begin{aligned}
&(x + x', y) - (x, y) - (x', y);\\
&(x, y + y') - (x, y) - (x, y');\\
&(xr, y) - (x, ry)
\end{aligned}$$

where $x, x' \in M$, $y, y' \in N$, $r \in R$. Define the *tensor product* of M and N over R as

$$T = M \otimes_R N = F/G$$

and denote the element $(x, y) + G$ of T by $x \otimes y$. Thus the general element of T is a finite sum of the form

$$t = \sum_i x_i \otimes y_i. \tag{1}$$

The relations force the map $h\colon (x, y) \to x \otimes y$ of $M \times N$ into T to be biadditive and R-balanced, so that

$$x \otimes (y + y') = x \otimes y + x \otimes y', \ (x + x') \otimes y = x \otimes y + x' \otimes y, \tag{2}$$

$$xr \otimes y = x \otimes ry. \tag{3}$$

If f is a biadditive, R-balanced mapping from $M \times N$ to the abelian group P, then f extends uniquely to an abelian group homomorphism from F to P, also called f. Since f is biadditive and R-balanced, the kernel of f contains G, so by the factor theorem there is a unique abelian group homomorphism $g\colon T \to P$ such that $g(x \otimes y) = f(x, y)$ for all $x \in M$, $y \in N$. Consequently, $gh = f$ and we have the universal mapping property:

Every biadditive, R-balanced mapping on $M \times N$ can be factored through T.

As before, any two abelian groups that satisfy the universal mapping property are isomorphic.

8.8.3 Bimodules

Let R and S be arbitrary rings. We say that M is an $S - R$ *bimodule* if M is both a left S-module and a right R-module, and in addition a compatibility condition is satisfied: $(sx)r = s(xr)$ for all $s \in S$, $r \in R$. We often abbreviate this as ${}_SM_R$.

If $f\colon R \to S$ is a ring homomorphism, then S is a left S-module, and also a right R-module by restriction of scalars, as in (8.7.1). The compatibility condition is satisfied: $(sx)r = sxf(r) = s(xr)$. Therefore S is an $S - R$ bimodule.

8.8.4 Proposition

If ${}_SM_R$ and ${}_RN_T$, then $M \otimes_R N$ is an $S - T$ bimodule.

Proof. Fix $s \in S$. The map $(x, y) \to sx \otimes y$ of $M \times N$ into $M \otimes_R N$ is biadditive and R-balanced. The latter holds because by the compatibility condition in the bimodule property of M, along with (3) of (8.8.2),

$$s(xr) \otimes y = (sx)r \otimes y = sx \otimes ry.$$

Thus there is an abelian group endomorphism on $M \otimes_R N$ such that $x \otimes y \to sx \otimes y$, and we use this to define scalar multiplication on the left by s. A symmetrical argument yields scalar multiplication on the right by t. To check the compatibility condition,

$$[s(x \otimes y)]t = (sx \otimes y)t = sx \otimes yt = s(x \otimes yt) = s[(x \otimes y)t]. \quad \clubsuit$$

8.8.5 Corollary

If ${}_SM_R$ and ${}_RN$, then $M \otimes_R N$ is a left S-module. If M_R and ${}_RN_T$, then $M \otimes_R N$ is a right T-module.

Proof. The point is that every module is, in particular, an abelian group, hence a $\mathbb{Z}$-module. Thus for the first statement, take $T = \mathbb{Z}$ in (8.8.4), and for the second statement, take $S = \mathbb{Z}$. $\clubsuit$

8.8.6 Extensions

As in Section 8.7, we can define the tensor product of any finite number of modules using multiadditive maps (additive in each variable) that are balanced. For example, suppose that M_R, ${}_RN_S$ and ${}_SP$. If $f\colon M \times N \times P \to G$, where G is an abelian group, the condition of balance is

$$f(xr, y, z) = f(x, ry, z) \text{ and } f(x, ys, z) = f(x, y, sz)$$

for all $x \in M$, $y \in N$, $z \in P$, $r \in R$, $s \in S$. An argument similar to the proof of (8.7.4) shows that

(a) $M \otimes_R N \otimes_S P \cong (M \otimes_R N) \otimes_S P \cong M \otimes_R (N \otimes_S P)$.

If M is a right R-module, and N and P are left R-modules, then

(b) $M \otimes_R (N \oplus P) \cong (M \otimes_R N) \oplus (M \otimes_R P)$.

This is proved as in (8.7.5), in fact the result can be extended to the direct sum of an arbitrary (not necessarily finite) number of left R-modules.

If M is a left R-module, then exactly as in (8.7.6) and (8.7.7), we have

(c) $R \otimes_R M \cong M$ and

(d) $R^m \otimes M \cong M^m$.

Let M_1 and M_2 be right R-modules, and let N_1 and N_2 be left R-modules. If $f_1 \colon M_1 \to N_1$ and $f_2 \colon M_2 \to N_2$ are R-module homomorphisms, the tensor product $f_1 \otimes f_2$ can be defined exactly as in (8.7.9). As before, the key property is

$$(f_1 \otimes f_2)(x_1 \otimes x_2) = f_1(x_1) \otimes f_2(x_2)$$

for all $x_1 \in M_1$, $x_2 \in M_2$.

8.8.7 Tensor Product of Algebras

If A and B are algebras over the commutative ring R, then the tensor product $A \otimes_R B$ becomes an R-algebra if we define multiplication appropriately. Consider the map of $A \times B \times A \times B$ into $A \otimes_R B$ given by

$$(a, b, a', b') \to aa' \otimes bb', \quad a, a' \in A, \quad b, b' \in B.$$

The map is 4-linear, so it factors through the tensor product to give an R-module homomorphism $g \colon A \otimes B \otimes A \otimes B \to A \otimes B$ such that

$$g(a \otimes b \otimes a' \otimes b') = aa' \otimes bb'.$$

Now let $h \colon (A \otimes B) \times (A \otimes B) \to A \otimes B \otimes A \otimes B$ be the bilinear map given by

$$h(u, v) = u \otimes v.$$

If we apply h followed by g, the result is a bilinear map $f \colon (A \otimes B) \times (A \otimes B) \to A \otimes B$ with

$$f(a \otimes b, a' \otimes b') = aa' \otimes bb',$$

and this defines our multiplication $(a \otimes b)(a' \otimes b')$ on $A \otimes B$. The multiplicative identity is $1_A \otimes 1_B$, and the distributive laws can be checked routinely. Thus $A \otimes_R B$ is a ring that is also an R-module. To check the compatibility condition, note that if $r \in R$, $a, a' \in A$, $b, b' \in B$, then

$$r[(a \otimes b)(a' \otimes b')] = [r(a \otimes b)](a' \otimes b') = (a \otimes b)[r(a' \otimes b')];$$

all three of these expressions coincide with $raa' \otimes bb' = aa' \otimes rbb'$.

Problems For Section 8.8

We will use the tensor product to define the *exterior algebra* of an R-module M, where R is a commutative ring. If p is a positive integer, we form the tensor product $M \otimes_R \cdots \otimes_R M$ of M with itself p times, denoted by $M^{\otimes p}$. Let N be the submodule of $M^{\otimes p}$ generated by those elements $x_1 \otimes \cdots \otimes x_p$, with $x_i \in M$ for all i, such that $x_i = x_j$ for some $i \neq j$. The pth *exterior power* of M is defined as

$$\Lambda^p M = M^{\otimes p}/N.$$

In most applications, M is a free R-module with a finite basis $x_1, \dots, x_n$ (with $1 \leq p \leq n$), and we will only consider this case. To simplify the notation, we write the element $a \otimes b \otimes \cdots \otimes c + N$ of $\Lambda^p M$ as $ab \cdots c$. (The usual notation is $a \wedge b \wedge \cdots \wedge c$.)

1. Let $y_1, \dots, y_p \in M$. Show that if y_i and y_j are interchanged in the product $y_1 \cdots y_p$, then the product is multiplied by -1.
2. Show that the products $x_{i_1} \cdots x_{i_p}$, where $i_1 < \cdots < i_p$, span $\Lambda^p M$.
3. Let $f\colon M^p \to Q$ be a multilinear map from M^p to the R-module Q, and assume that f is *alternating*, that is, $f(m_1, \dots, m_p) = 0$ if $m_i = m_j$ for some $i \neq j$. Show that f can be factored through $\Lambda^p M$, in other words, there is a unique R-homomorphism $g\colon \Lambda^p M \to Q$ such that $g(y_1 \cdots y_p) = f(y_1, \dots, y_p)$.

Let $y_i = \sum_{j=1}^n a_{ij} x_j$, $i = 1, \dots, n$. Since $\{x_1, \dots, x_n\}$ is a basis for M, y_i can be identified with row i of A. By the basic properties of determinants, the map $f(y_1, \dots, y_n) = \det A$ is multilinear and alternating, and $f(x_1, \dots, x_n) = 1$, the determinant of the identity matrix.

4. Show that $x_1 \cdots x_n \neq 0$ in $\Lambda^n M$, and that $\{x_1 \cdots x_n\}$ is a basis for $\Lambda^n M$.

Let $I = \{i_1, \cdots, i_p\}$, where $i_1 < \cdots < i_p$, and write the product $x_{i_1} \cdots x_{i_p}$ as x_I. Let J be the complementary set of indices. (For example, if $n = 5$, $p = 3$, and $I = \{1, 2, 4\}$, then $J = \{3, 5\}$.) Any equation involving $x_I \in \Lambda^p M$ can be multiplied by x_J to produce a valid equation in $\Lambda^n M$.

5. Show that the products x_I of Problem 2 are linearly independent, so that $\Lambda^p M$ is a free R-module of rank $\binom{n}{p}$.

Roughly speaking, the exterior algebra of M consists of the $\Lambda^p M$ for all p. By construction, $\Lambda^1 M = M$ and $\Lambda^p M = 0$ for $p > n$, since some index must repeat in any element of $\Lambda^p M$. By convention, we take $\Lambda^0 M = R$. Formally, the exterior powers are assembled into a *graded R-algebra*

$$A_0 \oplus A_1 \oplus A_2 \oplus \cdots$$

where $A_p = \Lambda^p M$. Multiplication is defined as in the discussion after Problem 4, that is, if $y_1 \cdots y_p \in A_p$ and $z_1 \cdots z_q \in A_q$, then the *exterior product* $y_1 \cdots y_p z_1 \cdots z_q$ belongs to A_{p+q}.

A ring R is said to be *graded* if, as an abelian group, it is the direct sum of subgroups R_n, $n = 0, 1, 2, \dots$, with $R_m R_n \subseteq R_{n+m}$ for all $m, n \geq 0$. [Example: $R =$

$k[X_1, \ldots, X_n]$, R_n = all homogeneous polynomials of degree n.] By definition, R_0 is a subring of R (because $R_0 R_0 \subseteq R_0$), and each R_n is a module over R_0 (because $R_0 R_n \subseteq R_n$).

6. Suppose that the ideal $I = \oplus_{n \geq 1} R_n$ is generated over R by finitely many elements $x_1, \ldots, x_r$, with $x_i \in R_{n_i}$. Show that $R_n \subseteq S = R_0[x_1, \ldots, x_r]$ for all $n = 0, 1, \ldots$, so that $R = S$.

7. Show that R is a Noetherian ring if and only if R_0 is Noetherian and R is a finitely generated R_0-algebra.

Chapter 9

Introducing Noncommutative Algebra

We will discuss noncommutative rings and their modules, concentrating on two fundamental results, the Wedderburn structure theorem and Maschke's theorem. Further insight into the structure of rings will be provided by the Jacobson radical.

9.1 Semisimple Modules

A vector space is the direct sum of one-dimensional subspaces (each subspace consists of scalar multiples of a basis vector). A one-dimensional space is simple in the sense that it does not have a nontrivial proper subspace. Thus any vector space is a direct sum of simple subspaces. We examine those modules which behave in a similar manner.

9.1.1 Definition

An R-module M is *simple* if $M \neq 0$ and the only submodules of M are 0 and M.

9.1.2 Theorem

Let M be a nonzero R-module. The following conditions are equivalent, and a module satisfying them is said to be *semisimple* or *completely reducible.*

(a) M is a sum of simple modules;

(b) M is a direct sum of simple modules;

(c) If N is a submodule of M, then N is a direct summand of M, that is, there is a submodule N' of M such that $M = N \oplus N'$.

Proof. (a) implies (b). Let M be the sum of simple modules M_i, $i \in I$. If $J \subseteq I$, denote $\sum_{j \in J} M_j$ by $M(J)$. By Zorn's lemma, there is a maximal subset J of I such that the sum defining $N = M(J)$ is direct. We will show that $M = N$. First assume that $i \notin J$.

Then $N \cap M_i$ is a submodule of the simple module M_i, so it must be either 0 or M_i. If $N \cap M_i = 0$, then $M(J \cup \{i\})$ is direct, contradicting maximality of J. Thus $N \cap M_i = M_i$, so $M_i \subseteq N$. But if $i \in J$, then $M_i \subseteq N$ by definition of N. Therefore $M_i \subseteq N$ for all i, and since M is the sum of all the M_i, we have $M = N$.

(b) implies (c). This is essentially the same as (a) implies (b). Let N be a submodule of M, where M is the direct sum of simple modules M_i, $i \in I$. Let J be a maximal subset of I such that the sum $N + M(J)$ is direct. If $i \notin J$ then exactly as before, $M_i \cap (N \oplus M(J)) = M_i$, so $M_i \subseteq N \oplus M(J)$. This holds for $i \in J$ as well, by definition of $M(J)$. It follows that $M = N \oplus M(J)$. [Notice that the complementary submodule N' can be taken as a direct sum of some of the original M_i.]

(c) implies (a). First we make several observations.

(1) If M satisfies (c), so does every submodule N. [Let $N \leq M$, so that $M = N \oplus N'$. If V is a submodule of N, hence of M, we have $M = V \oplus W$. If $x \in N$, then $x = v + w$, $v \in V$, $w \in W$, so $w = x - v \in N$ (using $V \leq N$). But v also belongs to N, and consequently $N = (N \cap V) \oplus (N \cap W) = V \oplus (N \cap W)$.]

(2) If $D = A \oplus B \oplus C$, then $A = (A + B) \cap (A + C)$. [If $a + b = a' + c$, where $a, a' \in A$, $b \in B$, $c \in C$, then $a' - a = b - c$, and since D is a direct sum, we have $b = c = 0$ and $a = a'$. Thus $a + b \in A$.]

(3) If N is a nonzero submodule of M, then N contains a simple submodule.

[Choose a nonzero $x \in N$. By Zorn's lemma, there is a maximal submodule V of N such that $x \notin V$. By (1) we can write $N = V \oplus V'$, and $V' \neq 0$ by choice of x and V. If V' is simple, we are finished, so assume the contrary. Then V' contains a nontrivial proper submodule V_1, so by (1) we have $V' = V_1 \oplus V_2$ with the V_j nonzero. By (2), $V = (V + V_1) \cap (V + V_2)$. Since $x \notin V$, either $x \notin V + V_1$ or $x \notin V + V_2$, which contradicts the maximality of V.]

To prove that (c) implies (a), let N be the sum of all simple submodules of M. By (c) we can write $M = N \oplus N'$. If $N' \neq 0$, then by (3), N' contains a simple submodule V. But then $V \leq N$ by definition of N. Thus $V \leq N \cap N' = 0$, a contradiction. Therefore $N' = 0$ and $M = N$. ♣

9.1.3 Proposition

Nonzero submodules and quotient modules of a semisimple module are semisimple.

Proof. The submodule case follows from (1) of the proof of (9.1.2). Let $N \leq M$, where $M = \sum_i M_i$ with the M_i simple. Applying the canonical map from M to M/N, we have

$$M/N = \sum_i (M_i + N)/N.$$

This key idea has come up before; see the proofs of (1.4.4) and (4.2.3). By the second isomorphism theorem, $(M_i + N)/N$ is isomorphic to a quotient of the simple module M_i. But a quotient of M_i is isomorphic to M_i or to zero, and it follows that M/N is a sum of simple modules. By (a) of (9.1.2), M/N is semisimple. ♣

Problems For Section 9.1

1. Regard a ring R as an R-module. Show that R is simple if and only if R is a division ring.
2. Let M be an R-module, with x a nonzero element of M. Define the R-module homomorphism $f\colon R \to Rx$ by $f(r) = rx$. Show that the kernel I of f is a proper ideal of R, and R/I is isomorphic to Rx.
3. If M is a nonzero R-module, show that M is simple if and only if $M \cong R/I$ for some maximal left ideal I.
4. If M is a nonzero R-module, show that M is simple if and only if M is cyclic (that is, M can be generated by a single element) and every nonzero element of M is a generator.
5. What do simple $\mathbb{Z}$-modules look like?
6. If F is a field, what do simple $F[X]$-modules look like?
7. Let V be an n-dimensional vector space over a field k. (Take $n \geq 1$ so that $V \neq 0$.) If f is an endomorphism (that is, a linear transformation) of V and $x \in V$, define $fx = f(x)$. This makes V into a module over the endomorphism ring $\mathrm{End}_k(V)$. Show that the module is simple.
8. Show that a nonzero module M is semisimple if and only if every short exact sequence $0 \to N \to M \to P \to 0$ splits.

9.2 Two Key Theorems

If M is a simple R-module, there are strong restrictions on a homomorphism either into or out of M. A homomorphism from one simple R-module to another is very severely restricted, as Schur's lemma reveals. This very useful result will be important in the proof of Wedderburn's structure theorem. Another result that will be needed is a theorem of Jacobson that gives some conditions under which a module homomorphism f amounts to multiplication by a fixed element of a ring, at least on part of the domain of f.

9.2.1 Schur's Lemma

(a) If $f \in \mathrm{Hom}_R(M, N)$ where M and N are simple R-modules, then f is either identically 0 or an isomorphism.

(b) If M is a simple R-module, then $\mathrm{End}_R(M)$ is a division ring.

Proof. (a) The kernel of f is either 0 or M, and the image of f is either 0 or N. If f is not the zero map, then the kernel is 0 and the image is N, so f is an isomorphism.

(b) Let $f \in \mathrm{End}_R(M)$, f not identically 0. By (a), f is an isomorphism, and therefore is invertible in the endomorphism ring of M. ♣

The next result prepares for Jacobson's theorem.

9.2.2 Lemma

Let M be a semisimple R-module, and let A be the endomorphism ring $\text{End}_R(M)$. [Note that M is an A-module; if $g \in A$ we take $g \bullet x = g(x)$, $x \in M$.] If $m \in M$ and $f \in \text{End}_A(M)$, then there exists $r \in R$ such that $f(m) = rm$.

Before proving the lemma, let's look more carefully at $\text{End}_A(M)$. Suppose that $f \in \text{End}_A(M)$ and $x \in M$. If $g \in A$ then $f(g(x)) = g(f(x))$. Thus $\text{End}_A(M)$ consists of those abelian group endomorphisms of M that commute with everything in $\text{End}_R(M)$. In turn, by the requirement that $f(rx) = rf(x)$, $\text{End}_R(M)$ consists of those abelian group endomorphisms of M that commute with R, more precisely with multiplication by r, for each $r \in R$. For this reason, $\text{End}_A(M)$ is sometimes called the *double centralizer* of R.

We also observe that the map taking $r \in R$ to multiplication by r is a ring homomorphism of R into $\text{End}_A(M)$. [Again use $rf(x) = f(rx)$.] Jacobson's theorem will imply that given any f in $\text{End}_A(M)$ and any finite set $S \subseteq M$, some g in the image of this ring homomorphism will agree with f on S. Thus in (9.2.2), we can replace the single element m by an arbitrary finite subset of M.

Proof. By (9.1.2) part (c), we can express M as a direct sum $Rm \oplus N$. Now if we have a direct sum $U = V \oplus W$ and $u = v + w$, $v \in V$, $w \in W$, there is a natural projection of U on V, namely $u \to v$. In the present case, let π be the natural projection of M on Rm. Then $\pi \in A$ and $f(m) = f(\pi m) = \pi f(m) \in Rm$. The result follows. ♣

Before proving Jacobson's theorem, we review some ideas that were introduced in the exercises in Section 4.4.

9.2.3 Comments

To specify an R-module homomorphism ψ from a direct sum $V^* = \oplus_{j=1}^n V_j$ to a direct sum $W^* = \oplus_{i=1}^m W_i$, we must give, for every i and j, the ith component of the image of $v_j \in V_j$. Thus the homomorphism is described by a matrix $[\psi_{ij}]$, where ψ_{ij} is a homomorphism from V_j to W_i. The ith component of $\psi(v_j)$ is $\psi_{ij}(v_j)$, so the ith component of $\psi(v_1 + \cdots + v_n)$ is $\sum_{j=1}^n \psi_{ij}(v_j)$. Consequently,

$$\psi(v_1 + \cdots + v_n) = [\psi_{ij}] \begin{bmatrix} v_1 \\ \vdots \\ v_n \end{bmatrix}. \tag{1}$$

This gives an abelian group isomorphism between $\text{Hom}_R(V^*, W^*)$ and $[\text{Hom}_R(V_j, W_i)]$, the collection of all m by n matrices whose ij entry is an R-module homomorphism from V_j to W_i. If we take $m = n$ and $V_i = W_j = V$ for all i and j, then $V^* = W^* = V^n$, the direct sum of n copies of V. Then the abelian group isomorphism given by (1) becomes

$$\text{End}_R(V^n) \cong M_n(\text{End}_R(V)), \tag{2}$$

the collection of all n by n matrices whose entries are$R$-endomorphisms of V. Since composition of endomorphisms corresponds to multiplication of matrices, (2) gives a ring isomorphism as well.

9.2.4 Theorem (Jacobson)

Let M be a semisimple R-module, and let A be the endomorphism ring $\mathrm{End}_R(M)$. If $f \in \mathrm{End}_A(M)$ and $m_1, \dots, m_n \in M$, then there exists $r \in R$ such that $f(m_i) = rm_i$ for all $i = 1, \dots, n$.

Proof. f induces an endomorphism $f^{(n)}$ of M^n, the direct sum of n copies of M, via

$$f^{(n)}(m_1 + \cdots + m_n) = f(m_1) + \cdots + f(m_n)$$

where $f(m_i)$ belongs to the ith copy of M. Thus the matrix that represents $f^{(n)}$ is the scalar matrix fI, where I is an n by n identity matrix. If $B = \mathrm{End}_R(M^n)$, then since a scalar matrix commutes with everything, $f^{(n)} \in \mathrm{End}_B(M^n)$. If $m_1, \dots, m_n \in M$, then by (9.2.2), there exists $r \in R$ such that $f^{(n)}(m_1 + \cdots m_n) = r(m_1 + \cdots m_n)$. [Note that since M is semisimple, so is M^n.] This is equivalent to $f(m_i) = rm_i$ for all i. ♣

Before giving a corollary, we must mention that the standard results that every vector space over a field has a basis, and any two bases have the same cardinality, carry over if the field is replaced by a division ring. Also recall that a module is said to be faithful if its annihilator is 0.

9.2.5 Corollary

Let M be a faithful, simple R-module, and let $D = \mathrm{End}_R(M)$, a division ring by (9.2.1(b)). If M is a finite-dimensional vector space over D, then $\mathrm{End}_D(M) \cong R$, a ring isomorphism.

Proof. Let $\{x_1, \dots, x_n\}$ be a basis for M over D. By (9.2.4), if $f \in \mathrm{End}_D(M)$, there exists $r \in R$ such that $f(x_i) = rx_i$ for all $i = 1, \dots, n$. Since the x_i form a basis, we have $f(x) = rx$ for every $x \in M$. Thus the map h from R to $\mathrm{End}_D(M)$ given by $r \to g_r =$ multiplication by r is surjective. If $rx = 0$ for all $x \in M$, then since M is faithful, we have $r = 0$ and h is injective. Since $h(rs) = g_r \circ g_s = h(r)h(s)$, h is a ring isomorphism. ♣

Problems For Section 9.2

1. Criticize the following argument. Let M be a simple R-module, and let $A = \mathrm{End}_R(M)$. "Obviously" M is also a simple A-module. For any additive subgroup N of M that is closed under the application of all R-endomorphisms of M is, in particular, closed under multiplication by an element $r \in R$. Thus N is an R-submodule of M, hence is 0 or M.

2. Let M be a nonzero cyclic module. Show that M is simple if and only if ann M, the annihilator of M, is a maximal left ideal.

3. In Problem 2, show that the hypothesis that M is cyclic is essential.

4. Let $V = F^n$ be the n-dimensional vector space of all n-tuples with components in the field F. If T is a linear transformation on V, then V becomes an $F[X]$-module via

$f(X)v = f(T)v$. For example, if $n = 2$, $T(a,b) = (0,a)$, and $f(X) = a_0 + a_1X + \cdots + a_nX^n$, then

$$\begin{aligned} f(X)(1,0) &= a_0(1,0) + a_1T(1,0) + a_2T^2(1,0) + \cdots + a_nT^n(1,0) \\ &= (a_0,0) + (0,a_1) \\ &= (a_0,a_1). \end{aligned}$$

Show that in this case, V is cyclic but not simple.

5. Suppose that M is a finite-dimensional vector space over an algebraically closed field F, and in addition M is a module over a ring R containing F as a subring. If M is a simple R-module and f is an R-module homomorphism, in particular an F-linear transformation, on M, show that f is multiplication by some fixed scalar $\lambda \in F$. This result is frequently given as a third part of Schur's lemma.

6. Let I be a left ideal of the ring R, so that R/I is an R-module but not necessarily a ring. Criticize the following statement: "Obviously", I annihilates R/I.

9.3 Simple and Semisimple Rings

9.3.1 Definitions and Comments

Since a ring is a module over itself, it is natural to call a ring R *semisimple* if it is semisimple as an R-module. Our aim is to determine, if possible, how semisimple rings are assembled from simpler components. A plausible idea is that the components are rings that are simple as modules over themselves. But this turns out to be too restrictive, since the components would have to be division rings (Section 9.1, Problem 1).

When we refer to a *simple left ideal* I of R, we will always mean that I is simple as a left R-module. We say that *the ring R is simple* if R is semisimple and all simple left ideals of R are isomorphic. [The definition of simple ring varies in the literature. An advantage of our choice (also favored by Lang and Bourbaki) is that we avoid an awkward situation in which a ring is simple but not semisimple.] Our goal is to show that the building blocks for semisimple rings are rings of matrices over a field, or more generally, over a division ring.

The next two results give some properties of modules over semisimple rings.

9.3.2 Proposition

If R is a semisimple ring, then every nonzero R-module M is semisimple.

Proof. By (4.3.6), M is a quotient of a free R-module F. Since F is a direct sum of copies of R (see (4.3.4)), and R is semisimple by hypothesis, it follows from (9.1.2) that F is semisimple. By (9.1.3), M is semisimple. ♣

9.3.3 Proposition

Let I be a simple left ideal in the semisimple ring R, and let M be a simple R-module. Denote by IM the R-submodule of M consisting of all finite linear combinations $\sum_i r_i x_i$, $r_i \in I$, $x_i \in M$. Then either $IM = M$ and I is isomorphic to M, or $IM = 0$.

Proof. If $IM \neq 0$, then since M is simple, $IM = M$. Thus for some $x \in M$ we have $Ix \neq 0$, and again by simplicity of M, we have $Ix = M$. Map I onto M by $r \to rx$, and note that the kernel cannot be I because $Ix \neq 0$. Since I is simple, the kernel must be 0, so $I \cong M$. ♣

9.3.4 Beginning the Decomposition

Let R be a semisimple ring. We regard two simple left ideals of R as equivalent if they are isomorphic (as R-modules), and we choose a representative I_i, $i \in T$ from each equivalence class. We define the basic building blocks of R as

$$B_i = \text{the sum of all left ideals of } R \text{ that are isomorphic to } I_i.$$

We have a long list of properties of the B_i to establish, and for the sake of economy we will just number the statements and omit the words "Lemma" and "Proof" in each case. We will also omit the end of proof symbol, except at the very end.

9.3.5

If $i \neq j$, then $B_i B_j = 0$. [The product of two left ideals is defined exactly as in (9.3.3).]

Apply (9.3.3) with I replaced by B_i and M by B_j.

9.3.6

$R = \sum_{i \in T} B_i$

If $r \in R$, then (r) is a left ideal, which by (9.1.2) and (9.1.3) (or (9.3.2)) is a sum of simple left ideats.

9.3.7

Each B_i is a two-sided ideal.

Using (9.3.5) and (9.3.6) we have

$$B_i \subseteq B_i R = B_i \sum_j B_j = B_i B_i \subseteq R B_i \subseteq B_i.$$

Thus $RB_i = B_i R = B_i$.

9.3.8

R has only finitely many isomorphism classes of simple left ideals $I_1, \dots, I_t$.

By (9.3.6), we can write the identity 1 of R as a finite sum of elements $e_i \in B_i$, $i \in T$. Adjusting the notation if necessary, let $1 = \sum_{i=1}^t e_i$. If $r \in B_j$ where $j \notin \{1, \dots, t\}$, then by (9.3.5), $re_i = 0$ for all $i = 1, \dots, t$, so $r = r1 = 0$. Thus $B_j = 0$ for $j \notin \{1, \dots, t\}$.

9.3.9

$R = \oplus_{i=1}^t B_i$. Thus 1 has a unique representation as $\sum_{i=1}^t e_i$, with $e_i \in B_i$.

By (9.3.6) and (9.3.8), R is the sum of the B_i. If $b_1 + \cdots + b_t = 0$, with $b_i \in B_i$, then

$$0 = e_i(b_1 + \cdots + b_t) = e_ib_1 + \cdots e_ib_t = e_ib_i = (e_1 + \cdots + e_t)b_i = 1b_i = b_i.$$

Therefore the sum is direct.

9.3.10

If $b_i \in B_i$, then $e_ib_i = b_i = b_ie_i$. Thus e_i is the identity on B_i and $B_i = Re_i = e_iR$.

The first assertion follows from the computation in (9.3.9), along with a similar computation with e_i multiplying on the right instead of the left. Now $B_i \subseteq Re_i$ because $b_i = b_ie_i$, and $Re_i \subseteq B_i$ by (9.3.7) and the fact that $e_i \in B_i$. The proof that $B_i = e_iR$ is similar.

9.3.11

Each B_i is a simple ring.

By the computation in (9.3.7), along with (9.3.10), B_i is a ring (with identity e_i). Let J be a simple left ideal of B_i. By (9.3.5) and (9.3.6), $RJ = B_iJ = J$, so J is a left ideal of R, necessarily simple. Thus J is isomorphic to some I_j, and we must have $j = i$. [Otherwise, J would appear in the sums defining both B_i and B_j, contradicting (9.3.9).] Therefore B_i has only one isomorphism class of simple left ideals. Now B_i is a sum of simple left ideals of R, and a subset of B_i that is a left ideal of R must be a left ideal of B_i. Consequently, B_i is semisimple and the result follows.

9.3.12

If M is a simple R-module, then M is isomorphic to some I_i. Thus there are only finitely many isomorphism classes of simple R-modules. In particular, if R is a simple ring, then all simple R-modules are isomorphic.

By (9.3.9),

$$R = \sum_{i=1}^t B_i = \sum_{i=1}^t \sum \{J \colon J \cong I_i\}$$

where the J are simple left ideals. Therefore

$$M = RM = \sum_{i=1}^t B_iM = \sum_{i=1}^t \sum \{JM \colon J \cong I_i\}.$$

By (9.3.3), $JM = 0$ or $J \cong M$. The former cannot hold for all J, since $M \neq 0$. Thus $M \cong I_i$ for some i. If R is a simple ring, then there is only one i, and the result follows.

9.3.13

Let M be a nonzero R-module, so that M is semisimple by (9.3.2). Define M_i as the sum of all simple submodules of M that are isomorphic to I_i, so that by (9.3.12), $M = \sum_{i=1}^t M_i$. Then

$$M = \bigoplus_{i=1}^{t} B_i M \text{ and } B_i M = e_i M = M_i, \; i = 1, \dots, t.$$

By definition of B_i,

$$B_i M_j = \sum \{JM_j \colon J \cong I_i\}$$

where the J's are simple left ideals. If N is any simple module involved in the definition of M_j, then JN is 0 or N, and by (9.3.3), $JN = N$ implies that $N \cong J \cong I_i$. But all such N are isomorphic to I_j, and therefore $B_i M_j = 0, i \neq j$. Thus

$$M_i = RM_i = \sum_j B_j M_i = B_i M_i$$

and

$$B_i M = \sum_j B_i M_j = B_i M_i.$$

Consequently, $M_i = B_i M = e_i RM = e_i M$ (using (9.3.10)), and all that remains is to show that the sum of the M_i is direct. Let $x_1 + \cdots + x_t = 0$, $x_i \in M_i$. Then

$$0 = e_i(x_1 + \cdots + x_t) = e_i x_i$$

since $e_i x_j \in B_i M_j = 0$ for $i \neq j$. Finally, by (9.3.9),

$$e_i x_i = (e_1 + \cdots + e_t) x_i = x_i.$$

9.3.14

A semisimple ring R is *ring-isomorphic* to a direct product of simple rings.

This follows from (9.3.9) and (9.3.5). For if $a_i, b_i \in B_i$, then

$$(a_1 + \cdots + a_t)(b_1 + \cdots + b_t) = a_1 b_1 + \cdots + a_t b_t. \quad \clubsuit$$

Problems For Section 9.3

In Problems 1 and 2, let M be a semisimple module, so that M is the direct sum of simple modules M_i, $i \in I$. We are going to show that M is a finite direct sum of simple modules if and only if M is finitely generated.

1. Suppose that $x_1, \dots, x_n$ generate M. It will follow that M is the direct sum of finitely many of the M_i. How would you determine which M_i's are involved?

2. Conversely, assume that M is a finite sum of simple modules. Show that M is finitely generated.
3. A left ideal I is said to be *minimal* if $I \neq 0$ and I has no proper subideal except 0. Show that the ring R is semisimple if and only if R is a direct sum of minimal left ideals.
4. Is $\mathbb{Z}$ semisimple?
5. Is $\mathbb{Z}_n$ semisimple?
6. Suppose that R is a ring with the property that every nonzero R-module is semisimple. Show that every R-module M is projective, that is, every exact sequence $0 \to A \to B \to M \to 0$ splits. Moreover, M is injective, that is, every exact sequence $0 \to M \to A \to B \to 0$ splits. [Projective and injective modules will be studied in Chapter 10.]
7. For any ring R, show that the following conditions are equivalent.
   (a) R is semisimple;
   (b) Every nonzero R-module is semisimple;
   (c) Every R-module is projective;
   (d) Every R-module is injective.

9.4 Further Properties of Simple Rings, Matrix Rings, and Endomorphisms

To reach the Wedderburn structure theorem, we must look at simple rings in more detail, and supplement what we already know about matrix rings and rings of endomorphisms.

9.4.1 Lemma

Let R be any ring, regarded as a left module over itself. If $h\colon R \to M$ is an R-module homomorpbism, then for some $x \in M$ we have $h(r) = rx$ for every $r \in R$. Moreover, we may choose $x = h(1)$, and the map $h \to h(1)$ is an isomorphism of $\operatorname{Hom}_R(R, M)$ and M. This applies in particular when $M = R$, in which case $h \in \operatorname{End}_R(R)$.

Proof. The point is that h is determined by what it does to the identity. Thus

$$h(r) = h(r1) = rh(1)$$

so we may take $x = h(1)$. If $s \in R$ and $h \in \operatorname{Hom}_R(R, M)$, we take $(sh)(r) = h(rs) = rsx$. This makes $\operatorname{Hom}_R(R, M)$ into a left R-module isomorphic to M. (For further discussion of this idea, see the exercises in Section 10.7.) ♣

Notice that although all modules are *left* R-modules, h is given by multiplication on the *right* by x.

9.4.2 Corollary

Let I and J be simple left ideals of the simple ring R. Then for some $x \in R$ we have $J = Ix$.

Proof. By the definition of a simple ring (see (9.3.1)), R is semisimple, so by (9.1.2), $R = I \oplus L$ for some left ideal L. Again by the definition of a simple ring, I and J are isomorphic (as R-modules). If $\tau\colon I \to J$ is an isomorphism and π is the natural projection of R on I, then $\tau\pi \in \operatorname{End}_R(R)$, so by (9.4.1), there exists $x \in R$ such that $\tau\pi(r) = rx$ for every $r \in R$. Allow r to range over I to conclude that $J = Ix$. ♣

A semisimple ring can be expressed as a direct sum of simple left ideals, by (9.1.2). If the ring is simple, only finitely many simple left ideals are needed.

9.4.3 Lemma

A simple ring R is a finite direct sum of simple left ideals.

Proof. Let $R = \oplus_j I_j$ where the I_j are simple left ideals. Changing notation if necessary, we have $1 = y_1 + \cdots + y_m$ with $y_j \in I_j$, $j = 1, \dots, m$. If $x \in R$, then

$$x = x1 = \sum_{j=1}^{m} xy_j \in \sum_{j=1}^{m} I_j.$$

Therefore R is a finite sum of the I_j, and the sum is direct because the original decomposition of R is direct. ♣

9.4.4 Corollary

If I is a simple left ideal of the simple ring R, then $IR = R$.

Proof. If J is any simple left ideal of R, then by (9.4.2), $J \subseteq IR$. By (9.4.3), R is a finite (direct) sum of simple left ideals, so $R \subseteq IR$. The reverse inclusion always holds, and the result follows. ♣

We now have some insight into the structure of simple rings.

9.4.5 Proposition

If R is a simple ring, then the only two-sided ideals of R are 0 and R.

Proof. Let J be a nonzero 2-sided ideal of R. By (9.1.3), J is a semisimple left R-module, so by (9.1.2), J is a sum of simple left ideals of J, hence of R. In particular, J contains a simple left ideal I. Since J is a right ideal, it follows that $J = JR$. Using (9.4.4), we have

$$J = JR \supseteq IR = R$$

so $J = R$. ♣

In the literature, a simple ring is often defined as a ring R whose only two-sided ideals are 0 and R, but then extra hypotheses must be added to force R to be semisimple. See the exercises for further discussion.

9.4.6 Corollary

Let I be a simple left ideal of the simple ring R, and let M be a simple R-module. Then $IM = M$ and M is faithful.

Proof. The first assertion follows from a computation that uses associativity of scalar multiplication in a module, along with (9.4.4):

$$M = RM = (IR)M = I(RM) = IM. \tag{1}$$

Now let b belong to the annihilator of M, so that $bM = 0$. We must show that $b = 0$. By a computation similar to (1) (using in addition the associativity of ring multiplication),

$$RbRM = RbM = R0 = 0. \tag{2}$$

But RbR is a two-sided ideal of R (see (2.2.7)), so by (9.4.5), $RbR = 0$ or R. In the latter case, $M = RM = RbRM = 0$ by (2), contradicting the assumption that M is simple. Therefore $RbR = 0$, in particular, $b = 1b1 = 0$. ♣

We are now ready to show that a simple ring is isomorphic to a ring of matrices. Let R be a simple ring, and V a simple R-module. [V exists because R is a sum of simple left ideals, and V is unique up to isomorphism by (9.3.12).] Let $D = \text{End}_R(V)$, a division ring by Schur's lemma (9.2.1(b)). Then (see (9.2.2)), V is a D-module, in other words, a vector space over D. V is a faithful R-module by (9.4.6), and if we can prove that V is finite-dimensional as a vector space over D, then by (9.2.5), R is ring-isomorphic to $\text{End}_D(V)$. If n is the dimension of V over D, then by (4.4.1), $\text{End}_D(V) \cong M_n(D^o)$, the ring of n by n matrices with entries in the opposite ring D^o.

9.4.7 Theorem

Let R be a simple ring, V a simple R-module, and D the endomorphism ring $\text{End}_R(V)$. Then V is a finite-dimensional vector space over D. If the dimension of this vector space is n, then (by the above discussion),

$$R \cong \text{End}_D(V) \cong M_n(D^o).$$

Proof. Assume that we have infinitely many linearly independent elements $x_1, x_2, \ldots$. Let I_m be the left ideal $\{r \in R\colon rx_i = 0 \text{ for all } i = 1, \ldots, m\}$. Then the I_m decrease as m increases, in fact they decrease strictly. [Given any m, let f be a D-linear transformation on V such that $f(x_i) = 0$ for $1 \le i \le m$ and $f(x_{m+1}) \neq 0$. By Jacobson's theorem (9.2.4), there exists $r \in R$ such that $f(x_i) = rx_i$, $i = 1, \ldots, m+1$. But then $rx_1 = \cdots = rx_m = 0$, $rx_{m+1} \neq 0$, so $r \in I_m \setminus I_{m+1}$.] Write $I_m = J_m \oplus I_{m+1}$, as in (9.1.2) part (c). [Recall from (9.1.3) that since R is semisimple, so are all left ideals.] Iterating this process, we construct a left ideal $J_1 \oplus J_2 \oplus \cdots$, and again by (9.1.2(c)),

$$R = J_0 \oplus J_1 \oplus J_2 \oplus \cdots .$$

Therefore 1 is a finite sum of elements $y_i \in J_i$, $i = 0, 1, \dots, t$. But then

$$R = J_0 \oplus J_1 \oplus \cdots \oplus J_t$$

and it follows that J_{t+1} must be 0, a contradiction. ♣

Problems For Section 9.4

Problems 1–5 are the key steps in showing that a ring R is simple if and only if R is Artinian and has no two-sided ideals except 0 and R. Thus if a simple ring is defined as one with no nontrivial two-sided ideals, then the addition of the Artinian condition gives our definition of simple ring; in particular, it forces the ring to be semisimple. The result that an Artinian ring with no nontrivial two-sided ideals is isomorphic to a matrix ring over a division ring (Theorem 9.4.7) is sometimes called the *Wedderburn-Artin theorem.*

In Problems 1–5, "simple" will always mean simple in our sense.

1. By (9.4.5), a simple ring has no nontrivial two-sided ideals. Show that a simple ring must be Artinian.
2. If R is an Artinian ring, show that there exists a simple R-module.
3. Let R be an Artinian ring with no nontrivial two-sided ideals. Show that R has a faithful, simple R-module.
4. Continuing Problem 3, if V is a faithful, simple R-module, and $D = \operatorname{End}_R(V)$, show that V is a finite-dimensional vector space over D.
5. Continuing Problem 4, show that R is ring-isomorphic to $\operatorname{End}_D(V)$, and therefore to a matrix ring $M_n(D^o)$ over a division ring.

In the next section, we will prove that a matrix ring over a division ring is simple; this concludes the proof that R is simple iff R is Artinian with no nontrivial two-sided ideals. (In the "if" part, semisimplicity of R follows from basic properties of matrix rings; see Section 2.2, Problems 2, 3 and 4.)

6. If an R-module M is a direct sum $\oplus_{i=1}^n M_i$ of finitely many simple modules, show that M has a composition series. (Equivalently, by (7.5.12), M is Artinian and Noetherian.)
7. Conversely, if M is semisimple and has a composition series, show that M is a finite direct sum of simple modules. (Equivalently, by Section 9.3, Problems 1 and 2, M is finitely generated.)

9.5 The Structure of Semisimple Rings

We have now done all the work needed for the fundamental theorem.

9.5.1 Wedderburn Structure Theorem

Let R be a semisimple ring.

(1) R is ring-isomorphic to a direct product of simple rings $B_1, \dots, B_t$.

(2) There are t isomorphism classes of simple R-modules. If $V_1, \dots, V_t$ are representatives of these classes, let D_i be the division ring $\operatorname{End}_R(V_i)$. Then V_i is a finite-dimensional vector space over D_i. If n_i is the dimension of this vector space, then there is a ring isomorphism

$$B_i \cong \operatorname{End}_{D_i}(V_i) \cong M_{n_i}(D_i^o).$$

Consequently, R is isomorphic to the direct product of matrix rings over division rings. Moreover,

(3) $B_iV_j = 0$, $i \neq j$; $B_iV_i = V_i$.

Proof. Assertion (1) follows from (9.3.5), (9.3.9) and (9.3.14). By (9.3.8) and (9.3.12), there are t isomorphism classes of simple R-modules. The remaining statements of (2) follow from (9.4.7). The assertions of (3) follow from (9.3.13) and its proof. ♣

Thus a semisimple ring can always be assembled from matrix rings over division rings. We now show that such matrix rings can never combine to produce a ring that is not semisimple.

9.5.2 Theorem

The ring $M_n(R)$ of all n by n matrices with entries in the division ring R is simple.

Proof. We have done most of the work in the exercises for Section 2.2. Let C_k be the set of matrices whose entries are 0 except perhaps in column k, $k = 1 \dots, n$. Then C_k is a left ideal of $M_n(R)$, and if any nonzero matrix in C_k belongs to a left ideal I, then $C_k \subseteq I$. (Section 2.2, Problems 2, 3, 4.) Thus each C_k is a simple left ideal, and $M_n(R)$, the direct sum of $C_1, \dots, C_n$, is semisimple.

Now let I be a nonzero simple left ideal. A nonzero matrix in I must have a nonzero entry in some column, say column k. Define $f\colon I \to C_k$ by $f(A) = A_k$, the matrix obtained from A by replacing every entry except those in column k by 0. Then f is an $M_n(R)$-module homomorphism, since

$$f(BA) = (BA)_k = BA_k = Bf(A).$$

By construction, f is not identically 0, so by Schur's lemma, f is an isomorphism. Since the C_k are mutually isomorphic, all simple left ideals are isomorphic, proving that $M_n(R)$ is simple. ♣

9.5.3 Informal Introduction to Group Representations

A major application of semisimple rings and modules occurs in group representation theory, and we will try to indicate the connection. Let k be any field, and let G be a finite group. We form the *group algebra* kG, which is a vector space over k with basis vectors corresponding to the elements of G. In general, if $G = \{x_1, \dots, x_m\}$, the elements of kG are of the form $\alpha_1 x_1 + \cdots + \alpha_m x_m$, where the α_i belong to k. Multiplication in kG is defined in the natural way; we set

$$(\alpha x_i)(\beta x_j) = \alpha\beta x_i x_j$$

and extend by linearity. Then kG is a ring (with identity $1_k 1_G$) that is also a vector space over k, and $\alpha(xy) = (\alpha x)y = x(\alpha y)$, $\alpha \in k$, $x, y \in G$, so kG is indeed an algebra over k. [This construction can be carried out with an arbitrary ring R in place of k, and with an arbitrary (not necessarily finite) group G. The result is the *group ring* RG, a free R-module with basis G.]

Now let V be an n-dimensional vector space over k. We want to describe the situation in which "G acts linearly on V". We are familiar with group action (Section 5.1), but we now add the condition that each $g \in G$ determines a linear transformation $\rho(g)$ on V. We will write $\rho(g)(v)$ as simply gv or $g(v)$, so that $g(\alpha v + \beta w) = \alpha g(v) + \beta g(w)$. Thus we can multiply vectors in V by scalars in G. Since elements of kG are linear combinations of elements of G with coefficients in k, we can multiply vectors in V by scalars in kG. To summarize very compactly,

$$V \text{ is a } kG\text{-module.}$$

Now since G acts on V, $(hg)v = h(gv)$ and $1_G v = v$, $g, h \in G$, $v \in V$. Thus $\rho(hg) = \rho(h)\rho(g)$, and each $\rho(g)$ is invertible since $\rho(g)\rho(g^{-1}) = \rho(1_G) =$ the identity on V. Therefore

$$\rho \text{ is a homomorphism from } G \text{ to } GL(V),$$

the group of invertible linear transformations on V. Multiplication in $GL(V)$ corresponds to composition of functions.

The homomorphism ρ is called a *representation* of G in V,

and n, the dimension of V, is called the *degree* of the representation. If we like, we can replace $GL(V)$ by the group of all nonsingular n by n matrices with entries in k. In this case, ρ is called a *matrix representation.*

The above process can be reversed. Given a representation ρ, we can define a linear action of G on V by $gv = \rho(g)(v)$, and thereby make V a kG-module. Thus representations can be identified with kG-modules.

9.5.4 The Regular Representation

If G has order n, then kG is an n-dimensional vector space over k with basis G. We take V to be kG itself, with gv the product of g and v in kG. As an example, let $G = \{e, a, a^2\}$, a cyclic group of order 3. V is a 3-dimensional vector space with basis e, a, a^2, and the action of G on V is determined by

$$ee = e, \; ea = a, \; ea^2 = a^2;$$
$$ae = a, \; aa = a^2, \; aa^2 = e;$$
$$a^2e = a^2, \; a^2a = e, \; a^2a^2 = a.$$

Thus the matrices $\rho(g)$ associated with the elements $g \in G$ are

$$[e] = \begin{bmatrix} 1 & 0 & 0 \\ 0 & 1 & 0 \\ 0 & 0 & 1 \end{bmatrix}, \; [a] = \begin{bmatrix} 0 & 0 & 1 \\ 1 & 0 & 0 \\ 0 & 1 & 0 \end{bmatrix}, \; [a^2] = \begin{bmatrix} 0 & 1 & 0 \\ 0 & 0 & 1 \\ 1 & 0 & 0 \end{bmatrix}.$$

9.5.5 The Role of Semisimplicity

Suppose that ρ is a representation of G in V. Assume that the basis vectors of V can be decomposed into two subsets $v(A)$ and $v(B)$ such that matrix of *every* $g \in G$ has the form

$$[g] = \begin{bmatrix} A & 0 \\ 0 & B \end{bmatrix}.$$

(The elements of A and B will depend on the particular g, but the dimensions of A and B do not change.) The corresponding statement about V is that

$$V = V_A \oplus V_B$$

where V_A and V_B are kG-submodules of V. We can study the representation by analyzing its behavior on the simpler spaces V_A and V_B. Maschke's theorem, to be proved in the next section, says that under wide conditions on the field k, this decomposition process can be continued until we reach subspaces that have no nontrivial kG-submodules. In other words, every kG-module is semisimple. In particular, kG is a semisimple ring, and the Wedderburn structure theorem can be applied to get basic information about representations.

We will need some properties of projection operators, and it is convenient to take care of this now.

9.5.6 Definitions and Comments

A linear transformation π on a vector space V [or more generally, a module homomorphism] is called a *projection* of V (on $\pi(V)$) if π is *idempotent*, that is, $\pi^2 = \pi$. We have already met the natural projection of a direct sum onto a component, but there are other possibilities. For example, let p be the projection of $\mathbb{R}^2 = \mathbb{R} \oplus \mathbb{R}$ given by $p(x, y) = \left(\frac{x-y}{2}, \frac{-x+y}{2}\right)$. Note that π must be the identity on $\pi(V)$, since $\pi(\pi(v)) = \pi(v)$.

If we choose subspaces carefully, we can regard any projection as natural.

9.5.7 Proposition

If π is a projection on V, then V is the direct sum of the image of π and the kernel of π.

Proof. Since $v = \pi(v) + (v - \pi(v))$ and $\pi(v - \pi(v)) = 0$, $V = \operatorname{im} V + \ker V$. To show that the sum is direct, let $v = \pi(w) \in \ker \pi$. Then $0 = \pi(v) = \pi^2(w) = \pi(w) = v$, so $\operatorname{im} \pi \cap \ker \pi = 0$. ♣

9.5.8 Example

For real numbers x and y, we have $(x, y) = (x - cy)(1, 0) + y(c, 1)$, where c is any fixed real number. Thus $\mathbb{R}^2 = \mathbb{R}(1, 0) \oplus \mathbb{R}(c, 1)$, and if we take $p(x, y) = (x - cy, 0)$, then p is a projection of $\mathbb{R}^2$ onto $\mathbb{R}(1, 0)$. By varying c we can change the complementary subspace $\mathbb{R}(c, 1)$. Thus we have many distinct projections onto the same subspace $\mathbb{R}(1, 0)$.

Problems For Section 9.5

1. Show that the regular representation is *faithful*, that is, the homomorphism ρ is injective.
2. Let G be a subgroup of S_n and let V be an n-dimensional vector space over k with basis $v(1), \dots, v(n)$. Define the action of G on V by

$$g(v(i)) = v(g(i)), \; i = 1, \dots, n.$$

   Show that the action is legal. (V is called a *permutation module.*)
3. Continuing Problem 2, if $n = 4$, find the matrix of $g = (1, 4, 3)$.
4. Here is an example of how a representation can arise in practice. Place an equilateral triangle in the plane V, with the vertices at $v_1 = (1, 0)$, $v_2 = \left(-\frac{1}{2}, \frac{1}{2}\sqrt{3}\right)$ and $v_3 = \left(-\frac{1}{2}, -\frac{1}{2}\sqrt{3}\right)$; note that $v_1 + v_2 + v_3 = 0$. Let $G = D_6$ be the group of symmetries of the triangle, with $g =$ counterclockwise rotation by 120 degrees and $h =$ reflection about the horizontal axis. Each member of D_6 is of the form $g^i h^j$, $i = 0, 1, 2$, $j = 0, 1$, and induces a linear transformation on V. Thus we have a representation of G in V (the underlying field k can be taken as $\mathbb{R}$).

With v_1 and v_2 taken as a basis for V, find the matrices $[g]$ and $[h]$ associated with g and h.

5. Continue from Problem 4, and switch to the standard basis $e_1 = v_1 = (1, 0)$, $e_2 = (0, 1)$. Changing the basis produces an *equivalent matrix representation.* The matrix representing the element $a \in G$ is now of the form

$$[a]' = P^{-1}[a]P$$

   where the similarity matrix P is the same for every $a \in G$ (the key point).

   Find the matrix P corresponding to the switch from $\{v_1, v_2\}$ to $\{e_1, e_2\}$, and the matrices $[g]'$ and $[h]'$.

6. Consider the dihedral group D_8, generated by elements R (rotation) and F (reflection). We assign to R the 2 by 2 matrix

$$A = \begin{bmatrix} 0 & 1 \\ -1 & 0 \end{bmatrix}$$

and to F the 2 by 2 matrix

$$B = \begin{bmatrix} 1 & 0 \\ 0 & -1 \end{bmatrix}.$$

Show that the above assignment determines a matrix representation of D_8 of degree 2.

7. Is the representation of Problem 6 faithful?

A very accessible basic text on group representation theory is "Representations and Characters of Groups" by James and Liebeck.

9.6 Maschke's Theorem

We can now prove the fundamental theorem on decomposition of representations. It is useful to isolate the key ideas in preliminary lemmas.

9.6.1 Lemma

Let G be a finite group, and k a field whose characteristic does not divide $|G|$ (so that division by $|G|$ is legal). Let V be a kG-module, and ψ a linear transformation on V as a vector space over k. Define $\theta\colon V \to V$ by

$$\theta(v) = \frac{1}{|G|} \sum_{g \in G} g^{-1}\psi g(v).$$

Then not only is θ a linear transformation on the vector space V, but it is also a kG-homomorphism.

Proof. Since ψ is a linear transformation and G acts linearly on V (see (9.5.3)), θ is linear. Now if $h \in G$, then

$$\theta(hv) = \frac{1}{|G|} \sum_{g \in G} g^{-1}\psi g(hv).$$

As g ranges over all of G, so does gh. Thus we can let $x = gh$, $g^{-1} = hx^{-1}$, to obtain

$$\theta(hv) = \frac{1}{|G|} \sum_{x \in G} hx^{-1}\psi(xv) = h\theta(v)$$

and the result follows. ♣

9.6.2 Lemma

In (9.6.1), suppose that ψ is a projection of V on a subspace W that is also a kG-submodule of V. Then θ is also a projection of V on W.

Proof. If $v \in W$, then $g(v) \in W$ since W is a kG-submodule of V. Thus $\psi g(v) = g(v)$ since ψ is a projection on W. By definition of θ we have $\theta(v) = v$. To prove that $\theta^2 = \theta$, note that since ψ maps V into the kG-submodule W, it follows from the definition of θ that θ also maps V into W. But θ is the identity on W, so

$$\theta^2(v) = \theta(\theta(v)) = \theta(v)$$

and θ is a projection. Since θ maps into W and is the identity on W, θ is a projection of V on W. ♣

9.6.3 Maschke's Theorem

Let G be a finite group, and k a field whose characteristic does not divide $|G|$. If V is a kG-module, then V is semisimple.

Proof. Let W be a kG-submodule of V. Ignoring the group algebra for a moment, we can write $V = W \oplus U$ as vector spaces over k. Let ψ be the natural projection of V on W, and define θ as in (9.6.1). By (9.6.1) and (9.6.2), θ is a kG-homomorphism and also a projection of V on W. By (9.5.7), $V = \operatorname{im}\theta \oplus \ker\theta = W \oplus \ker\theta$ *as kG-modules.* By (9.1.2), V is semisimple. ♣

We have been examining the decomposition of a semisimple module into a direct sum of simple modules. Suppose we start with an *arbitrary* module M, and ask whether M can be expressed as $M_1 \oplus M_2$, where M_1 and M_2 are nonzero submodules. If so, we can try to decompose M_1 and M_2, and so on. This process will often terminate in a finite number of steps.

9.6.4 Definition

The module M is *decomposable* if $M = M_1 \oplus M_2$, where M_1 and M_2 are nonzero submodules. Otherwise, M is *indecomposable.*

9.6.5 Proposition

Let M be a module with a composition series; equivalently, by (7.5.12), M is Noetherian and Artinian. Then M can be expressed as a finite direct sum $\oplus_{i=1}^n M_i$ of indecomposable submodules.

Proof. If the decomposition process does not terminate, infinite ascending and descending chains are produced, contradicting the hypothesis. ♣

As the above argument shows, the hypothesis can be weakened to M Noetherian *or* Artinian. But (9.6.5) is usually stated along with a uniqueness assertion which uses the stronger hypothesis:

If M has a composition series and $M = \oplus_{i=1}^n M_i = \oplus_{j=1}^m N_j$, where the M_i and N_j are indecomposable submodules, then $n = m$ and the M_i are, up to isomorphism, just a rearrangement of the N_i.

The full result (existence plus uniqueness) is most often known as the **Krull-Schmidt Theorem.** [One or more of the names Remak, Azumaya and Wedderburn are sometimes added.] The uniqueness proof is quite long (see, for example, Jacobson's Basic Algebra II), and we will not need the result.

Returning to semisimple rings, there is an asymmetry in the definition in that a ring is regarded as a left module over itself, so that submodules are left ideals. We can repeat the entire discussion using right ideals, so that we should distinguish between left-semisimple and right-semisimple rings. However, this turns out to be unnecessary.

9.6.6 Theorem

A ring R is left-semisimple if and only if it is right-semisimple.

Proof. If R is left-semisimple, then by (9.5.1), R is isomorphic to a direct product of matrix rings over division rings. But a matrix ring over a division ring is right-simple by (9.5.2) with left ideals replaced by right ideals. Therefore R is right-semisimple. The reverse implication is symmetrical. ♣

Problems For Section 9.6

1. Let V be the permutation module for $G = S_3$ (see Section 9.5, Problem 2), with basis v_1, v_2, v_3. Give an example of a nontrivial kG-submodule of V.

In Problems 2–4, we show that Maschke's theorem can fail if the characteristic of k divides the order of G. Let $G = \{1, a, \dots, a^{p-1}\}$ be a cyclic group of prime order p, and let V be a two-dimensional vector space over the field $\mathbb{F}_p$, with basis v_1, v_2. Take the matrix of a as

$$[a] = \begin{bmatrix} 1 & 1 \\ 0 & 1 \end{bmatrix}$$

so that

$$[a^r] = \begin{bmatrix} 1 & r \\ 0 & 1 \end{bmatrix}$$

and $[a^p]$ is the identity.

2. Show that W, the one-dimensional subspace spanned by v_1, is a kG-submodule of V.
3. Continuing Problem 2, show that W is the only one-dimensional kG-submodule of V.
4. Continuing Problem 3, show that V is not a semisimple kG-module.
5. Show that a semisimple module is Noetherian iff it is Artinian.

6. Let M be a decomposable R-module, so that M is the direct sum of nonzero submodules M_1 and M_2. Show that $\mathrm{End}_R(M)$ contains a nontrivial idempotent e (that is, $e^2 = e$ with e not the zero map and not the identity).
7. Continuing from Problem 6, suppose conversely that $\mathrm{End}_R(M)$ contains a nontrivial idempotent e. Show that M is decomposable. (Suggestion: use e to construct idempotents e_1 and e_2 that are *orthogonal*, that is, $e_1e_2 = e_2e_1 = 0$.)

9.7 The Jacobson Radical

There is a very useful device that will allow us to look deeper into the structure of rings.

9.7.1 Definitions and Comments

The *Jacobson radical* $J(R)$ of a ring R is the intersection of all maximal left ideals of R. More generally, the Jacobson radical $J(M) = J_R(M)$ of an R-module M is the intersection of all maximal submodules of M. ["Maximal submodule" will always mean "maximal proper submodule".] If M has no maximal submodule, take $J(M) = M$.

If M is finitely generated, then every submodule N of M is contained in a maximal submodule, by Zorn's lemma. [If the union of a chain of proper submodules is M, then the union contains all the generators, hence some member of the chain contains all the generators, a contradiction.] Taking $N = 0$, we see that $J(M)$ is a proper submodule of M. Since R is finitely generated (by 1_R), $J(R)$ is always a proper left ideal.

Semisimplicity of M imposes a severe constraint on $J(M)$.

9.7.2 Proposition

If M is semisimple, then $J(M) = 0$. Thus in a sense, the Jacobson radical is an "obstruction" to semisimplicity.

Proof. Let N be any simple submodule of M. By (9.1.2), $M = N \oplus N'$ for some submodule N'. Now $M/N' \cong N$, which is simple, so by the correspondence theorem, N' is maximal. Thus $J(M) \subseteq N'$, and therefore $J(M) \cap N = 0$. Since M is a sum of simple modules (see (9.1.2)), $J(M) = J(M) \cap M = 0$ ♣

Here is another description of the Jacobson radical.

9.7.3 Proposition

$J(R)$ is the intersection of all annihilators of simple R-modules.

Proof. By Section 9.1, Problem 3, simple modules are isomorphic to R/I for maximal left ideals I. If r annihilates all simple R-modules, then for every maximal left ideal I, r annihilates R/I, in particular, r annihilates $1 + I$. Thus $r(1 + I) = I$, that is, $r \in I$. Consequently, $r \in J(R)$.

Conversely, assume $r \in J(R)$. If M is a simple R-module, choose any nonzero element $x \in M$. The map $f_x\colon R \to M$ given by $f_x(s) = sx$ is an epimorphism by simplicity of

M. The kernel of f_x is the annihilator of x, denoted by $\operatorname{ann}(x)$. By the first isomorphism theorem, $M \cong R/\operatorname{ann}(x)$. By simplicity of M, $\operatorname{ann}(x)$ is a maximal left ideal, so by hypothesis, $r \in \bigcap_{x \in M} \operatorname{ann}(x) = \operatorname{ann}(M)$. Thus r annihilates all simple R-modules. ♣

9.7.4 Corollary

$J(R)$ is a two-sided ideal.

Proof. We noted in (4.2.6) that $\operatorname{ann}(M)$ is a two-sided ideal, and the result follows from (9.7.3). ♣

In view of (9.7.4), one might suspect that the Jacobson radical is unchanged if right rather than left ideals are used in the definition. This turns out to be the case.

9.7.5 Definitions and Comments

The element $a \in R$ is *left quasi-regular* (lqr) if $1 - a$ has a left inverse, *right quasi-regular* (rqr) if $1 - a$ has a right inverse, and *quasi-regular* (qr) if $1 - a$ is invertible. Note that if a is both lqr and rqr, it is qr, because if $b(1 - a) = (1 - a)c = 1$, then

$$b = b1 = b(1 - a)c = 1c = c.$$

9.7.6 Lemma

Let I be a left ideal of R. If every element of I is lqr, then every element of I is qr.

Proof. If $a \in I$, then we have $b(1 - a) = 1$ for some $b \in R$. Let $c = 1 - b$, so that $(1 - c)(1 - a) = 1 - a - c + ca = 1$. Thus $c = ca - a = (c - 1)a \in I$. By hypothesis, c is lqr, so $1 - c$ has a left inverse. But we know that $(1 - c)$ has a right inverse $(1 - a)$ [see above], so c is rqr. By (9.7.5), c is qr and $1 - c$ is the two-sided inverse of $1 - a$. ♣

9.7.7 Proposition

The Jacobson radical $J(R)$ is the largest two-sided ideal consisting entirely of quasi-regular elements.

Proof. First, we show that each $a \in J(R)$ is lqr, so by (9.7.6), each $a \in J(R)$ is qr. If $1 - a$ has no left inverse, then $R(1 - a)$ is a proper left ideal, which is contained in a maximal left ideal I (as in (2.4.2) or (9.7.1)). But then $a \in I$ and $1 - a \in I$, and therefore $1 \in I$, a contradiction.

Now we show that every left ideal (hence every two-sided ideal) I consisting entirely of quasi-regular elements is contained in $J(R)$. If $a \in I$ but $a \notin J(R)$, then for some maximal left ideal L we have $a \notin L$. By maximality of L, we have $I + L = R$, so $1 = b + c$ for some $b \in I$, $c \in L$. But then b is quasi-regular, so $c = 1 - b$ has an inverse, and consequently $1 \in L$, a contradiction. ♣

9.7.8 Corollary

$J(R)$ is the intersection of all maximal right ideals of R.

Proof. We can reproduce the entire discussion beginning with (9.7.1) with left and right ideals interchanged, and reach exactly the same conclusion, namely that the "right" Jacobson radical is the largest two-sided ideal consisting entirely of quasi-regular elements. It follows that the "left" and "right" Jacobson radicals are identical. ♣

We can now use the Jacobson radical to sharpen our understanding of semisimple modules and rings.

9.7.9 Theorem

If M is a nonzero R-module, the following conditions are equivalent:

(1) M is semisimple and has finite length, that is, has a composition series;

(2) M is Artinian and $J(M) = 0$.

Proof. (1) implies (2) by (7.5.12) and (9.7.2), so assume M Artinian with $J(M) = 0$. The Artinian condition implies that the collection of all finite intersections of maximal submodules of M has a minimal element N. If S is any maximal submodule of M, then $N \cap S$ is a finite intersection of maximal submodules, so by minimality of N, $N \cap S = N$, so $N \subseteq S$. Since $J(M)$ is the intersection of all such S, the hypothesis that $J(M) = 0$ implies that $N = 0$. Thus for some positive integer n we have maximal submodules $M_1, \dots, M_n$ such that $\cap_{i=1}^n M_i = 0$.

Now M is isomorphic to a submodule of $M' = \oplus_{i=1}^n (M/M_i)$. To see this, map $x \in M$ to $(x + M_1, \dots, x + M_n)$ and use the first isomorphism theorem. Since M' is a finite direct sum of simple modules, it is semisimple and has a composition series. (See Section 9.4, Problem 6.) By (9.1.3) and (7.5.7), the same is true for M. ♣

9.7.10 Corollary

The ring R is semisimple if and only if R is Artinian and $J(R) = 0$.

Proof. By (9.7.9), it suffices to show that if R is semisimple, then it has a composition series. But this follows because R is finitely generated, hence is a finite direct sum of simple modules (see Section 9.3, Problem 1). ♣

The Jacobson radical of an Artinian ring has some special properties.

9.7.11 Definitions and Comments

An ideal (or left ideal or right ideal) I of the ring R is *nil* if each element $x \in I$ is nilpotent, that is, $x^m = 0$ for some positive integer m; I is *nilpotent* if $I^n = 0$ for some positive integer n. Every nilpotent ideal is nil, and the converse holds if R is Artinian, as we will prove.

9.7.12 Lemma

If I is a nil left ideal of R, then $I \subseteq J(R)$.

Proof. If $x \in I$ and $x^m = 0$, then x is quasi-regular; the inverse of $1-x$ is $1+x+x^2+\cdots+x^{m-1}$. The result follows from the proof of (9.7.7). ♣

9.7.13 Proposition

If R is Artinian, then $J(R)$ is nilpotent. Thus by (9.7.11) and (9.7.12), $J(R)$ is the largest nilpotent ideal of R, and every nil ideal of R is nilpotent.

Proof. Let $J = J(R)$. The sequence $J \supseteq J^2 \supseteq \cdots$ stabilizes, so for some n we have $J^n = J^{n+1} = \cdots$, in particular, $J^n = J^{2n}$. We claim that $J^n = 0$. If not, then the collection of all left ideals Q of R such that $J^nQ \neq 0$ is nonempty (it contains J^n), hence has a minimal element N. Choose $x \in N$ such that $J^n x \neq 0$. By minimality of N, $J^n x = N$. Thus there is an element $c \in J^n$ such that $cx = x$, that is, $(1-c)x = 0$. But $c \in J^n \subseteq J$, so by (9.7.7), $1-c$ is invertible, and consequently $x = 0$, a contradiction. ♣

Problems For Section 9.7

1. Show that an R-module is M cyclic if and only if M is isomorphic to R/I for some left ideal I, and in this case we can take I to be $\text{ann}(M)$, the annihilator of M.
2. Show that the Jacobson radical of an R-module M is the intersection of all kernels of homomorphisms from M to simple R-modules.
3. If $I = J(R)$, show that $J(R/I) = 0$.
4. If f is an R-module homomorphism from M to N, show that $f(J(M)) \subseteq J(N)$.
5. Assume R commutative, so that $J(R)$ is the intersection of all maximal ideals of R. If $a \in R$, show that $a \in J(R)$ if and only if $1 + ab$ is a unit for every $b \in R$.
6. If N is a submodule of the Jacobson radical of the R-module M, show that $J(M)/N = J(M/N)$.

9.8 Theorems of Hopkins-Levitzki and Nakayama

From Section 7.5, we know that a Noetherian ring need not be Artinian, and an Artinian module need not be Noetherian. But the latter situation can never arise for rings, because of the following result.

9.8.1 Theorem (Hopkins and Levitzki)

Let R be an Artinian ring, and M a finitely generated R-module. Then M is both Artinian and Noetherian. In particular, with $M = R$, an Artinian ring is Noetherian.

Proof. By (7.5.9), M is Artinian. Let J be the Jacobson radical of R. By Section 9.7, Problem 3, the Jacobson radical of R/J is zero, and since R/J is Artinian by (7.5.7), it is semisimple by (9.7.9). Now consider the sequence

$$M_0 = M, \ M_1 = JM, \ M_2 = J^2M, \ldots.$$

By (9.7.13), J is nilpotent, so $M_n = 0$ for some n. Since $JM_i = M_{i+1}$, J annihilates M_i/M_{i+1}, so by Section 4.2, Problem 6, M_i/M_{i+1} is an R/J-module.

We claim that each M_i/M_{i+1} has a composition series.

We can assume that $M_i/M_{i+1} \neq 0$, otherwise there is nothing to prove. By (9.3.2), M_i/M_{i+1} is semisimple, and by (7.5.7), M_i/M_{i+1} is Artinian. [Note that submodules of M_i/M_{i+1} are the same, whether we use scalars from R or from R/J; see Section 4.2, Problem 6.] By Section 9.6, Problem 5, M_i/M_{i+1} is Noetherian, hence has a composition series by (7.5.12). Now intuitively, we can combine the composition series for the M_i/M_{i+1} to produce a composition series for M, proving that M is Noetherian. Formally, $M_{n-1} \cong M_{n-1}/M_n$ has a composition series. Since M_{n-2}/M_{n-1} has a composition series, so does M_{n-2}, by (7.5.7). Iterate this process until we reach M. ♣

We now proceed to a result that has many applications in both commutative and noncommutative algebra.

9.8.2 Nakayama's Lemma, Version 1

Let M be a finitely generated R-module, and I a two-sided ideal of R. If $I \subseteq J(R)$ and $IM = M$, then $M = 0$.

Proof. Assume $M \neq 0$, and let $x_1, \ldots, x_n$ generate M, where n is as small as possible. (Then $n \geq 1$ and the x_i are nonzero.) Since $x_n \in M = IM$, we can write $x_n = \sum_{i=1}^m b_i y_i$ for some $b_i \in I$ and $y_i \in M$. But y_i can be expressed in terms of the generators as $y_i = \sum_{j=1}^n a_{ij} x_j$ with $a_{ij} \in R$. Thus

$$x_n = \sum_{i,j} b_i a_{ij} x_j = \sum_{j=1}^{n} c_j x_j$$

where $c_j = \sum_{i=1}^m b_i a_{ij}$. Since I is a right ideal, $c_j \in I \subseteq J(R)$. (We need I to be a left ideal to make IM a legal submodule of M.) The above equation can be written as

$$(1 - c_n)x_n = \sum_{j=1}^{n-1} c_j x_j$$

and by (9.7.7), $1 - c_n$ is invertible. If $n > 1$, then x_n is a linear combination of the other x_i's, contradicting the minimality of n. Thus $n = 1$, in which case $(1 - c_1)x_1 = 0$, so $x_1 = 0$, again a contradiction. ♣

There is another version of Nakayama's lemma, which we prove after a preliminary result.

9.8.3 Lemma

Let N be a submodule of the R-module M, I a left ideal of R. Then $M = N + IM$ if and only if $M/N = I(M/N)$.

Proof. Assume $M = N + IM$, and let $x + N \in M/N$. Then $x = y + z$ for some $y \in N$ and $z \in IM$. Write $z = \sum_{i=1}^{t} a_i w_i$, $a_i \in I$, $w_i \in M$. It follows that

$$x + N = a_1(w_1 + N) + \cdots + a_t(w_t + N) \in I(M/N).$$

Conversely, assume $M/N = I(M/N)$, and let $x \in M$. Then

$$x + N = \sum_{i=1}^{t} a_i(w_i + N)$$

with $a_i \in I$ and $w_i \in M$. Consequently, $x - \sum_{i=1}^{t} a_i w_i \in N$, so $x \in N + IM$. ♣

9.8.4 Nakayama's Lemma, Version 2

Let N be a submodule of the R-module M, with M/N finitely generated over R. [This will be satisfied if M is finitely generated over R.] If I is a two-sided ideal contained in $J(R)$, and $M = N + IM$, then $M = N$.

Proof. By (9.8.3), $I(M/N) = M/N$, so by (9.8.2), $M/N = 0$, hence $M = N$. ♣

Here is an application of Nakayama's lemma.

9.8.5 Proposition

Let R be a commutative local ring with maximal ideal J (see (8.5.8)). Let M be a finitely generated R-module, and let $V = M/JM$. Then:

(i) V is a finite-dimensional vector space over the *residue field* $k = R/J$.

(ii) If $\{x_1 + JM, \dots, x_n + JM\}$ is a basis for V over k, then $\{x_1, \dots, x_n\}$ is a minimal set of generators for M.

(iii) Any two minimal generating sets for M have the same cardinality.

Proof. (i) Since J annihilates M/JM, it follows from Section 4.2, Problem 6, that V is a k-module, that is, a vector space over k. Since M is finitely generated over R, V is finite-dimensional over k.

(ii) Let $N = \sum_{i=1}^{n} Rx_i$. Since the $x_i + JM$ generate $V = M/JM$, we have $M = N + JM$. By (9.8.4), $M = N$, so the x_i generate M. If a proper subset of the x_i were to generate M, then the corresponding subset of the $x_i + JM$ would generate V, contradicting the assumption that V is n-dimensional.

(iii) A generating set S for M with more than n elements determines a spanning set for V, which must contain a basis with exactly n elements. By (ii), S cannot be minimal. ♣

Problems For Section 9.8

1. Let a be a nonzero element of the integral domain R. If $(a^t) = (a^{t+1})$ for some positive integer t, show that a is invertible.
2. Continuing Problem 1, show that every Artinian integral domain is a field.
3. If R is a commutative Artinian ring, show that every prime ideal of R is maximal.
4. Let R be a commutative Artinian ring. If $\mathcal{S}$ is the collection of all finite intersections of maximal ideals of R, then $\mathcal{S}$ is not empty, hence contains a minimal element $I = I_1 \cap I_2 \cap \cdots \cap I_n$, with the I_j maximal. Show that if P is any maximal ideal of R, then P must be one of the I_j. Thus R has only finitely many maximal ideals.
5. An R-module is *projective* if it is a direct summand of a free module. We will study projective modules in detail in Section 10.5. We bring up the subject now in Problems 5 and 6 to illustrate a nice application of Nakayama's lemma.

   Let R be a commutative local ring, and let M be a finitely generated projective module over R, with a minimal set of generators $\{x_1, \dots, x_n\}$ (see (9.8.5)). We can assume that for some free module F of rank n,

$$F = M \oplus N.$$

   To justify this, let F be free with basis $e_1, \dots, e_n$, and map F onto M via $e_i \to x_i$, $i = 1, \dots, n$. If the kernel of the mapping is K, then we have a short exact sequence

$$0 \to K \to F \to M \to 0,$$

   which splits since M is projective. [This detail will be covered in (10.5.3).]

   Let J be the maximal ideal of R, and $k = R/J$ the residue field. Show that

$$F/JF \cong M/JM \oplus N/JN.$$

6. Continue from Problem 5 and show that $N/JN = 0$. It then follows from Nakayama's lemma (9.8.2) that $N = 0$, and therefore $M = F$. We conclude that a finitely generated projective module over a commutative local ring is free.
7. We showed in (9.6.6) that there is no distinction between a left and a right-semisimple ring. This is not the case for Noetherian (or Artinian) rings.

Let X and Y be *noncommuting indeterminates*, in other words, $XY \neq YX$, and let $\mathbb{Z} < X, Y >$ be the set of all polynomials in X and Y with integer coefficients. [Elements of $\mathbb{Z}$ do commute with the indeterminates.] We impose the relations $Y^2 = 0$ and $YX = 0$ to produce the ring R; formally, $R = \mathbb{Z} < X, Y > /(Y^2, YX)$.

Consider $I = \mathbb{Z}[X]Y$, the set of all polynomials $f(X)Y$, $f(X) \in \mathbb{Z}[X]$. Then I is a two-sided ideal of R. Show that if I is viewed as a right ideal, it is not finitely generated. Thus R is not right-Noetherian.

8. Viewed as a left R-module, $R = \mathbb{Z}[X] \oplus \mathbb{Z}[X]Y$. Show that R is left-Noetherian.

9. Assume the hypothesis of (9.8.5). If $\{x_1, \ldots, x_n\}$ is a minimal generating set for M, show that $\{\overline{x}_1, \ldots, \overline{x}_n\}$, where $\overline{x}_i = x_i + JM$, is a basis for $M/JM = V$.
10. Continuing Problem 9, suppose that $\{x_1, \ldots, x_n\}$ and $\{y_1, \ldots, y_n\}$ are minimal generating sets for M, with $y_i = \sum_j a_{ij} x_j$, $a_{ij} \in R$. If A is the matrix of the a_{ij}, show that the determinant of A is a unit in R.

Chapter 10

Introducing Homological Algebra

Roughly speaking, homological algebra consists of (A) that part of algebra that is fundamental in building the foundations of algebraic topology, and (B) areas that arise naturally in studying (A).

10.1 Categories

We have now encountered many algebraic structures and maps between these structures. There are ideas that seem to occur regardless of the particular structure under consideration. Category theory focuses on principles that are common to all algebraic systems.

10.1.1 Definitions and Comments

A *category* $\mathcal{C}$ consists of *objects* $A, B, C, \dots$ and *morphisms* $f\colon A \to B$ (where A and B are objects). If $f\colon A \to B$ and $g\colon B \to C$ are morphisms, we have a notion of *composition*, in other words, there is a morphism $gf = g \circ f\colon A \to C$, such that the following axioms are satisfied.

(i) *Associativity*: If $f\colon A \to B$, $g\colon B \to C$, $h\colon C \to D$, then $(hg)f = h(gf)$;

(ii) *Identity*: For each object A there is a morphism $1_A\colon A \to A$ such that for each morphism $f\colon A \to B$, we have $f1_A = 1_B f = f$.

A remark for those familiar with set theory: For each pair (A, B) of objects, the collection of morphisms $f\colon A \to B$ is required to be a set rather than a proper class.

We have seen many examples:

1. **Sets**: The objects are sets and the morphisms are functions.
2. **Groups**: The objects are groups and the morphisms are group homomorphisms.
3. **Rings**: The objects are rings and the morphisms are ring homomorphisms.

4. **Fields**: The objects are fields and the morphisms are field homomorphisms [= field monomorphisms; see (3.1.2)].
5. **R-mod**: The objects are left R-modules and the morphisms are R-module homomorphisms. If we use right R-modules, the corresponding category is called **mod-R**.
6. **Top**: The objects are topological spaces and the morphisms are continuous maps.
7. **Ab**: The objects are abelian groups and the the morphisms are homomorphisms from one abelian group to another.

A morphism $f\colon A \to B$ is said to be an *isomorphism* if there is an inverse morphism $g\colon B \to A$, that is, $gf = 1_A$ and $fg = 1_B$. In **Sets**, isomorphisms are bijections, and in **Top**, isomorphisms are homeomorphisms. For the other examples, an isomorphism is a bijective homomorphism, as usual.

In the category of sets, a function f is injective iff $f(x_1) = f(x_2)$ implies $x_1 = x_2$. But in an abstract category, we don't have any elements to work with; a morphism $f\colon A \to B$ can be regarded as simply an arrow from A to B. How do we generalize injectivity to an arbitrary category? We must give a definition that does not depend on elements of a set. Now in **Sets**, f is injective iff it has a left inverse; equivalently, f is *left cancellable*, i.e. if $fh_1 = fh_2$, then $h_1 = h_2$. This is exactly what we need, and a similar idea works for surjectivity, since f is surjective iff f is *right cancellable*, i.e., $h_1 f = h_2 f$ implies $h_1 = h_2$.

10.1.2 Definitions and Comments

A morphism f is said to be *monic* if it is left cancellable, *epic* if it is right cancellable.

In all the categories listed in (10.1.1), a morphism f is monic iff f is injective as a mapping of sets. If f is surjective, then it is epic, but the converse can fail. See Problems 2 and 7–10 for some of the details.

In the category **R-mod**, the zero module $\{0\}$ has the property that for any R-module M, there is a unique module homomorphism from M to $\{0\}$ and a unique module homomorphism from $\{0\}$ to M. Here is a generalization of this idea.

10.1.3 Definitions and Comments

Let A be an object in a category. If for every object B, there is a unique morphism from A to B, then A is said to be an *initial object*. If for every object B there is a unique morphism from B to A, then A is said to be a *terminal object*. A *zero object* is both initial and terminal.

In the category of sets, there is only one initial object, the empty set. The terminal objects are singletons $\{x\}$, and consequently there are no zero objects. In the category of groups, the trivial group consisting of the identity alone is a zero object. We are going to prove that any two initial objects are isomorphic, and similarly for terminal objects. This will be a good illustration of the duality principle, to be discussed next.

10.1.4 Duality

If $\mathcal{C}$ is a category, the *opposite* or *dual* category $\mathcal{C}^{op}$ has the same objects as $\mathcal{C}$. The morphisms are those of $\mathcal{C}$ with arrows reversed; thus $f\colon A \to B$ is a morphism of $\mathcal{C}^{op}$

iff $f\colon B \to A$ is a morphism of $\mathcal{C}$. If the composition gf is permissible in $\mathcal{C}$, then fg is permissible in $\mathcal{C}^{\text{op}}$. To see how the duality principle works, let us first prove that if A and B are initial objects of $\mathcal{C}$, then A and B are isomorphic. There is a unique morphism $f\colon A \to B$ and a unique morphism $g\colon B \to A$. But $1_A\colon A \to A$ and $1_B\colon B \to B$, and it follows that $gf = 1_A$ and $fg = 1_B$. The point is that we need not give a separate proof that any two terminal objects are isomorphic. We have just proved the following:

If A and B are objects in a category $\mathcal{C}$, and for every object D of $\mathcal{C}$, there is a unique morphism from A to D and there is a unique morphism from B to D, then A and B are isomorphic.

Our statement is completely general; it does not involve the properties of any specific category. If we go through the entire statement and reverse all the arrows, equivalently, if we replace $\mathcal{C}$ by $\mathcal{C}^{\text{op}}$, we get:

If A and B are objects in a category $\mathcal{C}$, and for every object D of $\mathcal{C}$, there is a unique morphism from D to A and there is a unique morphism from D to B, then A and B are isomorphic.

In other words, any two terminal objects are isomorphic. If this is unconvincing, just go through the previous proof, reverse all the arrows, and interchange fg and gf. We say that *initial and terminal objects are dual.* Similarly, monic and epic morphisms are dual.

If zero objects exist in a category, then we have zero morphisms as well. If Z is a zero object and A and B arbitrary objects, there is a unique $f\colon A \to Z$ and a unique $g\colon Z \to B$. The *zero morphism* from A to B, denoted by 0_{AB}, is defined as gf, and it is independent of the particular zero object chosen (Problem 3). Note that since a zero morphism goes through a zero object, it follows that for an arbitrary morphism h, we have $h0 = 0h = 0$.

10.1.5 Kernels and Cokernels

If $f\colon A \to B$ is an R-module homomorphism, then its kernel is, as we know, $\{x \in A\colon f(x) = 0\}$. The *cokernel* of f is defined as the quotient group $B/\operatorname{im}(f)$. Thus f is injective iff its kernel is 0, and f is surjective iff its cokernel is 0. We will generalize these notions to an arbitrary category that contains zero objects. The following diagram indicates the setup for kernels.

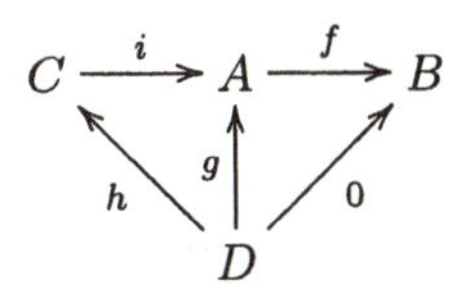

We take C to be the kernel of the module homomorphism f, with i the inclusion map. If $fg = 0$, then the image of g is contained in the kernel of f, so that g actually maps into C. Thus there is a unique module homomorphism $h\colon D \to C$ such that $g = ih$; simply take $h(x) = g(x)$ for all x. The key to the generalization is to think of the kernel as the *morphism* i. This is reasonable because C and i essentially encode the same information. Thus a *kernel* of the morphism $f\colon A \to B$ is a morphism $i\colon C \to A$ such that:

(1) $fi = 0$.

(2) If $g\colon D \to A$ and $fg = 0$, then there is a unique morphism $h\colon D \to C$ such that $g = ih$.

Thus any map killed by f can be factored through i.

If we reverse all the arrows in the above diagram and change labels for convenience, we get an appropriate diagram for cokernels.

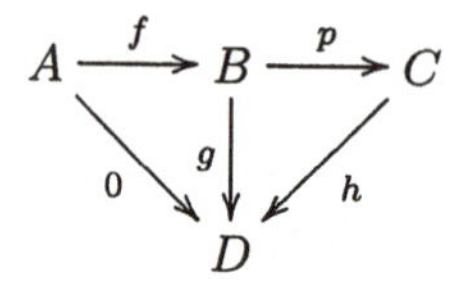

We take p to be the canonical map of B onto the cokernel of f, so that $C = B/\operatorname{im}(f)$. If $gf = 0$, then the image of f is contained in the kernel of g, so by the factor theorem, there is a unique homomorphism h such that $g = hp$. In general, a *cokernel* of a morphism $f\colon A \to B$ is a morphism $p\colon B \to C$ such that:

(1′) $pf = 0$.

(2′) If $g\colon B \to D$ and $gf = 0$, then there is a unique morphism $h\colon C \to D$ such that $g = hp$.

Thus any map that kills f can be factored through p.

Since going from kernels to cokernels simply involves reversing arrows, kernels and cokernels are dual. Note, however, that in an arbitrary category with 0, kernels and cokernels need not exist for arbitrary morphisms. But every monic has a kernel and (by duality) every epic has a cokernel; see Problem 5.

Problems For Section 10.1

1. Show that in any category, the identity and inverse are unique.
2. In the category of rings, the inclusion map $i\colon \mathbb{Z} \to \mathbb{Q}$ is not surjective. Show, however, that i is epic.
3. Show that the zero morphism 0_{AB} is independent of the particular zero object chosen in the definition.
4. Show that a kernel must be monic (and by duality, a cokernel must be epic). [In the definition of kernel, we can assume that i is monic in (1) of (10.1.5), and drop the uniqueness assumption on h. For i monic forces uniqueness of h, by definition of monic. Conversely, uniqueness of h forces i to be monic, by Problem 4.]
5. Show that in a category with 0, every monic has a kernel and every epic has a cokernel.
6. Show that if $i\colon C \to A$ and $j\colon D \to A$ are kernels of $f\colon A \to B$, then C and D are isomorphic. (By duality, a similar statement holds for cokernels.)
7. Let $f\colon A \to B$ be a group homomorphism with kernel K, and assume f not injective, so that $K \neq \{1\}$. Let g be the inclusion map of K into A. Find a homomorphism h such that $fg = fh$ but $g \neq h$.

8. It follows from Problem 7 that in the category of groups, f monic is equivalent to f injective as a mapping of sets, and a similar proof works in the category of modules. Why does the argument fail in the category of rings?
9. Continue from Problem 8 and give a proof that does work in the category of rings.
10. Let $f\colon M \to N$ be a module homomorphism with nonzero cokernel, so that f is not surjective. Show that f is not epic; it follows that epic is equivalent to surjective in the category of modules.

10.2 Products and Coproducts

We have studied the direct product of groups, rings, and modules. It is natural to try to generalize the idea to an arbitrary category, and a profitable approach is to forget (temporarily) the algebraic structure and just look at the cartesian product $A = \prod_i A_i$ of a family of sets $A_i, i \in I$. The key property of a product is that if we are given maps f_i from a set S *into* the factors A_i, we can *lift* the f_i into a single map $f\colon S \to \prod_i A_i$. The commutative diagram below will explain the terminology.

$$\begin{array}{ccc} & & A_i \\ & {\scriptstyle f_i}\nearrow & \uparrow {\scriptstyle p_i} \\ S & \xrightarrow[f]{} & A \end{array} \qquad (1)$$

In the picture, p_i is the projection of A onto the ith factor A_i. If $f_i(x) = a_i, i \in I$, we take $f(x) = (a_i, i \in I)$. It follows that $p_i \circ f = f_i$ for all i; this is what we mean by lifting the f_i to f. (Notice that there is only one possible lifting, i.e., f is unique.) If A is the direct product of groups A_i and the f_i are group homomorphisms, then f will also be a group homomorphism. Similar statements can be made for rings and modules. We can now give a generalization to an arbitrary category.

10.2.1 Definition

A *product* of objects A_i in a category $\mathcal{C}$ is an object A, along with morphisms $p_i\colon A \to A_i$, with the following *universal mapping property*. Given any object S of $\mathcal{C}$ and morphisms $f_i\colon S \to A_i$, there is a unique morphism $f\colon S \to A$ such that $p_i f = f_i$ for all i.

In a definition via a universal mapping property, we use a condition involving morphisms, along with a uniqueness statement, to specify an object and morphisms associated with that object. We have already seen this idea in connection with kernels and cokernels in the previous section, and in the construction of the tensor product in Section 8.7.

Not every category has products (see Problems 1 and 2), but if they do exist, they are essentially unique. (The technique for proving uniqueness is also essentially unique.)

10.2.2 Proposition

If $(A, p_i, i \in I)$ and $(B, q_i, i \in I)$ are products of the objects A_i, then A and B are isomorphic.

Proof. We use the above diagram (1) with $S = B$ and $f_i = q_i$ to get a morphism $f\colon B \to A$ such that $p_i f = q_i$ for all i. We use the diagram with $S = A$, A replaced by B, p_i replaced by q_i, and $f_i = p_i$, to get a morphism $h\colon A \to B$ such that $q_i h = p_i$. Thus

$$p_i f h = q_i h = p_i \quad \text{and} \quad q_i h f = p_i f = q_i.$$

But

$$p_i 1_A = p_i \quad \text{and} \quad q_i 1_B = q_i$$

and it follows from the uniqueness condition in (10.2.1) that $fh = 1_A$ and $hf = 1_B$. Formally, we are using the diagram two more times, once with $S = A$ and $f_i = p_i$, and once with $S = B$, A replaced by B, p_i replaced by q_i, and $f_i = q_i$. Thus A and B are isomorphic. ♣

The discussion of diagram (1) indicates that in the categories of groups, abelian groups, rings, and R-modules, products coincide with direct products. But a category can have products that have no connection with a cartesian product of sets; see Problems 1 and 2. Also, in the category of torsion abelian groups (torsion means that every element has finite order), products exist but do not coincide with direct products; see Problem 5.

The dual of a product is a coproduct, and to apply duality, all we need to do is reverse all the arrows in (1). The following diagram results.

$$\begin{array}{ccc} & & M_j \\ & {\scriptstyle f_j}\swarrow & \downarrow{\scriptstyle i_j} \\ N & \underset{f}{\longleftarrow} & M \end{array} \tag{2}$$

We have changed the notation because it is now profitable to think about modules. Suppose that M is the direct sum of the submodules M_j, and i_j is the inclusion map of M_j into M. If the f_j are module homomorphisms *out of* the factors M_j and into a module N, the f_j can be lifted to a single map f. If $x_{j1} \in M_{j1}, \dots, x_{jr} \in M_{jr}$, we take

$$f(x_{j1} + \cdots + x_{jr}) = f_{j1}(x_{j1}) + \cdots + f_{jr}(x_{jr}).$$

Lifting means that $f \circ i_j = f_j$ for all j. We can now give the general definition of coproduct.

10.2.3 Definition

A *coproduct* of objects M_j in a category $\mathcal{C}$ is an object M, along with morphisms $i_j\colon M_j \to M$, with the following universal mapping property. Given any object N of $\mathcal{C}$ and morphisms $f_j\colon M_j \to N$, there is a unique morphism $f\colon M \to N$ such that $f i_j = f_j$ for all j

Exactly as in (10.2.2), any two coproducts of a given collection objects are isomorphic.

The discussion of diagram (2) shows that in the category of R-modules, the coproduct is the direct sum, which is isomorphic to the direct product if there are only finitely many factors. In the category of sets, the coproduct is the *disjoint union*. To explain what this means, suppose we have sets $A_j, j \in J$. We can disjointize the A_j by replacing A_j by $A'_j = \{(x, j) : x \in A_j\}$. The coproduct is $A = \bigcup_{j \in J} A'_j$, with morphisms $i_j : A_j \to A$ given by $i_j(a_j) = (a_j, j)$. If for each j we have $f_j : A_j \to B$, we define $f : A \to B$ by $f(a_j, j) = f_j(a_j)$.

The coproduct in the category of groups will be considered in the exercises.

Problems For Section 10.2

1. Let S be a preordered set, that is, there is a reflexive and transitive relation $\leq$ on S. Then S can be regarded as a category whose objects are the elements of S. If $x \leq y$, there is a unique morphism from x to y, and if $x \not\leq y$, there are no morphisms from x to y. Reflexivity implies that there is an identity morphism on x, and transitivity implies that associativity holds. Show that a product of the objects x_i, if it exists, must be a greatest lower bound of the x_i. The greatest lower bound will be unique (not just essentially unique) if S is a partially ordered set, so that $\leq$ is antisymmetric.
2. Continuing Problem 1, do products always exist?
3. Continuing Problem 2, what can be said about coproducts?
4. If A is an abelian group, let $T(A)$ be the set of torsion elements of A. Show that $T(A)$ is a subgroup of A.
5. Show that in the category of torsion abelian groups, the product of groups A_i is $T(\prod A_i)$, the subgroup of torsion elements of the direct product.
6. Assume that we have a collection of groups G_i, pairwise disjoint except for a common identity 1. The *free product* of the G_i (notation $*_i G_i$) consists of all words (finite sequences) $a_1 \cdots a_n$ where the a_j belong to distinct groups. Multiplication is by concatenation with cancellation. For example, with the subscript j indicating membership in G_j,

$$(a_1a_2a_3a_4)(b_4b_2b_6b_1b_3) = a_1a_2a_3(a_4b_4)b_2b_6b_1b_3$$

   and if $b_4 = a_4^{-1}$, this becomes $a_1a_2a_3b_2b_6b_1b_3$. The empty word is the identity, and inverses are calculated in the usual way, as with free groups (Section 5.8). In fact a free group on S is a free product of infinite cyclic groups, one for each element of S. Show that in the category of groups, the coproduct of the G_i is the free product.
7. Suppose that products exist in the category of finite cyclic groups, and suppose that the cyclic group C with generator a is the product of the cyclic groups C_1 and C_2 with generators a_1 and a_2 respectively. Show that the projections p_1 and p_2 associated with the product of C_1 and C_2 are surjective.
8. By Problem 7, we may assume without loss of generality that $p_i(a) = a_i$, $i = 1, 2$. Show that for some positive integer n, $na_1 = a_1$ and $na_2 = 0$. [Take $f_1 : C_1 \to C_1$ to be the identity map, and let $f_2 : C_1 \to C_2$ be the zero map (using additive notation). Lift f_1 and f_2 to $f : C_1 \to C$.]
9. Exhibit groups C_1 and C_2 that can have no product in the category of finite cyclic groups.

10.3 Functors

We will introduce this fundamental concept with a concrete example. Let $\mathrm{Hom}_R(M, N)$ be the set of R-module homomorphisms from M to N. As pointed out at the beginning of Section 4.4, $\mathrm{Hom}_R(M, N)$ is an abelian group. It will also be an R-module if R is a commutative ring, but not in general. We are going to look at $\mathrm{Hom}_R(M, N)$ as a function of N, with M fixed.

10.3.1 The Functor $\mathrm{Hom}_R(M, -)$

We are going to construct a mapping from the category of R-modules to the category of abelian groups. Since a category consists of both objects and morphisms, our map will have two parts:

(i) Associate with each R-module N the abelian group $\mathrm{Hom}_R(M, N)$.

(ii) Associate with each R-module homomorphism $h\colon N \to P$ a homomorphism h_* from the abelian group $\mathrm{Hom}_R(M, N)$ to the abelian group $\mathrm{Hom}_R(M, P)$. The following diagram suggests how h_* should be defined.

$$M \xrightarrow{f} N \xrightarrow{h} P$$

Take

$$h_*(f) = hf.$$

Note that if h is the identity on N, then h_* is the identity on $\mathrm{Hom}_R(M, N)$.

Now suppose we have the following situation:

$$M \xrightarrow{f} N \xrightarrow{g} P \xrightarrow{h} Q$$

Then $(hg)_*(f) = (hg)f = h(gf) = h_*(g_*(f))$, so that

$$(hg)_* = h_*g_* \ (= h_* \circ g_*).$$

To summarize, we have a mapping F called a *functor* that takes an object A in a category $\mathcal{C}$ to an object $F(A)$ in a category $\mathcal{D}$; F also takes a morphism $h\colon A \to B$ in $\mathcal{C}$ to a morphism $h_* = F(h)\colon F(A) \to F(B)$ in $\mathcal{D}$. The key feature of F is the *functorial property*:

$$F(hg) = F(h)F(g) \text{ and } F(1_A) = 1_{F(A)}.$$

Thus a functor may be regarded as a homomorphism of categories.

10.3.2 The Functor $\mathrm{Hom}_R(-, N)$

We now look at $\mathrm{Hom}_R(M, N)$ as a function of M, with N fixed. Here is an appropriate diagram:

$$K \xrightarrow{g} L \xrightarrow{h} M \xrightarrow{f} N$$

If M is an R-module, we take $F(M)$ to be the abelian group $\text{Hom}_R(M, N)$. If $h\colon L \to M$ is an R-module homomorphism, we take $h^* = F(h)$ to be a homomorphism from the abelian group $\text{Hom}_R(M, N)$ to the abelian group $\text{Hom}_R(L, N)$, given by

$$h^*(f) = fh.$$

It follows that

$$(hg)^*(f) = f(hg) = (fh)g = g^*(fh) = g^*(h^*(f))$$

hence

$$(hg)^* = g^*h^*,$$

and if h is the identity on M, then h^* is the identity on $\text{Hom}_R(M, N)$.

Thus F does not quite obey the functorial property; we have $F(hg) = F(g)F(h)$ instead of $F(hg) = F(h)F(g)$. However, F is a legal functor on the *opposite category* of **R-mod**. In the literature, $\text{Hom}_R(-, N)$ is frequently referred to as a *contravariant functor* on the original category **R-mod**, and $\text{Hom}_R(M, -)$ as a *covariant functor* on **R-mod**.

If we replace the category of R-modules by an arbitrary category, we can still define functors (called *hom functors*) as in (10.3.1) and (10.3.2). But we must replace the category of abelian groups by the category of sets.

10.3.3 The Functors $\mathbf{M} \otimes_R -$ and $- \otimes_R \mathbf{N}$

To avoid technical complications, we consider tensor products of modules over a commutative ring R. First we discuss the tensor functor $T = M \otimes_R -$ The relevant diagram is given below.

$$N \xrightarrow{g} P \xrightarrow{f} Q$$

If N is an R-module, we take $T(N) = M \otimes_R N$. If $g\colon N \to P$ is an R-module homomorphism, we set $T(g) = 1_M \otimes g\colon M \otimes_R N \to M \otimes_R P$, where 1_M is the identity mapping on M [Recall that $(1_M \otimes g)(x \otimes y) = x \otimes g(y)$.] Then

$$T(fg) = 1_M \otimes fg = (1_M \otimes f)(1_M \otimes g) = T(f)T(g)$$

and

$$T(1_N) = 1_{T(N)}$$

so T is a functor from **R-mod** to **R-mod**.

The functor $S = - \otimes_R N$ is defined in a symmetrical way. If M is an R-module, then $S(M) = M \otimes_R N$, and if $f\colon L \to M$ is an R-module homomorphism, then $S(f)\colon L \otimes_R N \to M \otimes_R N$ is given by $S(f) = f \otimes 1_N$.

10.3.4 Natural Transformations

Again we will introduce this idea with an explicit example. The diagram below summarizes the data.

$$\begin{array}{ccc} F(A) & \xrightarrow{t_A} & G(A) \\ {\scriptstyle Ff}\downarrow & & \downarrow{\scriptstyle Gf} \\ F(B) & \xrightarrow[t_B]{} & G(B) \end{array} \qquad (1)$$

We start with abelian groups A and B and a homomorphism $f\colon A \to B$. We apply the *forgetful functor*, also called the *underlying functor* $\mathcal{U}$. This is a fancy way of saying that we forget the algebraic structure and regard A and B simply as sets, and f as a mapping between sets. Now we apply the *free abelian group functor* $\mathcal{F}$ to produce $F(A) = \mathcal{FU}(A)$, the free abelian group with A as basis (and similarly for $F(B)$). Thus $F(A)$ is the direct sum of copies of $\mathbb{Z}$, one copy for each $a \in A$. The elements of $F(A)$ can be represented as $\sum_a n(a)x(a)$, $n(a) \in \mathbb{Z}$, where $x(a)$ is the member of the direct sum that is 1 in the ath position and 0 elsewhere. [Similarly, we represent elements of B as $\sum_b n(b)y(b)$.] The mapping f determines a homomorphism $Ff\colon F(A) \to F(B)$, via $\sum n(a)x(a) \to \sum n(a)y(f(a))$.

Now let G be the identity functor, so that $G(A) = A, G(B) = B, Gf = f$. We define an abelian group homomorphism $t_A\colon F(A) \to A$ by $t_A(\sum n(a)x(a)) = \sum n(a)a$, and similarly we define $t_B(\sum n(b)y(b)) = \sum n(b)b$. (Remember that we began with *abelian groups* A and B.) The diagram (1) is then commutative, because

$$ft_A(x(a)) = f(a) \text{ and } t_B[(Ff)(x(a))] = t_B(y(f(a)) = f(a).$$

To summarize, we have two functors, F and G from the category $\mathcal{C}$ to the category $\mathcal{D}$. (In this case, $\mathcal{C} = \mathcal{D} =$ the category of abelian groups.) For all objects $A, B \in \mathcal{C}$ and morphisms $f\colon A \to B$, we have morphisms $t_A\colon F(A) \to G(A)$ and $t_B\colon F(B) \to G(B)$ such that the diagram (1) is commutative. We say that t is a *natural transformation* from F to G. If for every object $C \in \mathcal{C}$, t_C is an isomorphism (not the case in this example), t is said to be a *natural equivalence.*

The key intuitive point is that the process of going from $F(A)$ to $G(A)$ is "natural" in the sense that as we move from an object A to an object B, the essential features of the process remains the same.

Problems For Section 10.3

1. Let $F\colon S \to T$, where S and T are preordered sets. If we regard S and T as categories, as in Section 10.2, Problem 1, what property must F have in order to be a functor?
2. A group may be regarded as a category with a single object 0, with a morphism for each element $g \in G$. The composition of two morphisms is the morphism associated with the product of the elements. If $F\colon G \to H$ is a function from a group G to a group H, and we regard G and H as categories, what property must F have in order to be a functor?

3. We now look at one of the examples that provided the original motivation for the concept of a natural transformation. We work in the category of vector spaces (over a given field) and linear transformations. If V is a vector space, let V^* be the dual space, that is, the space of linear maps from V to the field of scalars, and let V^{**} be the dual of V^*. If $v \in V$, let $\overline{v} \in V^{**}$ be defined by $\overline{v}(f) = f(v), f \in V^*$. The mapping from v to $\overline{v}$ is a linear transformation, and in fact an isomorphism if V is finite-dimensional.

   Now suppose that $f\colon V \to W$ and $g\colon W \to X$ are linear transformations. Define $f^*\colon W^* \to V^*$ by $f^*(\alpha) = \alpha f$, $\alpha \in W^*$. Show that $(gf)^* = f^*g^*$.

4. The *double dual functor* takes a vector space V into its double dual V^{**}, and takes a linear transformation $f\colon V \to W$ to $f^{**}\colon V^{**} \to W^{**}$, where $f^{**}(v^{**}) = v^{**}f^*$. Show that the double dual functor is indeed a functor.

5. Now consider the following diagram.

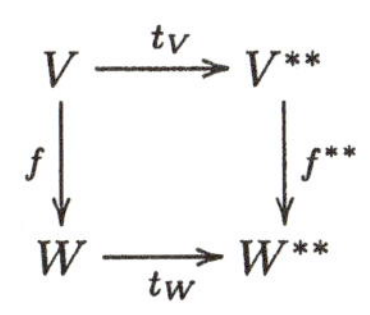

   We take $t_V(v) = \overline{v}$, and similarly for t_W. Show that the diagram is commutative, so that t is a natural transformation from the identity functor to the double dual functor.

   In the finite-dimensional case, we say that there is a natural isomorphism between a vector space and its double dual. "Natural" means coordinate-free in the sense that it is not necessary to choose a specific basis. In contrast, the isomorphism of V and its single dual V^* is not natural.

6. We say that $\mathcal{D}$ is a *subcategory* of $\mathcal{C}$ if the objects of $\mathcal{D}$ are also objects of $\mathcal{C}$, and similarly for morphisms (and composition of morphisms). The subcategory $\mathcal{D}$ is *full* if every $\mathcal{C}$-morphism $f\colon A \to B$, where A and B are *objects of* $\mathcal{D}$ (the key point) is also a $\mathcal{D}$-morphism. Show that the category of groups is a full subcategory of the category of monoids.

7. A functor $F\colon \mathcal{C} \to \mathcal{D}$ induces a map from $\mathcal{C}$-morphisms to $\mathcal{D}$-morphisms; $f\colon A \to B$ is mapped to $Ff\colon FA \to FB$. If this map is injective for all objects A, B of $\mathcal{C}$, we say that F is *faithful*. If the map is surjective for all objects A, B of $\mathcal{C}$, we say that F is *full*.

   (a) The *forgetful functor* from groups to sets assigns to each group its underlying set, and to each group homomorphism its associated map of sets. Is the forgetful functor faithful? full?

   (b) We can form the product $\mathcal{C} \times \mathcal{D}$ of two arbitrary categories; objects in the product are pairs (A, A') of objects, with $A \in \mathcal{C}$ and $A' \in \mathcal{D}$. A morphism from (A, A') to (B, B') is a pair (f, g), where $f\colon A \to B$ and $g\colon A' \to B'$. The *projection functor* from $\mathcal{C} \times \mathcal{D}$ to $\mathcal{C}$ takes (A, A') to A and (f, g) to f. Is the projection functor faithful? full?

10.4 Exact Functors

10.4.1 Definitions and Comments

We are going to investigate the behavior of the hom and tensor functors when presented with an exact sequence. We will be working in the categories of modules and abelian groups, but exactness properties can be studied in the more general setting of *abelian categories*, which we now describe very informally.

In any category $\mathcal{C}$, let $\text{Hom}_{\mathcal{C}}(A, B)$ (called a "hom set") be the set of morphisms in $\mathcal{C}$ from A to B. [As remarked in (10.1.1), the formal definition of a category requires that $\text{Hom}_{\mathcal{C}}(A, B)$ be a set for all objects A and B. The collection of all objects of $\mathcal{C}$ is a class but need not be a set.] For $\mathcal{C}$ to be an abelian category, the following conditions must be satisfied.

1. Each hom set is an abelian group.
2. The distributive laws $f(g+h) = fg + fh$, $(f+g)h = fh + gh$ hold.
3. $\mathcal{C}$ has a zero object.
4. Every finite set of objects has a product and a coproduct. (The existence of finite coproducts can be deduced from the existence of finite products, along with the requirements listed so far.)
5. Every morphism has a kernel and a cokernel.
6. Every monic is the kernel of its cokernel.
7. Every epic is the cokernel of its kernel.
8. Every morphism can be factored as an epic followed by a monic.

Exactness of functors can be formalized in an abelian category, but we are going to return to familiar ground by assuming that each category that we encounter is **R-mod** for some R. When $R = \mathbb{Z}$, we have the category of abelian groups.

10.4.2 Left Exactness of $\text{Hom}_R(M, -)$

Suppose that we have a short exact sequence

$$0 \longrightarrow A \xrightarrow{f} B \xrightarrow{g} C \longrightarrow 0 \tag{1}$$

We apply the covariant hom functor $F = \text{Hom}_R(M, -)$ to the sequence, dropping the last term on the right. We will show that the sequence

$$0 \longrightarrow FA \xrightarrow{Ff} FB \xrightarrow{Fg} FC \tag{2}$$

is exact. A functor that behaves in this manner is said to be *left exact*.

We must show that the transformed sequence is exact at FA and FB. We do this in three steps.

(a) Ff is monic.

Suppose that $(Ff)(\alpha) = f\alpha = 0$. Since f is monic (by exactness of the sequence (1)), $\alpha = 0$ and the result follows.

(b) im $Ff \subseteq \ker Fg$.

If $\beta \in \operatorname{im} Ff$, then $\beta = f\alpha$ for some $\alpha \in \operatorname{Hom}_R(M, A)$. By exactness of (1), $\operatorname{im} f \subseteq \ker g$, so $g\beta = gf\alpha = 0\alpha = 0$. Thus $\beta \in \ker g$.

(c) $\ker Fg \subseteq \operatorname{im} Ff$.

If $\beta \in \ker Fg$, then $g\beta = 0$, with $\beta \in \operatorname{Hom}_R(M, B)$. Thus if $y \in M$, then $\beta(y) \in \ker g = \operatorname{im} f$, so $\beta(y) = f(x)$ for some $x = \alpha(y) \in A$. Note that x is unique since f is monic, and $\alpha \in \operatorname{Hom}_R(M, A)$. Thus $\beta = f\alpha \in \operatorname{im} Ff$. ♣

10.4.3 Left Exactness of $\operatorname{Hom}_R(-, N)$

The contravariant hom functor $G = \operatorname{Hom}_R(-, N)$ is a functor on the opposite category, so before applying it to the sequence (1), we must reverse all the arrows. Thus left-exactness of G means that the sequence

$$0 \longrightarrow GC \xrightarrow{Gg} GB \xrightarrow{Gf} GA \tag{3}$$

is exact. Again we have three steps.

(a) Gg is monic.

If $(Gg)\alpha = \alpha g = 0$, then $\alpha = 0$ since g is epic.

(b) im $Gg \subseteq \ker Gf$.

If $\beta \in \operatorname{im} Gg$, then $\beta = \alpha g$ for some $\alpha \in \operatorname{Hom}_R(C, N)$. Thus $(Gf)\beta = \beta f = \alpha g f = 0$, so $\beta \in \ker Gf$.

(c) $\ker Gf \subseteq \operatorname{im} Gg$.

Let $\beta \in \operatorname{Hom}_R(B, N)$ with $\beta \in \ker Gf$, that is, $\beta f = 0$. If $y \in C$, then since g is epic, we have $y = g(x)$ for some $x \in B$. If $g(x_1) = g(x_2)$, then $x_1 - x_2 \in \ker g = \operatorname{im} f$, hence $x_1 - x_2 = f(z)$ for some $z \in A$. Therefore $\beta(x_1) - \beta(x_2) = \beta(f(z)) = 0$, so it makes sense to define $\alpha(y) = \beta(x)$. Then $\alpha \in \operatorname{Hom}_R(C, N)$ and $\alpha g = \beta$, that is, $(Gg)\alpha = \beta$. ♣

10.4.4 Right Exactness of the Functors $M \otimes_R -$ and $- \otimes_R N$

If we apply the functor $H = M \otimes_R -$ to the exact sequence

$$0 \longrightarrow A \xrightarrow{f} B \xrightarrow{g} C \longrightarrow 0$$

[see (1) of (10.4.2)], we will show that the sequence

$$HA \xrightarrow{Hf} HB \xrightarrow{Hg} HC \longrightarrow 0 \tag{4}$$

is exact. A similar result holds for $- \otimes_R N$. A functor that behaves in this way is said to be *right exact.* Once again, there are three items to prove.

(i) Hg is epic.

An element of $M \otimes C$ is of the form $t = \sum_i x_i \otimes y_i$ with $x_i \in M$ and $y_i \in C$. Since g is epic, there exists $z_i \in B$ such that $g(z_i) = y_i$. Thus $(1 \otimes g)(\sum_i x_i \otimes z_i) = \sum_i x_i \otimes g(z_i) = t$.

(ii) $\operatorname{im} Hf \subseteq \ker Hg$.

This is a brief computation: $(1 \otimes g)(1 \otimes f) = 1 \otimes gf = 1 \otimes 0 = 0$.

(iii) $\ker Hg \subseteq \operatorname{im} Hf$.

By (ii), the kernel of $1 \otimes g$ contains $L = \operatorname{im}(1 \otimes f)$, so by the factor theorem, there is a homomorphism $\overline{g}\colon (M \otimes_R B)/L \to M \otimes_R C$ such that $\overline{g}(m \otimes b + L) = m \otimes g(b)$, $m \in M$, $b \in B$.

Let π be the canonical map of $M \otimes_R B$ onto $(M \otimes_R B)/L$. Then $\overline{g}\pi(m \otimes b) = \overline{g}(m \otimes b + L) = m \otimes g(b)$, so

$$\overline{g}\pi = 1 \otimes g.$$

If we can show that $\overline{g}$ is an isomorphism, then

$$\ker(1 \otimes g) = \ker(\overline{g}\pi) = \ker \pi = L = \operatorname{im}(1 \otimes f)$$

and we are finished. To show that $\overline{g}$ is an isomorphism, we will display its inverse. First let h be the bilinear map from $M \times C$ to $(M \otimes_R B)/L$ given by $h(m, c) = m \otimes b + L$, where $g(b) = c$. [Such a b exists because g is epic. If $g(b) = g(b') = c$, then $b - b' \in \ker g = \operatorname{im} f$, so $b - b' = f(a)$ for some $a \in A$. Then $m \otimes b - m \otimes b' = m \otimes f(a) = (1 \otimes f)(m \otimes a) \in L$, and h is well-defined.] By the universal mapping property of the tensor product, there is a homomorphism $\overline{h}\colon M \otimes_R C \to (M \otimes_R B)/L$ such that

$$\overline{h}(m \otimes c) = h(m, c) = m \otimes b + L, \text{where } g(b) = c.$$

But $\overline{g}\colon (M \otimes_R B)/L \to M \otimes_R C$ and

$$\overline{g}(m \otimes b + L) = m \otimes g(b) = m \otimes c.$$

Thus $\overline{h}$ is the inverse of $\overline{g}$. ♣

10.4.5 Definition

A functor that is both left and right exact is said to be *exact*. Thus an exact functor is one that maps exact sequences to exact sequences. We have already seen one example, the localization functor (Section 8.5, Problems 4 and 5).

If we ask under what conditions the hom and tensor functors become exact, we are led to the study of projective, injective and flat modules, to be considered later in the chapter.

Problems For Section 10.4

In Problems 1–3, we consider the exact sequence (1) of (10.4.2) with $R = \mathbb{Z}$, so that we are in the category of abelian groups. Take $A = \mathbb{Z}$, $B = \mathbb{Q}$, the additive group of rational numbers, and $C = \mathbb{Q}/\mathbb{Z}$, the additive group of rationals mod 1. Let f be inclusion, and g the canonical map. We apply the functor $F = \mathrm{Hom}_R(M, -)$ with $M = \mathbb{Z}_2$. [We will omit the subscript R when $R = \mathbb{Z}$, and simply refer to $\mathrm{Hom}(M, -)$.]

1. Show that $\mathrm{Hom}(\mathbb{Z}_2, \mathbb{Q}) = 0$.
2. Show that $\mathrm{Hom}(\mathbb{Z}_2, \mathbb{Q}/\mathbb{Z}) \neq 0$.
3. Show that $\mathrm{Hom}(\mathbb{Z}_2, -)$ is not right exact.

In Problems 4 and 5, we apply the functor $G = \mathrm{Hom}(-, N)$ to the above exact sequence, with $N = \mathbb{Z}$.

4. Show that $\mathrm{Hom}(\mathbb{Q}, \mathbb{Z}) = 0$.
5. Show that $\mathrm{Hom}(-, \mathbb{Z})$ is not right exact.

Finally, in Problem 6 we apply the functor $H = M \otimes -$ to the above exact sequence, with $M = \mathbb{Z}_2$.

6. Show that $\mathbb{Z}_2 \otimes -$ (and similarly $- \otimes \mathbb{Z}_2$) is not left exact.
7. Refer to the sequences (1) and (2) of (10.4.2). If (2) is exact for all possible R-modules M, show that (1) is exact.
8. State an analogous result for the sequence (3) of (10.4.3), and indicate how the result is proved.

10.5 Projective Modules

Projective modules are direct summands of free modules, and are therefore images of natural projections. Free modules are projective, and projective modules are sometimes but not always free. There are many equivalent ways to describe projective modules, and we must choose one of them as the definition. In the diagram below and the definition to follow, all maps are R-module homomorphisms. The bottom row is exact, that is, g is surjective.

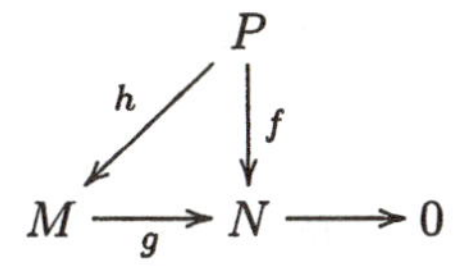

10.5.1 Definition

The R-module P is *projective* if given $f\colon P \to N$, and $g\colon M \to N$ surjective, there exists $h\colon P \to M$ (not necessarily unique) such that the diagram is commutative, that is, $f = gh$. We sometimes say that we have "lifted" f to h.

The definition may look obscure, but the condition described is a familiar property of free modules.

10.5.2 Proposition

Every free module is projective.

Proof. Let S be a basis for the free module P. By (4.3.6), f is determined by its behavior on basis elements $s \in S$. Since g is surjective, there exists $a \in M$ such that $g(a) = f(s)$. Take $h(s) = a$ and extend by linearity from S to P. Since $f = gh$ on S, the same must be true on all of P. ♣

Here is the list of equivalences.

10.5.3 Theorem

The following conditions on the R-module P are equivalent.

(1) P is projective.

(2) The functor $\mathrm{Hom}_R(P, -)$ is exact.

(3) Every short exact sequence $0 \to M \to N \to P \to 0$ splits.

(4) P is a direct summand of a free module.

Proof. (1) is equivalent to (2). In view of the left exactness of $F = \mathrm{Hom}_R(P, -)$ (see (10.4.2)), (2) says that if $g\colon M \to N$ is surjective, then so is $Fg\colon FM \to FN$. But Fg maps $h\colon P \to M$ to $gh\colon P \to N$, so what we must prove is that for an arbitrary morphism $f\colon P \to N$, there exists $h\colon P \to M$ such that $gh = f$. This is precisely the definition of projectivity of P.

(2) implies (3). Let $0 \to M \to N \to P \to 0$ be a short exact sequence, with $g\colon N \to P$ (necessarily surjective). Since P is projective, we have the following diagram.

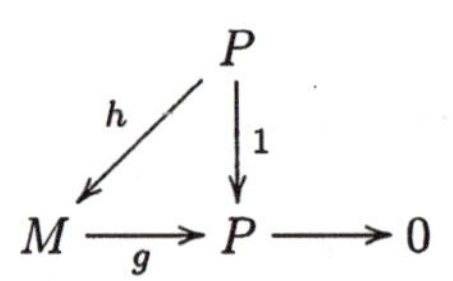

Thus there exists $h\colon P \to N$ such that $gh = 1_P$, which means that the exact sequence splits (see (4.7.1)).

(3) implies (4). By (4.3.6), P is a quotient of a free module, so there is an exact sequence $0 \to M \to N \to P \to 0$ with N free. By (3), the sequence splits, so by (4.7.4), P is a direct summand of N.

(4) implies (1). Let P be a direct summand of the free module F, and let π be the natural projection of F on P; see the diagram below.

$$\begin{array}{ccccc} F & \xrightarrow{\pi} & P & & \\ {\scriptstyle h}\downarrow & & \downarrow{\scriptstyle f} & & \\ M & \xrightarrow[g]{} & N & \longrightarrow & 0 \end{array}$$

Given $f\colon P \to N$, we have $f\pi\colon F \to N$, so by (10.5.2) there exists $h\colon F \to M$ such that $f\pi = gh$. If h' is the restriction of h to P, then $f = gh'$. ♣

10.5.4 Corollary

The direct sum $P = \oplus P_j$ is projective if and only if each P_j is projective.

Proof. If P is a direct summand of a free module, so is each P_j, and therefore the P_j are projective by (4) of (10.5.3). Conversely, assume that each P_j is projective. Let $f\colon P \to N$ and $g\colon M \to N$, with g surjective. If i_j is the inclusion map of P_j into P, then $fi_j\colon P_j \to N$ can be lifted to $h_j\colon P_j \to M$ such that $fi_j = gh_j$. By the universal mapping property of direct sum (Section 10.2), there is a morphism $h\colon P \to M$ such that $hi_j = h_j$ for all j. Thus $fi_j = ghi_j$ for every j, and it follows from the uniqueness part of the universal mapping property that $f = gh$. ♣

If we are searching for projective modules that are not free, the following result tells us where not to look.

10.5.5 Theorem

A module M over a principal ideal domain R is projective if and only if it is free.

Proof. By (10.5.2), free implies projective. If M is projective, then by (4) of (10.5.3), M is a direct summand of a free module. In particular, M is a submodule of a free module, hence is free by (4.6.2) and the discussion following it. ♣

10.5.6 Examples

1. A vector space over a field k is a free k-module, hence is projective.
2. A finite abelian group G is not a projective $\mathbb{Z}$-module, because it is not free. [If $g \in G$ and $n = |G|$, then $ng = 0$, so g can never be part of a basis.]
3. If p and q are distinct primes, then $R = \mathbb{Z}_{pq} = \mathbb{Z}_p \oplus \mathbb{Z}_q$. We claim that $\mathbb{Z}_p$ and $\mathbb{Z}_q$ are projective but not free R-modules. (As in Example 2, they are not projective $\mathbb{Z}$-modules.) This follows from (4) of (10.5.3) and the fact that any ring R is a free R-module (with basis $\{1\}$).

Problems For Section 10.5

In Problems 1–5, we are going to prove the *projective basis lemma*, which states that an R-module P is projective if and only if there are elements $x_i \in P$ ($i \in I$) and homomorphisms $f_i\colon P \to R$ such that for every $x \in P$, $f_i(x) = 0$ for all but finitely many i and

$$x = \sum_i f_i(x)x_i.$$

The set of x_i's is referred to as the projective basis.

1. To prove the "only if" part, let P be a direct summand of the free module F with basis $\{e_i\}$. Take f to be the inclusion map of P into F, and π the natural projection of F onto P. Show how to define the f_i and x_i so that the desired results are obtained.

2. To prove the "if" part, let F be a free module with basis $\{e_i, i \in I\}$, and define $\pi\colon F \to P$ by $\pi(e_i) = x_i$. Define $f\colon P \to F$ by $f(x) = \sum_i f_i(x)e_i$. Show that πf is the identity on P.
3. Continuing Problem 2, show that P is projective.
4. Assume that P is finitely generated by n elements. If R^n is the direct sum of n copies of R, show that P is projective iff P is a direct summand of R^n.
5. Continuing Problem 4, show that if P is projective and generated by n elements, then P has a projective basis with n elements.
6. Show that a module P is projective iff P is a direct summand of every module of which it is a quotient. In other words, if $P \cong M/N$, then P is isomorphic to a direct summand of M.
7. In the definition (10.5.1) of a projective module, give an explicit example to show that the mapping h need not be unique.

10.6 Injective Modules

If we reverse all arrows in the mapping diagram that defines a projective module, we obtain the dual notion, an injective module. In the diagram below, the top row is exact, that is, f is injective.

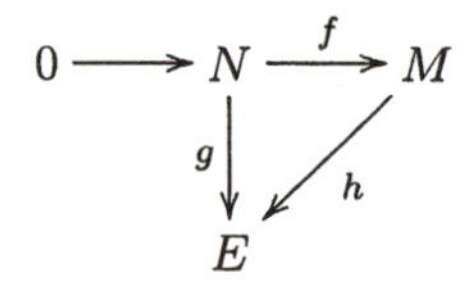

10.6.1 Definition

The R-module E is *injective* if given $g\colon N \to E$, and $f\colon N \to M$ injective, there exists $h\colon M \to E$ (not necessarily unique) such that $g = hf$. We sometimes say that we have "lifted" g to h.

As with projectives, there are several equivalent ways to characterize an injective module.

10.6.2 Theorem

The following conditions on the R-module E are equivalent.

(1) E is injective.

(2) The functor $\mathrm{Hom}_R(-, E)$ is exact.

(3) Every exact sequence $0 \to E \to M \to N \to 0$ splits.

Proof. (1) is equivalent to (2). Refer to (3) of (10.4.3), (1) of (10.4.2) and the definition of the contravariant hom functor in (10.3.2) to see what (2) says. We are to show that if $f\colon N \to M$ is injective, then $f^*\colon \mathrm{Hom}_R(M, E) \to \mathrm{Hom}_R(N, E)$ is surjective. But

$f^*(h) = hf$, so given $g\colon N \to E$, we must produce $h\colon M \to E$ such that $g = hf$. This is precisely the definition of injectivity.

(2) implies (3). Let $0 \to E \to M \to N \to 0$ be a short exact sequence, with $f\colon E \to M$ (necessarily injective). Since E is an injective module, we have the following diagram:

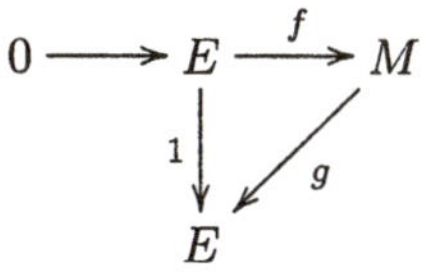

$$\begin{array}{ccccc} 0 & \longrightarrow & E & \xrightarrow{f} & M \\ & & {\scriptstyle 1}\downarrow & \swarrow{\scriptstyle g} & \\ & & E & & \end{array}$$

Thus there exists $g\colon M \to E$ such that $gf = 1_E$, which means that the exact sequence splits.

(3) implies (1). Given $g\colon N \to E$, and $f\colon N \to M$ injective, we form the *pushout* of f and g, which is a commutative square as indicated in the diagram below.

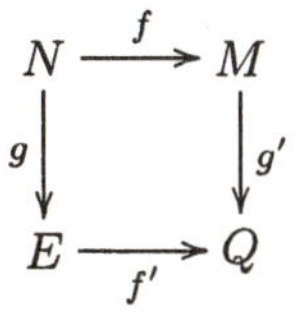

$$\begin{array}{ccc} N & \xrightarrow{f} & M \\ {\scriptstyle g}\downarrow & & \downarrow{\scriptstyle g'} \\ E & \xrightarrow[f']{} & Q \end{array}$$

Detailed properties of pushouts are developed in the exercises. For the present proof, all we need to know is that since f is injective, so is f'. Thus the sequence

$$0 \longrightarrow E \xrightarrow{f'} Q \longrightarrow Q/\operatorname{im} f' \longrightarrow 0$$

is exact. By (3), there exists $h\colon Q \to E$ such that $hf' = 1_E$. We now have $hg'\colon M \to E$ with $hg'f = hf'g = 1_E g = g$, proving that E is injective. ♣

We proved in (10.5.4) that a direct sum of modules is projective iff each component is projective. The dual result holds for injectives.

10.6.3 Proposition

A direct product $\prod_j E_j$ of modules is injective iff each E_j is injective. Consequently, a finite direct sum is injective iff each summand is injective.

Proof. If $f\colon N \to M$ is injective and $g\colon N \to \prod_i E_i$, let $g_i = p_i g$, where p_i is the projection of the direct product on E_i. Then finding $h\colon M \to \prod_i E_i$ such that $g = hf$ is equivalent to finding, for each i, a morphism $h_i\colon M \to E_i$ such that $g_i = h_i f$. [If $p_i g = h_i f = p_i hf$ for every i, then $g = hf$ by the uniqueness part of the universal mapping property for products.] The last assertion holds because the direct sum of finitely many modules coincides with the direct product. ♣

In checking whether an R-module E is injective, we are given $g\colon N \to E$, and $f\colon N \to M$, with f injective, and we must lift g to $h\colon M \to E$ with $g = hf$. The next result drastically reduces the collection of maps f and g that must be examined. We may take $M = R$ and restrict N to a left ideal I of R, with f the inclusion map.

10.6.4 Baer's Criterion

The R-module E is injective if and only if every R-homomorphism $f\colon I \to E$, where I is a left ideal of R, can be extended to an R-homomorphism $h\colon R \to E$.

Proof. The "only if" part follows from the above discussion, so assume that we are given $g\colon N \to E$ and $f\colon N \to M$, where (without loss of generality) f is an inclusion map. We must extend g to $h\colon M \to E$. A standard Zorn's lemma argument yields a maximal extension g_0 in the sense that the domain M_0 of g_0 cannot be enlarged. [The partial ordering is $(g_1, D_1) \leq (g_2, D_2)$ iff $D_1 \subseteq D_2$ and $g_1 = g_2$ on D_1.] If $M_0 = M$, we are finished, so assume $x \in M \setminus M_0$. Let I be the left ideal $\{r \in R\colon rx \in M_0\}$, and define $h_0\colon I \to E$ by $h_0(r) = g_0(rx)$. By hypothesis, h_0 can be extended to $h'_0\colon R \to E$. Let $M_1 = M_0 + Rx$ and define $h_1\colon M_1 \to E$ by

$$h_1(x_0 + rx) = g_0(x_0) + rh'_0(1).$$

To show that h_1 is well defined, assume $x_0 + rx = y_0 + sx$, with $x_0, y_0 \in M_0$ and $r, s \in R$. Then $(r - s)x = y_0 - x_0 \in M_0$, so $r - s \in I$. Using the fact that h'_0 extends h_0, we have

$$g_0(y_0 - x_0) = g_0((r - s)x) = h_0(r - s) = h'_0(r - s) = (r - s)h'_0(1)$$

and consequently, $g_0(x_0) + rh'_0(1) = g_0(y_0) + sh'_0(1)$ and h_1 is well defined. If $x_0 \in M_0$, take $r = 0$ to get $h_1(x_0) = g_0(x_0)$, so h_1 is an extension of g_0 to $M_1 \supset M_0$, contradicting the maximality of g_0. We conclude that $M_0 = M$. ♣

Since free modules are projective, we can immediately produce many examples of projective modules. The primary source of injective modules lies a bit below the surface.

10.6.5 Definitions and Comments

Let R be an integral domain. The R-module M is *divisible* if each $y \in M$ can be divided by any nonzero element $r \in R$, that is, there exists $x \in M$ such that $rx = y$. For example, the additive group of rational numbers is a divisible abelian group, as is $\mathbb{Q}/\mathbb{Z}$, the rationals mod 1. The quotient field of any integral domain (regarded as an abelian group) is divisible. A cyclic group of finite order $n > 1$ can never be divisible, since it is not possible to divide by n. The group of integers $\mathbb{Z}$ is not divisible since the only possible divisors of an arbitrary integer are ± 1. It follows that a nontrivial finitely generated abelian group, a direct sum of cyclic groups by (4.6.3), is not divisible.

It follows from the definition that a homomorphic image of a divisible module is divisible, hence a quotient or a direct summand of a divisible module is divisible. Also, a direct sum of modules is divisible iff each component is divisible.

10.6.6 Proposition

If R is any integral domain, then an injective R-module is divisible. If R is a PID, then an R-module is injective if and only if it is divisible.

Proof. Assume E is injective, and let $y \in E$, $r \in R$, $r \neq 0$. Let I be the ideal Rr, and define an R-homomorphism $f\colon I \to E$ by $f(tr) = ty$. If $tr = 0$, then since R is an integral domain, $t = 0$ and f is well defined. By (10.6.4), f has an extension to an R-homomorphism $h\colon R \to E$. Thus

$$y = f(r) = h(r) = h(r1) = rh(1)$$

so division by r is possible and E is divisible. Conversely, assume that R is a PID and E is divisible. Let $f\colon I \to E$, where I is an ideal of R. Since R is a PID, $I = Rr$ for some $r \in R$. We have no trouble extending the zero mapping, so assume $r \neq 0$. Since E is divisible, there exists $x \in E$ such that $rx = f(r)$. Define $h\colon R \to E$ by $h(t) = tx$. If $t \in R$, then

$$h(tr) = trx = tf(r) = f(tr)$$

so h extends f, proving E injective. ♣

Problems For Section 10.6

We now describe the construction of the pushout of two module homomorphisms $f\colon A \to C$ and $g\colon A \to B$; refer to Figure'10.6.1. Take

$$D = (B \oplus C)/W, \text{ where } W = \{(g(a), -f(a))\colon a \in A\},$$

and

$$g'(c) = (0, c) + W, \quad f'(b) = (b, 0) + W.$$

In Problems 1–6, we study the properties of this construction.

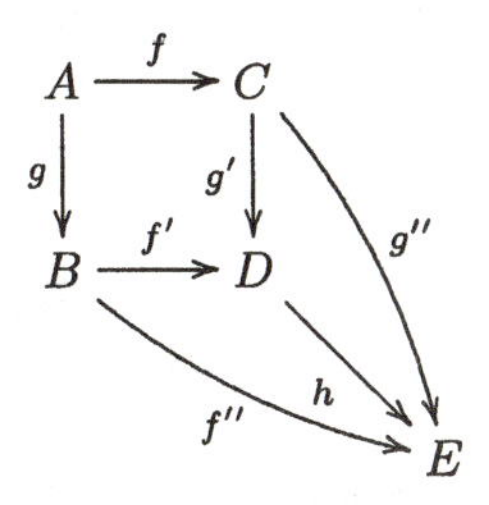

Figure 10.6.1

1. Show that the *pushout square* $ACDB$ is commutative, that is, $f'g = g'f$.
2. Suppose we have another commutative pushout square $ACEB$ with maps $f''\colon B \to E$ and $g''\colon C \to E$, as indicated in Figure 10.6.1. Define $h\colon D \to E$ by

$$h((b, c) + W) = g''(c) + f''(b).$$

Show that h is well defined.

3. Show that h makes the diagram commutative, that is, $hg' = g''$ and $hf' = f''$.
4. Show that if $h'\colon D \to E$ makes the diagram commutative, then $h' = h$.

   The requirements stated in Problems 1, 3 and 4 can be used to define the pushout via a universal mapping property. The technique of (10.2.2) shows that the pushout object D is unique up to isomorphism.
5. If f is injective, show that f' is also injective. By symmetry, the same is true for g and g'.
6. If f is surjective, show that f' is surjective. By symmetry, the same is true for g and g'.

Problems 7–10 refer to the dual construction, the *pullback*, defined as follows (see Figure 10.6.2). Given $f\colon A \to B$ and $g\colon C \to B$, take

$$D = \{(a,c) \in A \oplus C\colon f(a) = g(c)\}$$

and

$$g'(a,c) = a, \;\; f'(a,c) = c.$$

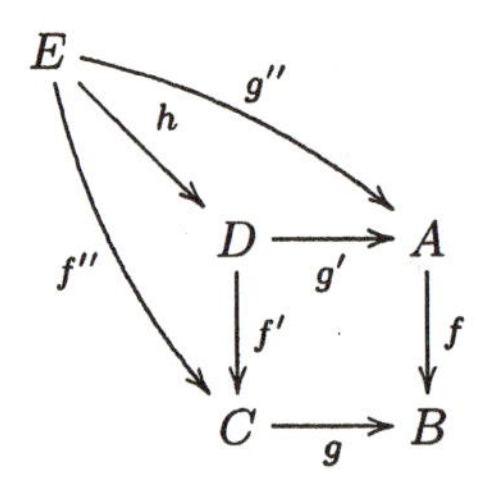

Figure 10.6.2

7. Show that the *pullback square* $DABC$ is commutative, that is, $fg' = gf'$.
8. If we have another commutative pullback square $EABC$ with maps $f''\colon E \to C$ and $g''\colon E \to A$, show that there is a unique $h\colon E \to D$ that makes the diagram commutative, that is, $g'h = g''$ and $f'h = f''$.
9. If f is injective, show that f' is injective. By symmetry, the same is true for g and g'.
10. If f is surjective, show that f' is surjective. By symmetry, the same is true for g and g'.
11. Let R be an integral domain with quotient field Q, and let f be an R-homomorphism from an ideal I of R to Q. Show that $f(x)/x$ is constant for all nonzero $x \in I$.
12. Continuing Problem 11, show that Q is an injective R-module.

10.7 Embedding into an Injective Module

We know that every module is a quotient of a projective (in fact free) module. In this section we prove the more difficult dual statement that every module can be embedded in an injective module. (To see that quotients and submodules are dual, reverse all the arrows in a short exact sequence.) First, we consider abelian groups.

10.7.1 Proposition

Every abelian group can be embedded in a divisible abelian group.

Proof. If A is an abelian group, then A is a quotient of a free abelian group F, say $A \cong F/B$. Now F is a direct sum of copies of $\mathbb{Z}$, hence F can be embedded in a direct sum D of copies of $\mathbb{Q}$, the additive group of rationals. It follows that F/B can be embedded in D/B; the embedding is just the inclusion map. By (10.6.5), D/B is divisible, and the result follows. ♣

10.7.2 Comments

In (10.7.1), we used $\mathbb{Q}$ as a standard divisible abelian group. It would be very desirable to have a canonical injective R-module. First, we consider $H = \mathrm{Hom}_{\mathbb{Z}}(R, A)$, the set of all abelian group homomorphisms from the additive group of the ring R to the abelian group A. If we are careful, we can make this set into a left R-module. The abelian group structure of H presents no difficulties, but we must also define scalar multiplication. If $f \in H$ and $s \in R$, we set

$$(sf)(r) = f(rs), \; r \in R.$$

Checking the module properties is routine except for associativity:

$$((ts)f)(r) = f(rts), \text{ and } (t(sf))(r) = (sf)(rt) = f(rts)$$

so $(ts)f = t(sf)$. Notice that sf is an *abelian group* homomorphism, not an R-module homomorphism.

Now if A is an R-module and B an abelian group, we claim that

$$\mathrm{Hom}_{\mathbb{Z}}(A, B) \cong \mathrm{Hom}_R(A, \mathrm{Hom}_{\mathbb{Z}}(R, B)), \tag{1}$$

equivalently,

$$\mathrm{Hom}_{\mathbb{Z}}(R \otimes_R A, B) \cong \mathrm{Hom}_R(A, \mathrm{Hom}_{\mathbb{Z}}(R, B)). \tag{2}$$

This is a special case of *adjoint associativity*:

If ${}_SM_R$, ${}_RN$, ${}_SP$, then

$$\mathrm{Hom}_S(M \otimes_R N, P) \cong \mathrm{Hom}_R(N, \mathrm{Hom}_S(M, P)). \tag{3}$$

Thus if F is the functor $M \otimes_R -$ and G is the functor $\mathrm{Hom}_S(M, -)$, then

$$\mathrm{Hom}_S(FN, P) \cong \mathrm{Hom}_R(N, GP) \tag{4}$$

which is reminiscent of the adjoint of a linear operator on an inner product space. We say that F and G are *adjoint functors*, with F *left adjoint* to G and G *right adjoint* to F. [There is a technical naturality condition that is added to the definition of adjoint functors, but we will not pursue this since the only adjoints we will consider are hom and tensor.]

Before giving the formal proof of (3), we will argue intuitively. The left side of the equation describes all biadditive, R-balanced maps from $M \times N$ to P. If $(x, y) \to f(x, y)$, $x \in M$, $y \in N$, is such a map, then f determines a map g from N to $\mathrm{Hom}_S(M, P)$, namely, $g(y)(x) = f(x, y)$. This is a variation of the familiar fact that a bilinear map amounts to a family of linear maps. Now if $s \in S$, then $g(y)(sx) = f(sx, y)$, which need not equal $sf(x, y)$, but if we factor f through the tensor product $M \otimes_R N$, we can then pull out the s. Thus $g(y) \in \mathrm{Hom}_S(M, P)$. Moreover, g is an R-module homomorphism, because $g(ry)(x) = f(x, ry)$, and we can factor out the r by the same reasoning as above. Since g determines f, the correspondence between f and g is an isomorphism of abelian groups.

To prove (3), let $f\colon M \otimes_R N \to P$ be an S-homomorphism. If $y \in N$, define $f_y\colon M \to P$ by $f_y(x) = f(x \otimes y)$, and define $\psi(f)\colon N \to \mathrm{Hom}_S(M, P)$ by $y \to f_y$. [$\mathrm{Hom}_S(M, P)$ is a left R-module by Problem 1.]

(a) $\psi(f)$ is an R-homomorphism:
$\psi(f)(y_1 + y_2) = f_{y_1+y_2} = f_{y_1} + f_{y_2} = \psi(f)(y_1) + \psi(f)(y_2)$;
$\psi(f)(ry) = f_{ry} = rf_y = (r\psi(f))(y)$.
$[f_{ry}(x) = f(x \otimes ry)$ and $(rf_y)(x) = f_y(xr) = f(xr \otimes y).]$

(b) ψ is an abelian group homomorphism:
We have $f_y(x) + g_y(x) = f(x \otimes y) + g(x \otimes y) = (f + g)(x \otimes y) = (f + g)_y(x)$ so $\psi(f + g) = \psi(f) + \psi(g)$.

(c) ψ is injective:
If $\psi(f) = 0$, then $f_y = 0$ for all $y \in N$, so $f(x \otimes y) = 0$ for all $x \in M$ and $y \in N$. Thus f is the zero map.

(d) If $g \in \mathrm{Hom}_R(N, \mathrm{Hom}_S(M, P))$, define $\varphi_g\colon M \times N \to P$ by $\varphi_g(x, y) = g(y)(x)$. Then φ_g is biadditive and R-balanced:
By definition, φ_g is additive in each coordinate, and [see Problem 1] $\varphi_g(xr, y) = g(y)(xr) = (rg(y))(x) = g(ry)(x) = \varphi_g(x, ry)$.

(e) ψ is surjective:

By (d), there is a unique S-homomorphism $\beta(g)\colon M \otimes_R N \to P$ such that $\beta(g)(x \otimes y) = \varphi_g(x, y) = g(y)(x)$, $x \in M$, $y \in N$. It follows that $\psi(\beta(g)) = g$, because $\psi(\beta(g))(y) = \beta(g)_y$, where $\beta(g)_y(x) = \beta(g)(x \otimes y) = g(y)(x)$. Thus $\beta(g)_y = g(y)$ for all y in N, so $\psi\beta$ is the identity and ψ is surjective.

It follows that ψ is an abelian group isomorphism. This completes the proof of (3).

Another adjointness result, which can be justified by similar reasoning, is that if N_R, ${}_RM_S$, P_S, then

$$\mathrm{Hom}_S(N \otimes_R M, P) \cong \mathrm{Hom}_R(N, \mathrm{Hom}_S(M, P)) \tag{5}$$

which says that $F = - \otimes_R M$ and $G = \mathrm{Hom}_S(M, -)$ are adjoint functors.

10.7.3 Proposition

If E is a divisible abelian group, then $\mathrm{Hom}_{\mathbb{Z}}(R, E)$ is an injective left R-module.

Proof. By (10.6.2), we must prove that $\text{Hom}_R(-, \text{Hom}_{\mathbb{Z}}(R, E))$ is exact. As in the proof of (1) implies (2) in (10.6.2), if $0 \to N \to M$ is exact, we must show that

$$\text{Hom}_R(M, \text{Hom}_{\mathbb{Z}}(R, E)) \to \text{Hom}_R(N, \text{Hom}_{\mathbb{Z}}(R, E)) \to 0$$

is exact. By (1) of (10.7.2), this is equivalent to showing that

$$\text{Hom}_{\mathbb{Z}}(M, E) \to \text{Hom}_{\mathbb{Z}}(N, E) \to 0$$

is exact. [As indicated in the informal discussion in (10.7.2), this replacement is allowable because a bilinear map can be regarded as a family of linear maps. A formal proof would invoke the naturality condition referred to in (10.7.2).] Since E is an injective $\mathbb{Z}$-module, the result now follows from (10.6.2). ♣

We can now prove the main result.

10.7.4 Theorem

If M is an arbitrary left R-module, then M can be embedded in an injective left R-module.

Proof. If we regard M as an abelian group, then by (10.7.1), we can assume that M is a subset of the divisible abelian group E. We will embed M in the injective left R-module $N = \text{Hom}_{\mathbb{Z}}(R, E)$ (see (10.7.3)). If $m \in M$, define $f(m)\colon R \to E$ by $f(m)(r) = rm$. Then $f\colon M \to N$, and we claim that f is an injective R-module homomorphism. If $f(m_1) = f(m_2)$, then $rm_1 = rm_2$ for every $r \in R$, and we take $r = 1$ to conclude that $m_1 = m_2$, proving injectivity. To check that f is an R-homomorphism, note that if $r, s \in R$ and $m \in M$, then

$$f(sm)(r) = rsm \text{ and } (sf(m))(r) = f(m)(rs) = rsm$$

by definition of scalar multiplication in the R-module N; see (10.7.2). ♣

It can be shown that every module M has an *injective hull*, that is, there is a smallest injective module containing M.

Problems For Section 10.7

1. If ${}_RM_S$ and ${}_RN$, show that $\text{Hom}_R(M, N)$ is a left S-module via

$$(sf)(m) = f(ms).$$

2. If ${}_RM_S$ and N_S, show that $\text{Hom}_S(M, N)$ is a right R-module via

$$(fr)(m) = f(rm).$$

3. If M_R and ${}_SN_R$, show that $\text{Hom}_R(M, N)$ is a left S-module via

$$(sf)(m) = sf(m).$$

4. If ${}_SM$ and ${}_SN_R$, show that $\mathrm{Hom}_S(M,N)$ is a right R-module via

$$(fr)(m) = f(m)r.$$

5. A useful mnemonic device for remembering the result of Problem 1 is that since M and N are *left* R-modules, we write the function f on the *right* of its argument. The result is $m(sf) = (ms)f$, a form of associativity. Give similar devices for Problems 2, 3 and 4.

Note also that in Problem 1, M is a right S-module, but $\mathrm{Hom}_R(M,N)$ is a left S-module. The reversal might be expected, because the hom functor is contravariant in its first argument. A similar situation occurs in Problem 2, but in Problems 3 and 4 there is no reversal. Again, this might be anticipated because the hom functor is covariant in its second argument.

6. Let R be an integral domain with quotient field Q. If M is a vector space over Q, show that M is a divisible R-module.
7. Conversely, if M is a torsion-free divisible R-module, show that M is a vector space over Q.
8. If R is an integral domain that is not a field, and Q is the quotient field of R, show that $\mathrm{Hom}_R(Q,R) = 0$.

10.8 Flat Modules

10.8.1 Definitions and Comments

We have seen that an R-module M is projective iff its covariant hom functor is exact, and M is injective iff its contravariant hom functor is exact. It is natural to investigate the exactness of the tensor functor $M \otimes_R -$, and as before we avoid complications by assuming all rings commutative. We say that M is *flat* if $M \otimes_R -$ is exact. Since the tensor functor is right exact by (10.4.4), an equivalent statement is that if $f\colon A \to B$ is an injective R-module homomorphism, then

$$1 \otimes f\colon M \otimes A \to M \otimes B$$

is injective. In fact it suffices to consider only R-modules A and B that are finitely generated. This can be deduced from properties of direct limits to be considered in the next section. [Any module is the direct limit of its finitely generated submodules (10.9.3, Example 2). The tensor product commutes with direct limits (Section 10.9, Problem 2). The direct limit is an exact functor (Section 10.9, Problem 4).] A proof that does not involve direct limits can also be given; see Rotman, "An Introduction to Homological Algebra", page 86.

10.8.2 Example

Since $\mathbb{Z}_2 \otimes_{\mathbb{Z}} -$ is not exact (Section 10.4, Problem 6), $\mathbb{Z}_2$ is not a flat $\mathbb{Z}$-module.

The next result is the analog for flat modules of property (10.5.4) of projective modules.

10.8.3 Proposition

The direct sum $\oplus_i M_i$ is flat if and only if each M_i is flat.

Proof. Let $f\colon A \to B$ be an injective R-homomorphism. In view of (8.8.6(b)), investigating the flatness of the direct sum amounts to analyzing the injectivity of the mapping

$$g\colon \oplus_i (M_i \otimes A) \to \oplus_i (M_i \otimes B)$$

given by

$$x_{i_1} \otimes a_1 + \cdots + x_{i_n} \otimes a_n \to x_{i_1} \otimes f(a_1) + \cdots + x_{i_n} \otimes f(a_n).$$

The map g will be injective if and only if all component maps $x_i \otimes a_i \to x_i \otimes f(a_i)$ are injective. This says that the direct sum is flat iff each component is flat. ♣

We now examine the relation between projectivity and flatness.

10.8.4 Proposition

R is a flat R-module.

Proof. If $f\colon A \to B$ is injective, we must show that $(1 \otimes f)\colon R \otimes_R A \to R \otimes_R B$ is injective. But by (8.7.6), $R \otimes_R M \cong M$ via $r \otimes x \to rx$. Thus the following diagram is commutative.

$$\begin{array}{ccc} R \otimes_R A & \xrightarrow{1 \otimes f} & R \otimes_R B \\ \| & & \| \\ A & \xrightarrow[f]{} & B \end{array}$$

Therefore injectivity of $1 \otimes f$ is equivalent to injectivity of f, and the result follows. ♣

10.8.5 Corollary

Every projective module, hence every free module, is flat.

Proof. By (10.8.3) and (10.8.4), every free module is flat. Since a projective module is a direct summand of a free module, it is flat by (10.8.3). ♣

Flat abelian groups can be characterized precisely.

10.8.6 Theorem

A $\mathbb{Z}$-module is flat iff it is torsion-free.

Proof. Suppose that M is a $\mathbb{Z}$-module that is not torsion-free. Let $x \in M$ be a nonzero element such that $nx = 0$ for some positive integer n. If $f\colon \mathbb{Z} \to \mathbb{Z}$ is multiplication by n, then $(1 \otimes f)\colon M \otimes \mathbb{Z} \to M \otimes \mathbb{Z}$ is given by

$$y \otimes z \to y \otimes nz = ny \otimes z$$

so that $(1 \otimes f)(x \otimes 1) = nx \otimes 1 = 0$. Since $x \otimes 1$ corresponds to x under the isomorphism between $M \otimes \mathbb{Z}$ and M, $x \otimes 1 \neq 0$, and $1 \otimes f$ is not injective. Therefore M is not flat.

The discussion in (10.8.1) shows that in checking flatness of M, we can restrict to finitely generated submodules of M. [We are examining equations of the form $(1 \otimes f)(t) = 0$, where $t = \sum_{i=1}^{n} x_i \otimes y_i$, $x_i \in M$, $y_i \in A$.]Thus without loss of generality, we can assume that M is a finitely generated abelian group. If M is torsion-free, then by (4.6.5), M is free and therefore flat by (10.8.5). ♣

10.8.7 Corollary

The additive group of rationals $\mathbb{Q}$ is a flat but not projective $\mathbb{Z}$-module.

Proof. Since $\mathbb{Q}$ is torsion-free, it is flat by (10.8.6). If $\mathbb{Q}$ were projective, it would be free by (10.5.5). This is a contradiction (see Section 4.1, Problem 5). ♣

Sometimes it is desirable to change the underlying ring of a module; the term *base change* is used in these situations.

10.8.8 Definitions and Comments

If $f\colon R \to S$ is a ring homomorphism and M is an S-module, we can create an R-module structure on M by $rx = f(r)x$, $r \in R$, $x \in M$. This is a base change by *restriction of scalars.*

If $f\colon R \to S$ is a ring homomorphism and M is an R-module, we can make $S \otimes_R M$ into an S-module via

$$s(s' \otimes x) = ss' \otimes x, \;\; s, s' \in S, \;\; x \in M.$$

This is a base change by *extension of scalars.* Note that S is an R-module by restriction of scalars, so the tensor product makes sense. What we are doing is allowing linear combinations of elements of M with coefficients in S. This operation is very common in algebraic topology.

In the exercises, we will look at the relation between base change and flatness. There will also be some problems on finitely generated algebras, so let's define these now.

10.8.9 Definition

The R-algebra A is *finitely generated* if there are elements $a_1, \ldots, a_n \in A$ such that every element of A is a polynomial in the a_i. Equivalently, the algebra homomorphism from the polynomial ring $R[X_1, \ldots, X_n] \to A$ determined by $X_i \to a_i$, $i = 1, \ldots, n$, is surjective. Thus A is a quotient of the polynomial ring.

It is important to note that if A is finitely generated *as an R-module*, then it is finitely generated as an R-algebra. [If $a = r_1a_1 + \cdots + r_na_n$, then a is certainly a polynomial in the a_i.]

Problems For Section 10.8

1. Give an example of a finitely generated R-algebra that is not finitely generated as an R-module.
2. Show that $R[X] \otimes_R R[Y] \cong R[X, Y]$.
3. Show that if A and B are finitely generated R-algebras, so is $A \otimes_R B$.
4. Let $f\colon R \to S$ be a ring homomorphism, and let M be an S-module, so that M is an R-module by restriction of scalars. If S is a flat R-module and M is a flat S-module, show that M is a flat R-module.
5. Let $f\colon R \to S$ be a ring homomorphism, and let M be an R-module, so that $S \otimes_R M$ is an S-module by extension of scalars. If M is a flat R-module, show that $S \otimes_R M$ is a flat S-module.
6. Let S be a multiplicative subset of the commutative ring R. Show that for any R-module M, $S^{-1}R \otimes_R M \cong S^{-1}M$ via $\alpha\colon (r/s) \otimes x \to rx/s$ with inverse $\beta\colon x/s \to (1/s) \otimes x$.
7. Continuing Problem 6, show that $S^{-1}R$ is a flat R-module.

10.9 Direct and Inverse Limits

If M is the direct sum of R-modules M_i, then R-homomorphisms $f_i\colon M_i \to N$ can be lifted uniquely to an R-homomorphism $f\colon M \to N$. The direct limit construction generalizes this idea. [In category theory, there is a further generalization called the *colimit*. The terminology is consistent because the direct sum is the coproduct in the category of modules.]

10.9.1 Direct Systems

A *directed set* is a partially ordered set I such that given any $i, j \in I$, there exists $k \in I$ such that $i \le k$ and $j \le k$. A typical example is the collection of finite subsets of a set, ordered by inclusion. If A and B are arbitrary finite subsets, then both A and B are contained in the finite set $A \cup B$.

Now suppose I is a directed set and we have a collection of objects $A_i, i \in I$, in a category $\mathcal{C}$. Assume that whenever $i \le j$, there is a morphism $h(i,j)\colon A_i \to A_j$. Assume further that the $h(i,j)$ are *compatible* in the sense that if $i \le j \le k$ and we apply $h(i,j)$

followed by $h(j,k)$, we get $h(i,k)$. We also require that for each i, $h(i,i)$ is the identity on A_i. The collection of objects and morphisms is called a *direct system.* As an example, take the objects to be the finitely generated submodules of a module, and the morphisms to be the natural inclusion maps. In this case, the directed set coincides with the set of objects, and the partial ordering is inclusion.

10.9.2 Direct Limits

Suppose that $\{A_i, h(i,j), i,j \in I\}$ is a direct system. The *direct limit* of the system will consist of an object A and morphisms $\alpha_i \colon A_i \to A$. Just as with coproducts, we want to lift morphisms $f_j \colon A_j \to B$ to a unique $f \colon A \to B$, that is, $f\alpha_j = f_j$ for all $j \in I$. But we require that the maps α_j be compatible with the $h(i,j)$, in other words, $\alpha_j h(i,j) = \alpha_i$ whenever $i \leq j$. A similar constraint is imposed on the f_j, namely, $f_j h(i,j) = f_i$, $i \leq j$. Thus the direct limit is an object A along with compatible morphisms $\alpha_j \colon A_j \to A$ such that given compatible morphisms $f_j \colon A_j \to B$, there is a unique morphism $f \colon A \to B$ such that $f\alpha_j = f_j$ for all j. Figure 10.9.1 summarizes the discussion.

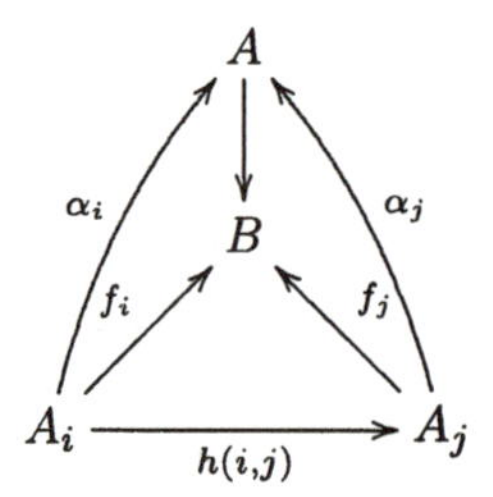

Figure 10.9.1

As in Section 10.2, any two direct limits of a given direct system are isomorphic.

If the ordering on I is the equality relation, then the only element j such that $i \leq j$ is i itself. Compatibility is automatic, and the direct limit reduces to a coproduct.

A popular notation for the direct limit is

$$A = \varinjlim A_i.$$

The direct limit is sometimes called an *inductive limit.*

10.9.3 Examples

1. A coproduct is a direct limit, as discussed above. In particular, a direct sum of modules is a direct limit.
2. Any module is the direct limit of its finitely generated submodules. [Use the direct system indicated in (10.9.1).]
3. The algebraic closure of a field F can be constructed (informally) by adjoining roots of all possible polynomials in $F[X]$; see (3.3.7). This suggests that the algebraic closure is the direct limit of the collection of all finite extensions of F. This can be proved with the aid of (3.3.9).

In the category of modules, direct limits always exist.

10.9.4 Theorem

If $\{M_i, h(i,j), i, j \in I\}$ is a direct system of R-modules, then the direct limit of the system exists.

Proof. Take M to be $(\oplus_i M_i)/N$, with N the submodule of the direct sum generated by all elements of the form

$$\beta_j h(i,j)x_i - \beta_i x_i, \; i \le j, \; x_i \in M_i \tag{1}$$

where β_i is the inclusion map of M_i into the direct sum. Define $\alpha_i \colon M_i \to M$ by

$$\alpha_i x_i = \beta_i x_i + N.$$

The α_i are compatible, because

$$\alpha_j h(i,j)x_i = \beta_j h(i,j)x_i + N = \beta_i x_i + N = \alpha_i x_i.$$

Given compatible $f_i \colon M_i \to B$, we define $f \colon M \to B$ by

$$f(\beta_i x_i + N) = f_i x_i,$$

the only possible choice. This forces $f\alpha_i = f_i$, provided we show that f is well-defined. But an element of N of the form (1) is mapped by our proposed f to

$$f_j h(i,j)x_i - f_i x_i$$

which is 0 by compatibility of the f_i. Thus f maps everything in N to 0, and the result follows. ♣

10.9.5 Inverse Systems and Inverse Limits

Inverse limits are dual to direct limits. An *inverse system* is defined as in (10.9.1), except that if $i \le j$, then $h(i,j)$ maps "backwards" from A_j to A_i. If we apply $h(j,k)$ followed by $h(i,j)$, we get $h(i,k)$; as before, $h(i,i)$ is the identity on A_i. The *inverse limit* of the inverse system $\{A_i, h(i,j), i, j \in I\}$ is an object A along with morphisms $p_i \colon A \to A_i$. As with products, we want to lift morphisms $f_i \colon B \to A_i$ to a unique $f \colon B \to A$. There is a compatibility requirement on the p_i and f_i: if $i \le j$, then $h(i,j)p_j = p_i$, and similarly $h(i,j)f_j = f_i$. Thus the inverse limit is an object A along with compatible morphisms $p_i \colon A \to A_i$ such that given compatible morphisms $f_i \colon B \to A_i$, there is a unique morphism $f \colon B \to A$ such that $p_i f = f_i$ for all i. See Figure 10.9.2.

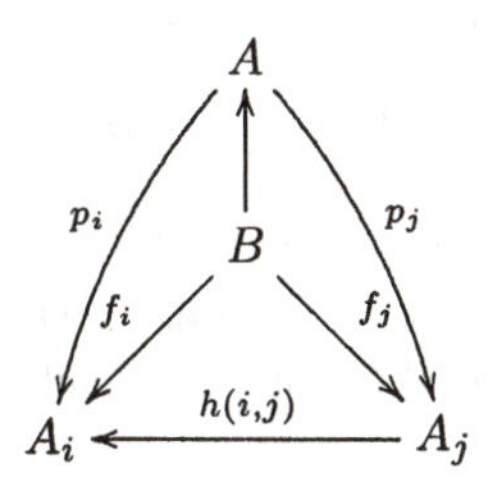

Figue 10.9.2

As in Section 10.2, the universal mapping property determines the inverse limit up to isomorphism.

If the ordering on I is the equality relation, the the inverse limit reduces to a product. In category theory, the *limit* is a generalization of the inverse limit.

A popular notation for the inverse limit is

$$A = \varprojlim A_i.$$

The inverse limit is sometimes called a *projective limit.*

We constructed the direct limit of a family of modules by forming a quotient of the direct sum. By duality, we expect that the inverse limit involves a submodule of the direct product.

10.9.6 Theorem

If $\{M_i, h(i,j), i,j \in I\}$ is an inverse system of R-modules, then the inverse limit of the system exists.

Proof. We take M to be the set of all $x = (x_i, i \in I)$ in the direct product $\prod_i M_i$ such that $h(i,j)x_j = x_i$ whenever $i \leq j$. Let p_i be the restriction to M of the projection on the ith factor. Then $h(i,j)p_jx = h(i,j)x_j = x_i = p_ix$, so the p_i are compatible. Given compatible $f_i \colon N \to M_i$, let f be the product of the f_i, that is, $fx = (f_ix, i \in I)$. By compatibility, $h(i,j)f_jx = f_ix$ for $i \leq j$, so f maps $\prod_i M_i$ into M. By definition of f we have $p_if = f_i$, and the result follows. ♣

10.9.7 Example

Recall from Section 7.9 that a p-adic integer can be represented as $a_0 + a_1p + a_2p^2 + \cdots$, where the a_i belong to $\{0, 1, \dots, p-1\}$. If we discard all terms after $a_{r-1}p^{r-1}$, $r = 1, 2, \dots$, we get the ring $\mathbb{Z}_{p^r}$. These rings form an inverse system; if $x \in \mathbb{Z}_{p^s}$ and $r \leq s$, we take $h(r,s)x$ to be the residue of x mod p^r. The inverse limit of this system is the ring of p-adic integers.

Problems For Section 10.9

1. In Theorem (10.9.6), why can't we say "obviously", since direct limits exist in the category of modules, inverse limits must also exist by duality.
2. Show that in the category of modules over a commutative ring, the tensor product commutes with direct limits. In other words,

$$\varinjlim (M \otimes N_i) = M \otimes \varinjlim N_i$$

assuming that the direct limit of the N_i exists.

3. For each $n = 1, 2, \dots$, let A_n be an R-module, with $A_1 \subseteq A_2 \subseteq A_3 \subseteq \cdots$. Take $h(i,j)$, $i \leq j$, to be the inclusion map. What is the direct limit of the A_n? (Be more explicit than in (10.9.4).)

4. Suppose that A, B and C are the direct limits of direct systems $\{A_i\}$, $\{B_i\}$ and $\{C_i\}$ of R-modules. Assume that for each i, the sequence

$$A_i \xrightarrow{f_i} B_i \xrightarrow{g_i} C_i$$

is exact. Give an intuitive argument to suggest that the sequence

$$A \xrightarrow{f} B \xrightarrow{g} C$$

is exact. Thus direct limit is an exact functor.

[A lot of formalism is being suppressed here. We must make the collection of direct systems into a category, and define a morphism in that category. This forces compatibility conditions on the f_i and g_i: $f_j h_A(i,j) = h_B(i,j) f_i$, $g_j h_B(i,j) = h_C(i,j) g_i$. The direct limit functor takes a direct system to its direct limit, but we must also specify what it does to morphisms.]

A possible strategy is to claim that since an element of a direct sum has only finitely many nonzero components, exactness at B is equivalent to exactness at each B_i. This is unconvincing because the direct limit is not simply a direct sum, but a quotient of a direct sum. Suggestions are welcome!

Problems 5 and 6 give some additional properties of direct products.

5. Show that

$$\operatorname{Hom}_R(\oplus A_i, B) \cong \prod_i \operatorname{Hom}_R(A_i, B).$$

6. Show that

$$\operatorname{Hom}_R(A, \prod_i B_i) \cong \prod_i \operatorname{Hom}_R(A, B_i).$$

7. If M is a nonzero R-module that is both projective and injective, where R is an integral domain that is not a field, show that $\operatorname{Hom}_R(M, R) = 0$.
8. Let R be an integral domain that is not a field. If M is an R-module that is both projective and injective, show that $M = 0$.

Appendix to Chapter 10

We have seen that an abelian group is injective if and only if it is divisible. In this appendix we give an explicit characterization of such groups.

A10.1 Definitions and Comments

Let G be an abelian group, and T the torsion subgroup of G (the elements of G of finite order). Then G/T is torsion-free, since $n(x+T) = 0$ implies $nx \in T$, hence $x \in T$. If p is a fixed prime, the *primary component* G_p associated with p consists of all elements whose order is a power of p. Note that G_p is a subgroup of G, for if $p^n a = p^m b = 0$, $n \geq m$, then $p^n(a-b) = 0$. (We use the fact that G is abelian; for example, $3(a-b) = a - b + a - b + a - b = a + a + a - b - b - b$.)

A10.2 Proposition

The torsion subgroup T is the direct sum of the primary components G_p, p prime.

Proof. Suppose x has order $m = \prod_{j=1}^k p_j^{r_j}$. If $m_i = m/p_i^{r_i}$, then the greatest common divisor of the m_i is 1, so there are integers $a_1, \dots, a_k$ such that $a_1m_1 + \cdots + a_km_k = 1$. Thus $x = 1x = \sum_{i=1}^k a_i(m_ix)$, and (by definition of m_i) m_ix has order $p_i^{r_i}$ and therefore belongs to the primary component G_{p_i}. This proves that G is the sum of the G_p. To show that the sum is direct, assume $0 \neq x \in G_p \cap \sum_{q \neq p} G_q$. Then the order of x is a power of p and also a product of prime factors unequal to p, which is impossible. For example, if y has order 9 and z has order 125, then $9(125)(y+z) = 0$, so the order of $y+z$ is of the form 3^r5^s. ♣

A10.3 Definitions and Comments

A *Prüfer group*, also called a *quasicyclic group* and denoted by $\mathbb{Z}(p^\infty)$, is a p-primary component of $\mathbb{Q}/\mathbb{Z}$, the rationals mod 1. Since every element of $\mathbb{Q}/\mathbb{Z}$ has finite order, it follows from (A10.2) that

$$\mathbb{Q}/\mathbb{Z} = \bigoplus_p \mathbb{Z}(p^\infty).$$

Now an element of $\mathbb{Q}/\mathbb{Z}$ whose order is a power of p must be of the form $a/p^r + \mathbb{Z}$ for some integer a and nonnegative integer r. It follows that the elements $a_r = 1/p^r + \mathbb{Z}$, $r = 1, 2, \dots$, generate $\mathbb{Z}(p^\infty)$. These elements satisfy the following relations:

$$pa_1 = 0, \ pa_2 = a_1, \dots, pa_{r+1} = a_r, \dots$$

A10.4 Proposition

Let H be a group defined by generators $b_1, b_2, \dots$ and relations $pb_1 = 0, pb_2 = b_1, \dots, pb_{r+1} = b_r, \dots$. Then H is isomorphic to $\mathbb{Z}(p^\infty)$.

Proof. First note that the relations imply that every element of H is an integer multiple of some b_i. Here is a typical computation:

$$\begin{aligned} 4b_7 + 6b_{10} + 2b_{14} &= 4(pb_8) + 6(pb_{11}) + 2b_{14} \\ &= \cdots = 4(p^7b_{14}) + 6(p^4b_{14}) + 2b_{14} = (4p^7 + 6p^4 + 2)b_{14}. \end{aligned}$$

By (5.8.5), there is an epimorphism $f\colon H \to \mathbb{Z}(p^\infty)$, and by the proof of (5.8.5), we can take $f(b_i) = a_i$ for all i. To show that f is injective, suppose $f(cb_i) = 0$ where $c \in \mathbb{Z}$. Then $cf(b_i) = ca_i = 0$, so $c/p^i \in \mathbb{Z}$, in other words, p^i divides c. (We can reverse this argument to conclude that $f(cb_i) = 0$ iff p^i divides c.) But the relations imply that $p^ib_i = 0$, and since c is a multiple of p^i, we have $cb_i = 0$. ♣

A10.5 Proposition

Let G be a divisible abelian group. Then its torsion subgroup T is also divisible. Moreover, G can be written as $T \oplus D$, where D is torsion-free and divisible.

Proof. If $x \in T$ and $0 \neq n \in \mathbb{Z}$, then for some $y \in G$ we have $ny = x$. Thus in the torsion-free group G/T we have $n(y+T) = x+T = 0$. But then $ny \in T$, so (as in (A10.1)) $y \in T$ and T is divisible, hence injective by (10.6.6). By (10.6.2), the exact sequence $0 \to T \to G \to G/T \to 0$ splits, so $G \cong T \oplus G/T$. Since G/T is torsion-free and divisible (see (10.6.5)), the result follows. ♣

We are going to show that an abelian group is divisible iff it is a direct sum of copies of $\mathbb{Q}$ (the additive group of rationals) and quasicyclic groups. To show that every divisible abelian group has this form, it suffices, by (A10.2), (A10.5) and the fact that a direct sum of divisible abelian groups is divisible, to consider only two cases, G torsion-free and G a p-group.

A10.6 Proposition

If G is a divisible, torsion-free abelian group, then G is isomorphic to a direct sum of copies of $\mathbb{Q}$.

Proof. The result follows from the observation that G can be regarded as a $\mathbb{Q}$-module, that is, a vector space over $\mathbb{Q}$; see Section 10.7, Problem 7. ♣

For any abelian group G, let $G[n] = \{x \in G\colon nx = 0\}$.

A10.7 Proposition

Let G and H be divisible abelian p-groups. Then any isomorphism φ of $G[p]$ and $H[p]$ can be extended to an isomorphism ψ of G and H.

Proof. Our candidate ψ arises from the injectivity of H, as the diagram below indicates.

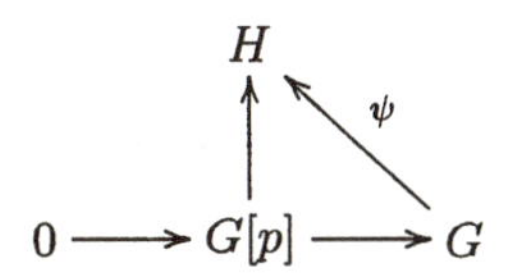

The map from $G[p]$ to G is inclusion, and the map from $G[p]$ to H is the composition of φ and the inclusion from $H[p]$ to H. Suppose that $x \in G$ and the order of x is $|x| = p^n$. We will prove by induction that $\psi(x) = 0$ implies $x = 0$. If $n = 1$, then $x \in G[p]$, so $\psi(x) = \varphi(x)$, and the result follows because φ is injective. For the inductive step, suppose $|x| = p^{n+1}$ and $\psi(x) = 0$. Then $|px| = p^n$ and $\psi(px) = p\psi(x) = 0$. By induction hypothesis, $px = 0$, which contradicts the assumption that x has order p^{n+1}.

Now we prove by induction that ψ is surjective. Explicitly, if $y \in H$ and $|y| = p^n$, then y belongs to the image of ψ. If $n = 1$, then $y \in H[p]$ and the result follows because φ

is surjective. If $|y| = p^{n+1}$, then $p^n y \in H[p]$, so for some $x \in G[p]$ we have $\varphi(x) = p^n y$. Since G is divisible, there exists $g \in G$ such that $p^n g = x$. Then

$$p^n(y - \psi(g)) = p^n y - \psi(p^n g) = p^n y - \psi(x) = p^n y - \varphi(x) = 0.$$

By induction hypothesis, there is an element $z \in G$ such that $\psi(z) = y - \psi(g)$. Thus $\psi(g + z) = y$. ♣

A10.8 Theorem

An abelian group G is divisible if and only if G is a direct sum of copies of $\mathbb{Q}$ and quasicyclic groups.

Proof. Suppose that G is such a direct sum. Since $\mathbb{Q}$ and $\mathbb{Z}(p^\infty)$ are divisible [$\mathbb{Z}(p^\infty)$ is a direct summand of the divisible group $\mathbb{Q}/\mathbb{Z}$], and a direct sum of divisible abelian groups is divisible, G must be divisible. Conversely, assume G divisible. In view of (A10.6) and the discussion preceding it, we may assume that G is a p-group. But then $G[p]$ is a vector space over the field $\mathbb{F}_p = \mathbb{Z}/p\mathbb{Z}$; the scalar multiplication is given by $(n + p\mathbb{Z})g = ng$. Since $pg = 0$, scalar multiplication is well-defined. If the dimension of $G[p]$ over $\mathbb{F}_p$ is d, let H be the direct sum of d copies of $\mathbb{Z}(p^\infty)$. An element of order p in a component of the direct sum is an integer multiple of $1/p + \mathbb{Z}$, and consequently $H[p]$ is also a d-dimensional vector space over $\mathbb{F}_p$. Thus $G[p]$ is isomorphic to $H[p]$, and it follows from (A10.7) that G is isomorphic to H. ♣

Supplement: The Long Exact Homology Sequence and Applications

S1. Chain Complexes

In the supplement, we will develop some of the building blocks for algebraic topology. As we go along, we will make brief comments [in brackets] indicating the connection between the algebraic machinery and the topological setting, but for best results here, please consult a text or attend lectures on algebraic topology.

S1.1 Definitions and Comments

A *chain complex* (or simply a *complex*) C_* is a family of R-modules C_n, $n \in \mathbb{Z}$, along with R-homomorphisms $d_n : C_n \to C_{n-1}$ called *differentials*, satisfying $d_n d_{n+1} = 0$ for all n. A chain complex with only finitely many C_n's is allowed; it can always be extended with the aid of zero modules and zero maps. [In topology, C_n is the abelian group of n-*chains*, that is, all formal linear combinations with integer coefficients of n-simplices in a topological space X. The map d_n is the *boundary operator*, which assigns to an n-simplex an $n-1$-chain that represents the oriented boundary of the simplex.]

The kernel of d_n is written $Z_n(C_*)$ or just Z_n; elements of Z_n are called *cycles* in dimension n. The image of d_{n+1} is written $B_n(C_*)$ or just B_n; elements of B_n are called *boundaries* in dimension n. Since the composition of two successive differentials is 0, it follows that $B_n \subseteq Z_n$. The quotient Z_n/B_n is written $H_n(C_*)$ or just H_n; it is called the nth *homology module* (or *homology group* if the underlying ring R is $\mathbb{Z}$).

[The key idea of algebraic topology is the association of an algebraic object, the collection of homology groups $H_n(X)$, to a topological space X. If two spaces X and Y are homeomorphic, in fact if they merely have the same homotopy type, then $H_n(X)$ and $H_n(Y)$ are isomorphic for all n. Thus the homology groups can be used to distinguish between topological spaces; if the homology groups differ, the spaces cannot be homeomorphic.]

Note that any exact sequence is a complex, since the composition of successive maps is 0.

S1.2 Definition

A *chain map* $f: C_* \to D_*$ from a chain complex C_* to a chain complex D_* is a collection of module homomorphisms $f_n: C_n \to D_n$, such that for all n, the following diagram is commutative.

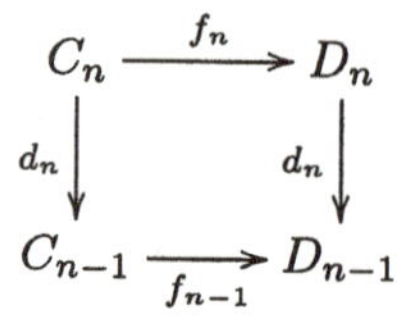

$$\begin{array}{ccc} C_n & \xrightarrow{f_n} & D_n \\ {\scriptstyle d_n}\downarrow & & \downarrow{\scriptstyle d_n} \\ C_{n-1} & \xrightarrow[f_{n-1}]{} & D_{n-1} \end{array}$$

We use the same symbol d_n to refer to the differentials in C_* and D_*.

[If $f: X \to Y$ is a continuous map of topological spaces and σ is a singular n-simplex in X, then $f_\#(\sigma) = f \circ \sigma$ is a singular n-simplex in Y, and $f_\#$ extends to a homomorphism of n-chains. If we assemble the $f_\#$'s for $n = 0, 1, \dots,$ the result is a chain map.]

S1.3 Proposition

A chain map f takes cycles to cycles and boundaries to boundaries. Consequently, the map $z_n + B_n(C_*) \to f_n(z_n) + B_n(D_*)$ is a well-defined homomorphism from $H_n(C_*)$ to $H_n(D_*)$. It is denoted by $H_n(f)$.

Proof. If $z \in Z_n(C_*)$, then since f is a chain map, $d_n f_n(z) = f_{n-1} d_n(z) = f_{n-1}(0) = 0$. Therefore $f_n(z) \in Z_n(D_*)$. If $b \in B_n(C_*)$, then $d_{n+1}c = b$ for some $c \in C_{n+1}$. Then $f_n(b) = f_n(d_{n+1}c) = d_{n+1} f_{n+1} c$, so $f_n(b) \in B_n(D_*)$. ♣

S1.4 The Homology Functors

We can create a category whose objects are chain complexes and whose morphisms are chain maps. The composition gf of two chain maps $f: C_* \to D_*$ and $g: D_* \to E_*$ is the collection of homomorphisms $g_n f_n$, $n \in \mathbb{Z}$. For any n, we associate with the chain complex C_* its nth homology module $H_n(C_*)$, and we associate with the chain map $f: C_* \to D_*$ the map $H_n(f): H_n(C_*) \to H_n(D_*)$ defined in (S1.3). Since $H_n(gf) = H_n(g)H_n(f)$ and $H_n(1_{C_*})$ is the identity on $H_n(C_*)$, H_n is a functor, called the nth *homology functor*.

S1.5 Chain Homotopy

Let f and g be chain maps from C_* to D_*. We say that f and g are *chain homotopic* and write $f \simeq g$ if there exist homomorphisms $h_n: C_n \to D_{n+1}$ such that $f_n - g_n = d_{n+1} h_n + h_{n-1} d_n$; see the diagram below.

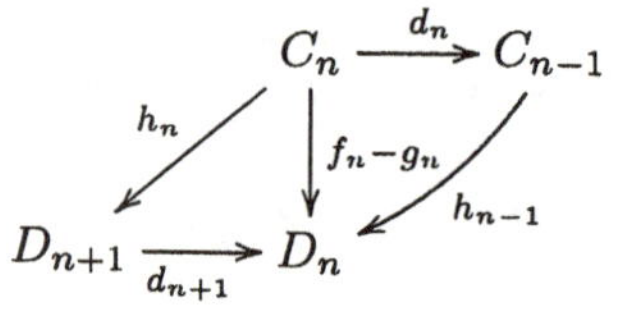

$$\begin{array}{ccccc} & & C_n & \xrightarrow{d_n} & C_{n-1} \\ & {\scriptstyle h_n}\swarrow & \downarrow{\scriptstyle f_n - g_n} & \swarrow{\scriptstyle h_{n-1}} & \\ D_{n+1} & \xrightarrow[d_{n+1}]{} & D_n & & \end{array}$$

[If f and g are homotopic maps from a topological space X to a topological space Y, then the maps $f_{\#}$ and $g_{\#}$ (see the discussion in (S1.2)) are chain homotopic,]

S1.6 Proposition

If f and g are chain homotopic, then $H_n(f) = H_n(g)$.

Proof. Let $z \in Z_n(C_*)$. Then

$$f_n(z) - g_n(z) = (d_{n+1}h_n + h_{n-1}d_n)z \in B_n(D_*)$$

since $d_n z = 0$. Thus $f_n(z) + B_n(D_*) = g_n(z) + B_n(D_*)$, in other words, $H_n(f) = H_n(g)$. ♣

S2. The Snake Lemma

We isolate the main ingredient of the long exact homology sequence. After an elaborate diagram chase, a homomorphism between two modules is constructed. The domain and codomain of the homomorphism are far apart in the diagram, and the arrow joining them tends to wiggle like a serpent. First, a result about kernels and cokernels of module homomorphisms.

S2.1 Lemma

Assume that the diagram below is commutative.

$$\begin{array}{ccc} A & \xrightarrow{f} & B \\ {\scriptstyle d}\downarrow & & \downarrow{\scriptstyle e} \\ C & \xrightarrow[g]{} & D \end{array}$$

(i) f induces a homomorphism on kernels, that is, $f(\ker d) \subseteq \ker e$.

(ii) g induces a homomorphism on cokernels, that is, the map $y + \operatorname{im} d \to g(y) + \operatorname{im} e$, $y \in C$, is a well-defined homomorphism from coker d to coker e.

(iii) If f is injective, so is the map induced by f, and if g is surjective, so is the map induced by g.

Proof. (i) If $x \in A$ and $d(x) = 0$, then $ef(x) = gd(x) = g0 = 0$.

(ii) If $y \in \operatorname{im} d$, then $y = dx$ for some $x \in A$. Thus $gy = gdx = efx \in \operatorname{im} e$. Since g is a homomorphism, the induced map is also.

(iii) The first statement holds because the map induced by f is simply a restriction. The second statement follows from the form of the map induced by g. ♣

Now refer to our *snake diagram*, Figure S2.1. Initially, we are given only the second and third rows (ABE0 and 0CDF), along with the maps d, e and h. Commutativity of the squares ABDC and BEFD is assumed, along with exactness of the rows. The diagram is now enlarged as follows. Take $A' = \ker d$, and let the map from A' to A be inclusion. Take $C' = \text{coker } d$, and let the map from C to C' be canonical. Augment columns 2 and 3 in a similar fashion. Let $A' \to B'$ be the map induced by f on kernels, and let $C' \to D'$ be the map induced by g on cokernels. Similarly, add $B' \to E'$ and $D' \to F'$. The enlarged diagram is commutative by (S2.1), and it has exact columns by construction.

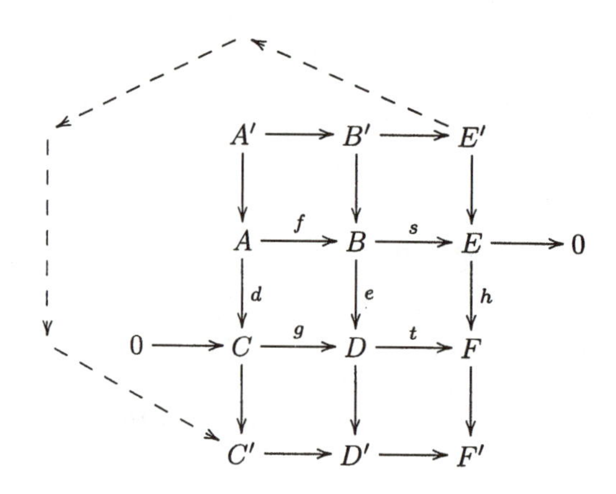

Figure S2.1

S2.2 Lemma

The first and fourth rows of the enlarged snake diagram are exact.

Proof. This is an instructive diagram chase, showing many standard patterns. Induced maps will be denoted by an overbar, and we first prove exactness at B'. If $x \in A' = \ker d$ and $y = \overline{f}x = fx$, then $sy = sfx = 0$, so $y \in \ker \overline{s}$. On the other hand, if $y \in B' \subseteq B$ and $\overline{s}y = sy = 0$, then $y = fx$ for some $x \in A$. Thus $0 = ey = efx = gdx$, and since g is injective, $dx = 0$. Therefore $y = fx$ with $x \in A'$, and $y \in \text{im } \overline{f}$.

Now to prove exactness at D', let $x \in C$. Then $\overline{t}(gx + \text{im } e) = tgx + \text{im } h = 0$ by exactness of the third row, so $\text{im } \overline{g} \subseteq \ker \overline{t}$. Conversely, if $y \in D$ and $\overline{t}(y + \text{im } e) = ty + \text{im } h = 0$, then $ty = hz$ for some $z \in E$. Since s is surjective, $z = sx$ for some $x \in B$. Now

$$ty = hz = hsx = tex$$

so $y - ex \in \ker t = \text{im } g$, say $y - ex = gw, w \in C$. Therefore

$$y + \text{im } e = \overline{g}(w + \text{im } d)$$

and $y + \text{im } e \in \text{im } \overline{g}$. ♣

S2.3 Remark

Sometimes an even bigger snake diagram is given, with column 1 assumed to be an exact sequence

$$0 \longrightarrow A' \longrightarrow A \xrightarrow{d} C \longrightarrow C' \longrightarrow 0$$

and similarly for columns 2 and 3. This is nothing new, because by replacing modules by isomorphic copies we can assume that A' is the kernel of d, C' is the cokernel of d, $A' \to A$ is inclusion, and $C \to C'$ is canonical.

S2.4 The Connecting Homomorphism

We will now connect E' to C' in the snake diagram while preserving exactness. The idea is to zig-zag through the diagram along the path $E'EBDCC'$.

Let $z \in E' \subseteq E$; Since s is surjective, there exists $y \in B$ such that $z = sy$. Then $tey = hsy = hz = 0$ since $E' = \ker h$. Thus $ey \in \ker t = \operatorname{im} g$, so $ey = gx$ for some $x \in C$. We define the *connecting homomorphism* $\partial \colon E' \to C'$ by $\partial z = x + \operatorname{im} d$. Symbolically,

$$\partial = [g^{-1} \circ e \circ s^{-1}]$$

where the brackets indicate that ∂z is the coset of x in $C' = C/\operatorname{im} d$.

We must show that ∂ is well-defined. Suppose that y' is another element of B with $sy' = z$. Then $y - y' \in \ker s = \operatorname{im} f$, so $y - y' = fu$ for some $u \in A$. Thus $e(y - y') = efu = gdu$. Now we know from the above computation that $ey = gx$ for some $x \in C$, and similarly $ey' = gx'$ for some $x' \in C$. Therefore $g(x - x') = gdu$, so $x - x' - du \in \ker g$. Since g is injective, $x - x' = du$, so $x + \operatorname{im} d = x' + \operatorname{im} d$. Thus ∂z is independent of the choice of the representatives y and x. Since every map in the diagram is a homomorphism, so is ∂.

S2.5 Snake Lemma

The sequence

$$A' \xrightarrow{\overline{f}} B' \xrightarrow{\overline{s}} E' \xrightarrow{\partial} C' \xrightarrow{\overline{g}} D' \xrightarrow{\overline{t}} F'$$

is exact.

Proof. In view of (S2.2), we need only show exactness at E' and C'. If $z = sy$, $y \in B' = \ker e$, then $ey = 0$, so $\partial z = 0$ by definition of ∂. Thus $\operatorname{im} \overline{s} \subseteq \ker \partial$. Conversely, assume $\partial z = 0$, and let x and y be as in the definition of ∂. Then $x = du$ for some $u \in A$, hence $gx = gdu = efu$. But $gx = ey$ by definition of ∂, so $y - fu \in \ker e = B'$. Since $z = sy$ by definition of ∂, we have

$$z = s(y - fu + fu) = s(y - fu) \in \operatorname{im} \overline{s}.$$

To show exactness at C', consider an element ∂z in the image of ∂. Then $\partial z = x + \operatorname{im} d$, so $\overline{g}\partial z = gx + \operatorname{im} e$. But $gx = ey$ by definition of ∂, so $\overline{g}\partial z = 0$ and $\partial z \in \ker \overline{g}$. Conversely, suppose $x \in C$ and $\overline{g}(x + \operatorname{im} d) = gx + \operatorname{im} e = 0$. Then $gx = ey$ for some $y \in B$. If $z = sy$, then $hsy = tey = tgx = 0$ by exactness of the third row. Thus $z \in E'$ and (by definition of ∂) we have $\partial z = x + \operatorname{im} d$. Consequently, $x + \operatorname{im} d \in \operatorname{im} \partial$. ♣

S3. The Long Exact Homology Sequence

S3.1 Definition

We say that

$$0 \longrightarrow C_* \xrightarrow{f} D_* \xrightarrow{g} E_* \longrightarrow 0$$

where f and g are chain maps, is a *short exact sequence of chain complexes* if for each n, the corresponding sequence formed by the component maps $f_n \colon C_n \to D_n$ and $g_n \colon D_n \to E_n$, is short exact. We will construct *connecting homomorphisms* $\partial_n \colon H_n(E_*) \to H_{n-1}(C_*)$ such that the sequence

$$\cdots \xrightarrow{g} H_{n+1}(E_*) \xrightarrow{\partial} H_n(C_*) \xrightarrow{f} H_n(D_*) \xrightarrow{g} H_n(E_*) \xrightarrow{\partial} H_{n-1}(C_*) \xrightarrow{f} \cdots$$

is exact. [We have taken some liberties with the notation. In the second diagram, f stands for the map induced by f_n on homology, namely, $H_n(f)$; similarly for g.] The second diagram is the *long exact homology sequence*, and the result may be summarized as follows.

S3.2 Theorem

A short exact sequence of chain complexes induces a long exact sequence of homology modules.

Proof. This is a double application of the snake lemma. The main ingredient is the following snake diagram.

$$\begin{array}{ccccccc}
 & & C_n/B_n(C_*) & \longrightarrow & D_n/B_n(D_*) & \longrightarrow & E_n/B_n(E_*) & \longrightarrow 0 \\
 & & \downarrow d & & \downarrow d & & \downarrow d & \\
0 & \longrightarrow & Z_{n-1}(C_*) & \longrightarrow & Z_{n-1}(D_*) & \longrightarrow & Z_{n-1}(E_*) &
\end{array}$$

The horizontal maps are derived from the chain maps f and g, and the vertical maps are given by $d(x_n + B_n) = dx_n$. The kernel of a vertical map is $\{x_n + B_n \colon x_n \in Z_n\} = H_n$, and the cokernel is $Z_{n-1}/B_{n-1} = H_{n-1}$. The diagram is commutative by the definition of a chain map. But in order to apply the snake lemma, we must verify that the rows are exact, and this involves another application of the snake lemma. The appropriate diagram is

$$\begin{array}{ccccccccc}
0 & \longrightarrow & C_n & \longrightarrow & D_n & \longrightarrow & E_n & \longrightarrow & 0 \\
 & & \downarrow d & & \downarrow d & & \downarrow d & & \\
0 & \longrightarrow & C_{n-1} & \longrightarrow & D_{n-1} & \longrightarrow & E_{n-1} & \longrightarrow & 0
\end{array}$$

where the horizontal maps are again derived from f and g. The exactness of the rows of the first diagram follows from (S2.2) and part (iii) of (S2.1), shifting indices from n to $n \pm 1$ as needed. ♣

S3.3 The connecting homomorphism explicitly

If $z \in H_n(E_*)$, then $z = z_n + B_n(E_*)$ for some $z_n \in Z_n(E_*)$. We apply (S2.4) to compute ∂z. We have $z_n + B_n(E_*) = g_n(y_n + B_n(D_*))$ for some $y_n \in D_n$. Then $dy_n \in Z_{n-1}(D_*)$ and $dy_n = f_{n-1}(x_{n-1})$ for some $x_{n-1} \in Z_{n-1}(C_*)$. Finally, $\partial z = x_{n-1} + B_{n-1}(C_*)$.

S3.4 Naturality

Suppose that we have a commutative diagram of short exact sequences of chain complexes, as shown below.

$$\begin{array}{ccccccccc} 0 & \longrightarrow & C & \longrightarrow & D & \longrightarrow & E & \longrightarrow & 0 \\ & & \downarrow & & \downarrow & & \downarrow & & \\ 0 & \longrightarrow & C' & \longrightarrow & D' & \longrightarrow & E' & \longrightarrow & 0 \end{array}$$

Then there is a corresponding commutative diagram of long exact sequences:

$$\begin{array}{ccccccccc} \cdots \xrightarrow{\partial} & H_n(C_*) & \longrightarrow & H_n(D_*) & \longrightarrow & H_n(E_*) & \xrightarrow{\partial} & H_{n-1}(C_*) & \longrightarrow \cdots \\ & \downarrow & & \downarrow & & \downarrow & & & \\ \cdots \xrightarrow{\partial} & H_n(C'_*) & \longrightarrow & H_n(D'_*) & \longrightarrow & H_n(E'_*) & \xrightarrow{\partial} & H_{n-1}(C'_*) & \longrightarrow \cdots \end{array}$$

Proof. The homology functor, indeed any functor, preserves commutative diagrams, so the two squares on the left commute. For the third square, an informal argument may help to illuminate the idea. Trace through the explicit construction of ∂ in (S3.3), and let f be the vertical chain map in the commutative diagram of short exact sequences. The first step in the process is

$$z_n + B_n(E_*) \to y_n + B_n(D_*).$$

By commutativity,

$$fz_n + B_n(E'_*) \to fy_n + B_n(D'_*).$$

Continuing in this fashion, we find that if $\partial z = x_{n-1} + B_{n-1}(C_*)$, then

$$\partial(fz) = fx_{n-1} + B_{n-1}(C'_*) = f(x_{n-1} + B_{n-1}(C_*)) = f(\partial z). \quad ♣$$

A formal proof can be found in "An Introduction to Algebraic Topology" by J. Rotman, page 95.

S4. Projective and Injective Resolutions

The functors Tor and Ext are developed with the aid of projective and injective resolutions of a module, and we will now examine these constructions.

S4.1 Definitions and Comments

A *left resolution* of a module M is an exact sequence

$$\cdots \longrightarrow P_2 \longrightarrow P_1 \longrightarrow P_0 \longrightarrow M \longrightarrow 0$$

A left resolution is a *projective resolution* if every P_i is projective, a *free resolution* if every P_i is free. By the first isomorphism theorem, M is isomorphic to the cokernel of the map $P_1 \to P_0$, so in a sense no information is lost if M is removed. A *deleted projective resolution* is of the form

$$\begin{array}{ccccccc} \cdots \longrightarrow & P_2 & \longrightarrow & P_1 & \longrightarrow & P_0 & \longrightarrow 0 \\ & & & & & \downarrow & \\ & & & & & M & \end{array}$$

and the deleted version turns out to be more convenient in computations. Notice that in a deleted projective resolution, exactness at P_0 no longer holds because the map $P_1 \to P_0$ need not be surjective. Resolutions with only finitely many P_n's are allowed, provided that the module on the extreme left is 0. The sequence can then be extended via zero modules and zero maps.

Dually, a *right resolution* of M is an exact sequence

$$0 \longrightarrow M \longrightarrow E_0 \longrightarrow E_1 \longrightarrow E_2 \cdots;$$

we have an *injective resolution* if every E_i is injective, . A *deleted injective resolution* has the form

$$\begin{array}{ccccccc} 0 \longrightarrow & E_0 & \longrightarrow & E_1 & \longrightarrow & E_2 & \longrightarrow \cdots \\ & \uparrow & & & & & \\ & M & & & & & \end{array}$$

Exactness at E_0 no longer holds because the map $E_0 \to E_1$ need not be injective.

We will use the notation $P_* \to M$ for a projective resolution, and $M \to E_*$ for an injective resolution.

S4.2 Proposition

Every module M has a free (hence projective) resolution.

Proof. By (4.3.6), M is a homomorphic image of a free module F_0. Let K_0 be the kernel of the map from F_0 onto M. In turn, there is a homomorphism with kernel K_1 from a free module F_1 onto K_0, and we have the following diagram:

$$0 \longrightarrow K_1 \longrightarrow F_1 \longrightarrow K_0 \longrightarrow F_0 \longrightarrow M \longrightarrow 0$$

Composing the maps $F_1 \to K_0$ and $K_0 \to F_0$, we get

$$0 \longrightarrow K_1 \longrightarrow F_1 \longrightarrow F_0 \longrightarrow M \longrightarrow 0$$

which is exact. But now we can find a free module F_2 and a homomorphism with kernel K_2 mapping F_2 onto K_1. The above process can be iterated to produce the desired free resolution. ♣

Specifying a module by generators and relations (see (4.6.6) for abelian groups) involves finding an appropriate F_0 and K_0, as in the first step of the above iterative process. Thus a projective resolution may be regarded as a generalization of a specification by generators and relations.

Injective resolutions can be handled by dualizing the proof of (S4.2).

S4.3 Proposition

Every module M has an injective resolution.

Proof. By (10.7.4), M can be embedded in an injective module E_0. Let C_0 be the cokernel of $M \to E_0$, and map E_0 canonically onto C_0. Embed C_0 in an injective module E_1, and let C_1 be the cokernel of the embedding map. We have the following diagram:

$$0 \longrightarrow M \longrightarrow E_0 \longrightarrow C_0 \longrightarrow E_1 \longrightarrow C_1 \longrightarrow 0$$

Composing $E_0 \to C_0$ and $C_0 \to E_1$, we have

$$0 \longrightarrow M \longrightarrow E_0 \longrightarrow E_1 \longrightarrow C_1 \longrightarrow 0$$

which is exact. Iterate to produce the desired injective resolution. ♣

S5. Derived Functors

S5.1 Left Derived Functors

Suppose that F is a right exact functor from modules to modules. (In general, the domain and codomain of F can be abelian categories, but the example we have in mind is $M \otimes_R -$.) Given a short exact sequence $0 \to A \to B \to C \to 0$, we form deleted projective resolutions $P_{A*} \to A$, $P_{B*} \to B$, $P_{C*} \to C$. It is shown in texts on homological algebra that it is possible to define chain maps to produce a short exact sequence of complexes as shown below.

$$\begin{array}{ccccccccc} 0 & \longrightarrow & A & \longrightarrow & B & \longrightarrow & C & \longrightarrow & 0 \\ & & \uparrow & & \uparrow & & \uparrow & & \\ 0 & \longrightarrow & P_{A*} & \longrightarrow & P_{B*} & \longrightarrow & P_{C*} & \longrightarrow & 0 \end{array}$$

The functor F will preserve exactness in the diagram, except at the top row, where we only have $FA \to FB \to FC \to 0$ exact. But remember that we are using deleted resolutions, so that the first row is suppressed. The *left derived functors* of F are defined by taking the homology of the complex $F(P)$, that is,

$$(L_nF)(A) = H_n[F(P_{A*})].$$

The word "left" is used because the L_nF are computed using left resolutions. It can be shown that up to natural equivalence, the derived functors are independent of the particular projective resolutions chosen. By (S3.2), we have the following long exact sequence:

$$\cdots \xrightarrow{\partial} (L_nF)(A) \longrightarrow (L_nF)(B) \longrightarrow (L_nF)(C) \xrightarrow{\partial} (L_{n-1}F)(A) \longrightarrow \cdots$$

S5.2 Right Derived Functors

Suppose now that F is a left exact functor from modules to modules, e.g., $\text{Hom}_R(M, -)$. We can dualize the discussion in (S5.1) by reversing the vertical arrows in the commutative diagram of complexes, and replacing projective resolutions such as P_{A*} by injective resolutions E_{A*}. The *right derived functors* of F are defined by taking the homology of $F(E)$. In other words,

$$(R^nF)(A) = H^n[F(E_{A*})]$$

where the superscript n indicates that we are using right resolutions and the indices are increasing as we move away from the starting point. By (S3.2), we have the following long exact sequence:

$$\cdots \xrightarrow{\partial} (R^nF)(A) \longrightarrow (R^nF)(B) \longrightarrow (R^nF)(C) \xrightarrow{\partial} (R^{n+1}F)(A) \longrightarrow \cdots$$

S5.3 Lemma

$(L_0F)(A) \cong F(A) \cong (R^0F)(A)$

Proof. This is a good illustration of the advantage of deleted resolutions. If $P_* \to A$, we have the following diagram:

$$\begin{array}{ccccc} F(P_1) & \longrightarrow & F(P_0) & \longrightarrow & 0 \\ & & \downarrow & & \\ & & F(A) & & \\ & & \downarrow & & \\ & & 0 & & \end{array}$$

The kernel of $F(P_0) \to 0$ is $F(P_0)$, so the 0th homology module $(L_0F)(A)$ is $F(P_0)$ mod the image of $F(P_1) \to F(P_0)$ [=the kernel of $F(P_0) \to F(A)$.] By the first isomorphism theorem and the right exactness of F, $(L_0F)(A) \cong F(A)$. To establish the other isomorphism, we switch to injective resolutions and reverse arrows:

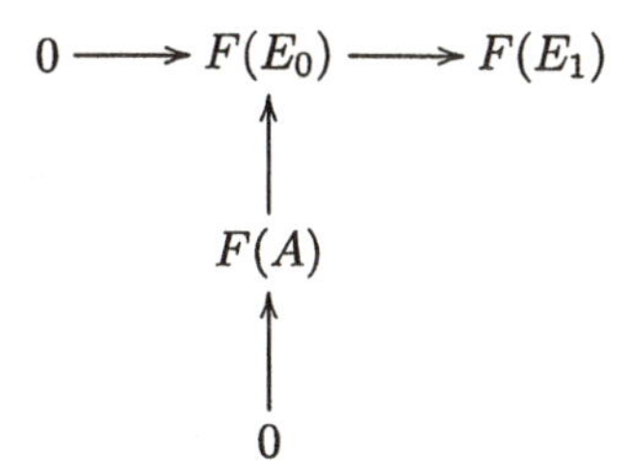

The kernel of $F(E_0) \to F(E_1)$ is isomorphic to $F(A)$ by left exactness of F, and the image of $0 \to F(E_0)$ is 0. Thus $(R^0F)(A) \cong F(A)$. ♣

S5.4 Lemma

If A is projective, then $(L_nF)(A) = 0$ for every $n > 0$; if A is injective, then $(R^nF)(A) = 0$ for every $n > 0$.

Proof. If A is projective [resp. injective], then $0 \to A \to A \to 0$ is a projective [resp. injective] resolution of A. Switching to a deleted resolution, we have $0 \to A \to 0$ in each case, and the result follows. ♣

S5.5 Definitions and Comments

If F is the right exact functor $M \otimes_R -$, the left derived functor L_nF is called $\text{Tor}_n^R(M, -)$. If F is the left exact functor $\text{Hom}_R(M, -)$, the right derived functor R^nF is called $\text{Ext}_R^n(M, -)$. It can be shown that the Ext functors can also be computed using projective resolutions and the contravariant hom functor. Specifically,

$$\text{Ext}_R^n(M, N) = [R^n \text{Hom}_R(-, N)](M).$$

A switch from injective to projective resolutions is a simplification, because projective resolutions are easier to find in practice.

The next three results sharpen Lemma S5.4. [The ring R is assumed fixed, and we write $\otimes_R$ simply as $\otimes$. Similarly, we drop the R in Tor^R and Ext_R. When discussing Tor, we assume R commutative.]

S5.6 Proposition

If M is an R-module, the following conditions are equivalent.

(i) M is flat;

(ii) $\text{Tor}_n(M, N) = 0$ for all $n \geq 1$ and all modules N;

(iii) $\text{Tor}_1(M, N) = 0$ for all modules N.

Proof. (i) implies (ii): Let $P_* \to N$ be a projective resolution of N. Since $M \otimes -$ is an exact functor (see (10.8.1)), the sequence

$$\cdots \to M \otimes P_1 \to M \otimes P_0 \to M \otimes N \to 0$$

is exact. Switching to a deleted resolution, we have exactness up to $M \otimes P_1$ but not at $M \otimes P_0$. Since the homology modules derived from an exact sequence are 0, the result follows.

(ii) implies (iii): Take $n = 1$.

(iii) implies (i): If $0 \to A \to B \to C \to 0$ is a short exact sequence, then by (S5.1), we have the following long exact sequence:

$$\cdots \operatorname{Tor}_1(M,C) \to \operatorname{Tor}_0(M,A) \to \operatorname{Tor}_0(M,B) \to \operatorname{Tor}_0(M,C) \to 0.$$

By hypothesis, $\operatorname{Tor}_1(M,C) = 0$, so by (S5.3),

$$0 \to M \otimes A \to M \otimes B \to M \otimes C \to 0$$

is exact, and therefore M is flat. ♣

S5.7 Proposition

If M is an R-module, the following conditions are equivalent.

(i) M is projective;

(ii) $\operatorname{Ext}^n(M,N) = 0$ for all $n \geq 1$ and all modules N;

(iii) $\operatorname{Ext}^1(M,N) = 0$ for all modules N.

Proof. (i) implies (ii): By (S5.4) and (S5.5), $\operatorname{Ext}^n(M,N) = [\operatorname{Ext}^n(-,N)](M) = 0$ for $n \geq 1$.

(ii) implies (iii): Take $n = 1$.

(iii) implies (i): Let $0 \to A \to B \to M \to 0$ be a short exact sequence. If N is any module, then using projective resolutions and the contravariant hom functor to construct Ext, as in (S5.5), we get the following long exact sequence:

$$0 \to \operatorname{Ext}^0(M,N) \to \operatorname{Ext}^0(B,N) \to \operatorname{Ext}^0(A,N) \to \operatorname{Ext}^1(M,N) \to \cdots$$

By (iii) and (S5.3),

$$0 \to \operatorname{Hom}(M,N) \to \operatorname{Hom}(B,N) \to \operatorname{Hom}(A,N) \to 0$$

is exact. Take $N = A$ and let g be the map from A to B. Then the map g^* from $\operatorname{Hom}(B,A)$ to $\operatorname{Hom}(A,A)$ is surjective. But $1_A \in \operatorname{Hom}(A,A)$, so there is a homomorphism $f\colon B \to A$ such that $g^*(f) = fg = 1_A$. Therefore the sequence $0 \to A \to B \to M \to 0$ splits, so by (10.5.3), M is projective. ♣

S5.8 Corollary

If N is an R-module, the following conditions are equivalent.

(a) N is injective;

(b) $\mathrm{Ext}^n(M,N)=0$ for all $n\geq 1$ and all modules M;

(c) $\mathrm{Ext}^1(M,N)=0$ for all modules M.

Proof. Simply saying "duality" may be unconvincing, so let's give some details. For (a) implies (b), we have $[\mathrm{Ext}^n(M,-)](N)=0$. For (c) implies (a), note that the exact sequence $0\to N\to A\to B\to 0$ induces the exact sequence

$$0\to \mathrm{Ext}^0(M,N)\to \mathrm{Ext}^0(M,A)\to \mathrm{Ext}^0(M,B)\to 0.$$

Replace $\mathrm{Ext}^0(M,N)$ by $\mathrm{Hom}(M,N)$, and similarly for the other terms. Then take $M=B$ and proceed exactly as in (S5.7). ♣

S6. Some Properties of Ext and Tor

We will compute $\mathrm{Ext}^n_R(A,B)$ and $\mathrm{Tor}^R_n(A,B)$ in several interesting cases.

S6.1 Example

We will calculate $\mathrm{Ext}^n_{\mathbb{Z}}(\mathbb{Z}_m,B)$ for an arbitrary abelian group B. To ease the notational burden slightly, we will omit the subscript $\mathbb{Z}$ in Ext and Hom, and use $=$ (most of the time) when we really mean $\cong$. We have the following projective resolution of $\mathbb{Z}_m$:

$$0\longrightarrow \mathbb{Z}\xrightarrow{m}\mathbb{Z}\longrightarrow \mathbb{Z}_m\longrightarrow 0$$

where the m over the arrow indicates multiplication by m. Switching to a deleted resolution and applying the contravariant hom functor, we get

$$\begin{array}{c} 0\longrightarrow \mathrm{Hom}(\mathbb{Z},B)\xrightarrow{m}\mathrm{Hom}(\mathbb{Z},B)\longrightarrow 0\\ \uparrow\\ \mathrm{Hom}(\mathbb{Z}_m,B)\end{array}$$

But by (9.4.1), we have

$$\mathrm{Hom}_R(R,B)\cong B \tag{1}$$

and the above diagram becomes

$$0\longrightarrow B\xrightarrow{m}B\longrightarrow 0 \tag{2}$$

By (S5.3), $\mathrm{Ext}^0(\mathbb{Z}_m,B)=\mathrm{Hom}(\mathbb{Z}_m,B)$. Now a homomorphism f from $\mathbb{Z}_m$ to B is determined by $f(1)$, and $f(m)=mf(1)=0$. If $B(m)$ is the set of all elements of B

that are annihilated by m, that is, $B(m) = \{x \in B: mx = 0\}$, then the map of $B(m)$ to $\operatorname{Hom}(\mathbb{Z}_m, B)$ given by $x \to f$ where $f(1) = x$, is an isomorphism. Thus

$$\operatorname{Ext}^0(\mathbb{Z}_m, B) = B(m).$$

It follows from (2) that

$$\operatorname{Ext}^n(\mathbb{Z}_m, B) = 0, \ n \geq 2$$

and

$$\operatorname{Ext}^1(\mathbb{Z}_m, B) = \ker(B \to 0)/\operatorname{im}(B \to B) = B/mB.$$

The computation for $n \geq 2$ is a special case of a more general result.

S6.2 Proposition

$\operatorname{Ext}^n_{\mathbb{Z}}(A, B) = 0$ for all $n \geq 2$ and all abelian groups A and B.

Proof. If B is embedded in an injective module E, we have the exact sequence

$$0 \to B \to E \to E/B \to 0.$$

This is an injective resolution of B since E/B is divisible, hence injective; see (10.6.5) and (10.6.6). Applying the functor $\operatorname{Hom}(A, -) = \operatorname{Hom}_{\mathbb{Z}}(A, -)$ and switching to a deleted resolution, we get the sequence

$$\begin{array}{ccccccc} 0 & \longrightarrow & \operatorname{Hom}(A, E) & \longrightarrow & \operatorname{Hom}(A, E/B) & \longrightarrow & 0 \\ & & \uparrow & & & & \\ & & \operatorname{Hom}(A, B) & & & & \end{array}$$

whose homology is 0 for all $n \geq 2$. ♣

S6.3 Lemma

$\operatorname{Ext}^0_{\mathbb{Z}}(\mathbb{Z}, B) = \operatorname{Hom}_{\mathbb{Z}}(\mathbb{Z}, B) = B$ and $\operatorname{Ext}^1_{\mathbb{Z}}(\mathbb{Z}, B) = 0$.

Proof. The first equality follows from (S5.3) and the second from (1) of (S6.1). Since $\mathbb{Z}$ is projective, the last statement follows from (S5.7). ♣

S6.4 Example

We will compute $\operatorname{Tor}^{\mathbb{Z}}_n(\mathbb{Z}_m, B)$ for an arbitrary abelian group B. As before, we drop the superscript $\mathbb{Z}$ and write $=$ for $\cong$. We use the same projective resolution of $\mathbb{Z}_m$ as in (S6.1), and apply the functor $- \otimes B$. Since $R \otimes_R B \cong B$ by (8.7.6), we reach diagram (2) as before. Thus

$$\mathrm{Tor}_n(\mathbb{Z}_m, B) = 0, \; n \geq 2;$$
$$\mathrm{Tor}_1(\mathbb{Z}_m, B) = \ker(B \to B) = \{x \in B \colon mx = 0\} = B(m);$$
$$\mathrm{Tor}_0(\mathbb{Z}_m, B) = \mathbb{Z}_m \otimes B = B/mB.$$

[To verify the last equality, use the universal mapping property of the tensor product to produce a map of $\mathbb{Z}_m \otimes B$ to B/mB such that $n \otimes x \to n(x + mB)$. The inverse of this map is $x + mB \to 1 \otimes x$.]

The result for $n \geq 2$ generalizes as in (S6.2):

S6.5 Proposition

$\mathrm{Tor}_n^{\mathbb{Z}}(A, B) = 0$ for all $n \geq 2$ and all abelian groups A and B.

Proof. B is the homomorphic image of a free module F. If K is the kernel of the homomorphism, then the exact sequence $0 \to K \to F \to B \to 0$ is a free resolution of B. [K is a submodule of a free module over a PID, hence is free.] Switching to a deleted resolution and applying the tensor functor, we get a four term sequence as in (S6.2), and the homology must be 0 for $n \geq 2$. ♣

S6.6 Lemma

$\mathrm{Tor}_1(\mathbb{Z}, B) = \mathrm{Tor}_1(A, \mathbb{Z}) = 0$; $\mathrm{Tor}_0(\mathbb{Z}, B) = \mathbb{Z} \otimes B = B$.

Proof. The first two equalities follow from (S5.6) since $\mathbb{Z}$ is flat. The other two equalities follow from (S5.3) and (8.7.6). ♣

S6.7 Finitely generated abelian groups

We will show how to compute $\mathrm{Ext}^n(A, B)$ and $\mathrm{Tor}_n(A, B)$ for arbitrary finitely generated abelian groups A and B. By (4.6.3), A and B can be expressed as a finite direct sum of cyclic groups. Now Tor commutes with direct sums:

$$\mathrm{Tor}_n(A, \oplus_{j=1}^r B_j) = \oplus_{j=1}^r \mathrm{Tor}_n(A, B_j).$$

[The point is that if P_{j*} is a projective resolution of B_j, then the direct sum of the P_{j*} is a projective resolution of $\oplus_j B_j$, by (10.5.4). Since the tensor functor is additive on direct sums, by (8.8.6(b)), the Tor functor will be additive as well. Similar results hold when the direct sum is in the first coordinate, and when Tor is replaced by Ext. (We use (10.6.3) and Problems 5 and 6 of Section 10.9, and note that the direct product and the direct sum are isomorphic when there are only finitely many factors.)]

Thus to complete the computation, we need to know $\mathrm{Ext}(A, B)$ and $\mathrm{Tor}(A, B)$ when $A = \mathbb{Z}$ or $\mathbb{Z}_m$ and $B = \mathbb{Z}$ or $\mathbb{Z}_n$. We have already done most of the work. By (S6.2) and

(S6.5), Ext^n and Tor_n are identically 0 for $n \geq 2$. By (S6.1),

$$\text{Ext}^0(\mathbb{Z}_m, \mathbb{Z}) = \mathbb{Z}(m) = \{x \in \mathbb{Z} \colon mx = 0\} = 0;$$
$$\text{Ext}^0(\mathbb{Z}_m, \mathbb{Z}_n) = \mathbb{Z}_n(m) = \{x \in \mathbb{Z}_n \colon mx = 0\} = \mathbb{Z}_d$$

where d is the greatest common divisor of m and n. [For example, $\mathbb{Z}_{12}(8) = \{0, 3, 6, 9\} \cong \mathbb{Z}_4$. The point is that the product of two integers is their greatest common divisor times their least common multiple.] By (S6.3),

$$\text{Ext}^0(\mathbb{Z}, \mathbb{Z}) = \text{Hom}(\mathbb{Z}, \mathbb{Z}) = \mathbb{Z}; \ \text{Ext}^0(\mathbb{Z}, \mathbb{Z}_n) = \mathbb{Z}_n.$$

By (S6.1),

$$\text{Ext}^1(\mathbb{Z}_m, \mathbb{Z}) = \mathbb{Z}/m\mathbb{Z} = \mathbb{Z}_m;$$
$$\text{Ext}^1(\mathbb{Z}_m, \mathbb{Z}_n) = \mathbb{Z}_n/m\mathbb{Z}_n = \mathbb{Z}_d$$

as above. By (S5.7),

$$\text{Ext}^1(\mathbb{Z}, \mathbb{Z}) = \text{Ext}^1(\mathbb{Z}, \mathbb{Z}_n) = 0$$

By (S6.4),

$$\text{Tor}_1(\mathbb{Z}_m, \mathbb{Z}) = \text{Tor}_1(\mathbb{Z}, \mathbb{Z}_m) = \mathbb{Z}(m) = 0;$$
$$\text{Tor}_1(\mathbb{Z}_m, \mathbb{Z}_n) = \mathbb{Z}_n(m) = \mathbb{Z}_d.$$

By (8.7.6) and (S6.4),

$$\text{Tor}_0(\mathbb{Z}, \mathbb{Z}) = \mathbb{Z};$$
$$\text{Tor}_0(\mathbb{Z}_m, \mathbb{Z}) = \mathbb{Z}/m\mathbb{Z} = \mathbb{Z}_m;$$
$$\text{Tor}_0(\mathbb{Z}_m, \mathbb{Z}_n) = \mathbb{Z}_n/m\mathbb{Z}_n = \mathbb{Z}_d.$$

Notice that $\text{Tor}_1(A, B)$ is a torsion group for all finitely generated abelian groups A and B. This is a partial explanation of the term "Tor". The Ext functor arises in the study of group extensions.

S7. Base Change in the Tensor Product

Let M be an A-module, and suppose that we have a ring homomorphism from A to B (all rings are assumed commutative). Then $B \otimes_A M$ becomes a B module (hence an A-module) via $b(b' \otimes m) = bb' \otimes m$. This is an example of base change, as discussed in (10.8.8). We examine some frequently occurring cases. First, consider $B = A/I$, where I is an ideal of A.

S7.1 Proposition

$(A/I) \otimes_A M \cong M/IM$.

Proof. Apply the (right exact) tensor functor to the exact sequence of A-modules

$$0 \to I \to A \to A/I \to 0$$

to get the exact sequence

$$I \otimes_A M \to A \otimes_A M \to (A/I) \otimes_A M \to 0.$$

Recall from (8.7.6) that $A \otimes_A M$ is isomorphic to M via $a \otimes m \to am$. By the first isomorphism theorem, $(A/I) \otimes_A M$ is isomorphic to M mod the image of the map from $I \otimes_A M$ to M. This image is the collection of all finite sums $\sum a_i m_i$ with $a_i \in I$ and $m_i \in M$, which is IM. ♣

Now consider $B = S^{-1}A$, where S is a multiplicative subset of A.

S7.2 Proposition

$(S^{-1}A) \otimes_A M \cong S^{-1}M$.

Proof. The map from $S^{-1}A \times M$ to $S^{-1}M$ given by $(a/s, m) \to am/s$ is A-bilinear, so by the universal mapping property of the tensor product, there is a linear map $\alpha\colon S^{-1}A \otimes_A M \to S^{-1}M$ such that $\alpha((a/s) \otimes m) = am/s$. The inverse map β is given by $\beta(m/s) = (1/s) \otimes m$. To show that β is well-defined, suppose that $m/s = m'/s'$. Then for some $t \in S$ we have $ts'm = tsm'$. Thus

$$1/s \otimes m = ts'/tss' \otimes m = 1/tss' \otimes ts'm = 1/tss' \otimes tsm' = 1/s' \otimes m'.$$

Now α followed by β takes $a/s \otimes m$ to am/s and then to $1/s \otimes am = a/s \otimes m$. On the other hand, β followed by α takes m/s to $1/s \otimes m$ and then to m/s. Consequently, α and β are inverses of each other and yield the desired isomorphism of $S^{-1}A \otimes_A M$ and $S^{-1}M$. ♣

Finally, we look at $B = A[X]$.

S7.3 Proposition

$$[X] \otimes_A M \cong M[X]$$

where the elements of $M[X]$ are of the form $a_0 m_0 + a_1 X m_1 + a_2 X^2 m_2 + \cdots + a_n X^n m_n$, $a_i \in A$, $m_i \in M$, $n = 0, 1, \ldots$.

Proof. This is very similar to (S7.2). In this case, the map α from $A[X] \otimes_A M$ to $M[X]$ takes $f(X) \otimes m$ to $f(X)m$, and the map β from $M[X]$ to $A[X] \otimes_A M$ takes $X^i m$ to $X^i \otimes m$. Here, there is no need to show that β is well-defined. ♣

Solutions Chapters 1–5

Section 1.1

1. Under multiplication, the positive integers form a monoid but not a group, and the positive even integers form a semigroup but not a monoid.
2. With $|a|$ denoting the order of a, we have $|0| = 1$, $|1| = 6$, $|2| = 3$, $|3| = 2$, $|4| = 3$, and $|5| = 6$.
3. There is a subgroup of order $6/d$ for each divisor d of 6. We have $\mathbb{Z}_6$ itself ($d = 1$), $\{0\}(d = 6), \{0, 2, 4\}(d = 2)$, and $\{0, 3\}(d = 3)$.
4. S forms a group under addition. The inverse operation is subtraction, and the zero matrix is the additive identity.
5. S^* does not form a group under multiplication, since a nonzero matrix whose determinant is 0 does not have a multiplicative inverse.
6. If d is the smallest positive integer in H, then H consists of all multiples of d. For if $x \in H$ we have $x = qd + r$ where $0 \leq r < d$. But then $r = x - qd \in H$, so r must be 0.
7. Consider the rationals with addition mod 1, in other words identify rational numbers that differ by an integer. Thus, for example, $1/3 = 4/3 = 7/3$, etc. The group is infinite, but every element generates a finite subgroup. For example, the subgroup generated by $1/3$ is $\{1/3, 2/3, 0\}$.
8. $(ab)^{mn} = (a^m)^n\,(b^n)^m = 1$, so the order of ab divides mn. Thus $|ab| = m_1 n_1$ where m_1 divides m and n_1 divides n. Consequently,

$$a^{m_1 n_1}\, b^{m_1 n_1} = 1 \tag{1}$$

   If $m = m_1 m_2$, raise both sides of (1) to the power m_2 to get $b^{mn_1} = 1$. The order of b, namely n, must divide mn_1, and since m and n are relatively prime, n must divide n_1. But n_1 divides n, hence $n = n_1$. Similarly, if $n = n_1 n_2$ we raise both sides of (1) to the power n_2 and conclude as above that $m = m_1$. But then $|ab| = m_1 n_1 = mn$, as asserted.

If c belongs to both $\langle a \rangle$ and $\langle b \rangle$ then since c is a power of a and also a power of b, we have $c^m = c^n = 1$. But then the order of c divides both m and n, and since m and n are relatively prime, c has order 1, i.e., $c = 1$.

9. Let $|a| = m$, $|b| = n$. If $[m, n]$ is the least common multiple, and (m, n) the greatest common divisor, of m and n, then $[m, n] = mn/(m, n)$. Examine the prime factorizations of m and n:

$$m = (p_1^{t_1} \cdots p_i^{t_i})(p_{i+1}^{t_{i+1}} \cdots p_j^{t_j}) = r\ r'$$
$$n = (p_1^{u_1} \cdots p_i^{u_i})(p_{i+1}^{u_{i+1}} \cdots p_j^{u_j}) = s'\ s$$

where $t_k \le u_k$ for $1 \le k \le i$, and $t_k \ge u_k$ for $i+1 \le k \le j$.
Now a^r has order m/r and b^s has order n/s, with m/r $(= r')$ and n/s $(= s')$ relatively prime. By Problem 8, $a^r b^s$ has order $mn/rs = mn/(m, n) = [m, n]$. Thus given elements of orders m and n, we can construct another element whose order is the least common multiple of m and n. Since the least common multiple of m, n and q is $[[m, n], q]$, we can inductively find an element whose order is the least common multiple of the orders of all elements of G.

10. Choose an element a that belongs to H but not K, and an element b that belongs to K but not H, where H and K are subgroups whose union is G. Then ab must belong to either H or K, say $ab = h \in H$. But then $b = a^{-1}h \in H$, a contradiction. If $ab = k \in K$, then $a = kb^{-1} \in K$, again a contradiction. To prove the last statement, note that if $H \cup K$ is a subgroup, the first result with G replaced by $H \cup K$ implies that $H = H \cup K$ or $K = H \cup K$, in other words, $K \subseteq H$ or $H \subseteq K$.

11. $a^{km} = 1$ if and only if km is a multiple of n, and the smallest such multiple occurs when km is the least common multiple of n and k. Thus the order of a^k is $[n, k]/k$. Examination of the prime factorizations of n and k shows that $[n, k]/k = n/(n, k)$.

12. We have $x \in A_i$ iff x is a multiple of p_i, and there are exactly n/p_i multiples of p_i between 1 and n. Similarly, x belongs to $A_i \cap A_j$ iff x is divisible by $p_i p_j$, and there are exactly $\frac{n}{p_i p_j}$ multiples of $p_i p_j$ between 1 and n. The same technique works for all other terms.

13. The set of positive integers in $\{1, 2, \dots, n\}$ and *not* relatively prime to n is $\cup_{i=1}^r A_i$, so $\varphi(n) = n - |\cup_{i=1}^r A_i|$. By the principle of inclusion and exclusion from basic combinatorics,

$$\Big|\bigcup_{i=1}^r A_i\Big| = \sum_{i=1}^r |A_i| - \sum_{i<j} |A_i \cap A_j| + \sum_{i<j<k} |A_i \cap A_j \cap A_k| - \cdots + (-1)^{r-1}|A_1 \cap A_2 \cap \cdots A_r|.$$

By Problem 12,

$$\varphi(n) = n\Big[1 - \sum_{i=1}^r \frac{1}{p_i} + \sum_{i<j} \frac{1}{p_i p_j} - \sum_{i<j<k} \frac{1}{p_i p_j p_k} + \cdots + (-1)^r 1 p_1 p_2 \cdots p_r\Big].$$

Thus $\varphi(n) = n(1 - \frac{1}{p_1})(1 - \frac{1}{p_2}) \cdots (1 - \frac{1}{p_r})$.

14. Let G be cyclic of prime order p. Since the only positive divisors of p are 1 and p, the only subgroups of G are G and $\{1\}$.

15. No. Any non-identity element of G generates a cyclic subgroup H. If $H \subset G$, we are finished. If $H = G$, then G is isomorphic to the integers, and therefore has many nontrivial proper subgroups. (See (1.1.4) and Problem 6 above.)

Section 1.2

1. The cycle decomposition is $(1,4)(2,6,5)$; there is one cycle of even length, so the permutation is odd.
2. The elements are I, $R = (A,B,C,D)$, $R^2 = (A,C)(B,D)$, $R^3 = (A,D,C,B)$, $F = (B,D)$, $RF = (A,B)(C,D)$, $R^2F = (A,C)$, $R^3F = (A,D)(B,C)$.
3. Such a permutation can be written as $(1,a_1,a_2,a_3,a_4)$ where (a_1,a_2,a_3,a_4) is a permutation of $\{2,3,4,5\}$. Thus the number of permutations is $4! = 24$.
4. Select two symbols from 5, then two symbols from the remaining 3, and divide by 2 since, for example, $(1,4)(3,5)$ is the same as $(3,5)(1,4)$. The number of permutations is $10(3)/2 = 15$.
5. For example, $(1,2,3)(1,2) = (1,3)$ but $(1,2)(1,2,3) = (2,3)$.
6. We have $V = \{I,(1,2)(3,4),(1,3)(2,4),(1,4)(2,3)\}$. Thus $V = \{I,a,b,c\}$ where the product of any two distinct elements from $\{a,b,c\}$ (in either order) is the third element, and the square of each element is I. It follows that V is an abelian group.
7. This follows because the inverse of the cycle $(a_1,a_2,\dots,a_k)$ is $(a_k,\dots,a_2,a_1)$.
8. Pick 3 symbols out of 4 to be moved, then pick one of two possible orientations, e.g., $(1,2,3)$ or $(1,3,2)$. The number of 3-cycles in S_4 is therefore $4(2) = 8$.
9. If π is a 3-cycle, then $\pi^3 = I$, so $\pi^4 = \pi$. But $\pi^4 = (\pi^2)^2$, and $\pi^2 \in H$ by hypothesis, so $(\pi^2)^2 \in H$ because H is a group. Thus $\pi \in H$.
10. There are 5 inversions, 21, 41, 51, 43 and 53. Thus we have an odd number of inversions and the permutation $\pi = (1,2,4)(3,5)$ is also odd.
11. This follows because a transposition of two adjacent symbols in the second row changes the number of inversions by exactly 1. Therefore such a transposition changes the parity of the number of inversions. Thus the parity of π coincides with the parity of the number of inversions. In the given example, it takes 5 transpositions of adjacent digits to bring 24513 into natural order 12345. It also takes 5 transpositions to create π:

$$\pi = (1,5)(1,4)(1,2)(3,5)(3,4)$$

Section 1.3

1. If $Ha = Hb$ then $a = 1a = hb$ for some $h \in H$, so $ab^{-1} = h \in H$. Conversely, if $ab^{-1} = h \in H$ then $Ha = Hhb = Hb$.
2. Reflexivity: $aa^{-1} = 1 \in H$.
   Symmetry: If $ab^{-1} \in H$ then $(ab^{-1})^{-1} = ba^{-1} \in H$.
   Transitivity: If $ab^{-1} \in H$ and $bc^{-1} \in H$ then $(ab^{-1})(bc^{-1}) = ac^{-1} \in H$.
3. $ab^{-1} \in H$ iff $(ab^{-1})^{-1} = ba^{-1} \in H$ iff $b \in Ha$.
4. $Ha^{-1} = Hb^{-1}$ iff $a^{-1}(b^{-1})^{-1} = a^{-1}b \in H$ iff $aH = bH$.
5. Since a_1 belongs to both aH and a_1H, we have $a_1H = aH$ because the left cosets partition G.

6. There are only two left cosets of H in G; one is H itself, and the other is, say, aH. Similarly, there are only two right cosets, H and Hb. Since the left cosets partition G, as do the right cosets, aH must coincide with Hb, so that every left coset if a right coset.
7. The permutations on the list are e, $(1,2,3)$, $(1,3,2)$, $(1,2)$, $(1,3)$, and $(2,3)$, which are in fact the 6 distinct permutations of $\{1,2,3\}$.
8. The left cosets of H are $H = \{e,b\}$, $aH = \{a,ab\}$, and $a^2H = \{a^2, a^2b\}$. The right cosets of H are $H = \{e,b\}$, $Ha = \{a,ba\} = \{a,a^2b\}$, and $Ha^2 = \{a^2, ba^2\} = \{a^2, ab\}$.
9. The computation of Problem 8 shows that the left cosets of H do not coincide with the right cosets. Explicitly, aH and a^2H are not right cosets (and similarly, Ha and Ha^2 are not left cosets).
10. $f(n) = f(1+1+\cdots 1) = f(1) + f(1) + \cdots f(1) = r + r + \cdots r = rn$.
11. In Problem 10, the image $f(\mathbb{Z})$ must coincide with $\mathbb{Z}$. But $f(\mathbb{Z})$ consists of all multiples of r, and the only way $f(\mathbb{Z})$ can equal $\mathbb{Z}$ is for r to be ± 1.
12. The automorphism group of $\mathbb{Z}$ is $\{I, -I\}$ where $(-I)^2 = I$. Thus the automorphisms of $\mathbb{Z}$ form a cyclic group of order 2. (There is only one such group, up to isomorphism.)
13. Reflexivity: $x = 1x1$. Symmetry: If $x = hyk$, then $y = h^{-1}xk^{-1}$. Transitivity: if $x = h_1yk_1$ and $y = h_2zk_2$, then $x = h_1h_2zk_2k_1$.
14. HxK is the union over all $k \in K$ of the right cosets $H(xk)$, and also the union over all $h \in H$ of the left cosets $(hx)K$.

Section 1.4

1. Define $f\colon \mathbb{Z} \to \mathbb{Z}_n$ by $f(x) =$ the residue class of x mod n. Then f is an epimorphism with kernel $n\mathbb{Z}$, and the result follows from the first isomorphism theorem.
2. Define $f\colon \mathbb{Z}_n \to \mathbb{Z}_{n/m}$ by $f(x) = x \bmod n/m$. Then f is an epimorphism with kernel $\mathbb{Z}_m$, and the result follows from the first isomorphism theorem. (In the concrete example with $n = 12$, $m = 4$, we have $f(0) = 0$, $f(1) = 1$, $f(2) = 2$, $f(3) = 0$, $f(4) = 1$, $f(5) = 2$, $f(6) = 0$, etc.)
3. $f(xy) = axya^{-1} = axa^{-1}aya^{-1} = f(x)f(y)$, so f is a homomorphism. If $b \in G$, we can solve $axa^{-1} = b$ for x, namely $x = a^{-1}ba$, so f is surjective. If $axa^{-1} = 1$ then $ax = a$, so $x = 1$ and f is injective. Thus f is an automorphism.
4. Note that $f_{ab}(x) = abx(ab)^{-1} = a(bxb^{-1})a^{-1} = f_a(f_b(x))$, and $y = f_a(x)$ iff $x = f_{a^{-1}}(y)$, so that $(f_a)^{-1} = f_{a^{-1}}$.
5. Define $\Psi\colon G \to \operatorname{Inn} G$, the group of inner automorphisms of G, by $\Psi(a) = f_a$. Then $\Psi(ab) = f_{ab} = f_a \circ f_b = \Psi(a)\Psi(b)$, so Ψ is a homomorphism (see the solution to Problem 4). Since a is arbitrary, Ψ is surjective. Now a belongs to $\ker \Psi$ iff f_a is the identity function, i.e., $axa^{-1} = x$ for all $x \in G$, in other words, a commutes with every x in G. Thus $\ker \Psi = Z(G)$, and the result follows from the first isomorphism theorem.
6. If f is an automorphism of $\mathbb{Z}_n$, then since 1 generates $\mathbb{Z}_n$, f is completely determined by $m = f(1)$, and since 1 has order n in $\mathbb{Z}_n$, m must have order n as well. But then m is a unit mod n (see (1.1.5)), and $f(r) = f(1+1+\cdots 1) = f(1)+f(1)+\cdots f(1) = rf(1) =$

rm. Conversely, any unit m mod n determines an automorphism $\theta(m)$ = multiplication by m. The correspondence between m and $\theta(m)$ is a group isomorphism because $\theta(m_1m_2) = \theta(m_1) \circ \theta(m_2)$.

7. The first assertion follows from the observation that HN is the subgroup generated by $H \cup N$ (see (1.3.6)). For the second assertion, note that if K is a subgroup of G contained in both H and N, then K is contained in $H \cap N$.
8. If $g(x) = y$, then $g \circ f_a \circ g^{-1}$ maps y to $g(axa^{-1}) = g(a)y[g(a)]^{-1}$.
9. If G is abelian, then $f_a(x) = axa^{-1} = aa^{-1}x = x$.

Section 1.5

1. $C_2 \times C_2$ has 4 elements $1 = (1,1)$, $\alpha = (a,1)$, $\beta = (1,a)$ and $\gamma = (a,a)$, and the product of any two distinct elements from $\{\alpha, \beta, \gamma\}$ is the third. Since each of α, β, γ has order 2 (and 1 has order 1), there is no element of order 4 and $C_2 \times C_2$ is not cyclic.
2. The four group is $V = \{I, a, b, c\}$ where the product of any two distinct elements from $\{a, b, c\}$ is the third. Therefore, the correspondence $1 \to I$, $\alpha \to a$, $\beta \to b$, $\gamma \to c$ is an isomorphism of $C_2 \times C_2$ and V.
3. Let $C_2 = \{1, a\}$ with $a^2 = 1$, and $C_3 = \{1, b, b^2\}$ with $b^3 = 1$. Then (a, b) generates $C_2 \times C_3$, since the successive powers of this element are (a,b), $(1,b^2)$, $(a,1)$, $(1,b)$, (a,b^2), and $(1,1)$. Therefore $C_2 \times C_3$ is cyclic of order 6, i.e., isomorphic to C_6.
4. Proceed as in Problem 3. If a has order n in C_n and b has order m in C_m, then (a,b) has order nm in $C_n \times C_m$, so that $C_n \times C_m$ is cyclic of order nm.
5. Suppose that (a,b) is a generator of the cyclic group $C_n \times C_m$. Then a must generate C_n and b must generate C_m (recall that $C_n \times \{1\}$ can be identified with C_n). But $(a,b)^k = 1$ iff $a^k = b^k = 1$, and it follows that the order of (a,b) is the least common multiple of the orders of a and b, i.e., the least common multiple of n and m. Since n and m are not relatively prime, the least common multiple is strictly smaller than nm, so that (a,b) cannot possibly generate $C_n \times C_m$, a contradiction.
6. By (1.3.3), G and H are both cyclic. Since p and q are distinct primes, they are relatively prime, and by Problem 4, $G \times H$ is cyclic.
7. Define $f\colon H \times K \to K \times H$ by $f(h,k) = (k,h)$. It follows from the definition of direct product that f is an isomorphism.
8. Define $f_1\colon G \times H \times K \to G \times (H \times K)$ by $f_1(g,h,k) = (g,(h,k))$, and define $f_2\colon G \times H \times K \to (G \times H) \times K$ by $f_2(g,h,k) = ((g,h),k)$. It follows from the definition of direct product that f_1 and f_2 are isomorphisms.

Section 2.1

1. Never. If f is a polynomial whose degree is at least 1, then f cannot have an inverse. For if $f(X)g(X) = 1$, then the leading coefficient of g would have to be 0, which is impossible.

2. If $f(X)g(X) = 1$, then (see Problem 1) f and g are polynomials of degree 0, in other words, elements of R. Thus the units of $R[X]$ are simply the nonzero elements of R.
3. (a) No element of the form $a_1X + a_2X^2 + \cdots$ can have an inverse.
   (b) For example, $1 - X$ is a unit because $(1 - X)(1 + X + X^2 + X^3 + \cdots) = 1$.
4. Since $\mathbb{Z}[i]$ is a subset of the field $\mathbb{C}$ of complex numbers, there can be no zero divisors in $\mathbb{Z}[i]$. If w is a nonzero Gaussian integer, then w has an inverse in $\mathbb{C}$, but the inverse need not belong to $\mathbb{Z}[i]$. For example, $(1+i)^{-1} = \frac{1}{2} - \frac{1}{2}i$.
5. If $z = a + bi$ with a and b integers, then $|z|^2 = a^2 + b^2$, so that if z is not zero, we must have $|z| \geq 1$. Thus if $zw = 1$, so that $|z||w| = 1$, we have $|z| = 1$, and the only possibilities are $a = 0, b = \pm 1$ or $a = \pm 1, b = 0$. Consequently, the units of $\mathbb{Z}[i]$ are 1, -1, i and $-i$.
6. All identities follow directly from the definition of multiplication of quaternions. Alternatively, (b) can be deduced from (a) by interchanging x_1 and x_2, y_1 and y_2, z_1 and z_2, and w_1 and w_2. Then the second identity of (c) can be deduced by noting that the quaternion on the right side of the equals sign in (a) is the conjugate of the quaternion on the right side of the equals sign in (b).
7. Multiply identities (a) and (b), and use (c). (This is not how Euler discovered the identity; quaternions were not invented until much later.)
8. The verification that End G is an abelian group under addition uses the fact that G is an abelian group. The additive identity is the zero function, and the additive inverse of f is given by $(-f)(a) = -f(a)$. Multiplication is associative because composition of functions is associative. To establish the distributive laws, note that the value of $(f+g)h$ at the element $a \in G$ is $f(h(a)) + g(h(a))$, so that $(f+g)h = fh + gh$. Furthermore, the value of $f(g+h)$ at a is $f(g(a) + h(a)) = f(g(a)) + f(h(a))$ since f is an endomorphism. Therefore $f(g+h) = fh + gh$. The multiplicative identity is the identity function, given by $E(a) = a$ for all a.
9. An endomorphism that has an inverse must be an isomorphism of G with itself. Thus the units of the ring End G are the automorphisms of G.
10. Use Euler's identity with $x_1 = 1, y_1 = 2, z_1 = 2, w_1 = 5$ $(34 = 1^2 + 2^2 + 2^2 + 5^2)$ and $x_2 = 1, y_2 = 1, z_2 = 4, w_2 = 6$ $(54 = 1^2 + 1^2 + 4^2 + 6^2)$. The result is $1836 = (34)(54) = 41^2 + 9^2 + 5^2 + 7^2$. The decomposition is not unique; another possibility is $x_1 = 0, y_1 = 0, z_1 = 3, w_1 = 5, x_2 = 0, y_2 = 1, z_2 = 2, w_2 = 7$.
11. In all four cases, sums and products of matrices of the given type are also of that type. But in (b), there is no matrix of the given form that can serve as the multiplicative identity.. Thus the sets (a), (c) and (d) are rings, but (b) is not.

Section 2.2

1. By Section 1.1, Problem 6, the additive subgroups of $\mathbb{Z}$ are of the form (n) = all multiples of n. But if $x \in (n)$ and $r \in \mathbb{Z}$ then $rx \in (n)$, so each (n) is an ideal as well.
2. If the n by n matrix A is 0 except perhaps in column k, and B is any n by n matrix, then BA is 0 except perhaps in column k. Similarly, if A is 0 off row k, then so is AB.

3. (a) This follows from the definition of matrix multiplication.
   (b) In (a) we have $a_{jr} = 0$ for $r \neq k$, and the result follows.
   (c) By (b), the ith term of the sum is a matrix with c_{ik} in the ik position, and 0's elsewhere. The sum therefore coincides with C.
4. The statement about left ideals follows from the formula of Problem 3(c). The result for right ideals is proved in a similar fashion. Explicitly, AE_{ij} has column i of A as its jth column, with 0's elsewhere. If $A \in R_k$ then AE_{ij} has a_{ki} in the kj position, with 0's elsewhere, so if $a_{ki} \neq 0$ we have $AE_{ij}a_{ki}^{-1} = E_{kj}$. Thus if $C \in R_k$ then
$$\sum_{j=1}^{n} AE_{ij}a_{ki}^{-1}c_{kj} = C.$$
5. If I is a two-sided ideal and $A \in I$ with $a_{rs} \neq 0$, then by considering products of the form $a_{rs}^{-1}E_{pq}AE_{kl}$ (which have the effect of selecting an entry of A and sliding it from one row or column to another), we can show that every matrix E_{ij} belongs to I. Since every matrix is a linear combination of the E_{ij}, it follows that $I = M_n(R)$.
6. A polynomial with no constant term is of the form $a_1X + a_2X^2 + \cdots a_nX^n = Xg(X)$. Conversely, a polynomial expressible as $Xg(X)$ has no constant term. Thus we may take $f = X$.
7. Let a be a nonzero element of R. Then the principal ideal (a) is not $\{0\}$, so $(a) = R$. Thus $1 \in (a)$, so there is an element $b \in R$ such that $ab = 1$.
8. Since an ideal I is a finite set in this case, it must have a finite set of generators $x_1, \dots, x_k$. Let d be the greatest common divisor of the x_i. Every element of I is of the form $a_1x_1 + \cdots + a_kx_k$, and hence is a multiple of d. Thus $I \subseteq (d)$. But $d \in I$, because there are integers a_i such that $\sum_i a_ix_i = d$. Consequently, $(d) \subseteq I$. [Technically, arithmetic is modulo n, but we get around this difficulty by noting that if $ab = c$ as integers, then $ab \equiv c$ modulo n.]

Section 2.3

1. Use the same maps as before, and apply the first isomorphism theorem for rings.
2. If I_n is the set of multiples of $n > 1$ in the ring of integers, then I_n is an ideal but not a subring (since $1 \notin I_n$). $\mathbb{Z}$ is a subring of the rational numbers $\mathbb{Q}$ but not an ideal, since a rational number times an integer need not be an integer.
3. In parts (2) and (3) of the Chinese remainder theorem, take $R = \mathbb{Z}$ and $I_i =$ the set of multiples of m_i.
4. Apply part (4) of the Chinese remainder theorem with $R = \mathbb{Z}$ and $I_i =$ the set of multiples of m_i.
5. To prove the first statement, define $f\colon R \to R_2$ by $f(r_1, r_2) = r_2$. Then f is a ring homomorphism with kernel R_1' and image R_2. By the first isomorphism theorem for rings, $R/R_1' \cong R_2$. A symmetrical argument proves the second statement. In practice, we tend to forget about the primes and write $R/R_1 \cong R_2$ and $R/R_2 \cong R_1$. There is

also a tendency to identify a ring with its isomorphic copy, and write $R/R_1 = R_2$ and $R/R_2 = R_1$ This should not cause any difficulty if you add, mentally at least, "up to isomorphism".

6. The product is always a subset of the intersection, by definition. First consider the case of two ideals. Then $1 = a_1 + a_2$ for some $a_1 \in I_1, a_2 \in I_2$. If $b \in I_1 \cap I_2$, then $b = b1 = ba_1 + ba_2 \in I_1 I_2$. The case of more than two ideals is handled by induction. Note that $R = (I_1 + I_n)(I_2 + I_n) \cdots (I_{n-1} + I_n) \subseteq (I_1 \cdots I_{n-1}) + I_n$. Therefore $(I_1 \cdots I_{n-1}) + I_n = R$. By the $n = 2$ case and the induction hypothesis, $I_1 \cdots I_{n-1} I_n = (I_1 \cdots I_{n-1}) \cap I_n = I_1 \cap I_2 \cap \cdots \cap I_n$.
7. Let $a + \cap_i I_i$ map to $(1 + I_1, 0 + I_2, c_3 + I_3, \ldots, c_n + I_n)$, where the c_j are arbitrary. Then $1 - a \in I_1$ and $a \in I_2$, so $1 = (1 - a) + a \in I_1 + I_2$. Thus $I_1 + I_2 = R$, and similarly $I_i + I_j = R$ for all $i \neq j$.

Section 2.4

1. If $n = rs$ with $r, s > 1$ then $r \notin \langle n \rangle, s \notin \langle n \rangle$, but $rs \in \langle n \rangle$, so that $\langle n \rangle$ is not prime. But $\mathbb{Z}/\langle n \rangle$ is isomorphic to $\mathbb{Z}_n$, the ring of integers modulo n (see Section 2.3, Problem 1). If n is prime, then $\mathbb{Z}_n$ is a field, in particular an integral domain, hence $\langle n \rangle$ is a prime ideal by (2.4.5).
2. By Problem 1, I is of the form $\langle p \rangle$ where p is prime. Since $\mathbb{Z}/\langle p \rangle$ is isomorphic to $\mathbb{Z}_p$, which is a field, $\langle p \rangle$ is maximal by (2.4.3).
3. The epimorphism $a_0 + a_1 X + a_2 X^2 + \cdots \to a_0$ of $F[[X]]$ onto F has kernel $\langle X \rangle$, and the result follows from (2.4.7).
4. The ideal $I = \langle 2, X \rangle$ is not proper; since $2 \in I$ and $\frac{1}{2} \in F \subseteq F[[X]]$, we have $1 \in I$ and therefore $I = F[[X]]$. The key point is that F is a field, whereas $\mathbb{Z}$ is not.
5. Suppose that $f(X) = a_0 + a_1 X + \cdots$ belongs to I but not to $\langle X \rangle$. Then a_0 cannot be 0, so by ordinary long division we can find $g(X) \in F[[X]]$ such that $f(X)g(X) = 1$. But then $1 \in I$, contradicting the assumption that I is proper.
6. Let $f(X) = a_n X^n + a_{n+1} X^{n+1} + \cdots, a_n \neq 0$, be an element of the ideal I, with n as small as possible. Then $f(X) \in (X^n)$, and if $g(X)$ is any element of I, we have $g(X) \in (X^m)$ for some $m \geq n$. Thus $I \subseteq (X^n)$. Conversely, if $f(X) = X^n g(X) \in I$, with $g(X) = a_n + a_{n+1} X + \cdots, a_n \neq 0,$, then as in Problem 5, $g(X)$ is a unit, and therefore $X^n \in I$. Thus $(X^n) \subseteq I$, so that $I = (X^n)$, as claimed.
7. $f^{-1}(P)$ is an additive subgroup by (1.3.15), part (ii). If $a \in f^{-1}(P)$ and $r \in R$, then $f(ra) = f(r)f(a) \in P$, so $ra \in f^{-1}(P)$. Thus $f^{-1}(P)$ is an ideal. If $ab \in f^{-1}(P)$, then $f(a)f(b) \in P$, so either a or b must belong to $f^{-1}(P)$. If $f^{-1}(P) = R$, then f maps eveything in R, including 1, into P; thus $f^{-1}(P)$ is proper. (Another method: As a proper ideal, P is the kernel of some ring homomorphism π. Consequently, $f^{-1}(P)$ is the kernel of $\pi \circ f$, which is also a ring homomorphism. Therefore $f^{-1}(P)$ is a proper ideal.) Consequently, $f^{-1}(P)$ is prime.
8. Let S be a field, and R an integral domain contained in S, and assume that R is not a field. For example, let $R = \mathbb{Z}$, $S = \mathbb{Q}$. Take f to be the inclusion map. Then $\{0\}$ is a maximal ideal of S, but $f^{-1}(\{0\}) = \{0\}$ is a prime but not maximal ideal of R.

9. If $P = I \cap J$ with $P \subset I$ and $P \subset J$, choose $a \in I \backslash P$ and $b \in J \backslash P$. Then $ab \in I \cap J = P$, contradicting the assumption that P is prime.

Section 2.5

1. Any number that divides a and b divides b and r_1, and conversely, any number that divides b and r_1 divides a and b. Iterating this argument, we find that $\gcd(a,b) = \gcd(b,r_1) = \gcd(r_1,r_2) = \cdots = \gcd(r_{j-1},r_j) = r_j$.
2. This follows by successive substitution. We start with $r_j = r_{j-2} - r_{j-1}q_j$, continue with $r_{j-1} = r_{j-3} - r_{j-2}q_{j-1}$, $r_{j-2} = r_{j-4} - r_{j-3}q_{j-2}$, and proceed up the ladder until we have expressed d as a linear combination of a and b. There is an easier way, as Problem 3 shows.
3. The first equation of the three describes the steps of the algorithm. We wish to prove that $ax_i + by_i = r_i$, that is,

$$a(x_{i-2} - q_i x_{i-1}) + b(y_{i-2} - q_i y_{i-1}) = r_i. \tag{1}$$

   But this follows by induction: if $ax_{i-2} + by_{i-2} = r_{i-2}$ and $ax_{i-1} + by_{i-1} = r_{i-1}$, then the left side of (1) is $r_{i-2} - q_i r_{i-1}$, which is r_i by definition of the Euclidean algorithm.
4. We have the following table:

i	q_{i+1}	r_i	x_i	y_i
−1	—	123	1	0
0	2	54	0	1
1	3	15	1	−2
2	1	9	−3	7
3	1	6	4	−9
4	2	3	−7	16

   For example, to go from $i = 1$ to $i = 2$ we have $x_2 = x_0 - q_2x_1 = 0 - 3(1) = -3$, $y_2 = y_0 - q_2y_1 = 1 - 3(-2) = 7$, and $r_2 = r_0 - q_2r_1 = 54 - 3(15) = 9$; also, $q_3 = \lfloor 15/9 \rfloor = 1$. We have $ax_2 + by_2 = 123(-3) + 54(7) = 9 = r_2$, as expected. The process terminates with $123(-7) + 54(16) = 3 = d$.
5. If p is composite, say $p = rs$ with $1 < r < p$, $1 < s < p$, then rs is 0 in $\mathbb{Z}_p$ but r and s are nonzero, so $\mathbb{Z}_p$ is not a field. If p is prime and a is not zero in $\mathbb{Z}_p$ then the greatest common divisor of a and p is 1, and consequently there are integers x and y such that $ax + py = 1$. In $\mathbb{Z}_p$ this becomes $ax = 1$, so that every nonzero element in $\mathbb{Z}_p$ has an inverse in $\mathbb{Z}_p$, proving that $\mathbb{Z}_p$ is a field.
6. Since $f(X)$ and $g(X)$ are multiples of $d(X)$, so are all linear combinations $a(X)f(X) + b(X)g(X)$, and consequently $I \subseteq J$. By Problem 2, there are polynomials $a(X)$ and $b(X)$ such that $a(X)f(X) + b(X)g(X) = d(X)$, so that $d(X)$ belongs to I. Since I is an ideal, every multiple of $d(X)$ belongs to I, and therefore $J \subseteq I$.
7. Take $f(X) = \sum_{i=0}^{n} b_i P_i(X)$.

8. If $g(X)$ is another polynomial such that $g(a_i) = f(a_i)$ for all i, then f and g agree at $n+1$ points, so that $f(X)-g(X)$ has more than n roots in F. By (2.5.3), $f(X)-g(X)$ must be the zero polynomial.
9. If F has only finitely many elements $a_1, \ldots, a_n$, take $f(X) = (X - a_1) \cdots (X - a_n)$.
10. Let F be the complex numbers $\mathbb{C}$. Then every polynomial of degree n has exactly n roots, counting multiplicity. Thus if $f(a) = 0$ at more than n points a, in particular if f vanishes at every point of $\mathbb{C}$, then $f = 0$. More generally, F can be any infinite field (use (2.5.3)).

Section 2.6

1. If $r = 0$ then I contains a unit, so that $1 \in I$ and $I = R$.
2. If $b \notin \langle p_1 \rangle$ then $b + \langle p_1 \rangle \neq \langle p_1 \rangle$, so $b + \langle p_1 \rangle$ has an inverse in $R/\langle p_1 \rangle$, say $c + \langle p_1 \rangle$. Thus $(b + \langle p_1 \rangle)(c + \langle p_1 \rangle) = 1 + \langle p_1 \rangle$, hence $(bc - 1) + \langle p_1 \rangle = \langle p_1 \rangle$, so $bc - 1 \in \langle p_1 \rangle$.
3. If $bc - dp_1 = 1$ then $bcp_2 \cdots p_n - dp_1 \cdots p_n = p_2 \cdots p_n$, and since b and $p_1 \cdots p_n$ belong to I, so does $p_2 \cdots p_n$, contradicting the minimality of n. (If $n = 1$, then $1 \in I$, so $I = R$.)
4. If $a, b \in R$ and $x, y \in J$ then $(ax + by)p_1 = xp_1 a + yp_1 b$. Since $x, y \in J$ we have $xp_1 \in I$ and $yp_1 \in I$, so that $(ax + by)p_1 \in I$, hence $ax + by \in J$.
5. If $x \in J$ then $xp_1 \in I$, so $Jp_1 \subseteq I$. Now $I \subseteq \langle p_1 \rangle$ by Problem 3, so if $a \in I$ then $a = xp_1$ for some $x \in R$. But then $x \in J$ by definition of J, so $a = xp_1 \in Jp_1$.
6. Since J contains a product of fewer than n primes, J is principal by the induction hypothesis. If $J = \langle d \rangle$ then by Problem 5, $I = J\langle p_1 \rangle$. But then $I = \langle dp_1 \rangle$, and the result follows. (If $n = 1$, then $p_1 \in I$, hence $1 \in J$, so $J = R$ and $I = J\langle p_1 \rangle = \langle p_1 \rangle$.)
7. Assume that $P \subseteq Q$. Then $p = aq$ for some $a \in R$, so $aq \in P$. Since P is prime, either a or q belongs to P. In the second case, $Q \subseteq P$ and we are finished. Thus assume $a \in P$, so that $a = bp$ for some $b \in R$. Then $p = aq = bpq$, and since R is an integral domain and $p \neq 0$, we have $bq = 1$, so q is a unit and $Q = R$, a contradiction of the assumption that Q is prime.
8. Let x be a nonzero element of P, with $x = up_1 \cdots p_n$, u a unit and the p_i prime. Then $p_1 \cdots p_n = u^{-1}x \in P$, and since P is prime, some p_i belongs to P. Thus P contains the nonzero principal prime ideal $\langle p_i \rangle$.

Section 2.7

1. If m is a generator of the indicated ideal then m belongs to all $\langle a_i \rangle$, so each a_i divides m. If each a_i divides b then b is in every $\langle a_i \rangle$, so $b \in \cap_{i=1}^n \langle a_i \rangle = \langle m \rangle$, so m divides b. Thus m is a least common multiple of A. Now suppose that m is an lcm of A, and let $\cap_{i=1}^n \langle a_i \rangle = \langle c \rangle$. Then c belongs to every $\langle a_i \rangle$, so each a_i divides c. Since $m = \text{lcm}(A)$, m divides c, so $\langle c \rangle$ is a subset of $\langle m \rangle$. But again since $m = \text{lcm}(A)$, each a_i divides m, so $m \in \cap_{i=1}^n \langle a_i \rangle = \langle c \rangle$. Therefore $\langle m \rangle \subseteq \langle c \rangle$, hence $\langle m \rangle = \langle c \rangle$, and m is a generator of $\cap_{i=1}^n \langle a_i \rangle$.

2. Let $a = 11 + 3i$, $b = 8 - i$. Then $a/b = (11 + 3i)(8 + i)/65 = 85/65 + i35/65$. Thus we may take $x_0 = y_0 = 1$, and the first quotient is $q_1 = 1 + i$. The first remainder is $r_1 = a - bq_1 = (11 + 3i) - (8 - i)(1 + i) = 2 - 4i$. The next step in the Euclidean algorithm is $(8-i)/(2-4i) = (8-i)(2+4i)/20 = 1+(3i/2)$. Thus the second quotient is $q_2 = 1 + i$ ($q_2 = 1 + 2i$ would be equally good). The second remainder is $r_2 = (8-i)-(2-4i)(1+i) = 2+i$. The next step is $(2-4i)/(2+i) = (2-4i)(2-i)/5 = -2i$, so $q_3 = -2i$, $r_3 = 0$. The gcd is the last divisor, namely $2 + i$.
3. We have $\Psi(1) \le \Psi(1(a)) = \Psi(a)$ for every nonzero a. If a is a unit with $ab = 1$, then $\Psi(a) \le \Psi(ab) = \Psi(1)$, so $\Psi(a) = \Psi(1)$. Conversely, suppose that $a \ne 0$ and $\Psi(a) = \Psi(1)$. Divide 1 by a to get $1 = aq + r$, where $r = 0$ or $\Psi(r) < \Psi(a) = \Psi(1)$. But if $r \ne 0$ then $\Psi(r)$ must be greater than or equal to $\Psi(1)$, so we must have $r = 0$. Therefore $1 = aq$, and a is a unit.
4. $\Psi((a_1 + b_1\sqrt{d})(a_2 + b_2\sqrt{d}))$
$$= \psi(a_1a_2 + b_1b_2d + (a_1b_2 + a_2b_1)\sqrt{d})$$
$$= \left|a_1a_2 + b_1b_2d + (a_1b_2 + a_2b_1)\sqrt{d}\right|\left|a_1a_2 + b_1b_2d - (a_1b_2 + a_2b_1)\sqrt{d}\right|;$$
$\Psi(a_1 + b_1\sqrt{d})\Psi(a_2 + b_2\sqrt{d})$
$$= \left|a_1 + b_1\sqrt{d}\right|\left|a_2 + b_2\sqrt{d}\right|\left|a_1 - b_1\sqrt{d}\right|\left|a_2 - b_2\sqrt{d}\right|$$
and it follows that $\Psi(\alpha\beta) = \Psi(\alpha)\Psi(\beta)$. Now $\Psi(\alpha) \ge 1$ for all nonzero α, for if $\Psi(\alpha) = |a^2 - db^2| = 0$, then $a^2 = db^2$. But if $b \ne 0$ then $d = (a/b)^2$, contradicting the assumption that d is not a perfect square. Thus $b = 0$, so a is 0 as well, and $\alpha = 0$, a contradiction. Thus $\Psi(\alpha\beta) = \Psi(\alpha)\Psi(\beta) \ge \Psi(\alpha)$.
5. Either d or $d - 1$ is even, so 2 divides $d(d - 1) = d^2 - d = (d + \sqrt{d})(d - \sqrt{d})$. But 2 does not divide $d + \sqrt{d}$ or $d - \sqrt{d}$. For example, if $2(a + b\sqrt{d}) = d + \sqrt{d})$ for integers a, b then $2a - d = (1 - 2b)\sqrt{d}$, which is impossible since $\sqrt{d}$ is irrational. (If $\sqrt{d} = r/s$ then $r^2 = ds^2$, which cannot happen if d is not a perfect square.)
6. Define Ψ as in Problem 4 (and Example (2.7.5)). Suppose $2 = \alpha\beta$ where α and β are nonunits in $\mathbb{Z}[\sqrt{d}]$. Then $4 = \Psi(2) = \Psi(\alpha)\Psi(\beta)$, with $\Psi(\alpha)$, $\Psi(\beta) > 1$ by Problems 3 and 4. But then $\Psi(\alpha) = \Psi(\beta) = 2$. If $\alpha = a + b\sqrt{d}$ then $|a^2 - db^2| = 2$, so $a^2 - db^2$ is either 2 or -2. Therefore if $b \ne 0$ (so that $b^2 \ge 1$), then since $d \le -3$ we have
$$a^2 - db^2 \ge 0 + 3(1) = 3,$$
a contradiction. Thus $b = 0$, so $\alpha = a$, and $2 = \Psi(a) = a^2$, an impossibility for $a \in \mathbb{Z}$.
7. This follows from Problems 5 and 6, along with (2.6.4).
8. Just as with ordinary integers, the product of two Gaussian integers is their greatest common divisor times their least common multiple. Thus by Problem 2, the lcm is $(11 + 3i)(8 - i)/(2 + i) = 39 - 13i$.
9. If $\alpha = \beta\gamma$, then $\Psi(\alpha) = \Psi(\beta)\Psi(\gamma)$. By hypothesis, either $\Psi(\beta)$ or $\Psi(\gamma)$ is $1 (= \Psi(1))$. By Problem 3, either β or γ is a unit.

Section 2.8

1. If D is a field, then the quotient field F, which can be viewed as the smallest field containing D, is D itself. Strictly speaking, F is isomorphic to D; the embedding map

$f(a) = a/1$ is surjective, hence an isomorphism. To see this, note that if $a/b \in F$, then $a/b = ab^{-1}/1 = f(ab^{-1})$.

2. The quotient field consists of all rational functions $f(X)/g(X)$, where $f(X)$ and $g(X)$ are polynomials in $F[X]$ and $g(X)$ is not the zero polynomial. To see this, note that the collection of rational functions is in fact a field, and any field containing $F[X]$ must contain all such rational functions.

3. $\frac{a}{b} + \left(\frac{c}{d} + \frac{e}{f}\right)$ and $\left(\frac{a}{b} + \frac{c}{d}\right) + \frac{e}{f}$ both compute to be $\frac{adf + bcf + bde}{bdf}$.

4. $\frac{a}{b}\left(\frac{c}{d} + \frac{e}{f}\right) = \frac{a}{b}\left(\frac{cf + de}{df}\right) = \frac{acf + ade}{bdf}$ and

   $\frac{ac}{bd} + \frac{ae}{bf} = \frac{acbf + bdae}{b^2df} = \frac{acf + dae}{bdf} = \frac{acf + ade}{bdf}$.

5. If g is any extension of h and $a/b \in F$, there is only one possible choice for $g(a/b)$, namely $h(a)/h(b)$. (Since $b \neq 0$ and h is a monomorphism, $h(b) \neq 0$.) If we define g this way, then $g(a) = g(a/1) = h(a)/h(1) = h(a)$, so g is in fact an extension of f. Furthermore, if $a/b = c/d$ then since h is a monomorphism, $h(a)/h(b) = h(c)/h(d)$. Therefore g is well-defined. Again since h is a monomorphism, it follows that $g(\frac{a}{b}+\frac{c}{d}) = g(\frac{a}{b}) + g(\frac{c}{d})$ and $g(\frac{a}{b}\frac{c}{d}) = g(\frac{a}{b})g(\frac{c}{d})$. Since g is an extension of h, we have $g(1) = 1$, so g is a homomorphism. Finally, if $g(a/b) = 0$, then $h(a) = 0$, so $a = 0$ by injectivity of h. Thus g is a monomorphism.

6. The problem is that h is not injective. As before, if g is to be an extension of h, we must have $g(a/b) = h(a)/h(b)$. But if b is a multiple of p, then $h(b)$ is zero, so no such g can exist.

7. We must have $\overline{g}(a/b) = \overline{g}(a/1)\overline{g}((b/1)^{-1}) = g(a)g(b)^{-1}$.

8. If $a/b = c/d$, then for some $s \in S$ we have $s(ad - bc) = 0$. So $g(s)[g(a)g(d) - g(b)g(c)] = 0$. Since $g(s)$ is a unit, we may multiply by its inverse to get $g(a)g(d) = g(b)g(c)$, hence $g(a)g(b)^{-1} = g(c)g(d)^{-1}$, proving that $\overline{g}$ is well-defined. To show that $\overline{g}$ is a homomorphism, we compute

$$\overline{g}\left(\frac{a}{b} + \frac{c}{d}\right) = \overline{g}\left(\frac{ad + bc}{bd}\right) = g(ad + bc)g(bd)^{-1}$$

$$= [g(a)g(d) + g(b)g(c)]g(b)^{-1}g(d)^{-1} = \overline{g}\left(\frac{a}{b}\right) + \overline{g}\left(\frac{c}{d}\right)$$

Similarly, we have $\overline{g}(\frac{a}{b}\frac{c}{d}) = \overline{g}(\frac{a}{b})\overline{g}(\frac{c}{d})$ and $\overline{g}(1) = 1$.

Section 2.9

1. We have $a_n(u/v)^n + a_{n-1}(u/v)^{n-1} + \cdots + a_1(u/v) + a_0 = 0$; multiply by v^n to get

$$a_nu^n + a_{n-1}u^{n-1}v + \cdots + a_1uv^{n-1} + a_0v^n = 0.$$

Therefore

$$a_nu^n = -a_{n-1}u^{n-1}v - \cdots - a_1uv^{n-1} - a_0v^n.$$

Since v divides the right side of this equation, it must divide the left side as well, and since u and v are relatively prime, v must divide a_n. Similarly,

$$a_0 v^n = -a_n u^n - a_{n-1} u^{n-1} v - \cdots - a_1 u v^{n-1},$$

so u divides a_0.

2. $X^n - p$ satisfies Eisenstein's criterion, and since the polynomial is primitive, it is irreducible over $\mathbb{Z}$.

3. $f_3(X) = X^3 + 2X + 1$, which is irreducible over $\mathbb{Z}_3$. For if $f_3(X)$ were reducible over $\mathbb{Z}_3$, it would have a linear factor (since it is a cubic), necessarily $X - 1$ or $X + 1 (= X - 2)$. But then 1 or 2 would be a root of f_3, a contradiction since $f_3(1) = 1$ and $f_3(2) = 1 \pmod 3$.

4. By Eisenstein, $X^4 + 3$ is irreducible over $\mathbb{Z}$. The substitution $X = Y + 1$ yields $Y^4 + 4Y^3 + 6Y^2 + 4Y + 4$, which is therefore irreducible in $\mathbb{Z}[Y]$. Thus $X^4 + 4X^3 + 6X^2 + 4X + 4$ is irreducible in $\mathbb{Z}[X]$, i.e., irreducible over $\mathbb{Z}$.

5. Note that $\langle n, X\rangle$ is a proper ideal since it cannot contain 1. If $\langle n, X\rangle = \langle f\rangle$ then $n \in \langle f\rangle$, so n is a multiple of f. Thus f is constant ($\neq 1$), in which case $X \notin \langle f\rangle$.

6. Since $1 \notin \langle X, Y\rangle$, $\langle X, Y\rangle$ is a proper ideal. Suppose $\langle X, Y\rangle = \langle f\rangle$. Then Y is a multiple of f, so f is a polynomial in Y alone (in fact $f = cY$). But then $X \notin \langle f\rangle$, a contradiction.

7. If $p = X + i$, then p is irreducible since $X + i$ is of degree 1. Furthermore, p divides $X^2 + 1$ but p^2 does not. Take the ring R to be $\mathbb{C}[X, Y] = (\mathbb{C}[X])[Y]$ and apply Eisenstein's criterion.

8. Write $f(X, Y)$ as $Y^3 + (X^3 + 1)$ and take $p = X + 1$. Since $X^3 + 1 = (X + 1)(X^2 - X + 1)$ and $X + 1$ does not divide $X^2 - X + 1$, the result follows as in Problem 7.

Section 3.1

1. $F(S)$ consists of all quotients of finite linear combinations (with coefficients in F) of finite products of elements of S. To prove this, note first that the set A of all such quotients is a field. Then observe that any field containing F and S must contain A, in particular, $A \subseteq F(S)$. But $F(S)$ is the smallest subfield containing F and S, so $F(S) \subseteq A$.

2. The composite consists of all quotients of finite sums of products of the form $x_{i_1} x_{i_2} \cdots x_{i_n}$, $n = 1, 2, \ldots$, where $i_1, i_2, \ldots, i_n \in I$ and $x_{i_j} \in K_{i_j}$. As in Problem 1, the set A of all such quotients is a field, and any field that contains all the K_i must contain A.

3. By (3.1.9), $[F[\alpha] : F] = [F[\alpha] : F[\beta]][F[\beta] : F]$, and since the degree of any extension is at least 1, the result follows.

4. Let $\min(-1 + \sqrt{2}, \mathbb{Q}) = a_0 + a_1 X + X^2$ (a polynomial of degree 1 cannot work because $-1 + \sqrt{2} \notin \mathbb{Q}$). Then $a_0 + a_1(-1 + \sqrt{2}) + (-1 + \sqrt{2})^2 = 0$. Since $(-1 + \sqrt{2})^2 = 3 - 2\sqrt{2}$, we have $a_0 - a_1 + 3 = 0$ and $a_1 - 2 = 0$, so $a_0 = -1, a_1 = 2$. Therefore $\min(-1 + 2\sqrt{2}, \mathbb{Q}) = X^2 + 2X - 1$.

5. Let $\beta = b_0 + b_1\alpha + \cdots + b_{n-1}\alpha^{n-1}$. Then for some $a_0, \dots, a_n \in F$ we have $a_0 + a_1\beta + \cdots + a_n\beta^n = 0$. Substituting the expression for β in terms of α into this equation, reducing to a polynomial in α of degree at most $n-1$ (as in the proof of (3.1.7)), and setting the coefficients of the α^i, $i = 0, 1, \dots, n-1$ equal to zero (remember that the α^i form a basis for $F[\alpha]$ over F), we get n linear equations in the $n+1$ unknowns a_i, $i = 0, \dots, n$. We know that a solution exists because β is algebraic over F. By brute force (try $a_i = 1$, $a_j = 0$, $j > i$ for $i = 1, 2, \dots, n$), we will eventually arrive at the minimal polynomial.
6. Define $\varphi\colon F(X) \to E$ by $\varphi(f(X)/g(X)) = f(\alpha)/g(\alpha)$. Note that φ is well-defined, since if g is a nonzero polynomial, then $g(\alpha) \neq 0$ (because α is transcendental over F). By (3.1.2), φ is a monomorphism. Since $\varphi(F(X)) = F(\alpha)$, it follows that $F(X)$ and $F(\alpha)$ are isomorphic.
7. The kernel of φ is I, and as in (3.1.3), $F[X]/I$ is a field. The image of φ is $F[\alpha]$, and by the first isomorphism theorem for rings, $F[\alpha]$ is isomorphic to $F[X]/I$. Therefore $F[\alpha]$ is a field, and consequently $F[\alpha] = F(\alpha)$.
8. If $f = gh$, then $(g + I)(h + I) = 0$ in $F[X]/I$, so $F[X]/I$ is not a field. By (2.4.3), I is not maximal.
9. The minimal polynomial over F belongs to $F[X] \subseteq E[X]$, and has α as a root. Thus $\min(\alpha, E)$ divides $\min(\alpha, F)$.
10. The result is true for $n = 1$; see (3.1.7). Let $E = F[\alpha_1, \dots, \alpha_{n-1}]$, so that $[F[\alpha_1, \dots, \alpha_n] : F] = [F[\alpha_1, \dots, \alpha_n] : E][E : F] = [E[\alpha_n] : E][E : F]$. But $[E[\alpha_n] : E]$ is the degree of the minimal polynomial of α_n over E, which is at most the degree of the minimal polynomial of α_n over F, by Problem 9. An application of the induction hypothesis completes the proof.

Section 3.2

1. $f(X) = (X - 2)^2$, so we may take the splitting field K to be Q itself.
2. $f(X) = (X - 1)^2 + 3$, with roots $1 \pm i\sqrt{3}$, so $K = \mathbb{Q}(i\sqrt{3})$. Now $i\sqrt{3} \notin \mathbb{Q}$ since $(i\sqrt{3})^2 = -3 < 0$, so $[K : \mathbb{Q}] \geq 2$. But $i\sqrt{3}$ is a root of $X^2 + 3$, so $[K : \mathbb{Q}] \leq 2$. Therefore $[K : \mathbb{Q}] = 2$.
3. Let α be the positive 4th root of 2. The roots of $f(X)$ are $\alpha, i\alpha, -\alpha$ and $-i\alpha$. Thus $K = \mathbb{Q}(\alpha, i)$. Now $f(X)$ is irreducible by Eisenstein, so $[\mathbb{Q}(\alpha) : \mathbb{Q}] = 4$. Since $i \notin \mathbb{Q}(\alpha)$ and i is a root of $X^2 + 1 \in \mathbb{Q}(\alpha)[X]$, we have $[K : \mathbb{Q}(\alpha)] = 2$. By (3.1.9), $[K : \mathbb{Q}] = 2 \times 4 = 8$.
4. The argument of (3.2.1) may be reproduced, with the polynomial f replaced by the family C of polynomials, and the roots $\alpha_1, \dots, \alpha_k$ of f by the collection of all roots of the polynomials in the family C.
5. Take $f = f_1 \cdots f_r$. Since α is a root of f iff α is a root of some f_i, the result follows.
6. If the degree is less than 4, it must be 2 (since $\sqrt{m} \notin \mathbb{Q}$). In this case, $\sqrt{n} = a + b\sqrt{m}$, so $n = a^2 + b^2m + 2ab\sqrt{m}$. Since m is square-free, we must have $a = 0$ or $b = 0$, and the latter is impossible because n is square-free. Thus $\sqrt{n} = b\sqrt{m}$, so $n = b^2m$, a contradiction of the hypothesis that m and n are distinct and square-free.

Section 3.3

1. If $\alpha_1, \dots, \alpha_n$ form a basis for E over F, then E is generated over F by the α_i. Each α_i is algebraic over F because $F(\alpha_i) \subseteq E$, and (3.1.10) applies.
2. There are only countably many polynomials with rational coefficients, and each such polynomial has only finitely many roots. Since an algebraic number must be a root of one of these polynomials, the set of algebraic numbers is countable. Since the complex field is uncountably infinite, there are uncountably many transcendental numbers.
3. The complex field $\mathbb{C}$ is algebraically closed, and $\mathbb{C}$ is an extension of the rational field $\mathbb{Q}$. But $\mathbb{C}$ is not algebraic over $\mathbb{Q}$, by Problem 2.
4. The algebraic numbers A form a field by (3.3.4), and A is algebraic over $\mathbb{Q}$ by definition. But it follows from Section 2.9, Problem 2, that A contains subfields of arbitrarily high degree (in fact subfields of every degree) over $\mathbb{Q}$, so that $A/\mathbb{Q}$ is not finite.
5. This can be verified by transfinite induction. A splitting field is always an algebraic extension (see (3.2.2)), and the field $F_{<f}$ is algebraic over F by the induction hypothesis. The result follows from (3.3.5).
6. By definition of algebraic number, A is an algebraic extension of $\mathbb{Q}$. If α is algebraic over A, then as in (3.3.5), α is algebraic over $\mathbb{Q}$, so $\alpha \in A$. Thus A has no proper algebraic extensions, so by (3.3.1), A is algebraically closed.
7. Since E is an extension of F we have $|F| \le |E|$. Suppose that $\alpha \in E$ and the minimal polynomial f of α has roots $\alpha_1, \dots, \alpha_n$, with $\alpha = \alpha_i$. Then the map $\alpha \to (f, i)$ is injective, since f and i determine α. It follows that $|E| \le |F[X]|\aleph_0 = |F[X]|$. But for each n, the set of polynomials of degree n over F has cardinality $|F|^{n+1} = |F|$, so $|F[X]| = |F|\aleph_0 = |F|$. Thus $|E| = |F|$.
8. Let C be an algebraic closure of F, and let A be the set of roots in C of all polynomials in S. Then $F(A)$, the field generated over F by the elements of A, is a splitting field for S over F; see Section 3.2, Problem 4.
9. If F is a finite field with elements $a_1, \dots, a_n$, the polynomial $f(X) = 1 + \prod_{i=1}^{n}(X - a_i)$ has no root in F, so F cannot be algebraically closed.

Section 3.4

1. Let $f(X) = (X - 1)^p$ over $\mathbb{F}_p$.
2. α is a root of $X^p - \alpha^p = (X - \alpha)^p$, so $m(X)$ divides $(X - \alpha)^p$.
3. By Problem 2, $m(X) = (X - \alpha)^r$ for some r. We are assuming that α is separable over $F(\alpha^p)$, so $m(X)$ must be simply $X - \alpha$. But then $\alpha \in F(\alpha^p)$.
4. The "if" part follows from the proof of (3.4.5), so assume that F is perfect and let $b \in F$. Let $f(X) = X^p - b$ and adjoin a root α of f. Then $\alpha^p = b$, so $F(\alpha^p) = F(b) = F$. By hypothesis, α is separable over $F = F(\alpha^p)$, so by Problem 3, $\alpha \in F$. But then b is the p^{th}power of an element of F.
5. If $\alpha_1, \dots, \alpha_n$ is a basis for E over F, then by the binomial expansion mod p, $K = F(\alpha_1^p, \dots, \alpha_n^p)$. Now since E/F is algebraic, the elements of $F(\alpha_1^p)$ can be expressed as polynomials in α_1^p with coefficients in F. Continuing, α_2^p is algebraic over

F, hence over $F(\alpha_1^p)$, so each element of $F(\alpha_1^p, \alpha_2^p)$ can be written as a polynomial in α_2^p with coefficients in $F(\alpha_1^p)$. Such an element has the form

$$\sum_s \Big(\sum_r b_{rs}\alpha_1^{pr}\Big)\alpha_2^{ps}$$

with the $b_{rs} \in F$. An induction argument completes the proof.

6. Extend the y_i to a basis $y_1, \dots, y_n$ for E over F. By Problem 5, every element of $E(= F(E^p))$ has the form $y = a_1 y_1^p + \cdots + a_n y_n^p$ with the $a_i \in F$. Thus $\{y_1^p, \dots, y_n^p\}$ spans E over F. It follows that this set contains a basis, hence (since there are exactly n vectors in the set) the set *is* a basis for E over F. The result follows.
7. Assume the extension is separable, and let $\alpha \in E$. Then α is separable over F, hence over $F(\alpha^p)$, so by Problem 3, $\alpha \in F(E^p)$. Thus $E = F(E^p)$. Conversely, suppose that $E = F(E^p)$ and the element $\alpha \in E$ has an inseparable minimal polynomial $m(X)$. By (3.4.3), $m(X)$ is of the form $b_0 + b_1 X^p + \cdots + b_{r-1}X^{(r-1)p} + X^{rp}$. Since $m(\alpha) = 0$, the elements $1, \alpha^p, \dots, \alpha^{rp}$ are linearly dependent over F. But by minimality of $m(X)$, $1, \alpha, \dots, \alpha^{rp-1}$ are linearly independent over F, hence $1, \alpha, \dots, \alpha^r$ are linearly independent over F. (Note that $rp - 1 \geq 2r - 1 \geq r$.) By Problem 6, $1, \alpha^p, \dots, \alpha^{rp}$ are linearly independent over F, which is a contradiction. Thus E/F is separable.
8. We may assume that F has prime characteristic p. By Problem 7, $E = K(E^p)$ and $K = F(K^p)$. Thus $E = F(K^p, E^p) = F(E^p)$ since $K \leq E$. Again by Problem 7, E/F is separable.
9. If g can be factored, so can f , and therefore g is irreducible. If $f(X) = g(X^{p^m})$ with m maximal, then $g \notin F[X^p]$. By (3.4.3) part (2), g is separable.
10. Suppose that the roots of g in a splitting field are $c_1, \dots, c_r$. Then $f(X) = g(X^{p^m}) = (X^{p^m} - c_1)\cdots(X^{p^m} - c_r)$. By separability of g, the c_j must be distinct, and since $f(\alpha) = 0$, we have $\alpha^{p^m} = c_j$ for all j. This is impossible unless $r = 1$, in which case $f(X) = X^{p^m} - c_1$. But $f \in F[X]$, so $\alpha^{p^m} = c_1 \in F$.
11. If $\alpha^{p^n} = c \in F$, then α is a root of $X^{p^n} - c = (X - \alpha)^{p^n}$, so $\min(\alpha, F)$ is a power of $X - \alpha$, and therefore has only one distinct root α. The converse follows from Problem 10 with $f = \min(\alpha, F)$.

Section 3.5

1. Take $F = \mathbb{Q}$, $K = \mathbb{Q}(\sqrt[3]{2})$ (see (3.5.3)), and let E be any extension of K that is normal over F, for example, $E = \mathbb{C}$.
2. The polynomial $f(X) = X^2 - a$ is irreducible, else it would factor as $(X - b)(X - c)$ with $b + c = 0$, $bc = a$, i.e., $(X - b)(X + b)$ with $b^2 = a$, contradicting the hypothesis. Thus E is obtained from $\mathbb{Q}$ by adjoining a root of f. The other root of f is $-\sqrt{a}$, so that E is a splitting field of f over $\mathbb{Q}$. By (3.5.7), $E/\mathbb{Q}$ is normal.
3. Take $F = \mathbb{Q}$, $K = \mathbb{Q}(\sqrt{2})$, $E = \mathbb{Q}(\sqrt[4]{2})$. Then K/F is normal by Problem 2, and E/K is normal by a similar argument. But E/F is not normal, since the two complex roots of $X^4 - 2$ do not belong to E. The same argument works with 2 replaced by any positive integer that is not a perfect square.

4. There are *at most* n embeddings of E in C extending σ. The proof is the same, except that now g has at most r distinct roots in C, so there are at most r possible choices of β. The induction hypothesis yields at most n/r extensions from $F(\alpha)$ to E, and the result follows.

5. Since the rationals have characteristic zero, the extension is separable. Since E is the splitting field of $(X^2-2)(X^2-3)$ over $\mathbb{Q}$, the extension is normal, hence Galois.

6. Since $\sqrt{3} \notin \mathbb{Q}(\sqrt{2})$, the extension has degree 4. By (3.5.9), there are exactly four $\mathbb{Q}$-automorphisms in the Galois group. By (3.5.1), each such $\mathbb{Q}$-automorphism must permute the roots of X^2-2 and must also permute the roots of X^2-3. There are only four possible ways this can be done. Since a $\mathbb{Q}$-automorphism is completely specified by its action on $\sqrt{2}$ and $\sqrt{3}$, the Galois group may be described as follows:

   (1) $\sqrt{2} \to \sqrt{2}$, $\quad \sqrt{3} \to \sqrt{3}$;
   (2) $\sqrt{2} \to \sqrt{2}$, $\quad \sqrt{3} \to -\sqrt{3}$;
   (3) $\sqrt{2} \to -\sqrt{2}$, $\quad \sqrt{3} \to \sqrt{3}$;
   (4) $\sqrt{2} \to -\sqrt{2}$, $\quad \sqrt{3} \to -\sqrt{3}$.

   Since the product (composition) of any two of automorphisms (2),(3),(4) is the third, the Galois group is isomorphic to the four group (Section 1.2, Problem 6).

7. Yes, up to isomorphism. If f is the polynomial given in (3.5.11), any normal closure is a splitting field for f over F, and the result follows from (3.2.5).

8. If f is irreducible over F and has a root in $E_1 \cap E_2$, then f splits over both E_1 and E_2, hence all roots of f lie in $E_1 \cap E_2$. Thus f splits over $E_1 \cap E_2$, and the result follows.

Section 4.1

1. If $x \in R$, take $r(x+I)$ to be $rx+I$ to produce a left R-module, and $(x+I)r = xr+I$ for a right R-module. Since I is an ideal, the scalar multiplication is well-defined, and the requirements for a module can be verified using the basic properties of quotient rings.

2. If A is an algebra over F, the map $x \to x1$ of F into A is a homomorphism, and since F is a field, it is a monomorphism (see (3.1.2)). Thus A contains a copy of F. Conversely, if F is a subring of A, then A is a vector space over F, and the compatibility conditions are automatic since A is commutative.

3. Let $R = \mathbb{Z}$ and let M be the additive group of integers mod m, where m is composite, say $m = ab$ with $a, b > 1$. Take $x = a \pmod{m}$ and $r = b$.

4. Any set containing 0 is linearly dependent, so assume a/b and c/d are nonzero rationals. Since $\frac{a/b}{c/d}$ is rational, the result follows.

5. In view of Problem 4, the only hope is that a single nonzero rational number a/b spans M over $\mathbb{Z}$. But this cannot happen, since an integer multiple of a/b must be a fraction whose denominator is a divisor of b.

6. If $a \in A \subseteq C$ and $x \in B \cap C$, then $ax \in (AB) \cap C$. Conversely, let $c = ab \in (AB) \cap C$. Then $b = a^{-1}c \in C$ since $A \subseteq C$. Thus $ab \in A(B \cap C)$.

7. If $f(X) = a_0 + a_1X + \cdots + a_nX^n$ and $v \in V$, take

$$f(X)v = f(T)v = a_0Iv + a_1Tv + \cdots + a_nT^nv$$

where I is the identity transformation and T^i is the composition of T with itself i times.

Section 4.2

1. Let W be a submodule of M/N. By the correspondence theorem, $W = L/N$ for some submodule L of M with $L \geq N$. Since $L = L + N$, we have $W = (L + N)/N$.
2. No. If S is any submodule of M, then $S + N$ is a submodule of M containing N, so $S + N$ corresponds to $W = (S + N)/N$. We know that W can also be written as $(L+N)/N$ where $L \geq N$. (For example, $L = S+N$.) By the correspondence theorem, $S + N = L + N$, and there is no contradiction.
3. If $A \in M_n(R)$, then AE_{11} retains column 1 of A, with all other columns zero.
4. To identify the annihilator of E_{11}, observe that by Problem 4, $AE_{11} = 0$ iff column 1 of A is zero. For the annihilator of M, note that $E_{j1} \in M$ for every j, and AE_{j1} has column j of A as column 1, with zeros elsewhere. (See Section 2.2, Problem 4.) Thus if A annihilates everything in M, then column j of A is zero for every j, so that A is the zero matrix.
5. R/I is an R-module by Problem 1 of Section 4.1 If $r \in R$ then $r + I = r(1 + I)$, so R/I is cyclic with generator $1 + I$.
6. We must show that scalar multiplication is well-defined, that is, if $r \in I$, then $rm = 0$ for all $m \in M$. Thus I must annihilate M, in other words, $IM = 0$, where the submodule IM is the set of all finite sums $\sum r_jm_j, r_j \in R, m_j \in M$.
7. No, since $(r + I)m$ coincides with rm.

Section 4.3

1. Essentially the same proof as in (4.3.3) works. If $z_1 + \cdots + z_n = 0$, with $z_i \in M_i$, then z_n is a sum of terms from previous modules, and is therefore 0. Inductively, every z_i is 0. (In the terminology of (4.3.3), z_i is $x_i - y_i$.)
2. Only when $A = \{0\}$. If A has n elements, then by Lagrange's theorem, $nx = 0$ for every $x \in A$, so there are no linearly independent sets (except the empty set).
3. This follows because $(-s)r + rs = 0$.
4. If I is not a principal ideal, then I can never be free. For if I has a basis consisting of a single element, then I is principal, a contradiction. But by Problem 3, there cannot be a basis with more than one element. If $I = \langle a \rangle$ is principal, then I is free if and only if a is not a zero-divisor.
5. $\mathbb{Z}$, or any direct sum of copies of $\mathbb{Z}$, is a free $\mathbb{Z}$-module. The additive group of rational numbers is not a free $\mathbb{Z}$-module, by Problem 5 of Section 4.1

6. The "only if" part was done in (4.3.6), so assume that M has the given property. Construct a free module $M' = \oplus_{i \in S} R_i$ where $R_i = R$ for all i. Then the map $f: S \to M'$ with $f(i) = e_i$ (where e_i has 1 in its ith component and zeros elsewhere) extends to a homomorphism (also called f) from M to M'. Let $g: M' \to M$ be the module homomorphism determined by $g(e_i) = i$. Then $g \circ f$ is the identity on S, hence on M, by the uniqueness assumption. Similarly, $f \circ g = 1$.
7. An element of M is specified by choosing a finite subset F of α, and then selecting an element $b_i \in R$ for each $i \in F$. The first choice can be made in α ways, and the second in $|R|^{|F|} = |R|$ ways. Thus $|M| = \alpha|R| = \max(\alpha, |R|)$.
8. We may take B to the set of "vectors" (e_i) with 1 in position i and zeros elsewhere. Thus there is a basis element for each copy of R, so $|B| = \alpha$.

Section 4.4

1. To prove that the condition is necessary, take the determinant of the equation $PP^{-1} = I$. Sufficiency follows from Cramer's rule.
2. A homomorphism $f: V \to W$ is determined by its action on elements of the form $(0, \dots, 0, x_j, 0, \dots, 0)$. Thus we must examine homomorphisms from V_j to $\oplus_{i=1}^{m} W_i$. Because of the direct sum, such mappings are assembled from homomorphisms from V_j to W_i, $i = 1, \dots, m$. Thus f may be identified with an $m \times n$ matrix whose ij element is a homomorphism from V_j to W_i. Formally, we have an abelian group isomorphism
$$\operatorname{Hom}_R(V, W) \cong [\operatorname{Hom}_R(V_j, W_i)].$$
3. In Problem 2, replace V and W by V^n and take all W_i and V_j to be V. This gives an abelian group isomorphism of the desired form. Now if f corresponds to $[f_{ij}]$ where $f_{ij}: V_j \to V_i$, and g corresponds to $[g_{ij}]$, then the composition $g \circ f$ is assembled from homomorphisms $g_{ik} \circ f_{kj}: V_j \to V_k \to V_i$. Thus composition of homomorphisms corresponds to multiplication of matrices, and we have a ring isomorphism.
4. In (4.4.1), take $n = m = 1$ and $M = R$.
5. Since $f(x) = f(x1) = xf(1)$, we may take $r = f(1)$.
6. This follows from Problems 3 and 4, with $V = R$.
7. If the endomorphism f is represented by the matrix A and g by B, then for any $c \in R$, we have $c(g \circ f) = (cg) \circ f = g \circ (cf)$, so $\operatorname{End}_R(M)$ is an R-algebra. Furthermore, cf is represented by cA, so the ring isomorphism is also an R-module homomorphism, hence an R-algebra isomorphism.

Section 4.5

1. Add column 2 to column 1, then add -3 times column 1 to column 2, then add -4 times row 2 to row 3. The Smith normal form is
$$S = \begin{bmatrix} 1 & 0 & 0 \\ 0 & 3 & 0 \\ 0 & 0 & 6 \end{bmatrix}$$

2. The matrix P^{-1} is the product of the elementary column matrices in the order in which they appeared. Thus

$$P^{-1} = \begin{bmatrix} 1 & 0 & 0 \\ 1 & 1 & 0 \\ 0 & 0 & 1 \end{bmatrix} \begin{bmatrix} 1 & -3 & 0 \\ 0 & 1 & 0 \\ 0 & 0 & 1 \end{bmatrix} = \begin{bmatrix} 1 & -3 & 0 \\ 1 & -2 & 0 \\ 0 & 0 & 1 \end{bmatrix}$$

$$P = \begin{bmatrix} -2 & 3 & 0 \\ -1 & 1 & 0 \\ 0 & 0 & 1 \end{bmatrix}$$

The matrix Q is the product of the elementary row matrices in opposite order (i.e., if R_1 appears first, followed by R_2 and R_3, then $Q = R_3R_2R_1$). In this case there is only one matrix, so

$$Q = \begin{bmatrix} 1 & 0 & 0 \\ 0 & 1 & 0 \\ 0 & -4 & 1 \end{bmatrix}$$

A direct computation shows that $QAP^{-1} = S$.

3. The new basis is given by $Y = PX$, i.e., $y_1 = -2x_1 + 3x_2$, $y_2 = -x_1 + x_2$, $y_3 = x_3$. The new set of generators is given by $V = SY$, i.e., $v_1 = y_1$, $v_2 = 3y_2$, $v_3 = 6y_3$.

4. Let $d_i = a_1 \cdots a_i$. Then d_i is the gcd of the $i \times i$ minors of S, and hence of A. The a_i are recoverable from the d_i via $a_1 = d_1$ and $a_i = d_i/d_{i-1}, i > 1$. Thus the a_i are determined by the matrix A and do not depend on any particular sequence leading to a Smith normal form.

5. If A and B have the same Smith normal form S, then A and B are each equivalent to S and therefore equivalent to each other. If A and B are equivalent, then by the result stated before Problem 4, they have the same gcd of $i \times i$ minors for all i. By Problem 4, they have the same invariant factors and hence the same Smith normal form.

6. Here are the results, in sequence:

    1. The second row is now (3 2 −13 2)
    2. The first row is (3 2 −13 2) and the second row is (6 4 13 5)
    3. The second row is (0 0 39 1) and the third row is (0 0 51 4)
    4. The third row becomes (0 0 12 3)
    5. The second row is (0 0 12 3) and the third row is (0 0 39 1)
    6. The third row is (0 0 3 −8)
    7. The second row is (0 0 3 −8) and the third row is (0 0 12 3)
    8. The third row is (0 0 0 35)
    9. The first row is now (3 2 2 −38)
    10. The final matrix is

$$\begin{bmatrix} 3 & 2 & 2 & 32 \\ 0 & 0 & 3 & 27 \\ 0 & 0 & 0 & 35 \end{bmatrix}.$$

7. We see from the Hermite normal form that we can take $x = 0, y = 7, z = 9$, provided 0 and 35 are congruent mod m. Thus m must be 5, 7 or 35.

Section 4.6

1. $441 = 3^2 \times 7^2$, and since there are two partitions of 2, there are $2 \times 2 = 4$ mutually nonisomorphic abelian groups of order 441, with the following invariant factors:

   (1) $a_1 = 3^2 7^2$, $G \cong \mathbb{Z}_{441}$

   (2) $a_1 = 3^0 7^1$, $a_2 = 3^2 7^1$, $G \cong \mathbb{Z}_7 \oplus \mathbb{Z}_{63}$

   (3) $a_1 = 3^1 7^0$, $a_2 = 3^1 7^2$, $G \cong \mathbb{Z}_3 \oplus \mathbb{Z}_{147}$

   (4) $a_1 = 3^1 7^1$, $a_2 = 3^1 7^1$, $G \cong \mathbb{Z}_{21} \oplus \mathbb{Z}_{21}$

2. $40 = 2^3 \times 5^1$, and since there are three partitions of 3 and one partition of 1, there are $3 \times 1 = 3$ mutually nonisomorphic abelian groups of order 40, with the following invariant factors:

   (1) $a_1 = 2^3 5^1, \quad G \cong \mathbb{Z}_{40}$

   (2) $a_1 = 2^1 5^0, \quad a_2 = 2^2 5^1, \quad G \cong \mathbb{Z}_2 \oplus \mathbb{Z}_{20}$

   (3) $a_1 = 2^1 5^0, \quad a_2 = 2^1 5^0, \quad a_3 = 2^1 5^1, \quad G \cong \mathbb{Z}_2 \oplus \mathbb{Z}_2 \oplus \mathbb{Z}_{10}$

3. The steps in the computation of the Smith normal form are

$$\begin{bmatrix} 1 & 5 & 3 \\ 2 & -1 & 7 \\ 3 & 4 & 2 \end{bmatrix} \to \begin{bmatrix} 1 & 5 & 3 \\ 0 & -11 & 1 \\ 0 & -11 & -7 \end{bmatrix} \to \begin{bmatrix} 1 & 0 & 0 \\ 0 & -11 & 1 \\ 0 & -11 & -7 \end{bmatrix} \to \begin{bmatrix} 1 & 0 & 0 \\ 0 & 1 & -11 \\ 0 & -7 & -11 \end{bmatrix}$$

$$\to \begin{bmatrix} 1 & 0 & 0 \\ 0 & 1 & -11 \\ 0 & 0 & -88 \end{bmatrix} \to \begin{bmatrix} 1 & 0 & 0 \\ 0 & 1 & 0 \\ 0 & 0 & 88 \end{bmatrix}$$

   Thus $G \cong \mathbb{Z}_1 \oplus \mathbb{Z}_1 \oplus \mathbb{Z}_{88} \cong \mathbb{Z}_{88}$.

4. Cancelling a factor of 2 is not appropriate. After the relations are imposed, the group is no longer free, so that $2y = 0$ does not imply that $y = 0$. Another difficulty is that the submodule generated by $2x_1 + 2x_2 + 8x_3$ is not the same as the submodule generated by $x_1 + x_2 + 4x_3$.

5. Take $M = \oplus_{n=1}^{\infty} M_n$, where each M_n is a copy of $\mathbb{Z}$. Take $N = \mathbb{Z}$ and $P = 0$. Since the union of a countably infinite set and a finite set is still countably infinite, we have the desired result.

6. If N and P are not isomorphic, then the decompositions of N and P will involve different sequences of invariant factors. But then the same will be true for $M \oplus N$ and $M \oplus P$, so $M \oplus N$ and $M \oplus P$ cannot be isomorphic.

Section 4.7

1. If u' is another solution, then $f'u = f'u'(= vf)$, and since f' is injective, $u = u'$.

2. By commutativity, $wgfa = g'vfa$, and by exactness, $gf = 0$. Thus $vfa \in \ker g' = \operatorname{im} f'$ by exactness.
3. For commutativity, we must have $f'ua = vfa = f'a'$, so $ua = a'$. Note that a' is unique because f' is injective. Checking that u is a homomorphism is routine, e.g., if $vfa_i = f'a'_i, i = 1, 2$, then $vf(a_1 + a_2) = f'(a'_1 + a'_2)$, so $u(a_1 + a_2) = ua_1 + ua_2$, etc.
4. $uc = ugb = g'vb$.
5. Suppose $c = gb_1 = gb_2$. Then $b_1 - b_2 \in \ker g = \operatorname{im} f$ by exactness, so $b_1 - b_2 = fa$. Then $f'wa = vfa = v(b_1 - b_2)$. By exactness, $0 = g'f'wa = g'v(b_1 - b_2)$, and the result follows.
6. Add a vertical identity map at the left side of the diagram and apply (ii) of the four lemma.
7. Add a vertical identity map at the right side of the diagram and apply (i) of the four lemma.
8. Add a vertical identity map w at the right side of the diagram and apply (ii) of the four lemma, shifting the notation $[s \to t,\ t \to u,\ u \to v,\ v \to w]$.
9. Since u and g' are surjective, v must be also, by commutativity.
10. Since f and u are injective, $f't$, hence t, must be also, by commutativity.
11. Add a vertical identity map s at the left side of the diagram, and apply (i) of the four lemma, shifting notation $[t \to s,\ u \to t,\ v \to u,\ w \to v]$.
12. If $vb = 0$, then $b = gm$, hence $0 = vgm = g'um$. Thus $um \in \ker g' = \operatorname{im} f'$, say $um = f'a'$. Since t is surjective, $a' = ta$, so $ufa = f'ta = f'a'$. Therefore um and ufa are both equal to $f'a'$. Since u is injective, $m = fa$, so $b = gm = gfa = 0$, proving that v is injective.
13. Let $a' \in A'$. Since u is surjective, $f'a' = um$, so $vgm = g'um = g'f'a' = 0$. Since v is injective, $gm = 0$, hence $m \in \ker g = \operatorname{im} f$, so $m = fa$. Thus $um = ufa = f'ta$. Therefore $f'a'$ and $f'ta$ are both equal to um. Since f' is injective, $a' = ta$, proving that t is surjective.

Section 5.1

1. The kernel of any homomorphism is a (normal) subgroup. If $g \in \ker \Phi$ then $g(xH) = xH$ for every $x \in G$, so by (1.3.1), $x^{-1}gx \in H$. Take $x = g$ to get $g \in H$.
2. By Problem 1, $\ker \Phi$ is a normal subgroup of G, necessarily proper since it is contained in H. Since G is simple, $\ker \Phi = \{1\}$, and hence Φ is injective. Since there are n left cosets of H, Φ maps into S_n.
3. If $[G : H] = n < \infty$, then by Problem 2, G can be embedded in S_n, so G is finite, a contradiction.
4. $g(xH) = xH$ iff $x^{-1}gx \in H$ iff $g \in xHx^{-1}$.
5. If $x \in G$, then $K = xKx^{-1} \subseteq xHx^{-1}$, and since x is arbitrary, $K \subseteq N$.
6. $g_1(H \cap K) = g_2(H \cap K)$ iff $g_2^{-1}g_1 \in H \cap K$ iff $g_1H = g_2H$ and $g_1K = g_2K$, proving both assertions.

7. Since $[G : H]$ and $[G : K]$ are relatively prime and divide $[G : H \cap K]$ by (1.3.5), their product divides, and hence cannot exceed, $[G : H \cap K]$.

8. By the first isomorphism theorem, G/N is isomorphic to a group of permutations of L, the set of left cosets of H. But $|L| = [G : H] = n$, so by Lagrange's theorem, $|G/N|$ divides $|S_n| = n!$.

9. Since $n > 1$, H is a proper subgroup of G, and since N is a subgroup of H, N is a proper subgroup of G as well. If $N = \{1\}$, then $|G| = [G : N]$, so by Problem 8, G divides $n!$, contradicting the hypothesis. Thus $\{1\} < N < G$, and G is not simple.

Section 5.2

1. For arbitrary σ and π, we have $\pi\sigma\pi^{-1}(\pi(i)) = \pi\sigma(i)$. In the cycle decomposition of σ, i is followed by $\sigma(i)$, and in the cycle decomposition of $\pi\sigma\pi^{-1}$, $\pi(i)$ is followed by $\pi\sigma(i)$, exactly as in the given numerical example.

2. If $g \in C_G(S)$ and $x \in S$ then $gxg^{-1} = x$, so $gSg^{-1} = S$, hence $C_G(S) \leq N_G(S)$. If $g \in N_G(S)$ and $x \in S$, then $gxg^{-1} \in S$, and the action is legal. As in (5.1.3), Example 3, the kernel of the action consists of all elements of $N_G(S)$ that commute with everything in S, that is, $N_G(S) \cap C_G(S) = C_G(S)$.

3. We have $z \in G(gx)$ iff $zgx = gx$ iff $g^{-1}zgx = x$ iff $g^{-1}zg \in G(x)$ iff $z \in gG(x)g^{-1}$.

4. We have $g_1G(x) = g_2G(x)$ iff $g_2^{-1}g_1 \in G(x)$ iff $g_2^{-1}g_1x = x$ iff $g_1x = g_2x$, proving that Ψ is well-defined and injective. If the action is transitive and $y \in X$, then for some x, $y = gx = \Psi(gG(x))$ and Ψ is surjective.

5. If $g, h \in G$, then h takes gx to hgx. In the coset action, the corresponding statement is that h takes $gG(x)$ to $hgG(x)$. The formal statement is that Ψ is a "G-set isomorphism". In other words, Ψ is a bijection of the space of left cosets of $G(x)$ and X, with $\Psi(hy) = h\Psi(y)$ for all $h \in G$ and y in the coset space. Equivalently, the following diagram is commutative.

$$\begin{array}{ccc} gx & \longrightarrow & hgx \\ \Psi\uparrow & & \uparrow\Psi \\ gG(x) & \longrightarrow & hgG(x) \end{array}$$

6. The two conjugacy classes are $\{1\}$ and $G \setminus \{1\}$. Thus if $|G| = n > 1$, the orbit sizes under conjugacy on elements are 1 and $n - 1$. But each orbit size divides the order of the group, so $n - 1$ divides n. Therefore $n = k(n - 1)$, where k is a positive integer. Since $k = 1$ is not possible, we must have $k \geq 2$, so $n \geq 2(n - 1)$, so $n \leq 2$.

7. If g_i is an element in the ith conjugacy class, $1 \leq i \leq k$, then by the orbit-stabilizer theorem, the size of this class is $|G|/|C_G(g_i)|$. Since the orbits partition G, the sum of the class sizes is $|G|$, and

$$\sum_{i=1}^{k} \frac{1}{x_i} = 1$$

where $x_i = |C_G(g_i)|$. If, say, $g_1 = 1$, so that $x_1 = |G|$, the result follows from the observation that each x_i, in particular x_1, is bounded by $N(k)$.

Section 5.3

1. The group elements are I, $R = (1,2,3,4)$, $R^2 = (1,3)(2,4)$, $R^3 = (1,4,3,2)$, $F = (1)(3)(2,4)$, $RF = (1,2)(3,4)$, $R^2F = (1,3)(2)(4)$, $R^3F = (1,4)(2,3)$. Thus the number of distinct colorings is

$$\frac{1}{8}\left(n^4 + n + n^2 + n + n^3 + n^2 + n^3 + n^2\right) = \frac{1}{8}\left(n^4 + 2n^3 + 3n^2 + 2n\right).$$

2. Yes. If the vertices of the square are $1,2,3,4$ in counterclockwise order, we can identify vertex 1 with side 12, vertex 2 with side 23, vertex 3 with side 34, and vertex 4 with side 41. This gives a one-to-one correspondence between colorings in one problem and colorings in the other.

3. If the vertices of the square are $1,2,3,4$ in counterclockwise order, then $WGGW$ will mean that vertices 1 and 4 are colored white, and vertices 2 and 3 green. The equivalence classes are

$$\{WWWW\}, \{GGGG\}, \{WGGG, GWGG, GGWG, GGGW\},$$
$$\{GWWW, WGWW, WWGW, WWWG\},$$
$$\{WWGG, GWWG, GGWW, WGGW\}, \{WGWG, GWGW\}.$$

4. Label $(-1,0)$ as vertex 1, $(0,0)$ as vertex 2, and $(1,0)$ as vertex 3. Then $I = (1)(2)(3)$ and $\sigma = (1,3)(2)$. Thus the number of distinct colorings is $\frac{1}{2}(n^3 + n^2)$.

5. We have free choice of color in two cycles of I and one cycle of σ. The number of distinct colorings is $\frac{1}{2}(n^2 + n)$.

6. We can generate a rotation by choosing a face of the tetrahedron to be placed on a table or other flat surface, and then choosing a rotation of 0,120 or 240 degrees. Thus there are 12 rotations, and we have enumerated all of them. By examining what each rotation does to the vertices, we can verify that all permutations are even. Since A_4 has $4!/2 = 12$ members, G must coincide with A_4, up to isomorphism.

7. The members of A_4 are $(1,2,3)$, $(1,3,2)$, $(1,2,4)$, $(1,4,2)$, $(1,3,4)$, $(1,4,3)$, $(2,3,4)$, $(2,4,3)$, $(1,2)(3,4)$, $(1,3)(2,4)$, $(1,4)(2,3)$, and the identity. Counting cycles of length 1, we have 11 permutations with 2 cycles and one permutation with 4 cycles. The number of distinct colorings is $\frac{1}{12}(n^4 + 11n^2)$.

8. In the above list of permutations, the first 8 have no fixed colorings. In the next 3, we can pick a cycle to be colored B, and pick a different color for the other cycle. This gives $2 \times 3 = 6$ fixed colorings. For the identity, we can pick two vertices to be colored B, and then choose a different color for each of the other two vertices. The number of fixed colorings is $\binom{4}{2}3^2 = 54$. The number of distinct colorings of the vertices is $[(6x3) + 54)]/12 = 6$.

9. As in Problem 6, a rotation can be generated by choosing a face of the cube to be placed on a table, and then choosing a rotation of 0,$\pm$90 or 180 degrees. Thus there are 24 rotations, and we have enumerated all of them. Alternatively, there is a one-to-one correspondence between rotations and permutations of the 4 diagonals of the cube. Since there are $4! = 24$ permutations of a set with 4 elements, there can be no additional rotations. The correspondence between rotations and permutations of the diagonals yields an isomorphism of G and S_4.

10. Any permutation of the faces except the identity has a cycle of length 2 or more, and each of the faces within that cycle must receive the same color, which is a contradiction. Thus $f(\pi) = 0$ for $\pi \neq I$. Now I fixes all legal colorings, and since there are 6 colors and 6 faces, the number of legal colorings is $6! = 720$. The number of distinct colorings is therefore $720/24 = 30$.

    Remark This problem can be solved directly without using the heavy machinery of this section. Without loss of generality, choose any particular color for a particular face, and move the cube so that this face is at the bottom. Choose one of the remaining 5 colors for the top face. The number of allowable colorings of the 4 remaining sides of the cube is the number of circular permutations of 4 objects, which is $3! = 6$. The number of distinct colorings is $5 \times 6 = 30$.

11. The group $G = \{1, R, R^2, \dots, R^{p-1}\}$ is cyclic of order p. Since p is prime, each R^i, $i = 1, \dots, p-1$, has order p, and therefore as a permutation of the vertices consists of a single cycle. Thus the number of distinct colorings is

$$\frac{1}{p}\left[n^p + (p-1)n\right].$$

12. Since the result of Problem 11 is an integer, $n^p + (p-1)n = n^p - n + np$ is a multiple of p, hence so is $n^p - n$. Thus for any positive integer n, $n^p \equiv n \bmod p$. It follows that if n is not a multiple of p, then $n^{p-1} \equiv 1 \bmod p$.

Section 5.4

1. Let G act on subgroups by conjugation. If P is a Sylow p-subgroup, then the stabilizer of P is $N_G(P)$ (see (5.2.2), Example 4). By (5.2.3), the index of $N_G(P)$ is n_p.

2. Since P is normal in $N_G(P)$ (see (5.2.2), Example 4), $PQ = QP \leq G$ by (1.4.3). By (5.2.4), PQ is a p-subgroup.

3. The Sylow p-subgroup P is contained in PQ, which is a p-subgroup by Problem 2. Since a Sylow p-subgroup is a p-subgroup of maximum possible size, we have $P = PQ$, and therefore $Q \subseteq P$.

4. (a) By definition of normalizer, we have $gPg^{-1} \leq gN_G(P)g^{-1} \leq gHg^{-1} = H$. Thus P and gPg^{-1} are subgroups of H, and since they are p-subgroups of maximum possible size, they are Sylow p-subgroups of H.

    (b) Since H is always a subgroup of its normalizer, let $g \in N_G(H)$. By (a), P and gPg^{-1} are conjugate in H, so for some $h \in H$ we have $gPg^{-1} = hPh^{-1}$. Thus $(h^{-1}g)P(h^{-1}g)^{-1} = P$, so $h^{-1}g \in N_G(P) \leq H$. But then $g \in H$, and the result follows.

5. By the second isomorphism theorem, $[N : P \cap N] = [PN : P] = |PN|/|P|$. Since $|P|$ is the largest possible power of p for p-subgroups of G, $[PN : P]$ and p must be relatively prime. Therefore $[N : P \cap N]$ and p are relatively prime, so $P \cap N$ is a p-subgroup of N of maximum possible size, i.e., a Sylow p-subgroup of N.
6. By the third isomorphism theorem, $[G/N : PN/N] = [G : PN] = |G|/|PN|$. Since $|G|/|P|$ and p are relatively prime and $P \leq PN$, it follows that $|G|/|PN|$ and p are relatively prime. The result follows as in Problem 5.
7. Since f is an automorphism, $f(P)$ is a subgroup of G and has the same number of elements as P, in other words, $f(P)$ is a Sylow p-subgroup. By hypothesis, $f(P) = P$.
8. By (1.3.5), $[G : N] = [G : H][H : N] = p[H : N]$, and since $[G : N]$ divides $p! = p(p-1)!$, the result follows.
9. If q is a prime factor of $[H : N]$, then by Problem 8, q is a divisor of some integer between 2 and $p-1$, in particular, $q \leq p-1$. But by Lagrange's theorem, q divides $|H|$, hence q divides $|G|$. This contradicts the fact that p is the smallest prime divisor of $|G|$. We conclude that there are no prime factors of $[H : N]$, which means that $[H : N] = 1$, Thus $H = N$, proving that H is normal in G.

Section 5.5

1. This follows from (5.5.6), part (iii), with $p = 3$ and $q = 5$.
2. Let $Z(G)a$ be a generator of $G/Z(G)$. If $g_1, g_2 \in G$, then $Z(G)g_1 = Z(G)a^i$ for some i, so $g_1a^{-i} = z_1 \in Z(G)$, and similarly $g_2a^{-j} = z_2 \in Z(G)$. Thus $g_1g_2 = a^iz_1a^jz_2 = z_1z_2a^{i+j} = z_2z_1a^{j+i} = z_2a^jz_1a^i = g_2g_1$.
3. By (5.5.3), the center $Z(G)$ is nontrivial, so has order p or p^2. In the latter case, $G = Z(G)$, so G is abelian. If $|Z(G)| = p$, then $|G/Z(G)| = p$, and $G/Z(G)$ has order p and is therefore cyclic. By Problem 2, G is abelian (and $|Z(G)|$ must be p^2, not p).
4. Each Sylow p-subgroup is of order p and therefore has $p-1$ elements of order p, with a similar statement for q and r. If we include the identity, we have $1 + n_p(p-1) + n_q(q-1) + n_r(r-1)$ distinct elements of G, and the result follows.
5. G cannot be abelian, for if so it would be cyclic of prime order. By (5.5.5), n_p, n_q and n_r are greater than 1. We know that n_p divides qr and $n_p > 1$. But n_p can't be q since $q \not\equiv 1 \bmod p$ (because $p > q$). Similarly, n_p can't be r, so $n_p = qr$. Now n_q divides pr and is greater than 1, so as above, n_q must be either p or pr (it can't be r because $q > r$, so $r \not\equiv 1 \bmod q$). Thus $n_q \geq p$. Finally, n_r divides pq and is greater than 1, so n_r is p, q, or pq. Since $p > q$, we have $n_r \geq q$.
6. Assume that G is simple. Substituting the inequalities of Problem 5 into the identity of Problem 4, we have

$$pqr \geq 1 + qr(p-1) + p(q-1) + q(r-1).$$

Thus

$$0 \geq pq - p - q + 1 = (p-1)(q-1),$$

a contradiction.

7. Since $|P| = p^r$ with $r \geq 1$ and $m > 1$, we have $1 < |P| < |G|$. Since G is simple, P is not normal in G. By (5.5.4), $n > 1$. By Problem 9 of Section 5.1, $|G|$ divides $n!$.
8. Assume G simple, and let $n = n_p$ with $p = 5$. By Sylow (2), n divides $2^4 = 16$ and $n \equiv 1 \bmod 5$. The only divisors of 16 that are congruent to 1 mod 5 are 1 and 16, and 1 is excluded by Problem 7. Thus the only possibility is $n = 16$, and by Problem 7, $2^4 5^6$ divides $16!$, hence 5^6 divides $16!$. But in the prime factorization of $16!$, 5 appears with exponent 3 (not 6), due to the contribution of 5,10 and 15. We have reached a contradiction, so G cannot be simple.

Section 5.6

1. Apply the Jordan-Hölder theorem to the series $1 \trianglelefteq N \trianglelefteq G$.
2. $\mathbb{Z}$ has no composition series. By Section 1.1, Problem 6, each nontrivial subgroup of $\mathbb{Z}$ consists of multiples of some positive integer, so the subgroup is isomorphic to $\mathbb{Z}$ itself. Thus $\mathbb{Z}$ has no simple subgroups, so if we begin with $\{0\}$ and attempt to build a composition series, we cannot even get started.
3. We have the composition series $1 \triangleleft \mathbb{Z}_2 \triangleleft \mathbb{Z}_2 \oplus \mathbb{Z}_3 \cong Z_6$ (or $1 \triangleleft \mathbb{Z}_3 \triangleleft \mathbb{Z}_2 \oplus \mathbb{Z}_3 \cong \mathbb{Z}_6$) and $1 \triangleleft A_3 \triangleleft S_3$.
4. A_n consists of products of an even number of transpositions, and the result follows from the observation that $(a, c)(a, b) = (a, b, c)$ and $(c, d)(a, b) = (a, d, c)(a, b, c)$.
5. If $(a, b, c) \in N$ and (d, e, f) is any 3-cycle, then for some permutation π we have $\pi(a, b, c)\pi^{-1} = (d, e, f)$. Explicitly, we can take $\pi(a) = d$, $\pi(b) = e$, $\pi(c) = f$; see Section 5.2, Problem 1. We can assume without loss of generality that π is even, for if it is odd, we can replace it by $(g, h)\pi$, where g and h are not in $\{d, e, f\}$. (We use $n \geq 5$ here.) Since N is normal, $(d, e, f) \in N$.
6. If N contains $(1, 2, 3, 4)$, then it contains $(1, 2, 3)(1, 2, 3, 4)(1, 3, 2) = (1, 4, 2, 3)$, and hence contains $(1, 4, 2, 3)(1, 4, 3, 2) = (1, 2, 4)$, contradicting Problem 5. If N contains $(1, 2, 3, 4, 5)$, then it contains $(1, 2, 3)(1, 2, 3, 4, 5)(1, 3, 2) = (1, 4, 5, 2, 3)$, and so contains $(1, 4, 5, 2, 3)(1, 5, 4, 3, 2) = (1, 2, 4)$, a contradiction. The analysis for longer cycles is similar. [Actually, we should have assumed that N contains a permutation π whose disjoint cycle decomposition is $\cdots(1, 2, 3, 4)\cdots$. But multiplication by $\pi^{-1} = \cdots(1, 4, 3, 2)\cdots$ cancels the other cycles.]
7. If N contains $(1, 2, 3)(4, 5, 6)$, then it must also contain $(3, 4, 5)(1, 2, 3)(4, 5, 6)(3, 5, 4) = (1, 2, 4)(3, 6, 5)$. Thus N also contains $(1, 2, 4)(3, 6, 5)(1, 2, 3)(4, 5, 6) = (1, 4, 3, 2, 6)$, which contradicts Problem 6. If the decomposition of a permutation σ in N contains a single 3-cycle, then σ^2 *is* a 3-cycle in N, because a transposition is its own inverse. This contradicts Problem 5.
8. If, $(1, 2)(3, 4) \in N$, then $(1, 5, 2)(1, 2)(3, 4)(1, 2, 5) = (1, 5)(3, 4)$ belongs to N, and so does $(1, 5)(3, 4)(1, 2)(3, 4) = (1, 2, 5)$, contradicting Problem 5.
9. If N contains $(1, 2)(3, 4)(5, 6)(7, 8)$, then it contains

$$(2, 3)(4, 5)(1, 2)(3, 4)(5, 6)(7, 8)(2, 3)(4, 5) = (1, 3)(2, 5)(4, 6)(7, 8).$$

Therefore N contains

$$(1,3)(2,5)(4,6)(7,8)(1,2)(3,4)(5,6)(7,8) = (1,5,4)(2,3,6),$$

contradicting Problem 7.

10. We can reproduce the analysis leading to the Jordan-Hölder theorem, with appropriate notational changes. For example, we replace the "subnormal" condition $G_i \trianglelefteq G_{i+1}$ by the "normal" condition $G_i \trianglelefteq G$.

11. We say that N is a *minimal normal subgroup* of H if $\{1\} < N \trianglelefteq H$ and there is no normal subgroup of H strictly between $\{1\}$ and N. In a chief series, there can be no normal subgroup of G strictly between G_i and G_{i+1}. Equivalently, by the correspondence theorem, there is no normal subgroup of G/G_i strictly between $G_i/G_i = \{1\}$ and G_{i+1}/G_i. Thus G_{i+1}/G_i is a minimal normal subgroup of G/G_i.

Section 5.7

1. S_3 is nonabelian and solvable ($1 \triangleleft A_3 \triangleleft S_3$).

2. Let $1 = G_0 \triangleleft G_1 \triangleleft \cdots \triangleleft G_r = G$ be a composition series, with all G_i/G_{i-1} cyclic of prime order (see (5.7.5)). Since $|G_i| = |G_i/G_{i-1}||G_{i-1}|$ and G_0 is finite, an induction argument shows that G is finite.

3. The factors of a composition series are simple p-groups P, which must be cyclic of prime order. For $1 \triangleleft Z(P) \trianglelefteq P$, so $Z(P) = P$ and P is abelian, hence cyclic of prime order by (5.5.1). [The trivial group is solvable with a derived series of length 0.]

4. S_3 is solvable by Problem 1, but is not nilpotent. Since S_3 is nonabelian, a central series must be of the form $1 \triangleleft H \triangleleft S_3$ with $H \subseteq Z(S_3) = 1$, a contradiction.

5. If S_n is solvable, then so is A_n by (5.7.4), and this contradicts (5.7.2).

6. By (5.5.3), P has a nontrivial center. Since $Z(P)$ is normal in P and P is simple, $Z(P) = P$ and P is abelian. By (5.5.1), P is cyclic of prime order, and since P is a p-group, the only possibility is $|P| = p$.

7. Let N be a maximal proper normal subgroup of P. (N exists because P is finite and nontrivial, and $1 \triangleleft P$.). Then the p-group P/N is simple (by the correspondence theorem). By Problem 6, $|P/N| = p$.

8. If P is a Sylow p-subgroup of G, then $|P| = p^r$ and by Problem 7, P has a subgroup Q_1 of index p, hence of order p^{r-1}. If Q_1 is nontrivial, the same argument shows that Q_1 has a subgroup Q_2 of order p^{r-2}. An induction argument completes the proof.

9. Let $G = D_6$, the group of symmetries of the equilateral triangle. Take $N = \{I, R, R^2\}$, where R is rotation by 120 degrees. Then N has index 2 in G and is therefore normal. (See Section 1.3, Problem 6, or Section 5.4, Problem 9.) Also, N has order 3 and G/N has order 2, so both N and G/N are cyclic, hence abelian. But G is not abelian, since rotations and reflections do not commute.

10. It follows from the splicing technique given in the proof of (5.7.4) that $\text{dl}(G) \leq \text{dl}(N) + dl(G/N)$.

Section 5.8

1. Let $H = \langle a \mid a^n \rangle$, and let C_n be a cyclic group of order n generated by a. Then $a^n = 1$ in C_n, and since $a^{n+j} = a^j$, we have $|H| \le n = |C_n|$. The result follows as in (5.8.6).
2. The discussion in Example 4 of (2.1.3), with $i = a$ and $j = b$, shows that the quaternion group Q satisfies all the relations. Since $ab = ba^{-1}$, it follows as in (1.2.4) that every element of the given presentation H is of the form $b^r a^s, r, s \in \mathbb{Z}$. Since $b^2 = a^2$, we can restrict r to 0 or 1, and since $a^4 = 1$, we can restrict s to 0, 1, 2 or 3. Thus $|H| \le 8$, and the result follows as in (5.8.6).
3. Take $a = (1,2,3)$ and $b = (1,2)$ to show that S_3 satisfies all the relations. Since $ba = a^{-1}b$, each element of H is of the form $a^r b^s, r, s \in \mathbb{Z}$. Since $a^3 = b^2 = 1$, $|H| \le 3 \times 2 = 6$, and the result follows as in (5.8.6).
4. No. There are many counterexamples; an easy one is $C_n = \langle a \mid a^n = 1, a^{2n} = 1 \rangle$, the cyclic group of order n.
5. $n_2^{-1} n_1 = h_2 h_1^{-1} \in N \cap H = 1$.
6. Take ψ to be the inclusion map. Then $\pi\psi(h) = \pi(h) = \pi(1h) = h$. To show that π is a homomorphism, note that $n_1 h_1 n_2 h_2 = n_1(h_1 n_2 h_1^{-1}) h_1 h_2$ and $h_1 n_2 h_1^{-1} \in N$.
7. If $g \in G$ then $g = (g\pi(g)^{-1})\pi(g)$ with $\pi(g) \in H$ and $\pi(g\pi(g)^{-1}) = \pi(g)\pi(g)^{-1} = 1$, so $g\pi(g)^{-1} \in N$. [Remember that since we are taking ψ to be inclusion, π is the identity on H.] Thus $G = NH$. If $g \in N \cap H$, then $g \in \ker \pi$ and $g \in H$, so $g = \pi(g) = 1$, proving that $H \cap N = 1$.
8. If we define $\pi(n, h) = (1, h)$, $i(n, 1) = (n, 1)$, and $\psi(1, h) = (1, h)$, then the sequence of Problem 6 is exact and splits on the right.
9. We have $(n_1 h_1)(n_2 h_2) = n_1(h_1 n_2 h_1^{-1}) h_1 h_2$, so we may take $f(h)$ to be the inner automorphism of N given by conjugation by $h \in H$.
10. Consider the sequence

$$1 \longrightarrow C_3 \xrightarrow{i} S_3 \xrightarrow{\pi} C_2 \longrightarrow 1$$

where C_3 consists of the identity 1 and the 3-cycles $(1,2,3)$ and $(1,3,2)$, and C_2 consists of the identity and the 2-cycle $(1,2)$. The map i is inclusion, and π takes each 2-cycle to $(1,2)$ and each 3-cycle to the identity. The identity map from C_2 to S_3 gives a right-splitting, but there is no left splitting. If g were a left-splitting map from S_3 to C_3, then $g(1,2) = (1,2,3)$ is not possible because $g(1) = g(1,2)g(1,2) = (1,2,3)(1,2,3) = (1,3,2)$, a contradiction. Similarly, $g(1,2) = (1,3,2)$ is impossible, so $g(1,2) = 1$, so $g \circ i$ cannot be the identity. Explicitly, $g(2,3) = g((1,2)(1,2,3)) = g(1,2,3) = (1,2,3)$, and $g(1,3) = g((1,2)(1,3,2)) = g(1,3,2) = (1,3,2)$. Consequently, $g(1,3,2) = g((1,3)(2,3)) = 1$, a contradiction.

11. In the exact sequence of Problem 6, take $G = \mathbb{Z}_{p^2}$, $N = \mathbb{Z}_p$, $H = G/N \cong \mathbb{Z}_p$, i the inclusion map, and π the canonical epimorphism. If f is a right-splitting map, its image must be a subgroup with p elements (since f is injective), and there is only one such subgroup, namely $\mathbb{Z}_p$. But then $\pi \circ f = 0$, a contradiction.

12. If $g \in G$, then $gPg^{-1} \subseteq gNg^{-1} = N$, so P and gPg^{-1} are both Sylow p-subgroups of N. By Sylow (3), they are *conjugate in* N (the key point). Thus for some $n \in N$ we have $P = n(gPg^{-1})n^{-1}$. But then by definition of normalizer we have $ng \in N_G(P)$, hence $g \in NN_G(P)$.

13. The multiplication table of the group is completely determined by the relations $a^n = 1$, $b^2 = 1$, and $bab^{-1} = a^{-1}$. The relations coincide with those of D_{2n}, with $a = R$ and $b = F$.

14. The relation $a^n = 1$ disappears, and we have $\langle a, b \mid b^2 = 1, bab^{-1} = a^{-1} \rangle$.

Solutions Chapters 6–10

Section 6.1

1. We have $r_1 = 2$, $r_2 = 1$, $r_3 = 1$ so $t_1 = 1$, $t_2 = 0$, $t_3 = 1$. The algorithm terminates in one step after after subtraction of $(X_1 + X_2 + X_3)(X_1X_2X_3)$. The given polynomial can be expressed as e_1e_3.
2. We have $r_1 = 2, r_2 = 1, r_3 = 0$ so $t_1 = 1, t_2 = 1, t_3 = 0$. At step 1, subtract $(X_1 + X_2 + X_3)(X_1X_2 + X_1X_3 + X_2X_3)$. The result is $-3X_1X_2X_3 + 4X_1X_2X_3 = X_1X_2X_3$. By inspection (or by a second step of the algorithm), the given polynomial can be expressed as $e_1e_2 + e_3$.
3. Equation (1) follows upon taking $\sigma_1(h)$ outside the summation and using the linear dependence. Equation (2) is also a consequence of the linear dependence, because $\sigma_i(h)\sigma_i(g) = \sigma_i(hg)$.
4. By hypothesis, the characters are distinct, so for some $h \in G$ we have $\sigma_1(h) \neq \sigma_2(h)$. Thus in (3), each a_i is nonzero and

$$\sigma_1(h) - \sigma_i(h) \begin{cases} = 0 & \text{if } i = 1; \\ \neq 0 & \text{if } i = 2. \end{cases}$$

This contradicts the minimality of r. (Note that the $i = 2$ case is important, since there is no contradiction if $\sigma_1(h) - \sigma_i(h) = 0$ for all i.)

5. By (3.5.10), the Galois group consists of the identity alone. Since the identity fixes all elements, the fixed field of G is $\mathbb{Q}(\sqrt[3]{2})$.
6. Since $\mathbb{C} = \mathbb{R}[i]$, an $\mathbb{R}$-automorphism σ of $\mathbb{C}$ is determined by its action on i. Since σ must permute the roots of $X^2 + 1$ by (3.5.1), we have $\sigma(i) = i$ or $-i$. Thus the Galois group has two elements, the identity automorphism and complex conjugation.
7. The complex number z is fixed by complex conjugation if and only if z is real, so the fixed field is $\mathbb{R}$.

Section 6.2

1. The right side is a subset of the left since both E_i and E_{i+1}^p are contained in E_{i+1}. Since E_i is contained in the set on the right, it is enough to show that $\alpha_{i+1} \in E_i(E_{i+1}^p)$. By

hypothesis, α_{i+1} is separable over F, hence over $E_i(\alpha_{i+1}^p)$. By Section 3.4, Problem 3, $\alpha_{i+1} \in E_i(\alpha_{i+1}^p) \subseteq E_i(E_{i+1}^p)$.

2. Apply Section 3.4, Problem 7, with $E = F(E^p)$ replaced by $E_{i+1} = E_i(E_{i+1}^p)$, to conclude that E_{i+1} is separable over E_i. By the induction hypothesis, E_i is separable over F. By transitivity of separable extensions (Section 3.4, Problem 8), E_{i+1} is separable over F. By induction, E/F is separable.

3. Let f_i be the minimal polynomial of α_i over F. Then E is a splitting field for $f = f_1 \cdots f_n$ over F, and the result follows.

4. This is a corollary of part 2 of the fundamental theorem, with F replaced by K_{i-1} and G replaced by $\operatorname{Gal}(E/K_{i-1}) = H_{i-1}$.

5. $E(A)$ is a field containing $E \geq F$ and A, hence $E(A)$ contains E and K, so that by definition of composite, $EK \leq E(A)$. But any field (in particular EK) that contains E and K contains E and A, hence contains $E(A)$. Thus $E(A) \leq EK$.

6. If $\sigma \in G$, define $\Psi(\sigma)(\tau(x)) = \tau\sigma(x)$, $x \in E$. Then $\psi(\sigma) \in G'$. [If $y = \tau(x) \in F'$ with $x \in F$, then $\Psi(\sigma)y = \Psi(\sigma)\tau x = \tau\sigma(x) = \tau(x) = y$.] Now $\Psi(\sigma_1\sigma_2)\tau(x) = \tau\sigma_1\sigma_2(x)$ and $\Psi(\sigma_1)\Psi(\sigma_2)\tau(x) = \Psi(\sigma_1)\tau\sigma_2(x) = \tau\sigma_1\sigma_2(x)$, so Ψ is a group homomorphism. The inverse of Ψ is given by $\Psi'(\sigma')\tau^{-1}y = \tau^{-1}\sigma'(y)$, $\sigma' \in G'$, $y \in E'$. To see this, we compute

$$\Psi'(\Psi(\sigma))\tau^{-1}y = \tau^{-1}\Psi(\sigma)y = \tau^{-1}\Psi(\sigma)\tau x = \tau^{-1}\tau\sigma(x) = \sigma(x) = \sigma(\tau^{-1}y).$$

Thus $\Psi'\Psi$ is the identity on G.

7. Since H' is a normal subgroup of G, its fixed field $L = \mathcal{F}(H')$ is normal over F, so by minimality of the normal closure, we have $N \subseteq L$. But all fixed fields are subfields of N, so $L \subseteq N$, and consequently $L = N$.

8. If $\sigma \in H'$, then σ fixes everything in the fixed field N, so σ is the identity. Thus the largest normal subgroup of G that is contained in H is trivial. But this largest normal subgroup is the core of H in G, and the resulting formula follows from Problems 4 and 5 of Section 5.1.

Section 6.3

1. $G = \{\sigma_1, \dots, \sigma_n\}$ where σ_i is the unique F-automorphism of E that takes α to α_i.

2. We must find an α such that $1, \alpha, \dots, \alpha^{n-1}$ is a basis for $E/\mathbb{Q}$. If $\alpha = b_1x_1 + \cdots + b_nx_n$, we can compute the various powers of α and write $\alpha^i = c_{i1}x_1 + \cdots + c_{in}x_n$, $i = 0, 1, \dots, n-1$, where each c_{ij} is a rational number. The powers of α will form a basis iff $\det[c_{ij}] \neq 0$. This will happen "with probability 1"; if a particular choice of the b_i yields $\det[c_{ij}] = 0$, a slight perturbation of the b_i will produce a nonzero determinant.

3. By (6.3.1), we may regard G as a group of permutations of the roots $\alpha_1, \dots, \alpha_n$ of f, and therefore G is isomorphic to a subgroup H of S_n. Since G acts transitively on the α_i (see (6.3.1)), the natural action of H on $\{1, 2, \dots, n\}$ is transitive. [For an earlier appearance of the natural action, see the discussion before (5.3.1).]

4. The Galois group G must be isomorphic to a transitive subgroup of S_2, which is cyclic of order 2. There is only one transitive subgroup of S_2, namely S_2 itself, so G is a cyclic group of order 2.

5. Since $[\mathbb{Q}(\sqrt{2}) : \mathbb{Q}] = 2$ and $[\mathbb{Q}(\sqrt{2}, \sqrt{3}) : \mathbb{Q}(\sqrt{2})] = 2$, the Galois group G has order 4. [Note that $\sqrt{3} \notin \mathbb{Q}(\sqrt{2})$ because $a + b\sqrt{2}$ can never be $\sqrt{3}$ for $a, b \in \mathbb{Q}$.] An automorphism σ in G must take $\sqrt{2}$ to $\pm\sqrt{2}$ and $\sqrt{3}$ to $\pm\sqrt{3}$. Thus σ is either the identity or has order 2. Now a group in which every element has order 1 or 2 must be abelian, regardless of the size of the group [$(ab)(ab) = 1$, so $ab = b^{-1}a^{-1} = ba$]. Since G is not cyclic, it must be isomorphic to the four group $\mathbb{Z}_2 \oplus \mathbb{Z}_2$. (See the analysis in (4.6.4).)

6. Let H be the subgroup generated by H_1 and H_2, that is, by $H_1 \cup H_2$. If $\sigma \in H_1 \cup H_2$, then σ fixes $K_1 \cap K_2 = K$. Since H consists of all finite products (= compositions) of elements in H_1 or H_2, everything in H fixes K, so that $K \subseteq \mathcal{F}(H)$. On the other hand, if $x \in \mathcal{F}(H)$ but $x \notin K$, say $x \notin K_1$. Then some $\tau \in H_1 \subseteq H$ fails to fix x, so $x \notin \mathcal{F}(H)$, a contradiction. Therefore $K = \mathcal{F}(H)$.

7. The fixed field is K_1K_2, the composite of K_1 and K_2. For if σ fixes K_1K_2, then it fixes both K_1 and K_2, so σ belongs to $H_1 \cap H_2$. Conversely, if $\sigma \in H_1 \cap H_2$, then σ is the identity on both K_1 and K_2. But by the explicit form of K_1K_2 (see Section 3.1, Problem 1 and Section 6.2, Problem 5), σ is the identity on K_1K_2. Thus $\mathcal{F}(H_1 \cap H_2) = K_1K_2$.

8. We have $E = F(\alpha_1, \dots, \alpha_n)$, where the α_i are the roots of f. Since $\min(\alpha_i, F)$ divides the separable polynomial f, each α_i is separable over F. By Section 6.2, Problem 1, E is separable over F.

9. Since $[\mathbb{Q}(\theta, i) : \mathbb{Q}] = [\mathbb{Q}(\theta) : \mathbb{Q}][\mathbb{Q}(\theta, i) : \mathbb{Q}(\theta)] = 4 \times 2 = 8$, we have $|G| = 8$. Any $\sigma \in G$ must map θ to a root of f (4 choices), and i to a root of $X^2 + 1$ (2 choices, i or $-i$). Since σ is determined by its action on θ and i, we have found all 8 members of G.

10. Let $\sigma(\theta) = i\theta$, $\sigma(i) = i$, and let $\tau(\theta) = \theta$, $\tau(i) = -i$. Then $\sigma^4 = 1$, $\tau^2 = 1$, and the automorphisms $1, \sigma, \sigma^2, \sigma^3, \tau, \sigma\tau, \sigma^2\tau, \sigma^3\tau$ are distinct (by direct verification). Also, we have $\sigma\tau = \tau\sigma^{-1} = \tau\sigma^3$. The result follows from the analysis of the dihedral group in Section 5.8.

11. By direct verification, every member of N fixes $i\theta^2 = i\sqrt{2}$. Since N has index 2 in G, the fixed field of N has degree 2 over $\mathbb{Q}$. But the minimal polynomial of $i\sqrt{2}$ over $\mathbb{Q}$ is $X^2 + 2$, and it follows that $\mathcal{F}(N) = \mathbb{Q}(i\sqrt{2}\}$. $\mathcal{F}(N)$ is the splitting field of $X^2 + 2$ over $\mathbb{Q}$ and is therefore normal over $\mathbb{Q}$, as predicted by Galois theory.

Section 6.4

1. We have $\alpha^4 = 1 + \alpha + \alpha^2 + \alpha^3$ and $\alpha^5 = 1$. Thus the powers of α do not exhaust the nonzero elements of $GF(16)$.

2. We may assume that $E = GF(p^n)$ and that E contains $F = GF(p^m)$, where $n = md$. Then $[E : F] = [E : \mathbb{F}_p]/[F : \mathbb{F}_p] = n/m = d$. Since E/F is separable, we have $E = F(\alpha)$ by the theorem of the primitive element. The minimal polynomial of α over F is an irreducible polynomial of degree d.

3. Exactly as in (6.4.5), carry out a long division of $X^n - 1$ by $X^m - 1$. The division will be successful iff m divides n.

4. Since the b_i belong to L, we have $K \subseteq L$, and since $h \in L[X]$, it follows that $g|h$. But $g \in K[X]$ by definition of K, so $h|g$. Since g and h are monic, they must be equal. In particular, they have the same degree, so $[E : L] = [E : K]$. Since $K \subseteq L$, we have $L = K$.

5. Since $L = K$, L is completely determined by g. But if $f = \min(\alpha, F)$, then g divides f. Since f has only finitely many irreducible divisors, there can only be finitely many intermediate fields L.

6. Since there are finitely many intermediate fields between E and F, the same is true between L and F. By induction hypothesis, $L = F(\beta)$ for some $\beta \in L$. Thus $E = L(\alpha_n) = F(\beta, \alpha_n)$.

7. By hypothesis, there are only finitely many fields of the form $F(c\beta + \alpha_n)$, $c \in F$. But there are infinitely many choices of c, and the result follows.

8. Since $E = F(\beta, \alpha_n)$, it suffices to show that $\beta \in F(c\beta + \alpha_n)$. This holds because

$$\beta = \frac{(c\beta + \alpha_n) - (d\beta + \alpha_n)}{c - d}.$$

9. Let $\sigma\colon F \to F$ be the Frobenius automorphism, given by $\sigma(x) = x^p$. Let $f = \min(\alpha, \mathbb{F}_p)$ and $g = \min(\alpha^p, \mathbb{F}_p)$. Then $f(\alpha^p) = f(\sigma(\alpha)) = \sigma(f(\alpha))$ since σ is a monomorphism, and $\sigma(f(\alpha)) = \sigma(0) = 0$. Thus g divides the monic irreducible polynomial f, so $g = f$.

10. By Problem 9, the subsets are $\{0\}, \{1,3,9\}, \{2,6,5\}, \{4,12,10\}$, and $\{7,8,11\}$. [For example, starting with 2, we have $2 \times 3 = 6$, $6 \times 3 = 18 \equiv 5 \bmod 13$, $5 \times 3 = 15 \equiv 2 \bmod 13$.] In the second case, we get

$$\{0\}, \{1,2,4,8\}, \{3,6,9,12\}, \{5,10\}, \{7,14,13,11\}.$$

Section 6.5

1. $\Psi_n(X^p) = \prod_i (X^p - \omega_i)$ where the ω_i are the primitive n^{th} roots of unity. But the roots of $X^p - \omega_i$ are the pth roots of ω_i, which must be primitive np^{th} roots of unity because p is prime and p divides n. The result follows. (The map $\theta \to \theta^p$ is a bijection between primitive np^{th} roots of unity and primitive n^{th} roots of unity, because $\varphi(np) = p\varphi(n)$.)

2. By (6.5.1) and (6.5.6), the Galois group of the nth cyclotomic extension of $\mathbb{Q}$ can be identified with the group of automorphisms of the cyclic group of nth roots of unity. By (6.5.6), the Galois group is isomorphic to U_n, and the result follows.

3. The powers of 3 mod 7 are 3, $9 \equiv 2$, 6, $18 \equiv 4$, $12 \equiv 5$, 1.

4. This follows from Problem 3 and (1.1.4).

5. $\sigma_6(\omega + \omega^6) = \omega^6 + \omega^{36} = \omega + \omega^6$, so $\omega + \omega^6 \in K$. Now $\omega + \omega^6 = \omega + \omega^{-1} = 2\cos 2\pi/7$, so ω satisfies a quadratic equation over $\mathbb{Q}(\cos 2\pi/7)$. By (3.1.9),

$$[\mathbb{Q}_7 : \mathbb{Q}] = [\mathbb{Q}_7 : K][K : \mathbb{Q}(\cos 2\pi/7)][\mathbb{Q}(\cos 2\pi/7) : \mathbb{Q}]$$

where the term on the left is 6, the first term on the right is $|\langle\sigma_6\rangle| = 2$, and the second term on the right is (by the above remarks) 1 or 2. But $[K : \mathbb{Q}(\cos 2\pi/7)]$ cannot be 2 (since 6 is not a multiple of 4), so we must have $K = \mathbb{Q}(\cos 2\pi/7)$.

6. $\sigma_2(\omega + \omega^2 + \omega^4) = \omega^2 + \omega^4 + \omega^8 = \omega + \omega^2 + \omega^4$, so $\omega + \omega^2 + \omega^4 \in L$; $\sigma_3(\omega + \omega^2 + \omega^4) = \omega^3 + \omega^6 + \omega^{12} = \omega^3 + \omega^5 + \omega^6 \neq \omega + \omega^2 + \omega^4$, so $\omega + \omega^2 + \omega^4 \notin \mathbb{Q}$. [If $\omega^3 + \omega^5 + \omega^6 = \omega + \omega^2 + \omega^4$, then we have two distinct monic polynomials of degree 6 satisfied by ω (the other is $\Psi_7(X)$), which is impossible.]
7. By the fundamental theorem, $[L : \mathbb{Q}] = [G : \langle\sigma_2\rangle] = 2$, so we must have $L = \mathbb{Q}(\omega + \omega^2 + \omega^4)$.
8. The roots of Ψ_q are the p^rth roots of unity that are not p^{r-1}th roots of unity. Thus

$$\Psi_q(X) = \frac{X^{p^r} - 1}{X^{p^{r-1}} - 1} = \frac{t^p - 1}{t - 1}$$

and the result follows.
9. By Problem 1,

$$\Psi_{18}(X) = \Psi_{(3)(6)}(X) = \Psi_6(X^3) = X^6 - X^3 + 1.$$

Section 6.6

1. f is irreducible by Eisenstein, and the Galois group is S_3. This follows from (6.6.7) or via the discriminant criterion of (6.6.3); we have $D(f) = -27(4) = -108$, which is not a square in $\mathbb{Q}$.
2. f is irreducible by the rational root test, and $D(f) = -4(-3)^3 - 27 = 108 - 27 = 81$, a square in $\mathbb{Q}$. Thus the Galois group is A_3.
3. f is irreducible by Eisenstein. The derivative is $f'(X) = 5X^4 - 40X^3 = 5X^3(X - 8)$. We have $f'(x)$ positive for $x < 0$ and for $x > 8$, and $f'(x)$ negative for $0 < x < 8$. Sincc $f(0) > 0$ and $f(8) < 0$, graphing techniques from calculus show that f has exactly 3 real roots. By (6.6.7), $G = S_5$.
4. f is irreducible by the rational root test. By the formula for the discriminant of a general cubic with $a = 3, b = -2, c = 1$, we have $D = 9(-8) - 4(-8) - 27 - 18(6) = -175$. Alternatively, if we replace X by $X - \frac{a}{3} = X - 1$, the resulting polynomial is $g(X) = X^3 - 5X + 5$, whose discriminant is $-4(-5)^3 - 27(25) = -175$. In any event, D is not a square in $\mathbb{Q}$, so $G = S_3$. (Notice also that g is irreducible by Eisenstein, so we could have avoided the rational root test at the beginning.)
5. If f is reducible, then it is the product of a linear factor and a quadratic polynomial g. If g is irreducible, then G is cyclic of order 2 (Section 6.3, Problem 4). If g is reducible, then all roots of f are in the base field, and G is trivial.
6. Let the roots be $a, b + ic$ and $b - ic$. Then

$$\Delta = (a - b - ic)(a - b + ic)2ic = ((a - b)^2 + c^2)2ic$$

and since $i^2 = -1$, we have $D < 0$. Since D cannot be a square in $\mathbb{Q}$, the Galois group is S_3. [This also follows from (6.6.7).]

7. If the roots are a, b and c, then $D = (a-b)^2(a-c)^2(b-c)^2 > 0$. The result follows from (6.6.3).

Section 6.7

1. By (6.7.2), the Galois group of $\mathbb{Q}(\sqrt{m})/\mathbb{Q}$ is $\mathbb{Z}_2$ for $m = 2, 3, 5, 7$. It follows that the Galois group of $\mathbb{Q}(\sqrt{2}, \sqrt{3}, \sqrt{5}, \sqrt{7})/\mathbb{Q}$ is $\mathbb{Z}_2 \times \mathbb{Z}_2 \times \mathbb{Z}_2 \times \mathbb{Z}_2$. See (6.7.5), and note that $\mathbb{Q}$ contains a primitive square root of unity, namely -1. (It is not so easy to prove that the Galois group has order 16. One approach is via the texhnique of Section 7.3, Problems 9 and 10.)
2. Yes. Let E be the pth cyclotomic extension of $\mathbb{Q}$, where p is prime. If $p > 2$, then $\mathbb{Q}$ does not contain a primitive p^{th} root of unity. By (6.5.6), the Galois group is isomorphic to the group of units mod$p$, which is cyclic.
3. Since the derivative of $X^n - a$ is $nX^{n-1} \neq 0$, it follows as in (6.5.1) that f has n distinct roots $\beta_1, \dots, \beta_n$ in E. Since $\beta_i^n = a$ and $\beta_i^{-n} = a^{-1}$, there are n distinct n^{th} roots of unity in E, namely $1 = \beta_1\beta_1^{-1}, \beta_2\beta_1^{-1}, \dots, \beta_n\beta_1^{-1}$. Since the group of nth roots of unity is cyclic, there must be a primitive n^{th} root of unity in E.
4. Each root of g is of the form $\omega^i\theta$, so $g_0 = \omega^k\theta^d$ for some k. Since $\omega^p = 1$, we have $g_0^p = \theta^{dp}$. But $c = \theta^p$ since θ is also a root of f, and the result follows.
5. By Problem 4 we have
$$c = c^1 = c^{ad}c^{bp} = g_0^{ap}c^{bp} = (g_0^a c^b)^p$$
with $g_0^a c^b \in F$. Thus $g_0^a c^b$ is a root of f in F.
6. $[E : F(\omega)]$ divides p and is less than p by (6.7.2); note that E is also a splitting field for f over $F(\omega)$. Thus $[E : F(\omega)]$ must be 1, so $E = F(\omega)$.
7. F contains a primitive p^{th} root of unity ω iff $E (= F(\omega)) = F$ iff $X^p - c$ splits over F.
8. By induction, $\sigma^j(\theta) = \theta + j$, $0 \leq j \leq p-1$. Thus the subgroup of G that fixes θ, hence fixes $F(\theta)$, consists only of the identity. By the fundamental theorem, $E = F(\theta)$.
9. We have $\sigma(\theta^p - \theta) = \sigma(\theta)^p - \sigma(\theta) = (\theta+1)^p - (\theta+1) = \theta^p - \theta$ in characteristic p. Thus $\theta^p - \theta$ belongs to the fixed field of G, which is F. Let $a = \theta^p - \theta$, and the result follows.
10. Since $f(\theta) = 0$, $\min(\theta, F)$ divides f. But the degree of the minimal polynomial is $[F(\theta) : F] = [E : F] = p = \deg f$. Thus $f = \min(\theta, F)$, which is irreducible.
11. Since $\theta^p - \theta = a$, we have $(\theta+1)^p - (\theta+1) = \theta^p - \theta = a$. Inductively, $\theta, \theta+1, \dots, \theta+p-1$ are distinct roots of f in E, and since f has degree p, we have found all the roots and f is separable. Since E is a splitting field for f over F, we have $E = F(\theta)$.
12. By Problem 11, every root of f generates the same extension of F, namely E. But any monic irreducible factor of f is the minimal polynomial of at least one of the roots of f, and the result follows.
13. $[E : F] = [F(\theta) : F] = \deg(\min(\theta, F)) = \deg f = p$. Thus the Galois group has prime order p and is therefore cyclic.

Section 6.8

1. Take the real part of each term of the identity to get

$$\cos 3\theta = \cos^3\theta + 3\cos\theta(i\sin\theta)^2 = \cos^3\theta - 3\cos\theta(1-\cos^2\theta);$$

thus $\cos 3\theta = 4\cos^3\theta - 3\cos\theta$. If $3\theta = \pi/3$, we have

$$\cos\pi/3 = 1/2 = 4\alpha^3 - 3\alpha$$

so $8\alpha^3 - 6\alpha - 1 = 0$. But $8X^3 - 6X - 1$ is irreducible over $\mathbb{Q}$ (rational root test), so α is algebraic over $\mathbb{Q}$ and $[\mathbb{Q}(\alpha):\mathbb{Q}] = 3$ (not a power of 2), a contradiction.

2. $X^3 - 2$ is irreducible by Eisenstein, so $[\mathbb{Q}(\sqrt[3]{2}):\mathbb{Q}] = 3$ and $\sqrt[3]{2}$ is not constructible.

3. The side of such a square would be $\sqrt{\pi}$, so $\sqrt{\pi}$, hence π, would be algebraic over $\mathbb{Q}$, a contradiction.

4. ω is a root of $X^2 - 2(\cos 2\pi/n)X + 1$ since $\cos 2\pi/n = \frac{1}{2}(\omega + \omega^{-1})$ and $\omega^2 - (\omega+\omega^{-1})\omega + 1 = 0$. The discriminant of the quadratic polynomial is negative, proving irreducibility over $\mathbb{R} \supseteq \mathbb{Q}(\cos 2\pi/n)$.

5. By (6.5.2), (6.5.5) and (3.1.9),

$$\varphi(n) = [\mathbb{Q}(\omega):\mathbb{Q}] = [\mathbb{Q}(\omega):\mathbb{Q}(\cos 2\pi/n)][\mathbb{Q}(\cos 2\pi/n):\mathbb{Q}].$$

By Problem 4, $[\mathbb{Q}(\omega):\mathbb{Q}(\cos 2\pi/n)] = 2$, and if the regular n-gon is constructible, then $[\mathbb{Q}(\cos 2\pi/n):\mathbb{Q}]$ is a power of 2. The result follows.

6. By hypothesis, $G = \mathrm{Gal}(\mathbb{Q}(\omega)/\mathbb{Q})$ is a 2-group since its order is $\varphi(n)$. Therefore every quotient group of G, in particular $\mathrm{Gal}(\mathbb{Q}(\cos 2\pi/n)/\mathbb{Q})$, is a 2-group. [Note that by (6.5.1), G is abelian, hence every subgroup of G is normal, and therefore every intermediate field is a Galois extension of $\mathbb{Q}$. Thus part 2c of the fundamental theorem (6.2.1) applies.]

7. By the fundamental theorem (specifically, by Section 6.2, Problem 4), there are fields $\mathbb{Q} = K_0 \le K_1 \le \cdots \le K_r = \mathbb{Q}(\cos 2\pi/n)$ with $[K_i : K_{i-1}] = 2$ for all $i = 1,\dots,r$. Thus $\cos 2\pi/n$ is constructible.

8. If $n = p_1^{e_1}\cdots p_r^{e_r}$, then (see Section 1.1, Problem 13)

$$\varphi(n) = p_1^{e_1-1}(p_1-1)\cdots p_r^{e_r-1}(p_r-1).$$

If $p_i \neq 2$, we must have $e_i = 1$, and in addition, $p_i - 1$ must be a power of 2. The result follows.

9. If m is not a power of 2, then m can be factored as ab where a is odd and $1 < b < m$. In the quotient $(X^a+1)/(X+1)$, set $X = 2^b$. It follows that $(2^m+1)/(2^b+1)$ is an integer. Since $1 < 2^b + 1 < 2^m + 1$, $2^m + 1$ cannot be prime.

10. The e_i belong to E and are algebraically independent over K, so the transcendence detree of E over K is at least n. It follows that the α_i are algebraically independent over K, and the transcendence degree is exactly n. Therefore any permutation of the α_i induces a K-automorphism of $E = K(\alpha_1,\dots,\alpha_n)$ which fixes each e_i, hence fixes F. Thus the Galois group of f consists of all permutations of n letters.

11. Since S_n is not solvable, the general equation of degree n is not solvable by radicals if $n \geq 5$. In other words, if $n \geq 5$, there is no sequence of operations on $e_1, \dots, e_n$ involving addition, subtraction, multiplication, division and extraction of mth roots, that will yield the roots of f.

Section 6.9

1. If S is not maximal, keep adding elements to S until a maximal algebraically independent set is obtained. If we go all the way to T, then T is algebraically independent and spans E algebraically, hence is a transcendence basis. (Transfinite induction supplies the formal details.)

2. For the first statement, take $T = E$ in Problem 1. For the second statement, take $S = \emptyset$.

3. (i) implies (ii): Suppose that t_i satisfies $f(t_i) = b_0 + b_1 t_i + \cdots + b_m t_i^m = 0$, with $b_j \in F(T \setminus \{t_i\})$. By forming a common denominator for the b_j, we may assume that the b_j are polynomials in $F[T \setminus \{t_i\}] \subseteq F[T]$. By (i), $b_j = 0$ for all j, so $f = 0$.

   (ii) implies (iii): Note that $F(t_1, \dots, t_{i-1}) \subseteq F(T \setminus \{t_i\})$.

   (iii) implies (i): Suppose that f is a nonzero polynomial in $F[X_1, \dots, X_m]$ such that $f(t_1, \dots, t_m) = 0$, where m is as small as possible. Then $f = h_0 + h_1 X_m + \cdots + h_r X_m^r$ where the h_j belong to $F[X_1, \dots, X_{m-1}]$. Now $f(t_1, \dots, t_m) = b_0 + b_1 t_m + \cdots + b_r t_m^r$ where $b_j = h_j(t_1, \dots, t_{m-1})$. If the b_j are not all zero, then t_m is algebraic over $F(t_1, \dots, t_{m-1})$, contradicting (iii). Thus $b_j \equiv 0$, so by minimality of m, $h_j \equiv 0$, so $f = 0$.

4. If $S \cup \{t\}$ is algebraically dependent over F, then there is a positive integer n and a nonzero polynomial f in $F[X_1, \dots, X_n, Z]$ such that $f(t_1, \dots, t_n, t) = 0$ for some $t_1, \dots, t_n \in S$. Since S is algebraically independent over F, f must involve Z. We may write $f = b_0 + b_1 Z + \cdots + b_m Z^m$ where $b_m \neq 0$ and the b_j are polynomials in $F[X_1, \dots, X_n]$. But then t is algebraic over $F(S)$.

   Conversely, if t is algebraic over $F(S)$, then for some positive integer n, there are elements $t_1, \dots, t_n \in S$ such that t is algebraic over $F(t_1, \dots, t_n)$. By Problem 3, $\{t_1, \dots, t_n, t\}$ is algebraically dependent over F, hence so is $S \cup \{t\}$.

5. Let $A = \{s_1, \dots, s_m, t_1, \dots, t_n\}$ be an arbitrary finite subset of $S \cup T$, with $s_i \in S$ and $t_j \in T$. By Problem 3, s_i is transcendental over $F(s_1, \dots, s_{i-1})$ and t_j is transcendental over $K(t_1, \dots, t_{j-1})$, hence over $F(s_1, \dots, s_m, t_1, \dots, t_{j-1})$ since $S \subseteq K$. Again by Problem 3, A is algebraically independent over F. Since A is arbitrary, $S \cup T$ is algebraically independent over F. Now if $t \in K$ then $\{t\}$ is algebraically dependent over K (t is a root of $X - t$). But if t also belongs to T, then T is algebraically dependent over K, contradicting the hypothesis. Thus K and T, hence S and T, are disjoint.

6. By Problem 5, $S \cup T$ is algebraically independent over F. By hypothesis, E is algebraic over $K(T)$ and K is algebraic over $F(S)$. Since each $t \in T$ is algebraic over $F(S)(T) =$

$F(S \cup T)$, it follows that $K(T)$ is algebraic over $F(S \cup T)$. By (3.3.5), E is algebraic over $F(S \cup T)$. Therefore $S \cup T$ is a transcendence basis for E/F.

7. If T is algebraically independent over F, the map $f(X_1, \dots, X_n) \to f(t_1, \dots, t_n)$ extends to an F-isomorphism of $F(X_1, \dots, X_n)$ and $F(t_1, \dots, t_n)$. Conversely, assume that $F(T)$ is F-isomorphic to the rational function field. By Problem 2, there is a transcendence basis B for $F(T)/F$ such that $B \subseteq T$. By (6.9.7), the transcendence degree of $F(T)/F$ is $|T| = n$. By (6.9.5) or (6.9.6), $B = T$, so T is algebraically independent over F.
8. The "if" part is clear since $[K(z) : K]$ can't be finite; if so, $[F : K] < \infty$. For the "only if" part, z is algebraic over $K(x)$, so let

$$z^n + \varphi_{n-1}(x)z^{n-1} + \cdots + \varphi_0(x) = 0, \ \varphi_i \in K(x).$$

Clear denominators to get a polynomial $f(z, x) = 0$, with coefficients of f in K. Now x must appear in f, otherwise z is not transcendental. Thus x is algebraic over $K(z)$, so $[K(z, x) : K(z)] < \infty$. Therefore

$$[F : K(z)] = [F : K(z, x)][K(z, x) : K(z)].$$

The first term on the right is finite since $K(x) \subseteq K(z, x)$, and the second term is finite, as we have just seen. Thus $[F : K(z)] < \infty$, and the result follows. ♣
9. We have $\operatorname{tr}\deg(\mathbb{C}/\mathbb{Q}) = c$, the cardinality of $\mathbb{C}$ (or $\mathbb{R}$). For if $\mathbb{C}$ has a countable transcendence basis $z_1, z_2, \dots$ over $\mathbb{Q}$, then $\mathbb{C}$ is algebraic over $\mathbb{Q}(z_1, z_2, \dots)$. Since a polynomial over $\mathbb{Q}$ can be identified with a finite sequence of rationals, it follows that $|\mathbb{C}| = |\mathbb{Q}|$, a contradiction.

Section 7.1

1. Replace (iii) by (iv) and the proof goes through as before. If R is a field, then in (iii) implies (i), x is an eigenvalue of C, so $\det(xI - C) = 0$.
2. Replace (iii) by (v) and the proof goes through as before. [Since B is an $A[x]$-module, in (iii) implies (i) we have $x\beta_i \in B$; when we obtain $[\det(xI - C)]b = 0$ for every $b \in B$, the hypothesis that B is faithful yields $\det(xI - C) = 0$.]
3. Multiply the equation by a^{n-1} to get

$$a^{-1} = -(c_{n-1} + \cdots + c_1 a^{n-2} + c_0 a^{n-1}) \in A.$$

4. Since $A[b]$ is a subring of B, it is an integral domain. Thus if $bz = 0$ and $b \neq 0$, then $z = 0$.
5. Any linear transformation on a finite-dimensional vector space is injective iff it is surjective. Thus if $b \in B$ and $b \neq 0$, there is an element $c \in A[b] \subseteq B$ such that $bc = 1$. Therefore B is a field.
6. P is the preimage of Q under the inclusion map of A into B, so P is a prime ideal. The map $a + P \to a + Q$ is a well-defined injection of A/P into B/Q, since $P = Q \cap A$. Thus A/P can be viewed as a subring of B/Q.

7. If $b+Q \in B/Q$, then b satisfies an equation of the form

$$x^n + a_{n-1}x^{n-1} + \cdots + a_1 x + a_0 = 0, \; a_i \in A.$$

By Problem 6, $b+Q$ satisfies the same equation with a_i replaced by $a_i + P$ for all i. Thus B/Q is integral over A/P.

8. By Problems 3–5, A/P is a field if and only if B/Q is a field, and the result follows. (Note that since Q is a prime ideal, B/Q is an integral domain, as required in the hypothesis of the result just quoted.)

Section 7.2

1. By the quadratic formula, $L = \mathbb{Q}(\sqrt{b^2-4c})$. Since $b^2 - 4c \in \mathbb{Q}$, we may write $b^2 - 4c = s/t = st/t^2$ for relatively prime integers s and t. We also have $s = uy^2$ and $t = vz^2$ where $u, v, y, z \in \mathbb{Z}$, with u and v relatively prime and square-free. Thus $L = \mathbb{Q}(\sqrt{uv}) = \mathbb{Q}(\sqrt{d})$.
2. If $\mathbb{Q}(\sqrt{d}) = \mathbb{Q}(\sqrt{e})$, then $\sqrt{d} = a + b\sqrt{e}$ for rational numbers a and b. Thus $d = a^2 + b^2 e + 2ab\sqrt{e}$, so $\sqrt{e}$ is rational, a contradiction (unless $a = 0$ and $b = 1$).
3. Any isomorphism of $\mathbb{Q}(\sqrt{d})$ and $\mathbb{Q}(\sqrt{e})$ must carry $\sqrt{d}$ into $a+b\sqrt{e}$ for rational numbers a and b. Thus d is mapped to $a^2 + b^2 e + 2ab\sqrt{e}$. But a $\mathbb{Q}$-isomorphism maps d to d, and we reach a contradiction as in Problem 2.
4. Since $\omega_n = \omega_{2n}^2$ we have $\omega_n \in \mathbb{Q}(\omega_{2n})$, so $\mathbb{Q}(\omega_n) \subseteq \mathbb{Q}(\omega_{2n})$. If n is odd then $n+1 = 2r$, so

$$\omega_{2n} = -\omega_{2n}^{2r} = -(\omega_{2n}^2)^r = -\omega_n^r.$$

Therefore $\mathbb{Q}(\omega_{2n}) \subseteq \mathbb{Q}(\omega_n)$.

5. Let f be a monic polynomial over $\mathbb{Z}$ with $f(x) = 0$. If f is factorable over $\mathbb{Q}$, then it is factorable over $\mathbb{Z}$ by (2.9.2). Thus $\min(x, \mathbb{Q})$ is the monic polynomial in $\mathbb{Z}[X]$ of least degree such that $f(x) = 0$.
6. $\mathbb{Q}(\sqrt{-3}) = \mathbb{Q}(\omega)$ where $\omega = -\frac{1}{2} + \frac{1}{2}\sqrt{-3}$ is a primitive cube root of unity.
7. If $n = [L : Q]$, then an integral basis consists of n elements of L that are linearly independent over $\mathbb{Z}$, hence over $\mathbb{Q}$. (A linear dependence relation over $\mathbb{Q}$ can be converted to one over $\mathbb{Z}$ by multiplying by a common denominator.)

Section 7.3

1. The Galois group of $E/\mathbb{Q}$ consists of the identity and the automorphism $\sigma(a+b\sqrt{d}) = a - b\sqrt{d}$. By (7.3.6), $T(x) = x + \sigma(x) = 2a$ and $N(x) = x\sigma(x) = a^2 + db^2$.
2. A basis for $E/\mathbb{Q}$ is $1, \theta, \theta^2$, and

$$\theta^2 1 = \theta^2, \quad \theta^2\theta = \theta^3 = 3\theta - 1, \quad \theta^2\theta^2 = \theta^4 = \theta\theta^3 = 3\theta^2 - \theta.$$

Thus

$$m(\theta^2) = \begin{bmatrix} 0 & -1 & 0 \\ 0 & 3 & -1 \\ 1 & 0 & 3 \end{bmatrix}$$

and we have $T(\theta^2) = 6$, $N(\theta^2) = 1$. Note that if we had already computed the norm of θ (the matrix of θ is

$$m(\theta) = \begin{bmatrix} 0 & 0 & -1 \\ 1 & 0 & 3 \\ 0 & 1 & 0 \end{bmatrix}$$

and $T(\theta) = 0$, $N(\theta) = -1$), it would be easier to calculate $N(\theta^2)$ as $[N(\theta)]^2 = (-1)^2 = 1$.

3. The cyclotomic polynomial Ψ_6 has only two roots, ω and its complex conjugate $\overline{\omega}$. By (7.3.5),

$$T(\omega) = \omega + \overline{\omega} = e^{i\pi/3} + e^{-i\pi/3} = 2\cos(\pi/3) = 1.$$

4. By (7.3.6), $N(x) = x\sigma(x)\cdots\sigma^{n-1}(x)$ and $T(x) = x + \sigma(x) + \cdots + \sigma^{n-1}(x)$. If $x = y/\sigma(y)$, then $\sigma(x) = \sigma(y)/\sigma^2(y), \ldots, \sigma^{n-1}(x) = \sigma^{n-1}(y)/\sigma^n(y) = \sigma^{n-1}(y)/y$, and the telescoping effect gives $N(x) = 1$. If $x = z - \sigma(z)$, then $\sigma(x) = \sigma(z) - \sigma^2(z)$, $\ldots, \sigma^{n-1}(x) = \sigma^{n-1}(z) - z$, and a similar telescoping effect gives $T(x) = 0$.

5. Choose $v \in E$ such that

$$y = v + x\sigma(v) + x\sigma(x)\sigma^2(v) + \cdots + x\sigma(x)\cdots\sigma^{n-2}(x)\sigma^{n-1}(v) \neq 0$$

and hence

$$\sigma(y) = \sigma(v) + \sigma(x)\sigma^2(v) + \sigma(x)\sigma^2(x)\sigma^3(v) + \cdots + \sigma(x)\sigma^2(x)\cdots\sigma^{n-1}(x)\sigma^n(v).$$

We are assuming that $N(x) = x\sigma(x)\cdots\sigma^{n-1}(x) = 1$, and it follows that the last summand in $\sigma(y)$ is $x^{-1}\sigma^n(v) = x^{-1}v$. Comparing the expressions for y and $\sigma(y)$, we have $x\sigma(y) = y$, as desired.

6. Since $T(x) = 0$, we have $-x = \sigma(x) + \cdots + \sigma^{n-1}(x)$, so the last summand of $\sigma(w)$ is $-xu$. Thus

$$w - \sigma(w) = x(u + \sigma(u) + \sigma^2(u) + \cdots + \sigma^{n-1}(u)) = xT(u).$$

7. We have

$$z - \sigma(z) = \frac{w}{T(u)} - \frac{\sigma(w)}{\sigma(T(u))} = \frac{w - \sigma(w)}{T(u)}$$

since $T(u)$ belongs to F and is therefore fixed by σ (see (7.3.3)). By Problem 6, $z - \sigma(z) = x$.

8. No. We have $y/\sigma(y) = y'/\sigma(y')$ iff $y/y' = \sigma(y/y')$ iff y/y' is fixed by all automorphisms in the Galois group G iff y/y' belongs to the fixed field of G, which is F. Similarly, $z - \sigma(z) = z' - \sigma(z')$ iff $z - z' \in F$.
9. We have $\min(\theta, \mathbb{Q}) = X^4 - 2$, $\min(\theta^2, \mathbb{Q}) = X^2 - 2$, $\min(\theta^3, \mathbb{Q}) = X^4 - 8$, $\min(\sqrt{3}\theta, \mathbb{Q}) = X^4 - 18$. (To compute the last two minimal polynomials, note that $(\theta^3)^4 = (\theta^4)^3 = 2^3 = 8$, and $(\sqrt{3}\theta)^4 = 18$.) Therefore, all 4 traces are 0.
10. Suppose that $\sqrt{3} = a + b\theta + c\theta^2 + d\theta^3$ with $a, b, c \in \mathbb{Q}$. Take the trace of both sides to conclude that $a = 0$. (The trace of $\sqrt{3}$ is 0 because its minimal polynomial is $X^2 - 3$.) Thus $\sqrt{3} = b\theta + c\theta^2 + d\theta^3$, so $\sqrt{3}\theta = b\theta^2 + c\theta^3 + 2d$. Again take the trace of both sides to get $d = 0$. The same technique yields $b = c = 0$, and we reach a contradiction.

Section 7.4

1. If $l(y) = 0$, then $(x, y) = 0$ for all x. Since the bilinear form is nondegenerate, we must have $y = 0$.
2. Since V and V^* have the same dimension n, the map $y \to l(y)$ is surjective.
3. We have $(x_i, y_j) = l(y_j)(x_i) = f_j(x_i) = \delta_{ij}$. Since the $f_j = l(y_j)$ form a basis, so do the y_j.
4. Write $x_i = \sum_{k=1}^{n} a_{ik} y_k$, and take the inner product of both sides with x_j to conclude that $a_{ij} = (x_i, x_j)$.
5. The "if" part was done in the proof of (7.4.10). If $\det C = \pm 1$, then C^{-1} has coefficients in $\mathbb{Z}$ by Cramer's rule.
6. If $d \not\equiv 1 \bmod 4$, then by (7.2.3), 1 and $\sqrt{d}$ form an integral basis. Since the trace of $a + b\sqrt{d}$ is $2a$ (Section 7.3, Problem 1), the field discriminant is

$$D = \det \begin{bmatrix} 2 & 0 \\ 0 & 2d \end{bmatrix} = 4d.$$

If $d \equiv 1 \bmod 4$, then 1 and $\frac{1}{2}(1 + \sqrt{d})$ form an integral basis, and

$$\left[\frac{1}{2}(1 + \sqrt{d})\right]^2 = \frac{1}{4} + \frac{d}{4} + \frac{1}{2}\sqrt{d}.$$

Thus

$$D = \det \begin{bmatrix} 2 & 1 \\ 1 & \frac{d+1}{2} \end{bmatrix} = d.$$

7. The first statement follows because multiplication of each element of a group G by a particular element $g \in G$ permutes the elements of G. The plus and minus signs are balanced in $P + N$ and PN, before and after permutation. We can work in a Galois extension of $\mathbb{Q}$ containing L, and each automorphism in the Galois group restricts to one of the σ_i on L. Thus $P + N$ and PN belong to the fixed field of the Galois group, which is $\mathbb{Q}$.

8. Since the x_j are algebraic integers, so are the $\sigma_i(x_j)$, as in the proof of (7.3.10). By (7.1.5), P and N, hence $P+N$ and PN, are algebraic integers. By (7.1.7), $\mathbb{Z}$ is integrally closed, so by Problem 7, $P+N$ and PN belong to $\mathbb{Z}$.
9. $D = (P-N)^2 = (P+N)^2 - 4PN \equiv (P+N)^2 \bmod 4$. But any square is congruent to 0 or 1 mod 4, and the result follows.
10. We have $y_i = \sum_{j=1}^{n} a_{ij}x_j$ with $a_{ij} \in \mathbb{Z}$. By (7.4.3), $D(y) = (\det A)^2 D(x)$. Since $D(y)$ is square-free, $\det A = \pm 1$, so A has an inverse with entries in $\mathbb{Z}$. Thus $x = A^{-1}y$, as claimed.
11. Every algebraic integer is a $\mathbb{Z}$-linear combination of the x_i, hence of the y_i by Problem 10. Since the y_i form a basis for L over $\mathbb{Q}$, they are linearly independent and the result follows.
12. No. For example, let $L = \mathbb{Q}(\sqrt{d})$, where d is a square-free integer with $d \not\equiv 1 \bmod 4$. (See Problem 6). The field discriminant is $4d$, which is not square-free.
13. This follows from the proof of (7.4.7).

Section 7.5

1. $A_0 \subset A_1 \subset A_2 \subset \cdots$
2. Let $a/p^n \in B$, where p does not divide a. There are integers r and s such that $ra + sp^n = 1$. Thus $ra/p^n = 1/p^n$ in $\mathbb{Q}/\mathbb{Z}$, and $A_n \subseteq B$. If there is no upper bound on n, then $1/p^n \in B$ for all n (note $1/p^n = p/p^{n+1} = p^2/p^{n+2}$, etc.), hence $B = A$. If there is a largest n, then for every $m > n$, $B \cap A_m \subseteq A_n$ by maximality of n. Therefore $B = A_n$.
3. Let $x_1, x_2, \ldots$ be a basis for V. Let M_r be the subspace spanned by $x_1, \ldots, x_r$, and L_r the subspace spanned by the $x_j, j > r$. If V is n-dimensional, then $V = L_0 > L_1 > \cdots > L_{n-1} > L_n = 0$ is a composition series since a one-dimensional subspace is a simple module. [$V = M_n > M_{n-1} > \cdots > M_1 > 0$ is another composition series.] Thus V is Noetherian and Artinian. If V is infinite-dimensional, then $M_1 < M_2 < \cdots$ violates the acc, and $L_0 > L_1 > L_2 > \cdots$ violates the dcc. Thus V is neither Noetherian nor Artinian. [Note that if V has an uncountable basis, there is no problem; just take a countably infinite subset of it.]
4. $l(M)$ is finite iff M has a composition series iff M is Noetherian and Artinian iff N and M/N are Noetherian and Artinian iff $l(N)$ and $l(M/N)$ are finite.
5. By Problem 4, the result holds when $l(M) = \infty$, so assume $l(M)$, hence $l(N)$ and $l(M/N)$, finite. Let $0 < N_1 < \cdots < N_r = N$ be a composition series for N, and let $N/N < (M_1+N)/N < \cdots < (M_s+N)/N = M/N$ be a composition series for M/N. Then

$$0 < N_1 < \cdots < N_r < M_1 + N < \cdots < M_s + N = M$$

is a composition series for M. (The factors in the second part of the series are simple by the third isomorphism theorem.) It follows that $l(M) = r + s = l(N) + l(M/N)$.

6. By (7.5.9), R is a Noetherian S-module, hence a Noetherian R-module. (Any R-submodule T of R is, in particular, an S-submodule of R. Therefore T is finitely generated.)
7. Yes. Map a polynomial to its constant term and apply the first isomorphism theorem to show that $R \cong R[X]/(X)$. Thus R is a quotient of a Noetherian R-module, so is Noetherian by (7.5.7).
8. If there is an infinite descending chain of submodules M_i of M, then the intersection $N = \cap_i M_i$ cannot be expressed as the intersection of finitely many M_i. By the correspondence theorem, $\cap_i(M_i/N) = 0$, but no finite intersection of the submodules M_i/N of M/N is 0. Thus M/N is not finitely cogenerated. Conversely, suppose that M/N is not finitely cogenerated. By the correspondence theorem, we have $\cap_\alpha M_\alpha = N$, but no finite intersection of the M_α is N. Pick any M_α and call it M_1. If $M_1 \subseteq M_\alpha$ for all α, then $M_1 = N$, a contradiction. Thus we can find $M_\alpha = M_2$ such that $M_1 \supset M_1 \cap M_2$. Continue inductively to produce an infinite descending chain.

Section 7.6

1. The "only if" part follows from (7.6.2). If the given condition is satisfied and $ab \in P$, then $(a)(b) \subseteq P$, hence $(a) \subseteq P$ or $(b) \subseteq P$, and the result follows.
2. If $x_i \notin P_i$ for some i, then $x_i \in I \setminus \cup_{j=1}^n P_j$ and we are finished.
3. Since I is an ideal, $x \in I$. Say $x \in P_1$. All terms in the sum that involve x_1 belong to P_1 by Problem 2. The remaining term $x_2 \cdots x_n$ is the difference of two elements in P_1, hence $x_2 \cdots x_n \in P_1$. Since P_1 is prime, $x_j \in P_1$ for some $j \neq 1$, contradicting the choice of x_j.
4. The product of ideals is always contained in the intersection. If I and J are relatively prime, then $1 = x + y$ with $x \in I$ and $y \in J$. If $z \in I \cap J$, then $z = z1 = zx + zy \in IJ$. The general result follows by induction, along with the computation

$$R = (I_1 + I_3)(I_2 + I_3) \subseteq I_1 I_2 + I_3.$$

Thus $I_1 I_2$ and I_3 are relatively prime.

5. See (2.6.9).
6. Assume that R is not a field, equivalently, $\{0\}$ is not a maximal ideal. Thus by (7.6.9), every maximal idea is invertible.
7. Let r be a nonzero element of R such that $rK \subseteq R$, hence $K \subseteq r^{-1}R \subseteq K$. Thus $K = r^{-1}R$. Since $r^{-2} \in K$ we have $r^{-2} = r^{-1}s$ for some $s \in R$. But then $r^{-1} = s \in R$, so $K \subseteq R$ and consequently $K = R$.
8. $R = R^r = (P_1 + P_2)^r \subseteq P_1^r + P_2$. Thus P_1^r and P_2 are relatively prime for all $r \geq 1$. Assuming inductively that P_1^r and P_2^s are relatively prime, we have

$$P_2^s = P_2^s R = P_2^s(P_1^r + P_2) \subseteq P_1^r + P_2^{s+1}$$

so

$$R = P_1^r + P_2^s \subseteq P_1^r + (P_1^r + P_2^{s+1}) = P_1^r + P_2^{s+1}$$

completing the induction.

Section 7.7

1. By Section 7.3, Problem 1, the norms are 6, 6, 4 and 9. Now if $x = a + b\sqrt{-5}$ and $x = yz$, then $N(x) = a^2 + 5b^2 = N(y)N(z)$. The only algebraic integers of norm 1 are ± 1, and there are no algebraic integers of norm 2 or 3. Thus there cannot be a nontrivial factorization of $1 \pm \sqrt{-5}$, 2 or 3.
2. If $(a+b\sqrt{-5})(c+d\sqrt{-5}) = 1$, take norms to get $(a^2+5b^2)(c^2+5d^2) = 1$, so $b = d = 0$, $a = \pm 1$, $c = \pm 1$.
3. By Problem 2, if two factors are associates, then the quotient of the factors is ± 1, which is impossible.
4. This is a nice application of the principle that divides means contains. The greatest common divisor is the smallest ideal containing both I and J, that is, $I + J$. The least common multiple is the largest ideal contained in both I and J, which is $I \cap J$.
5. If I is a fractional ideal, then by (7.7.1) there is a fractional ideal I' such that $II' = R$. By definition of fractional ideal, there is a nonzero element $r \in R$ such that rI' is an integral ideal. If $J = rI'$, then $IJ = Rr$, a principal ideal of R.
6. This is done just as in Problems 1–3, using the factorization $18 = (2)(3^2) = (1 + \sqrt{-17})(1 - \sqrt{-17})$.
7. By (7.2.2), the algebraic integers are of the form $a + b\sqrt{-3}, a, b \in \mathbb{Z}$, or $\frac{u}{2} + \frac{v}{2}\sqrt{-3}$ where u and v are odd integers. If we require that the norm be 1, we only get ± 1 in the first case. But in the second case, we have $u^2 + 3v^2 = 4$, so $u = \pm 1, v = \pm 1$. Thus if $\omega = e^{i2\pi/3}$, the algebraic integers of norm 1 are ± 1, $\pm\omega$, and $\pm\omega^2$.

Section 7.8

1. If Rx and Ry belong to $P(R)$, then $(Rx)(Ry)^{-1} = (Rx)(Ry^{-1}) = Rxy^{-1} \in P(R)$, and the result follows from (1.1.2).
2. If $C(R)$ is trivial, then every integral ideal I of R is a principal fractional ideal $Rx, x \in K$. But $I \subseteq R$, so $x = 1x$ must belong to R, proving that R is a PID. The converse holds because every principal ideal is a principal fractional ideal.
3. $1 - \sqrt{-5} = 2 - (1 + \sqrt{-5}) \in P_2$, so $(1 + \sqrt{-5})(1 - \sqrt{-5}) = 6 \in P_2^2$.
4. Since $2 \in P_2$, it follows that $4 \in P_2^2$, so by Problem 3, $2 = 6 - 4 \in P_2^2$.
5. $(2, 1+\sqrt{-5})(2, 1+\sqrt{-5}) = (4, 2(1+\sqrt{-5}), (1+\sqrt{-5})^2)$, and $(1+\sqrt{-5})^2 = -4+2\sqrt{-5}$. Thus each of the generators of the ideal P_2^2 is divisible by 2, hence belongs to (2). Therefore $P_2^2 \subseteq (2)$.
6. $x^2+5 \equiv (x+1)(x-1) \bmod 3$, which suggests that $(3) = P_3P_3'$, where $P_3 = (3, 1+\sqrt{-5})$ and $P_3' = (3, 1 - \sqrt{-5})$.
7. $P_3P_3' = (3, 3(1-\sqrt{-5}), 3(1+\sqrt{-5}), 6) \subseteq (3)$ since each generator of P_3P_3' is divisible by 3. But $3 \in P_3 \cap P_3'$, hence $9 \in P_3P_3'$, and therefore $9 - 6 = 3 \in P_3P_3'$. Thus $(3) \subseteq P_3P_3'$, and the result follows.

Section 7.9

1. Using (1), the product is $z = 4 + 2p + 4p^2 + p^3 + p^4$. But $4 = 3 + 1 = 1 + p$ and $4p^2 = p^2 + 3p^2 = p^2 + p^3$. Thus $z = 1 + 3p + p^2 + 2p^3 + p^4 = 1 + 2p^2 + 2p^3 + p^4$. Using (2), we are multiplying $x = \{2, 5, 14, 14, \dots\}$ by $y = \{2, 2, 11, 11, \dots\}$. Thus $z_0 = 4, z_1 = 10, z_2 = 154, z_3 = 154, z_4 = 154$, and so on. But $4 \equiv 1 \bmod 3$, $10 \equiv 1 \bmod 9$, $154 \equiv 19 \bmod 27$, $154 \equiv 73 \bmod 81$, $154 \equiv 154 \bmod 243$. The standard form is $\{1, 1, 19, 73, 154, 154, \dots\}$. As a check, the product is $(2+3+9)(2+9) = 154$, whose base 3 expansion is $1 + 0(3) + 2(9) + 2(27) + 1(81)$ as found above.
2. We have $a_0 = -1$ and $a_n = 0$ for $n \geq 1$; equivalently, $x_n = -1$ for all n. In standard form, $x_0 = p-1, x_1 = p^2-1, x_2 = p^3-1, \dots$.Since $(p^r-1)-(p^{r-1}-1) = (p-1)(p^{r-1})$, the series representation is

$$(p-1) + (p-1)p + (p-1)p^2 + \cdots + (p-1)p^n + \cdots .$$

The result can also be obtained by multiplying by -1 on each side of the equation

$$1 = (1-p)(1 + p + p^2 + \cdots).$$

3. Let x be a nonzero element of $GF(q)$. By (6.4.1), $x^{q-1} = 1$, so $|x|^{q-1} = 1$. Thus $|x|$ is a root of unity, and since absolute values are nonnegative real, we must have $|x| = 1$, and the result follows.
4. If the absolute value is nonarchimedian, then S is bounded by (7.9.6). If the absolute value is archimedian, then by (7.9.6), $|n| > 1$ for some n. But then $|n^k| = |n|^k \to \infty$ as $k \to \infty$. Therefore S is unbounded.
5. A field of prime characteristic p has only finitely many integers $0, 1, \dots, p-1$. Thus the set S of Problem 4 must be bounded, so the absolute value is nonarchimedian.
6. The "only if" part is handled just as in calculus. For the "if" part, note that by (iv) of (7.9.5), we have $|z_m + z_{m+1} + \cdots + z_n| \leq \max\{|z_i|: m \leq i \leq n\} \to 0$ as $m, n \to \infty$. Thus the nth partial sums form a Cauchy sequence, which must converge to an element in $\mathbb{Q}_p$.
7. Since $n! = 1 \cdot 2 \cdots p \cdots 2p \cdots 3p \cdots$, it follows from (7.9.2) and (7.9.3) that if $rp \leq n < (r+1)p$, then $|n!| = 1/p^r$. Thus $|n!| \to 0$ as $n \to \infty$.
8. No. Although $|p^r| = 1/p^r \to 0$ as $r \to \infty$, all integers n such that $rp < n < (r+1)p$ have p-adic absolute value 1, by (7.9.2). Thus the sequence of absolute values $|n|$ cannot converge, hence the sequence itself cannot converge.

Section 8.1

1. If $x \in V$ and $f_1(x) \neq 0$, then $f_2(x)$ must be 0 since $f_1 f_2 \in I(V)$; the result follows.
2. By Problem 1, $V \subseteq V(f_1) \cup V(f_2)$. Thus

$$V = (V \cap V(f_1)) \cup (V \cap V(f_2)) = V_1 \cup V_2.$$

Since $f_1 \notin I(V)$, there exists $x \in V$ such that $f_1(x) \neq 0$. Thus $x \notin V_1$, so $V_1 \subset V$; similarly, $V_2 \subset V$.

3. $I(V) \supseteq I(W)$ by (4). If $I(V) = I(W)$, let $V = V(S), W = V(T)$. Then $IV(S) = IV(T)$, and by applying V to both sides, we have $V = W$ by (6).
4. Let $x \in V$; if $f_1(x) \neq 0$, then since $f_1 \in I(V_1)$, we have $x \notin V_1$. But then $x \in V_2$, and therefore $f_2(x) = 0$ (since $f_2 \in I(V_2)$). Thus $f_1 f_2 = 0$ on V, so $f_1 f_2 \in I(V)$.
5. If V is reducible, then V is the union of proper subvarieties V_1 and V_2. If V_1 is reducible, then it too is the union of proper subvarieties. This decomposition process must terminate in a finite number of steps, for otherwise by Problems 1–4, there would be a strictly increasing infinite sequence of ideals, contradicting the fact that $k[X_1, \dots, X_n]$ is Noetherian.
6. If $V = \bigcup_i V_i = \bigcup_j W_j$, then $V_i = \bigcup_j (V_i \bigcap W_j)$, so by irreducibility, $V_i = V_i \bigcap W_j$ for some j. Thus $V_i \subseteq W_j$, and similarly $W_j \subseteq V_k$ for some k. But then $V_i \subseteq V_k$, hence $i = k$ (otherwise we would have discarded V_i). Thus each V_i can be paired with a corresponding W_j, and vice versa.
7. By hypothesis, $A^n = \cup(A^n \setminus V(I_i))$. Taking complements, we have $\cap V(I_i) = \emptyset$. But by (8.1.2), $\cap V(I_i) = V(\cup I_i) = V(I)$, so by the weak Nullstellensatz, $I = k[X_1, \dots, X_n]$. Thus the constant polynomial 1 belongs to I.
8. Suppose that the open sets $A^n \setminus V(I_i)$ cover A^n. By Problem 7, $1 \in I$, hence 1 belongs to a finite sum $\sum_{i \in F} I_i$. Since 1 never vanishes, $V(\sum_{i \in F} I_i) = \emptyset$. By (8.1.2), $\cap_{i \in F} V_i = \emptyset$, where $V_i = V(I_i)$. Taking complements, we have $\cup_{i \in F}(A^n \setminus V_i) = A^n$. Thus the original open covering of A^n has a finite subcovering, proving compactness.

Section 8.2

1. If $a \notin (a_1, \dots, a_k)$, then g or some other element of I would extend the inductive process to step $k+1$.
2. In going from d_i to d_{i+1} we are taking the minimum of a smaller set.
3. By minimality of m, $a \notin (a_1, \dots, a_{m-1})$, hence f_m and g satisfy conditions 1 and 2. By choice of f_m we have $d_m \leq d$. (If $m = 1$, then $d_1 \leq d$ by choice of f_1.)
4. Let f be the unique ring homomorphism from $R[X_1, \dots, X_n]$ to S such that f is the identity on R and $f(X_i) = x_i$, $i = 1, \dots, n$. (For example, if $a \in R$, then $aX_1^2 X_4^7 \to ax_1^2 x_4^7$.) Since the image of f contains R and $\{x_1, \dots, x_n\}$, f is surjective and the result follows.
5. By the Hilbert basis theorem, $R[X_1, \dots, X_n]$ is a Noetherian ring, hence a Noetherian $R[X_1, \dots, X_n]$-module. By (7.5.7), S is a Noetherian $R[X_1, \dots, X_n]$-module. But the submodules of S considered as an $R[X_1, \dots, X_n]$-module coincide with the submodules of S as an S-module. (See Section 4.2, Problems 6 and 7; note that the kernel of the homomorphism f of Problem 4 annihilates S.) Thus S is a Noetherian S-module, that is, a Noetherian ring.

Section 8.3

1. Suppose that $xy \in J$ with $x \notin J$ and $y \notin J$. By maximality of J, the ideal $J + (x)$ contains an element $s \in S$. Similarly, $J + (y)$ contains an element $t \in S$. But then

$st \in (J + (x))(J + (y)) \subseteq J + (xy) \subseteq J$, so $S \cap J \neq \emptyset$, a contradiction.

2. Let $S = \{1, f, f^2, \dots, f^r, \dots\}$. Then $I \cap S = \emptyset$ since $f \notin \sqrt{I}$. By Problem 1, I is contained in a prime ideal P disjoint from S. But $f \in S$, so f cannot belong to P, and the result follows.

3. The "if" part follows because f and f^r have the same zero-set. Conversely, if $V(f) = V(g)$, then by the Nullstellensatz, $\sqrt{(f)} = \sqrt{(g)}$, and the result follows.

4. $W \subseteq V$ since $(t^4)^2 = (t^3)(t^5)$ and $(t^5)^2 = (t^3)^2(t^4)$; $L \subseteq V$ by direct verification. Conversely, if $y^2 = xz$ and $z^2 = x^2y$, let $t = y/x$. (If $x = 0$, then $y = z = 0$ and we can take $t = 0$.) Then $z = y^2/x = (y/x)^2x = t^2x$, and $z^2 = x^2y$. Therefore $z^2 = t^4x^2 = x^2y$, hence $y = t^4$. If $t = 0$ then $y = 0$, hence $z = 0$ and $(x, y, z) \in L$. Thus assume $t \neq 0$. But then $x = y/t = t^3$ and $z = t^2x = t^5$.

5. We will show that $I(V)$ is a prime ideal (see the exercises in Section 8.1). If $fg \in I(V)$, then fg vanishes on V. Using the parametric form, we have $f(t, t^2, t^3)g(t, t^2, t^3) = 0$ for all complex numbers t. Since we are now dealing with polynomials in only one variable, either $f(t, t^2, t^3) = 0$ for all t or $g(t, t^2, t^3) = 0$ for all t. Thus $f \in I(V)$ or $g \in I(V)$.

6. (a) $x = 2t/(t^2 + 1)$, $y = (t^2 - 1)/(t^2 + 1)$

   (b) $x = t^2$, $y = t^3$

   (c) $x = t^2 - 1$, $y = t(t^2 - 1)$

7. (Following Shafarevich, Basic Algebraic Geometry, Vol.1, page 2.) We can assume that x appears in f with positive degree. Viewing f and g as polynomials in $k(y)[x]$, a PID, f is still irreducible because irreducibility over an integral domain implies irreducibility over the quotient field. If $g = fh$ where h is a polynomial in x with coefficients in $k(y)$, then by clearing denominators we see that f must divide g in $k[x, y]$, a contradiction. (Since f is irreducible, it must either divide g or a polynomial in y alone, and the latter is impossible because x appears in f.) Thus f does not divide g in $k(y)[x]$. Since f and g are relatively prime, there exist $s, t \in k(y)[x]$ such that $fs + gt = 1$. Clearing denominators, we get $u, v \in k[x, y]$ such that $fu + gv = a$, where a is a nonzero polynomial in y alone. Now if $\alpha, \beta \in k$ and $f(\alpha, \beta) = g(\alpha, \beta) = 0$, then $a(\beta) = 0$, and this can only happen for finitely many β. For any fixed β, consider $f(x, \beta) = 0$. If this polynomial in x is not identically 0, then there are only finitely many α such that $f(\alpha, \beta) = 0$, and we are finished. Thus assume $f(x, \beta) \equiv 0$. Then $f(x, y) = f(x, y) - f(x, \beta) = (y - \beta)h$ in $k(x)[y]$, contradicting the irreducibility of f.

Section 8.4

1. Since $f = 0$ iff some $f_i = 0$, $V(f)$ is the union of the $V(f_i)$. Since each f_i is irreducible, the ideal $I_i = (f_i)$ is prime by (2.6.1), hence $V(I_i) = V(f_i)$ is an irreducible subvariety of $V(f)$. [See the problems in Section 8.1, along with the Nullstellensatz and the fact that every prime ideal is a radical ideal (Section 8.3, Problem 2).] No other decomposition is possible, for if $V(f_i) \subseteq V(f_j)$, then $(f_i) \supseteq (f_j)$. This is impossible if f_i and f_j are distinct irreducible factors of f.

2. By the Nullstellensatz, $IV(f) = \sqrt{(f)}$, and we claim that $\sqrt{(f)} = (f_1 \cdots f_r)$. For if $g \in (f_1 \cdots f_r)$, then a sufficiently high power of g will belong to (f). Conversely, if $g^m = hf$, then each f_i divides g^m, and since the f_i are irreducible, each f_i divides g, so $(f_1 \cdots f_r)$ divides g.
3. By Problem 1, f is irreducible if and only if $V(f)$ is an irreducible hypersurface. If f and g are irreducible and $V(f) = V(g)$, then as in Problem 1, $(f) = (g)$, so $f = cg$ for some nonzero constant c (Section 2.1, Problem 2). Thus $f \to V(f)$ is a bijection between irreducible polynomials and irreducible hypersurfaces, if the polynomials f and $cf, c \neq 0$, are identified.
4. This follows from the definition of $I(X)$ in (8.1.3), and the observation that a function vanishes on a union of sets iff it vanishes on each of the sets.
5. By Section 8.1, Problem 5, every variety V is the union of finitely many irreducible subvarieties $V_1, \dots, V_r$. By Problem 4, $I(V) = \cap_{i=1}^r I(V_i)$. By the Problems in Section 8.1, each $I(V_i)$ is a prime ideal. By (8.4.3), every radical ideal is $I(V)$ for some variety V, and the result follows.
6. By Section 8.1, Problem 6, and the inclusion-reversing property of I (part (4) of (8.1.3)), the decomposition is unique if we discard any prime ideal that properly contains another one. In other words, we retain only the minimal prime ideals.
7. If f is any irreducible factor of any of the f_i, then f does not divide g. Thus for some $j \neq i$, f does not divide f_j. By Problem 7 of Section 8.3, the simultaneous equations $f = f_j = 0$ have only finitely many solutions, and consequently X is a finite set.
8. With notation as in Problem 7, $f_i = gh_i$, where the gcd of the h_i is constant. Thus X is the union of the algebraic curve defined by $g = 0$ and the finite set defined by $h_1 = \cdots = h_m = 0$. (This analysis does not apply when X is defined by the zero polynomial, in which case $X = A^2$.)
9. If $k = \mathbb{R}$, the zero-set of $x^2 + y^{2n}$ is $\{(0,0)\}$ for all $n = 1, 2, \dots$. If k is algebraically closed, then as a consequence of the Nullstellensatz, $V(f) = V(g)$ with f and g irreducible implies that $f = cg$ for some constant c. (See Problem 3).
10. Let $k = \mathbb{F}_2$, and let I be the ideal of $k[X]$ generated by $f(X) = X^2 + X + 1$. Since f is irreducible, I is a maximal ideal (Section 3.1, Problem 8), in particular, I is proper. But $f(0)$ and $f(1)$ are nonzero, so $V(I)$ is empty, contradicting the weak Nullstellensatz.

Section 8.5

1. If $x \notin M$, then the ideal generated by M and x is R, by maximality of M. Thus there exists $y \in M$ and $z \in R$ such that $y + zx = 1$. By hypothesis, zx, hence x, is a unit. Take the contrapositive to conclude that every nonunit belongs to M.
2. Any additive subgroup of the cyclic additive group of $\mathbb{Z}_{p^n}$ must consist of multiples of some power of p, and it follows that every ideal is contained in (p), which must therefore be the unique maximal ideal.
3. No. A can be nilpotent, that is, some power of A can be 0. The set will be multiplicative if A is invertible.

4. $S^{-1}(gf)$ takes m/s to $g(f(m))/s$, as does $S^{-1}gS^{-1}f$. If f is the identity on M, then $S^{-1}f$ is the identity on $S^{-1}M$.

5. By hypothesis, $gf = 0$, so $S^{-1}gS^{-1}f = S^{-1}gf = S^{-1}0 = 0$. Thus $\operatorname{im} S^{-1}f \subseteq \ker S^{-1}g$. Conversely, let $x \in N, s \in S$, with $x/s \in \ker S^{-1}g$. Then $g(x)/s = 0/1$, so for some $t \in S$ we have $tg(x) = g(tx) = 0$. Therefore $tx \in \ker g = \operatorname{im} f$, so $tx = f(y)$ for some $y \in M$. We now have $x/s = f(y)/st = (S^{-1}f)(y/st) \in \operatorname{im} S^{-1}f$.

6. The set of nonuits is $M = \{f/g\colon g(a) \neq 0, f(a) = 0\}$, which is an ideal. By (8.5.9), R is a local ring with maximal ideal M.

7. The sequence $0 \to N \to M \to M/N \to 0$ is exact, so by Problem 5, $0 \to N_S \to M_S \to (M/N)_S \to 0$ is exact. (If f is one of the maps in the first sequence, the corresponding map in the second sequence is $S^{-1}f$.) It follows from the definition of localization of a module that $N_S \leq M_S$, and by exactness of the second sequence we have $(M/N)_S \cong M_S/N_S$, as desired.

Section 8.6

1. If x^m belongs to the intersection of the I_i, then x belongs to each $\sqrt{I_i}$, so $x \in \cap_{i=1}^n \sqrt{I_i}$. Conversely, if $x \in \cap_{i=1}^n \sqrt{I_i}$, let $x^{m_i} \in I_i$. If m is the maximum of the m_i, then $x^m \in \cap_{i=1}^n I_i$, so $x \in \sqrt{\cap_{i=1}^n I_i}$.

2. We are essentially setting $X = Z = 0$ in R, and this collapses R down to $k[Y]$. Formally, map $f + I$ to $g + I$, where g consists of those terms in f that do not involve X or Z. Then $R/P \cong k[Y]$, an integral domain. Therefore P is prime.

3. $(X + I)(Y + I) = Z^2 + I \in P^2$, but $X + I \notin P^2$ and $Y + I \notin \sqrt{P^2} = P$.

4. P_1 is prime because $R/P_1 \cong k[Y]$, an integral domain. P_2 is maximal by (8.3.1), so P_2^2 is P_2-primary by (8.6.6). The radical of Q is P_2, so by (8.6.5), Q is P_2-primary.

5. The first assertion is that

$$(X^2, XY) = (X) \cap (X, Y)^2 = (X) \cap (X^2, XY, Y^2)$$

and the second is

$$(X^2, XY) = (X) \cap (X^2, Y).$$

In each case, the left side is contained in the right side by definition of the ideals involved. The inclusion from right to left follows because if $f(X,Y)X = g(X,Y)Y^2$ (or $f(X,Y)X = g(X,Y)Y$), then $g(X,Y)$ must involve X and $f(X,Y)$ must involve Y. Thus $f(X,Y)X$ is a polynomial multiple of XY.

6. By (8.6.9), a proper ideal I can be expressed as the intersection of finitely many primary ideals Q_i. If Q_i is P_i-primary, then by Problem 1,

$$I = \sqrt{I} = \cap_i \sqrt{Q_i} = \cap_i P_i.$$

7. Since X^3 and Y^n belong to I_n, we have $X, Y \in \sqrt{I_n}$, so $(X,Y) \subseteq \sqrt{I_n}$. By (8.3.1), (X,Y) is a maximal ideal. Since $\sqrt{I_n}$ is proper (it does not contain 1), we have $(X,Y) = \sqrt{I_n}$. By (8.6.5), I_n is primary.

Section 8.7

1. $[(x, y+y')+G] - [(x,y)+G] - [(x,y')+G] = 0$ since $(x, y+y') - (x,y) - (x,y') \in G$; $r[(x,y)+G] - [(rx,y)+G] = 0$ since $r(x,y) - (rx,y) \in G$; the other cases are similar.
2. Let a and b be integers such that $am + bn = 1$. If $x \in \mathbb{Z}_m$ and $y \in \mathbb{Z}_n$, then $x \otimes y = 1(x \otimes y) = a(mx \otimes y) + b(x \otimes ny) = 0$ since $z \otimes 0 = 0 \otimes z = 0$.
3. Let A be a torsion abelian group, that is, each element of A has finite order. If $x \in A$ and $y \in \mathbb{Q}$, then $nx = 0$ for some positive integer n. Thus $x \otimes y = n(x \otimes (y/n)) = nx \otimes (y/n) = 0 \otimes (y/n) = 0$.
4. We have $h = g'h' = g'gh$ and $h' = gh = gg'h'$. But if $P = T$ and $f = h$, then $g = 1_T$ makes the diagram commute, as does $g'g$. By the uniqueness requirement in the universal mapping property, we must have $g'g = 1_T$, and similarly $gg' = 1_{T'}$. Thus T and T' are isomorphic.
5. $n \otimes x = n(1 \otimes x) = 1 \otimes nx = 1 \otimes 0 = 0$.
6. $nZ \otimes \mathbb{Z}_n \cong \mathbb{Z} \otimes \mathbb{Z}_n \cong \mathbb{Z}_n$ by (8.7.6), with $n \otimes x \to 1 \otimes x \to x$, and since $x \neq 0$, $n \otimes x$ cannot be 0.
7. We have a bilinear map $(f,g) \to f \otimes g$ from $\operatorname{Hom}_R(M, M') \times \operatorname{Hom}_R(N, N')$ to $\operatorname{Hom}_R(M \otimes_R N, M' \otimes_R N')$, and the result follows from the universal mapping property of tensor products.
8. In terms of matrices, we are to prove that $M_m(R) \otimes M_n(R) \cong M_{mn}(R)$. This follows because $M_t(R)$ is a free R-module of rank t^2.

Section 8.8

1. $y_1 \cdots (y_i + y_j) \cdots (y_i + y_j) \cdots y_p = y_1 \cdots y_i \cdots y_i \cdots y_p + y_1 \cdots y_i \cdots y_j \cdots y_p + y_1 \cdots y_j \cdots y_i \cdots y_p + y_1 \cdots y_j \cdots y_j \cdots y_p$. The left side, as well as the first and last terms on the right, are zero by definition of N. Thus $y_1 \cdots y_i \cdots y_j \cdots y_p = -y_1 \cdots y_j \cdots y_i \cdots y_p$, as asserted.
2. If π is any permutation of $\{a, \dots, b\}$, then by Problem 1,

$$x_{\pi(a)} \cdots x_{\pi(b)} = (\operatorname{sgn} \pi) x_a \cdots x_b.$$

   The left side will be $\pm x_a \cdots x_b$, regardless of the particular permutation π, and the result follows.
3. The multilinear map f induces a unique $h\colon M^{\otimes p} \to Q$ such that $h(y_1 \otimes \cdots \otimes y_p) = f(y_1, \dots, y_p)$. Since f is alternating, the kernel of h contains N, so the existence and uniqueness of the map g follows from the factor theorem.
4. By Problem 3, there is a unique R-homomorphism $g\colon \Lambda^n M \to R$ such that $g(y_1 \cdots y_n) = f(y_1, \dots, y_n)$. In particular, $g(x_1 \cdots x_n) = f(x_1, \dots, x_n) = 1 \neq 0$. Thus $x_1 \cdots x_n \neq 0$. If r is any nonzero element of R, then $g(rx_1 x_2 \cdots x_n) = f(rx_1, \dots, x_n) = r$, so $rx_1 \cdots x_n \neq 0$. By Problem 2, $\{x_1 \cdots x_n\}$ is a basis for $\Lambda^n M$.
5. Fix the set of indices I_0 and its complementary set J_0. If $\sum_I a_I x_I = 0$, $x_I \in R$, multiply both sides on the right by x_{J_0}. If $I \neq I_0$, then $x_I x_{J_0} = 0$ by definition of N.

Thus $a_{I_0}x_{I_0}x_{J_0} = \pm a_{I_0}x_1 \cdots x_n = 0$. By Problem 4, $a_{I_0} = 0$. Since I_0 is arbitrary, the result follows.

6. We have $R_0 \subseteq S$ by definition of S. Assume that $R_m \subseteq S$ for $m = 0, 1, \dots, n-1$, and let $a \in R_n$ $(n > 0)$. Then $a \in I$, so $a = \sum_{i=1}^r c_i x_i$ where (since $x_i \in R_{n_i}$ and R is the direct sum of the R_m) $c_i \in R_{n-n_i}$. By induction hypothesis, $c_i \in S$, and since $x_i \in S$ by definition of S, we have $a \in S$, completing the induction.
7. The "if" part follows from Section 8.2, Problem 5, so assume R Noetherian. Since $R_0 \cong R/I$, it follows that R_0 is Noetherian. Since R is Noetherian, I is finitely generated, so by Problem 6, $R = S$, a finitely generated R_0-algebra.

Section 9.1

1. Assume R is simple, and let $x \in R, x \neq 0$. Then Rx coincides with R, so $1 \in Rx$. Thus there is an element $y \in R$ such that $yx = 1$. Similarly, there is an element $z \in R$ such that $zy = 1$. Therefore
$$z = z1 = zyx = 1x = x, \text{ so } xy = zy = 1$$
and y is a two-sided inverse of x. Conversely, assume that R is a division ring, and let x be a nonzero element of the left ideal I. If y is the inverse of x, then $1 = yx \in I$, so $I = R$ and R is simple.
2. I is proper because $f(1) = x \neq 0$, and $R/I \cong Rx$ by the first isomorphism theorem.
3. The "if" part follows from the correspondence theorem, so assume that M is simple. If x is a nonzero element of M, then $M = Rx$ by simplicity. If $I = \ker f$ as in Problem 2, then $M \cong R/I$, and I is maximal by the correspondence theorem.
4. The "only if" part was done in Problem 3, so assume that M is not simple. Let N be a submodule of M with $0 < N < M$. If x is a nonzero element of N, then $Rx \leq N < M$, so x cannot generate M.
5. By Problem 3, a simple $\mathbb{Z}$-module is isomorphic to $\mathbb{Z}/I$, where I is a maximal ideal of $\mathbb{Z}$. By Section 2.4, Problems 1 and 2, $I = (p)$ where p is prime.
6. As in Problem 5, a simple $F[X]$-module is isomorphic to $F[X]/(f)$, where f is an irreducible polynomial in $F[X]$. (See Section 3.1, Problem 8.)
7. If x is a nonzero element of V and y an arbitrary element of V, there is an endomorphism f such that $f(x) = y$. Therefore $V = (\operatorname{End}_k V)x$. By Problem 4, V is a simple $\operatorname{End}_k(V)$-module.
8. By (4.7.4), every such short exact sequence splits iff for any submodule $N \leq M$, $M \cong N \oplus P$, where the map $N \to M$ can be identified with inclusion and the map $M \to P$ can be identified with projection. In other words, every submodule of M is a direct summand. Equivalently, by (9.1.2), M is semisimple.

Section 9.2

1. Unfortunately, multiplication by r is not necessarily an R-endomorphism of M, since $r(sx) = (rs)x$, which need not equal $s(rx) = (sr)x$.

2. Let x be a generator of M, and define $f\colon R \to M$ by $f(r) = rx$. By the first isomorphism theorem, $M \cong R/\operatorname{ann} M$. The result follows from the correspondence theorem.
3. Let $M = \mathbb{Z}_p \oplus \mathbb{Z}_p$ where p is prime. Then M is not a simple $\mathbb{Z}$-module, but $\operatorname{ann} M = p\mathbb{Z}$ is a maximal ideal of $\mathbb{Z}$.
4. The computation given in the statement of the problem shows that (1,0) is a generator of V, hence V is cyclic. But $N = \{(0, b)\colon b \in F\}$ is a nontrivial proper submodule of V. (Note that $T(0, b) = (0, 0) \in N$.) Therefore V is not simple.
5. Since F is algebraically closed, f has an eigenvalue $\lambda \in F$. Thus the kernel of $f - \lambda I$ is not zero, so it must be all of M. Therefore $f(m) = \lambda m$ for all $m \in M$.
6. If $r \in I$ and $s+I \in R/I$, then $r(s+I) = rs+I$. But if I is not a right ideal, we cannot guarantee that rs belongs to I.

Section 9.3

1. For each $j = 1, \dots, n$, there is a finite subset $I(j)$ of I such that x_j belongs to the direct sum of the $M_i, i \in I(j)$. If J is the union of the $I(j)$, then $M \subseteq \bigoplus_{i\in J} M_i \subseteq M$, so M is the direct sum of the $M_i, i \in J$.
2. Each simple module $M_i, i = 1, \dots, n$, is cyclic (Section 9.1, Problem 4), and therefore can be generated by a single element x_i. Thus M is generated by $x_1, \dots, x_n$.
3. A left ideal is simple iff it is minimal, so the result follows from (9.1.2).
4. No. If it were, then by Section 9.1, Problem 5, $\mathbb{Z}$ would be a direct sum of cyclic groups of prime order. Thus each element of $\mathbb{Z}$ would have finite order, a contradiction.
5. By (4.6.4), every finite abelian group is the direct sum of various $\mathbb{Z}_p$, p prime. If p and q are distinct primes, then $\mathbb{Z}_p \oplus \mathbb{Z}_q \cong \mathbb{Z}_{pq}$ by the Chinese remainder theorem. Thus $\mathbb{Z}_n$ can be assembled from cyclic groups of prime order as long as no prime appears more than once in the factorization of n. (If $\mathbb{Z}_p \oplus \mathbb{Z}_p$ is part of the decomposition, the group cannot be cyclic.) Consequently, Z_n is semisimple if and only if n is square-free.
6. This follows from Section 9.1, Problem 8. (In the first case, B is semisimple by hypothesis, and in the second case A is semisimple. The degenerate case $M = 0$ can be handled directly.)
7. Conditions (a) and (b) are equivalent by (9.3.2) and the definition of semisimple ring. By Problem 6, (b) implies both (c) and (d). To show that (c) implies (b) and (d) implies (b), let M be a nonzero R-module, with N a submodule of M. By hypothesis, the sequence $0 \to N \to M \to M/N \to 0$ splits. (By hypothesis, M/N is projective in the first case and N is injective in the second case.) By Section 9.1, Problem 8, M is semisimple.

Section 9.4

1. If there is an infinite descending sequence $I_1 \supset I_2 \supset \cdots$ of left ideals, we can proceed exactly as in (9.4.7) to reach a contradiction.

2. Let I_1 be any nonzero left ideal. If I_1 is simple, we are finished. If not, there is a nonzero left ideal I_2 such that $I_1 \supset I_2$. If we continue inductively, the Artinian hypothesis implies that the process must terminate in a simple left ideal I_t.
3. By Problem 2, the ring R has a simple R-module M. The hypothesis that R has no nontrivial two-sided ideals implies that we can proceed exactly as in (9.4.6) to show that M is faithful.
4. If V is infinite-dimensional over D, then exactly as in (9.4.7), we find an infinite descending chain of left ideals, contradicting the assumption that R is Artinian.
5. By Problem 4, V is a finite-dimensional vector space over D, so we can reproduce the discussion preceding (9.4.7) to show that $R \cong \operatorname{End}_D(V) \cong M_n(D^o)$.
6. The following is a composition series:

$$0 < M_1 < M_1 \oplus M_2 < \cdots < M_1 \oplus M_2 \oplus \cdots \oplus M_n = M.$$

7. By (9.1.2), M is a direct sum of simple modules. If the direct sum is infinite, then we can proceed as in Problem 6 to construct an infinite ascending (or descending) chain of submodules of M, contradicting the hypothesis that M is Artinian and Noetherian.

Section 9.5

1. If $g \in \ker \rho$, then $gv = v$ for every $v \in V$. Take $v = 1_G$ to get $g1_G = 1_G$, so $g = 1_G$ and ρ is injective.
2. $(gh)(v(i)) = v(g(h(i)))$ and $g(h(v(i))) = g(v(h(i))) = v(g(h(i)))$. Also, $1_G(v(i)) = v(1_G(i)) = v(i)$.
3. We have $g(v(1)) = v(4)$, $g(v(2)) = v(2)$, $g(v(3)) = v(1)$, $g(v(4)) = v(3)$, so

$$[g] = \begin{bmatrix} 0 & 0 & 1 & 0 \\ 0 & 1 & 0 & 0 \\ 0 & 0 & 0 & 1 \\ 1 & 0 & 0 & 0 \end{bmatrix}.$$

4. We have $gv_1 = v_2$, $gv_2 = v_3 = -v_1 - v_2$; $hv_1 = v_1$, $hv_2 = v_3 = -v_1 - v_2$. Thus

$$[g] = \begin{bmatrix} 0 & -1 \\ 1 & -1 \end{bmatrix}, \quad [h] = \begin{bmatrix} 1 & -1 \\ 0 & -1 \end{bmatrix}.$$

5. We have $v_1 = e_1$ and $v_2 = -\frac{1}{2}e_1 + \frac{1}{2}\sqrt{3}e_2$. Thus

$$P^{-1} = \begin{bmatrix} 1 & -\frac{1}{2} \\ 0 & \frac{1}{2}\sqrt{3} \end{bmatrix}, \quad P = \begin{bmatrix} 1 & \frac{1}{3}\sqrt{3} \\ 0 & \frac{2}{3}\sqrt{3} \end{bmatrix}$$

and

$$[g]' = P^{-1}[g]P = \begin{bmatrix} -\frac{1}{2} & -\frac{1}{2}\sqrt{3} \\ \frac{1}{2}\sqrt{3} & -\frac{1}{2} \end{bmatrix},$$

$$[h]' = P^{-1}[h]P = \begin{bmatrix} 1 & 0 \\ 0 & -1 \end{bmatrix}.$$

6. Check by direct computation that the matrices A and B satisfy the defining relations of D_8: $A^4 = I$, $B^2 = I$, $AB = BA^{-1}$. (See Section 5.8.)
7. Yes. Again, check by direct computation that the matrices A^iB^j, $i = 0, 1, 2, 3$, $j = 0, 1$, are distinct. Thus if $g \in D_8$ and $\rho(g) = I$, then g is the identity element of D_8.

Section 9.6

1. Let W be the one-dimensional subspace spanned by $v_1+v_2+v_3$. Since any permutation in S_3 permutes the v_i, $v \in W$ implies $gv \in W$.
2. Multiplying $[a^r]$ by $\begin{bmatrix} 1 \\ 0 \end{bmatrix}$ and $\begin{bmatrix} 0 \\ 1 \end{bmatrix}$, we have $a^rv_1 = v_1$ and $a^rv_2 = rv_1 + v_2$. Since W is spanned by v_1, it is closed under the action of G and is therefore a kG-submodule.
3. If

$$[a^r]\begin{bmatrix} x \\ y \end{bmatrix} = c\begin{bmatrix} x \\ y \end{bmatrix}$$

then $x + ry = cx$ and $y = cy$. If $y \neq 0$ then $c = 1$, so ry, hence y, must be 0. Thus $c = 1$ and x is arbitrary, so that any one-dimensional kG-submodule must coincide with W.

4. If $V = W \oplus U$, where U is a kG-submodule, then U must be one-dimensional. By Problem 3, $W = U$, and since $W \neq 0$, this is impossible.
5. If M is semisimple and either Noetherian or Artinian, then M is the direct sum of finitely many simple modules. These simple modules are the factors of a composition series, and the result follows from (7.5.12).
6. Let e be the natural projection on M_1. Then e is an idempotent, $e \neq 0$ since $M_1 \neq 0$, and $e \neq 1$ since $e = 0$ on M_2 and $M_2 \neq 0$.
7. Let e be a nontrivial idempotent, and define $e_1 = e$, $e_2 = 1-e$. By direct computation, e_1 and e_2 are nontrivial idempotents that are orthogonal. Take $M_1 = e_1(M)$, $M_2 = e_2(M)$. Then M_1 and M_2 are nonzero submodules with $M = M_1 + M_2$. To show that the sum is direct, let $z = e_1x = e_2y \in M_1 \cap M_2$, with $x, y \in M$. Then $e_1z = e_1e_2y = 0$, and similarly $e_2z = 0$. Thus $z = 1z = e_1z + e_2z = 0$.

Section 9.7

1. If $M = Rx$, define $f\colon R \to M$ by $f(r) = rx$. By the first isomorphism theorem, $M \cong R/\ker f$. Moreover, $\ker f = \operatorname{ann}(M)$. Conversely, R/I is cyclic since it is generated by $1 + I$.
2. If N is a maximal submodule of M, then N is the kernel of the canonical map of M onto the simple module M/N. Conversely, if $f\colon M \to S$, S simple, then $f(M)$ is 0

or S, so $f = 0$ or $S \cong M/\ker f$. Thus $\ker f$ is either M or a maximal submodule of M. The intersection of all kernels therefore coincides with the intersection of all maximal submodules. [If there are no maximal submodules, then the intersection of all kernels is M.]

3. By the correspondence theorem, the intersection of all maximal left ideals of R/I is 0. This follows because the intersection of all maximal left ideals of R containing I is the intersection of all maximal left ideals of R. [Note that $J(R)$ is contained in every maximal left ideal, and $I = J(R)$.]

4. Let g be an R-module homomorphism from N to the simple R-module S. Then $gf\colon M \to S$, so by Problem 2, $J(M) \subseteq \ker(gf)$. But then $f(J(M)) \subseteq \ker g$. Take the intersection over all g to get $f(J(M)) \subseteq J(N)$.

5. Suppose $a \in J(R)$. If $1+ab$ is a nonunit, then it belongs to some maximal ideal M. But then a belongs to M as well, and therefore so does ab. Thus $1 \in M$, a contradiction. Now assume $a \notin J(R)$, so that for some maximal ideal M, $a \notin M$. By maximality, $M + Ra = R$, so $1 = x + ra$ for some $x \in M$ and $r \in R$. Since x belongs to M, it cannot be a unit, so if we set $b = -r$, it follows that $1 + ab$ is a nonunit.

6. By the correspondence theorem, there is a bijection, given by $\psi(A) = A/N$, between maximal submodules of M containing N and maximal submodules of M/N. Since $N \le J(M)$ by hypothesis, a maximal submodule of M containing N is the same thing as a maximal submodule of M. Thus $J(M)$ corresponds to $J(M/N)$, that is, $\psi(J(M)) = J(M/N)$. Since $\psi(J(M)) = J(M)/N$, the result follows.

Section 9.8

1. $a^t \in (a^{t+1})$, so there exists $b \in R$ such that $a^t = ba^{t+1}$. Since R is an integral domain we have $1 = ba$.

2. Let a be a nonzero element of the Artinian integral domain R. The sequence $(a) \supseteq (a^2) \supseteq \dots$ stabilizes, so for some t we have $(a^t) = (a^{t+1})$. By Problem 1, a has an inverse, proving that R is a field.

3. If P is a prime ideal of R, then R/P is an Artinian integral domain by (7.5.7) and (2.4.5). By Problem 2, R/P is a field, so by (2.4.3), P is maximal.

4. We have $P \cap I \in \mathcal{S}$ and $P \cap I \subseteq I$, so by minimality of I, $P \cap I = I$. Thus $P \supseteq I = \cap_{j=1}^n I_j$, so by (7.6.2), $P \supseteq I_j$ for some j. But P and I_j are maximal ideals, hence $P = I_j$.

5. If $z \in F$, with $z = x + y$, $x \in M$, $y \in N$, define $h\colon F \to M/JM \oplus N/JN$ by $h(z) = (x + JM) + (y + JN)$. Then h is an epimorphism with kernel $JM + JN = JF$, and the result follows from the first isomorphism theorem.

6. By (9.8.5), M/JM is an n-dimensional vector space over the residue field k. Since F, hence F/JF, is generated by n elements, F/JF has dimension at most n over k. Thus N/JN must have dimension zero, and the result follows.

7. Multiplication of an element of I on the right by a polynomial $f(X, Y)$ amounts to multiplication by the constant term of f. Thus I is finitely generated as a right R-module iff it is finitely generated as an abelian group. This is a contradiction, because

as an abelian group,

$$I = \oplus_{n=0}^{\infty} \mathbb{Z}X^n Y.$$

8. By the Hilbert basis theorem, $\mathbb{Z}[X]$ is a Noetherian ring and therefore a Noetherian left $\mathbb{Z}[X]$-module. The isomorphic copy $\mathbb{Z}[X]Y$ is also a Noetherian left $\mathbb{Z}[X]$-module, hence so is R, by (7.5.8). A left ideal of R that is finitely generated as a $\mathbb{Z}[X]$-module is finitely generated as an R-module, so R is left-Noetherian.
9. The set $\{\overline{x}_1, \dots, \overline{x}_n\}$ spans V, and therefore contains a basis. If the containment is proper, then by (9.8.5) part (ii), $\{x_1, \dots, x_n\}$ cannot be a minimal generating set, a contradiction.
10. In vector-matrix notation we have $y = Ax$ and therefore $\overline{y} = \overline{A}\overline{x}$, where $\overline{a}_{ij} = a_{ij} + J$. By Problem 9, $\overline{x}$ and $\overline{y}$ are bases, so that $\det \overline{A} \neq 0$. But under the canonical map of R onto k, $\det A$ maps to $\det \overline{A}$, and therefore $\det A$ cannot belong to the kernel of the map, namely J. But by (8.5.9), J is the ideal of nonunits of R, so $\det A$ is a unit.

Section 10.1

1. This is exactly the same as the proof for groups in (1.1.1).
2. Any ring homomorphism on $\mathbb{Q}$ is determined by its values on $\mathbb{Z}$ (write $m = (m/n)n$ and apply the homomorphism g to get $g(m/n) = g(m)/g(n)$. Thus if $gi = hi$, then g coincides with h on $\mathbb{Z}$, so $g = h$, proving that i is epic.
3. Let AZB be a shorthand notation for the composition of the unique morphism from A to the zero object Z, followed by the unique morphism from Z to B. If Z' is another zero object, then $AZB = (AZ'Z)B = A(Z'ZB) = A(Z'B) = AZ'B$, as claimed.
4. If $is = it$, then $fis = fit = 0$, so $is (= it) = ih$ where h is unique. Thus s and t must coincide with h, hence i is monic.
5. A kernel of a monic $f\colon A \to B$ is 0, realized as the zero map from a zero object Z to A. For $f0 = 0$, and if $fg = 0$, then $fg = f0$; since f is monic, $g = 0$. But then g can be factored through 0_{ZA}. Similarly, a cokernel of the epic $f\colon A \to B$ is the zero map from B to a zero object.
6. If $j = ih$ and $i = jh'$, then $i = ihh'$, so by uniqueness in part (2) of the definition of kernel (applied when $g = i$), hh', and similarly $h'h$, must be the identity.
7. Define $h\colon K \to A$ by $h(x) = 1$ for all x. Since K is nontrivial, h cannot be injective, so that $g \neq h$. But $fg = fh$, since both maps take everything in K to the identity of B.
8. The kernel of a ring homomorphism is an ideal, but not a subring (since it does not contains the multiplicative identity).
9. Let $f\colon A \to B$ be a noninjective ring homomorphism. Let C be the set of pairs (x, y) in the direct product $A \times A$ such that $f(x) = f(y)$. Since f is not injective, there is an element (x, y) of C with $x \neq y$. Thus if $D = \{(x, x)\colon x \in A\}$, then $D \subset C$. If g is the projection of $A \times A$ on the first coordinate, and h is the projection on the second coordinate, then $f(g(x, y)) = f(x)$ and $f(h(x, y)) = f(y)$, so $fg = fh$ on the ring C. But g and h disagree on the nonempty set $C \setminus D$, so f is not monic.

10. Let g be the canonical map of N onto $N/f(M)$, and let $h\colon N \to N/f(M)$ be identically zero. Since gf sends everything to 0, we have $gf = hf$ with $g \neq h$. Thus f is not epic.

Section 10.2

1. Suppose that y is the product of the x_i. By definition of product, if $f_i\colon x \to x_i$ for all i, there is a unique $f\colon x \to y$ such that $p_i f = f_i$. Since $p_i\colon y \to x_i$, we have $y \leq x_i$. Moreover, if $x \leq x_i$ for all i, then $x \leq y$. Therefore y is a greatest lower bound of the x_i.
2. No. For example, consider the usual ordering on the integers.
3. By duality, a coproduct of the x_i, if it exists, is a least upper bound.
4. If x has order r and y has order s, then $rs(x+y) = s(rx) + r(sy) = 0$. Thus the sum of two elements of finite order also has finite order, and the result follows.
5. The key point is that if f is a homomorphism of a torsion abelian group S, then $f(S)$ is a also torsion [since $nf(x) = f(nx)$]. Thus in diagram (1) with $A = \prod A_i$, we have $f(S) \subseteq T(A)$. Since $\prod A_i$ is the product in the category of abelian groups, it follows that $T(A)$ satisfies the universal mapping property and is therefore the product in the category of torsion abelian groups.
6. Given homomorphisms $f_j\colon G_j \to H$, we must lift the f_j to a homomorphism from the free product to H. This is done via $f(a_1 \cdots a_n) = f_1(a_1) \cdots f_n(a_n)$. If i_j is the inclusion map from G_j to $*_i G_i$, then $f(i_j(a_j)) = f(a_j) = f_j(a_j)$, as required.
7. We have $p_i f = f_i$, where $f_i\colon G \to C_i$. The f_i can be chosen to be surjective (e.g., take G to be the direct product of the C_i), and it follows that the p_i are surjective.
8. Since $f\colon C_1 \to C$, we have $f(a_1) = na$ for some positive integer n. Thus

$$a_1 = f_1(a_1) = p_1 f(a_1) = p_1(na) = np_1(a) = na_1;$$
$$0 = f_2(a_1) = p_2 f(a_1) = p_2(na) = np_2(a) = na_2.$$

9. By Problem 8, the order of C_1 divides $n-1$, and the order of C_2 divides n. There are many choices of C_1 and C_2 for which this is impossible. For example, let C_1 and C_2 be nontrivial p-groups for a fixed prime p.

Section 10.3

1. If $f\colon x \to y$, then $Ff\colon Fx \to Fy$. By definition of the category of preordered sets, this statement is equivalent to $x \leq y \implies Fx \leq Fy$. Thus functors are order-preserving maps.
2. F must take the morphism associated with xy to the composition of the morphism associated with Fx and the morphism associated with Fy. In other words, $F(xy) = F(x)F(y)$, that is, F is a homomorphism.
3. If $\beta \in X^*$, then $(gf)^*(\beta) = \beta gf$ and $f^* g^*(\beta) = f^*(\beta g) = \beta gf$.

4. To verify the functorial property, note that

$$(gf)^{**}(v^{**}) = v^{**}(gf)^* = v^{**}f^*g^* \text{ (by Problem 3)}$$

and

$$g^{**}f^{**}v^{**} = g^{**}(v^{**}f^*) = v^{**}f^*g^*.$$

Thus $(gf)^{**} = g^{**}f^{**}$. If f is the identity, then so is f^*, and consequently so is f^{**}.

5. $f^{**}t_V(v) = f^{**}(\overline{v}) = \overline{v}f^*$, and if $\beta \in W^*$, then $(\overline{v}f^*)(\beta) = \overline{v}(f^*\beta) = (f^*\beta)(v) = \beta f(v)$. But $t_W f(v) = \overline{f(v)}$ where $\overline{f(v)}(\beta) = \beta f(v)$.

6. Groups form a subcategory because every group is a monoid and every group homomorphism is, in particular, a monoid homomorphism. The subcategory is full because every monoid homomorphism from one group to another is also a group homomorphism.

7. (a) If two group homomorphisms are the same as set mappings, they are identical as homomorphisms as well. Thus the forgetful functor is faithful. But not every map of sets is a homomorphism, so the forgetful functor is not full.

   (b) Since (f, g) is mapped to f for arbitrary g, the projection functor is full but not faithful (except in some degenerate cases).

Section 10.4

1. If a homomorphism from $\mathbb{Z}_2$ to $\mathbb{Q}$ takes 1 to x, then $0 = 1 + 1 \to x + x = 2x$. But 0 must be mapped to 0, so $x = 0$.

2. A nonzero homomorphism can be constructed with $0 \to 0$, $1 \to \frac{1}{2}$. Then $1 + 1 \to \frac{1}{2} + \frac{1}{2} = 1 = 0$ in $\mathbb{Q}/\mathbb{Z}$.

3. Since a trivial group cannot be mapped onto a nontrivial group, there is no way that Fg can be surjective.

4. Let f be a homomorphism from $\mathbb{Q}$ to $\mathbb{Z}$. If r is any rational number and m is a positive integer, then

$$f(r) = f\left(\frac{r}{m} + \cdots + \frac{r}{m}\right) = mf\left(\frac{r}{m}\right)$$

so

$$f\left(\frac{r}{m}\right) = \frac{f(r)}{m}.$$

But if $f(r) \neq 0$, we can choose m such that $f(r)/m$ is not an integer, a contradiction. Therefore $f = 0$.

5. By Problem 4, $\text{Hom}(\mathbb{Q}, \mathbb{Z}) = 0$. But $\text{Hom}(\mathbb{Z}, \mathbb{Z}) \neq 0$, so as in Problem 3, Gf cannot be surjective.

6. We have $\mathbb{Z}_2 \otimes \mathbb{Z} \cong \mathbb{Z}_2$ and $\mathbb{Z}_2 \otimes \mathbb{Q} = 0$

$$\left[1 \otimes \frac{m}{n} = 1 \otimes \frac{2m}{2n} = 2 \otimes \frac{m}{2n} = 0 \otimes \frac{m}{2n} = 0.\right]$$

   Thus the map Hf cannot be injective.

7. Since $f_* = Ff$ is injective, $f\alpha = 0$ implies $\alpha = 0$, so f is monic and hence injective. Since $g_* f_* = 0$, we have $gf\alpha = 0$ for all $\alpha \in \text{Hom}(M, A)$. Take $M = A$ and $\alpha = 1_A$ to conclude that $gf = 0$, so that $\text{im } f \subseteq \ker g$. Finally, take $M = \ker g$ and $\alpha\colon M \to B$ the inclusion map. Then $g_*\alpha = g\alpha = 0$, so $\alpha \in \ker g_* = \text{im } f_*$. Thus $\alpha = f\beta$ for some $\beta \in \text{Hom}(M, A)$. Thus $\ker g = M = \text{im } \alpha \subseteq \text{im } f$.

8. If (3) is exact for all possible R-modules N, then (1) is exact. This is dual to the result of Problem 7, and the proof amounts to interchanging injective and surjective, monic and epic, inclusion map and canonical map, kernel and cokernel.

Section 10.5

1. If $x \in P$, then $f(x)$ can be expressed as $\sum_i t_i e_i$ (a finite sum), and we define $f_i(x) = t_i$ and $x_i = \pi(e_i)$. Then

$$x = \pi(f(x)) = \pi(\sum_i t_i e_i) = \sum_i t_i \pi(e_i) = \sum_i f_i(x) x_i.$$

2. $\pi(f(x)) = \sum_i f_i(x)\pi(e_i) = \sum_i f_i(x) x_i = x$.

3. By Problem 2, the exact sequence $0 \to \ker \pi \to F \to P \to 0$ (with $\pi\colon F \to P$) splits, and therefore P is a direct summand of the free module F and hence projective.

4. Since R^n is free, the "if" part follows from (10.5.3), part (4). If P is projective, then by the proof of (3) implies (4) in (10.5.3), P is a direct summand of a free module of rank n. [The free module can be taken to have a basis whose size is the same as that of a set of generators for P.]

5. This follows from Problem 1 with $F = R^n$.

6. If P is projective and isomorphic to M/N, we have an exact sequence $0 \to N \to M \to P \to 0$. Since P is projective, the sequence splits by (10.5.3), part (3), so P is isomorphic to a direct summand of M. Conversely, assume that P is a direct summand of every module of which it is a quotient. Since P is a quotient of a free module F, it follows that P is a direct summand of F. By (10.5.3) part (4), P is projective.

7. In the diagram above (10.5.1), take $M = R_1 \oplus R_2$, with $R_1 = R_2 = R$. Take $P = N = R_1$, $f = 1_{R_1}$, and let g be the natural projection of M on N. Then we can take $h(r) = r + s$, $r \in R_1, s \in R_2$, where s is either 0 or r. By replacing M by an arbitrary direct sum of copies of R, we can produce two choices for the component of $h(r)$ in each R_i, $i = 2, 3, \dots$ (with the restriction that only finitely many components are nonzero). Thus there will be infinitely many possible choices for h altogether.

Section 10.6

1. $f'g(a) = (g(a),0) + W$ and $g'f(a) = (0, f(a)) + W$, with $(g(a),0) - (0, f(a)) = (g(a), -f(a)) \in W$. Thus $f'g = g'f$.
2. If $(b,c) \in W$, then $b = g(a), c = -f(a)$ for some $a \in A$. Therefore

$$g''(c) + f''(b) = -g''f(a) + f''g(a) = 0$$

   and h is well-defined.
3. $hg'(c) = h((0,c) + W) = g''(c) + 0 = g''(c)$ and $hf'(b) = h((b,0) + W) = 0 + f''(b) = f''(b)$.
4. $h'((b,c) + W) = h'((0,c) + W + (b,0) + W) = h'(g'(c) + f'(b)) = h'g'(c) + h'f'(b) = g''(c) + f''(b) = h((b,c) + W)$.
5. If $f'(b) = 0$, then by definition of f', $(b,0) \in W$, so for some $a \in A$ we have $b = g(a)$ and $f(a) = 0$. Since f is injective, $a = 0$, hence $b = 0$ and f' is injective.
6. If $b \in B, c \in C$, then surjectivity of f gives $c = f(a)$ for some $a \in A$. Thus $f'(b + g(a)) = (b + g(a), 0) + W = (b + g(a), 0) + W + (-g(a), f(a)) + W$ [note that $(-g(a), f(a)) \in W$] $= (b, f(a)) + W = (b,c) + W$, proving that f' is surjective.
7. If $(a,c) \in D$, then $f(a) = g(c)$ and $fg'(a,c) = f(a)$, $gf'(a,c) = g(c)$. Thus $fg' = gf'$.
8. If $x \in E$, then $fg''(x) = gf''(x)$, so $(g''(x), f''(x)) \in D$. Take $h(x) = (g''(x), f''(x))$, which is the only possible choice that satisfies $g'h = g''$ and $f'h = f''$.
9. If $(a,c) \in D$ and $f'(a,c) = 0$, then $c = 0$, so $f(a) = g(c) = 0$. Since f is injective, $a = 0$. Consequently, $(a,c) = 0$ and f' is injective.
10. If $c \in C$, then there exists $a \in A$ such that $f(a) = g(c)$. Thus $(a,c) \in D$ and $f'(a,c) = c$, proving that f' is surjective.
11. If $x, y \in I$, then $f(xy) = xf(y)$ and $f(xy) = f(yx) = yf(x)$. Thus $xf(y) = yf(x)$, and if x and y are nonzero, the result follows upon division by xy.
12. We must extend $f\colon I \to Q$ to $h\colon R \to Q$. Let z be the common value $f(x)/x$, $x \in I$, $x \neq 0$. Define $h(r) = rz$, $r \in R$. Then h is an R-homomorphism, and if $x \in I$, $x \neq 0$, then $h(x) = xz = xf(x)/x = f(x)$. Since $h(0) = f(0) = 0$, h is an extension of f and the result follows from (10.6.4).

Section 10.7

1. The only nonroutine verification is the check that $sf \in \mathrm{Hom}_R(M,N)$:

$$(sf)(rm) = f(rms) = rf(ms) = r[(sf)(m)].$$

2. $(fr)(ms) = f(r(ms)) = f((rm)s) = f(rm)s = [(fr)(m)]s$.
3. $(sf)(mr) = s(f(mr)) = s(f(m)r) = (sf(m)r = [(sf)(m)]r$.
4. $(fr)(sm) = (f(sm))r = (sf(m))r = s(f(m)r) = s[(fr)(m)]$.

5. In Problem 2, M and N are right S-modules, so we write f on the left: $(fr)m = f(rm)$. In Problem 3, M and N are right R-modules, and we write f on the left: $(sf)m = s(fm)$. In Problem 4, M and N are left S-modules, and we write f on the right: $m(fr) = (mf)r$.
6. Let $y \in M$, $r \in R$ with $r \neq 0$. By hypothesis, $x = \frac{1}{r}y \in M$, so we have $x \in M$ such that $y = rx$, proving M divisible.
7. If $y \in M$, $r \in R$ with $r \neq 0$, we must define $\frac{1}{r}y$. Since M is divisible, there exists $x \in M$ such that $y = rx$, and we take $\frac{1}{r}y = x$. If $x' \in M$ and $y = rx'$, then $r(x - x') = 0$, and since M is torsion-free, $x = x'$. Thus x is unique and scalar multiplication is well-defined.
8. Let f be a nonzero R-homomorphism from Q to R. Then $f(u) = 1$ for some $u \in Q$. [If $f(x) = r \neq 0$, then $rf(x/r) = f(rx/r) = f(x) = r$, so we can take $u = x/r$.] Now if s is a nonzero element of R, then $sf(u/s) = f(su/s) = f(u) = 1$, so $f(u/s)$ is an inverse of s. Consequently, R is a field, contradicting the hypothesis.

Section 10.8

1. Let $A = R[X]$ where R is any commutative ring. As an R-algebra, A is generated by X, but A is not finitely generated as an R-module since it contains polynomials of arbitrarily high degree.
2. The bilinear map determined by $(X^i, Y^j) \to X^iY^j$ induces an R-homomorphism of $R[X] \otimes_R R[Y]$ onto $R[X, Y]$, with inverse determined by $X^iY^j \to X^i \otimes Y^j$.
3. Abbreviate $X_1, \dots, X_n$ by X and $Y_1, \dots, Y_m$ by Y. Let A be a homomorphic image of $R[X]$ under f, and B a homomorphic image of $R[Y]$ under g. Then $A \otimes_R B$ is a homomorphic image of $R[X] \otimes R[Y] (\cong R[X, Y]$ by Problem 2) under $f \otimes g$.
4. If $f\colon A \to B$ is an injective R-module homomorphism, then by hypothesis, $(1 \otimes f)\colon S \otimes_R A \to S \otimes_R B$ is injective. Also by hypothesis,

$$(1 \otimes (1 \otimes f))\colon M \otimes_S S \otimes_R A \to M \otimes_S S \otimes_R B$$

is injective. Since $M \otimes_S S \cong M$, the result follows.
5. Let $f\colon A \to B$ be injective. Since $A \otimes_S S \cong A$ and $B \otimes_S S \cong B$, it follows from the hypothesis that $(f \otimes 1)\colon A \otimes_S (S \otimes_R M) \to B \otimes_S (S \otimes_R M)$ is injective. Thus $S \otimes_R M$ is a flat S-module.
6. α is derived from the bilinear map $S^{-1}R \times M \to S^{-1}M$ given by $(r/s, x) \to rx/s$. We must also show that β is well-defined. If $x/s = y/t$, then there exists $u \in S$ such that $utx = usy$. Thus

$$\frac{1}{s} \otimes x = \frac{ut}{sut} \otimes x = \frac{1}{sut} \otimes utx = \frac{1}{sut} \otimes usy = \frac{1}{t} \otimes y$$

as required. By construction, α and β are inverses of each other and yield the desired isomorphism.
7. We must show that $S^{-1}R \otimes_R -$ is an exact functor. But in view of Problem 6, an equivalent statement is the localization functor S^{-1} is exact, and this has already been proved in Section 8.5, Problem 5.

Section 10.9

1. The proof of (10.9.4) uses the fact that we are working in the category of modules. To simply say "duality" and reverse all the arrows, we would need an argument that did not depend on the particular category.
2. Let N be the direct limit of the N_i. The direct system $\{N_i, h(i,j)\}$ induces a direct system $\{M \otimes N_i, 1 \otimes h(i,j)\}$. Compatibility in the new system reduces to compatibility in the old system; tensoring with 1 is harmless. Since compatible maps $f_i\colon N_i \to B$ can be lifted to $f\colon N \to B$, it follows that compatible maps $g_i\colon M \otimes N_i \to B$ can be lifted to $g\colon M \otimes N \to B$. Thus $M \otimes N$ satisfies the universal mapping property for $\{M \otimes N_i\}$.
3. The direct limit is $A = \cup_{n=1}^{\infty} A_n$, with $\alpha_n\colon A_n \to A$ the inclusion map
5. Each R-homomorphism f from the direct sum of the A_i to B induces an R-homomorphism $f_i\colon A_i \to B$. [f_i is the injection of A_i into the direct sum, followed by f]. Take $\alpha(f) = (f_i, i \in I) \in \prod_i \operatorname{Hom}_R(A_i, B)$. Conversely, given such a family $(f_i, i \in I)$, the f_i can be lifted uniquely to $\beta(f_i, i \in I) = f$. Since α and β are inverse R-homomorphisms, the result follows.
6. If $f\colon A \to \prod_i B_i$, define $\alpha(f) = (p_i f, i \in I) \in \prod_i \operatorname{Hom}_R(A, B_i)$, where p_i is the projection of the direct product onto the ith factor. Conversely, given $(g_i, i \in I)$, where $g_i\colon A \to B_i$, the g_i can be lifted to a unique $g\colon A \to \prod_i B_i$ such that $p_i g = g_i$ for all i. If we take $\beta(g_i, i \in I) = g$, then α and β are inverse R-homomorphisms, and the result follows.
7. There is a free module F such that $F = M \oplus M'$, and since F is torsion-free, so is M. Since M is injective, it is divisible, so by Problem 7 of Section 10.7, M is a vector space over the quotient field Q, hence a direct sum of copies of Q. Therefore, using Problem 5 above and Problem 8 of Section 10.7,

$$\operatorname{Hom}_R(M, R) = \operatorname{Hom}_R(\oplus Q, R) \cong \prod \operatorname{Hom}(Q, R) = 0.$$

8. By Problem 7, $\operatorname{Hom}_R(M, R) = 0$. Let M be a direct summand of the free module F with basis $\{r_i, i \in I\}$. If x is a nonzero element of M, then x has some nonzero coordinate with respect to the basis, say coordinate j. If p_j is the projection of F on coordinate j (the jth copy of R), then p_j restricted to M is a nonzero R-homomorphism from M to R. (Note that x does not belong to the kernel of p_j.) Thus the assumption that $M \neq 0$ leads to a contradiction.

BIBLIOGRAPHY

General

Cohn, P.M., *Algebra, Volumes 1 and 2*, John Wiley and Sons, New York, 1989
Dummit, D.S. and Foote, R.M., *Abstract Algebra*, Prentice-Hall, Upper Saddle River, NJ, 1999
Hungerford, T.M., *Algebra*, Springer-Verlag, New York, 1974
Isaacs, I.M., *Algebra, a Graduate Course*, Brooks-Cole, a division of Wadsworth, Inc., Pacific Grove, CA, 1994
Jacobson, N., *Basic Algebra I and II*, W.H. Freeman and Company, San Francisco, 1980
Lang, S., *Algebra*, Addison-Wesley, Reading, MA, 1993

Modules

Adkins, W.A., and Weintraub, S.H., *Algebra, An Approach via Module Theory*, Springer-Verlag, New York, 1992
Blyth, T.S., *Module Theory*, Oxford University Press, Oxford, 1990

Basic Group Theory

Alperin, J.L., and Bell, R.B., *Groups and Representations*, Springer-Verlag, New York, 1995
Humphreys, J.F., *A Course in Group Theory*, Oxford University Press, Oxford 1996
Robinson, D.S., *A Course in the Theory of Groups*, Springer-Verlag, New York, 1993
Rose, J.S., *A Course on Group Theory*, Dover, New York, 1994
Rotman, J.J., *An Introduction to the Theory of Groups*, Springer-Verlag, New York, 1998

Fields and Galois Theory

Adamson, I.T., *Introduction to Field Theory*, Cambridge University Press, Cambridge, 1982
Garling, D.J.H., *A Course in Galois Theory*, Cambridge University Press, Cambridge, 1986
Morandi, P., *Fields and Galois Theory*, Springer-Verlag, New York, 1996
Roman, S., *Field Theory*, Springer-Verlag, New York, 1995
Rotman, J.J., *Galois Theory*, Springer-Verlag, New York, 1998

Algebraic Number Theory

Borevich, Z.I., and Shafarevich, I.R., *Number Theory*, Academic Press, San Diego, 1966
Fröhlich, A., and Taylor, M.J., *Algebraic Number Theory*, Cambridge University Press, Cambridge, 1991
Gouvea, F.Q., *p-adic Numbers*, Springer-Verlag, New York, 1997
Janusz, G.J., *Algebraic Number Fields*, American Mathematical Society, Providence, 1996
Lang, S., *Algebraic Number Theory*, Springer-Verlag, New York, 1994
Marcus, D.A., *Number Fields*, Springer-Verlag, New York, 1977
Samuel, P., *Algebraic Theory of Numbers*, Hermann, Paris, 1970

Algebraic Geometry and Commutative Algebra

Atiyah, M.F., and Macdonald, I.G., *Introduction to Commutative Algebra*, Addison-Wesley, Reading, MA, 1969
Bump, D., *Algebraic Geometry*, World Scientific, Singapore, 1998

Cox, D., Little, J., and O'Shea, D., *Ideals, Varieties, and Algorithms*, Springer-Verlag, New York, 1992

Eisenbud, D., *Commutative Algebra with a View Toward Algebraic Geometry*, Springer-Verlag, New York, 1995

Fulton, W., *Algebraic Curves*, W.A. Benjamin, New York, 1969

Hartshorne, R., *Algebraic Geometry*, Springer-Verlag, New York, 1977

Kunz, E., *Introduction to Commutative Algebra and Algebraic Geometry*, Birkhäuser, Boston, 1985

Matsumura, H., *Commutative Ring Theory*, Cambridge University Press, Cambridge, 1986

Reid, M., *Undergraduate Algebraic Geometry*, Cambridge University Press, Cambridge, 1988

Reid, M., *Undergraduate Commutative Algebra*, Cambridge University Press, Cambridge, 1995

Shafarevich, I.R., *Basic Algebraic Geometry, Volumes 1 and 2*, Springer-Verlag, New York 1988

Ueno, K., *Algebraic Geometry 1*, American Mathematical Society, Providence 1999

Noncommutative Rings

Anderson, F.W., and Fuller, K.R., *Rings and Categories of Modules*, Springer-Verlag, New York, 1992

Beachy, J.A., *Introductory Lectures on Rings and Modules*, Cambridge University Press, Cambridge, 1999

Farb, B., and Dennis, R.K., *Noncommutative Algebra*, Springer-Verlag, New York, 1993

Herstein, I.N., *Noncommutative Rings*, Mathematical Association of America, Washington, D.C., 1968

Lam, T.Y., *A First Course in Noncommutative Rings*, Springer-Verlag, New York, 1991

Group Representation Theory

Curtis, C.W., and Reiner, I., *Methods of Representation Theory*, John Wiley and Sons, New York, 1981

Curtis, C.M., and Reiner, I., *Representation Theory of Finite Groups and Associative Algebras*, John Wiley and Sons, New York, 1966

Dornhoff, L., *Group Representation Theory*, Marcel Dekker, New York, 1971

James, G., and Liebeck, M., *Representations and Characters of Groups*, Cambridge University Press, Cambridge, 1993

Homological Algebra

Hilton, P.J., and Stammbach, U., *A Course in Homological Algebra*, Springer-Verlag, New York, 1970

Hilton, P., and Wu, Y-C., *A Course in Modern Algebra*, John Wiley and Sons, New York, 1974

Mac Lane, S., *Categories for the Working Mathematician*, Springer-Verlag, New York, 1971

Rotman, J.J., *An Introduction to Algebraic Topology*, Springer-Verlag, New York, 1988

Rotman, J.J., *An Introduction to Homological Algebra*, Springer-Verlag, New York, 1979

Weibel, C.A., *An Introduction to Homological Algebra*, Cambridge University Press, Cambridge, 1994

List of Symbols

Throughout the text, $\subseteq$ means subset, $\subset$ means proper subset

$\mathbb{Z}_n$	integers modulo n	1.1
$\mathbb{Z}$	integers	1.1
$\langle A \rangle$	subgroup generated by A	1.1
S_n	symmetric group	1.2
A_n	alternating group	1.2
D_{2n}	dihedral group	1.2
φ	Euler phi function	1.1, 1.3
$\trianglelefteq$	normal subgroup	1.3
$\triangleleft$	proper normal subgrouad	1.3
ker	kernel	1.3, 2.2
$\cong$	isomorphism	1.4
$Z(G)$	center of a group	1.4
$H \times K$	direct product	1.5
$\mathbb{Q}$	rationals	2.1
$M_n(R)$	matrix ring	2.1
$R[X]$	polynomial ring	2.1
$R[[X]]$	formal power series ring	2.
End	endomorphism ring	2.1
$\langle X \rangle$	ideal generated by X	2.2
UFD	unique factorization domain	2.6
PID	principal ideal domain	2.6
ED	Euclidean domain	2.7
$\min(\alpha, F)$	minimal polynomial	3.1
$\bigvee_i K_i$	composite of fields	3.1
$\mathrm{Gal}(E/F)$	Galois group	3.5
0	the module $\{0\}$ and the ideal $\{0\}$	4.1
$\oplus_i M_i$	direct sum of modules	4.3
$\sum_i M_i$	sum of modules	4.3
$\mathrm{Hom}_R(M, N)$	set of R-module homomorphisms from M to N	4.4
$\mathrm{End}_R(M)$	endomorphism ring	4.4
$g \bullet x$	group action	5.1
G'	commutator subgroup	5.7
$G^{(i)}$	derived subgroups	5.7
$\langle S \mid K \rangle$	presentation of a group	5.8
$\mathcal{F}(H)$	fixed field	6.1
$\mathcal{G}(K)$	fixing group	6.1
$GF(p^n)$	finite field with p^n elements	6.4
$\Psi_n(X)$	n^{th} cyclotomic polynomial	6.5
Δ	product of differences of roots	6.6
D	discriminant	6.6, 7.4
$N[E/F]$	norm	7.3
$T[E/F]$	trace	7.3
char	characteristic polynomial	7.3
$n_P(I)$	exponent of P in the factorization of I	7.7
v_p	p-adic valuation	7.9
$\mid\ \mid_p$	p-adic absolute value	7.9
$V(S)$	variety in affine space	8.1
$I(X)$	ideal of a set of points	8.1
$k[X_1, \dots, X_n]$	polynomial ring in n variables over the field k	8.1
$\sqrt{I}$	radical of an ideal	8.3
$k(X_1, \dots, X_n)$	rational function field over k	8.4

$S^{-1}R$	localization of the ring R by S	8.5
$S^{-1}M$	localization of the module M by S	8.5
$\mathcal{N}(R)$	nilradical of the ring R	8.6
$M \otimes_R N$	tensor product of modules	8.7
UMP	universal mapping property	8.7
$A \otimes B$	tensor (Kronecker) product of matrices	8.7
kG	group algebra	9.5
RG	group ring	9.5
$J(M), J(R)$	Jacobson radical	9.7
$\mathrm{Hom}_R(M, -)$, $\mathrm{Hom}_R(-, N)$	hom functors	10.3
$M \otimes_R -$, $- \otimes_R N$	tensor fuctors	10.3
$\mathbb{Q}/\mathbb{Z}$	additive group of rationals mod 1	10.6 (also 1.1, Problem 7)
$\mathbb{Z}(p^\infty)$	quasicyclic group	A10
$G[n]$	elements of G annihilated by n	A10
H_n	homology functor	S1
$f \simeq g$	chain homotopy	S1
∂	connecting homomorphism	S2, S3
$P_* \to M$	projective resolution	S4
$M \to E_*$	injective resolution	S4
L_nF	left derived functor	S5
R^nF	right derived functor	S5
Tor	derived functor of $\otimes$	S5
Ext	derived functor of Hom	S5

INDEX

A CATALOG OF SELECTED

DOVER BOOKS

IN SCIENCE AND MATHEMATICS

Mathematics

FUNCTIONAL ANALYSIS (Second Corrected Edition), George Bachman and Lawrence Narici. Excellent treatment of subject geared toward students with background in linear algebra, advanced calculus, physics and engineering. Text covers introduction to inner-product spaces, normed, metric spaces, and topological spaces; complete orthonormal sets, the Hahn-Banach Theorem and its consequences, and many other related subjects. 1966 ed. 544pp. 6⅛ x 9¼. 0-486-40251-7

ASYMPTOTIC EXPANSIONS OF INTEGRALS, Norman Bleistein & Richard A. Handelsman. Best introduction to important field with applications in a variety of scientific disciplines. New preface. Problems. Diagrams. Tables. Bibliography. Index. 448pp. 5⅜ x 8½. 0-486-65082-0

VECTOR AND TENSOR ANALYSIS WITH APPLICATIONS, A. I. Borisenko and I. E. Tarapov. Concise introduction. Worked-out problems, solutions, exercises. 257pp. 5⅝ x 8¼. 0-486-63833-2

AN INTRODUCTION TO ORDINARY DIFFERENTIAL EQUATIONS, Earl A. Coddington. A thorough and systematic first course in elementary differential equations for undergraduates in mathematics and science, with many exercises and problems (with answers). Index. 304pp. 5⅜ x 8½. 0-486-65942-9

FOURIER SERIES AND ORTHOGONAL FUNCTIONS, Harry F. Davis. An incisive text combining theory and practical example to introduce Fourier series, orthogonal functions and applications of the Fourier method to boundary-value problems. 570 exercises. Answers and notes. 416pp. 5⅜ x 8½. 0-486-65973-9

COMPUTABILITY AND UNSOLVABILITY, Martin Davis. Classic graduate-level introduction to theory of computability, usually referred to as theory of recurrent functions. New preface and appendix. 288pp. 5⅜ x 8½. 0-486-61471-9

ASYMPTOTIC METHODS IN ANALYSIS, N. G. de Bruijn. An inexpensive, comprehensive guide to asymptotic methods–the pioneering work that teaches by explaining worked examples in detail. Index. 224pp. 5⅜ x 8½ 0-486-64221-6

APPLIED COMPLEX VARIABLES, John W. Dettman. Step-by-step coverage of fundamentals of analytic function theory–plus lucid exposition of five important applications: Potential Theory; Ordinary Differential Equations; Fourier Transforms; Laplace Transforms; Asymptotic Expansions. 66 figures. Exercises at chapter ends. 512pp. 5⅜ x 8½. 0-486-64670-X

INTRODUCTION TO LINEAR ALGEBRA AND DIFFERENTIAL EQUATIONS, John W. Dettman. Excellent text covers complex numbers, determinants, orthonormal bases, Laplace transforms, much more. Exercises with solutions. Undergraduate level. 416pp. 5⅜ x 8½. 0-486-65191-6

RIEMANN'S ZETA FUNCTION, H. M. Edwards. Superb, high-level study of landmark 1859 publication entitled "On the Number of Primes Less Than a Given Magnitude" traces developments in mathematical theory that it inspired. xiv+315pp. 5⅜ x 8½. 0-486-41740-9

CALCULUS OF VARIATIONS WITH APPLICATIONS, George M. Ewing. Applications-oriented introduction to variational theory develops insight and promotes understanding of specialized books, research papers. Suitable for advanced undergraduate/graduate students as primary, supplementary text. 352pp. 5⅜ x 8½. 0-486-64856-7

COMPLEX VARIABLES, Francis J. Flanigan. Unusual approach, delaying complex algebra till harmonic functions have been analyzed from real variable viewpoint. Includes problems with answers. 364pp. 5⅜ x 8½. 0-486-61388-7

AN INTRODUCTION TO THE CALCULUS OF VARIATIONS, Charles Fox. Graduate-level text covers variations of an integral, isoperimetrical problems, least action, special relativity, approximations, more. References. 279pp. 5⅜ x 8½. 0-486-65499-0

COUNTEREXAMPLES IN ANALYSIS, Bernard R. Gelbaum and John M. H. Olmsted. These counterexamples deal mostly with the part of analysis known as "real variables." The first half covers the real number system, and the second half encompasses higher dimensions. 1962 edition. xxiv+198pp. 5⅜ x 8½. 0-486-42875-3

CATASTROPHE THEORY FOR SCIENTISTS AND ENGINEERS, Robert Gilmore. Advanced-level treatment describes mathematics of theory grounded in the work of Poincaré, R. Thom, other mathematicians. Also important applications to problems in mathematics, physics, chemistry and engineering. 1981 edition. References. 28 tables. 397 black-and-white illustrations. xvii + 666pp. 6⅛ x 9¼. 0-486-67539-4

INTRODUCTION TO DIFFERENCE EQUATIONS, Samuel Goldberg. Exceptionally clear exposition of important discipline with applications to sociology, psychology, economics. Many illustrative examples; over 250 problems. 260pp. 5⅜ x 8½. 0-486-65084-7

NUMERICAL METHODS FOR SCIENTISTS AND ENGINEERS, Richard Hamming. Classic text stresses frequency approach in coverage of algorithms, polynomial approximation, Fourier approximation, exponential approximation, other topics. Revised and enlarged 2nd edition. 721pp. 5⅜ x 8½. 0-486-65241-6

INTRODUCTION TO NUMERICAL ANALYSIS (2nd Edition), F. B. Hildebrand. Classic, fundamental treatment covers computation, approximation, interpolation, numerical differentiation and integration, other topics. 150 new problems. 669pp. 5⅜ x 8½. 0-486-65363-3

THREE PEARLS OF NUMBER THEORY, A. Y. Khinchin. Three compelling puzzles require proof of a basic law governing the world of numbers. Challenges concern van der Waerden's theorem, the Landau-Schnirelmann hypothesis and Mann's theorem, and a solution to Waring's problem. Solutions included. 64pp. 5⅜ x 8½. 0-486-40026-3

THE PHILOSOPHY OF MATHEMATICS: AN INTRODUCTORY ESSAY, Stephan Körner. Surveys the views of Plato, Aristotle, Leibniz & Kant concerning propositions and theories of applied and pure mathematics. Introduction. Two appendices. Index. 198pp. 5⅜ x 8½. 0-486-25048-2

TENSOR CALCULUS, J.L. Synge and A. Schild. Widely used introductory text covers spaces and tensors, basic operations in Riemannian space, non-Riemannian spaces, etc. 324pp. 5⅝ x 8¼. 0-486-63612-7

ORDINARY DIFFERENTIAL EQUATIONS, Morris Tenenbaum and Harry Pollard. Exhaustive survey of ordinary differential equations for undergraduates in mathematics, engineering, science. Thorough analysis of theorems. Diagrams. Bibliography. Index. 818pp. 5⅜ x 8½. 0-486-64940-7

INTEGRAL EQUATIONS, F. G. Tricomi. Authoritative, well-written treatment of extremely useful mathematical tool with wide applications. Volterra Equations, Fredholm Equations, much more. Advanced undergraduate to graduate level. Exercises. Bibliography. 238pp. 5⅜ x 8½. 0-486-64828-1

FOURIER SERIES, Georgi P. Tolstov. Translated by Richard A. Silverman. A valuable addition to the literature on the subject, moving clearly from subject to subject and theorem to theorem. 107 problems, answers. 336pp. 5⅜ x 8½. 0-486-63317-9

INTRODUCTION TO MATHEMATICAL THINKING, Friedrich Waismann. Examinations of arithmetic, geometry, and theory of integers; rational and natural numbers; complete induction; limit and point of accumulation; remarkable curves; complex and hypercomplex numbers, more. 1959 ed. 27 figures. xii+260pp. 5⅜ x 8½. 0-486-63317-9

POPULAR LECTURES ON MATHEMATICAL LOGIC, Hao Wang. Noted logician's lucid treatment of historical developments, set theory, model theory, recursion theory and constructivism, proof theory, more. 3 appendixes. Bibliography. 1981 edition. ix + 283pp. 5⅜ x 8½. 0-486-67632-3

CALCULUS OF VARIATIONS, Robert Weinstock. Basic introduction covering isoperimetric problems, theory of elasticity, quantum mechanics, electrostatics, etc. Exercises throughout. 326pp. 5⅜ x 8½. 0-486-63069-2

THE CONTINUUM: A CRITICAL EXAMINATION OF THE FOUNDATION OF ANALYSIS, Hermann Weyl. Classic of 20th-century foundational research deals with the conceptual problem posed by the continuum. 156pp. 5⅜ x 8½. 0-486-67982-9

CHALLENGING MATHEMATICAL PROBLEMS WITH ELEMENTARY SOLUTIONS, A. M. Yaglom and I. M. Yaglom. Over 170 challenging problems on probability theory, combinatorial analysis, points and lines, topology, convex polygons, many other topics. Solutions. Total of 445pp. 5⅜ x 8½. Two-vol. set. Vol. I: 0-486-65536-9 Vol. II: 0-486-65537-7

Paperbound unless otherwise indicated. Available at your book dealer, online at **www.doverpublications.com**, or by writing to Dept. GI, Dover Publications, Inc., 31 East 2nd Street, Mineola, NY 11501. For current price information or for free catalogues (please indicate field of interest), write to Dover Publications or log on to **www.doverpublications.com** and see every Dover book in print. Dover publishes more than 500 books each year on science, elementary and advanced mathematics, biology, music, art, literary history, social sciences, and other areas.